T0330976

# Difficult Dynamics Concepts Better Explained

## Principles and Applications

# Difficult
# Dynamics Concepts
# Better Explained

## Principles and Applications

Jay F Tu

North Carolina State University, USA

**World Scientific**

NEW JERSEY · LONDON · SINGAPORE · BEIJING · SHANGHAI · HONG KONG · TAIPEI · CHENNAI · TOKYO

*Published by*

World Scientific Publishing Co. Pte. Ltd.

5 Toh Tuck Link, Singapore 596224

*USA office:* 27 Warren Street, Suite 401-402, Hackensack, NJ 07601

*UK office:* 57 Shelton Street, Covent Garden, London WC2H 9HE

**Library of Congress Cataloging-in-Publication Data**
Names: Tu, Jay F., author.
Title: Difficult dynamics concepts better explained : principles and applications /
    Jay F. Tu, North Carolina State University, USA.
Description: Hackensack, N.J. : World Scientific, 2023. | Includes index.
Identifiers: LCCN 2022040124 | ISBN 9789811262470 (hardcover) |
    ISBN 9789811262487 (ebook for institutions) | ISBN 9789811262494 (ebook for individuals)
Subjects: LCSH: Dynamics.
Classification: LCC TA352 .T845 2023 | DDC 620.1/04--dc23/eng/20221110
LC record available at https://lccn.loc.gov/2022040124

**British Library Cataloguing-in-Publication Data**
A catalogue record for this book is available from the British Library.

For any available supplementary material, please visit
https://www.worldscientific.com/worldscibooks/10.1142/13029#t=suppl

Desk Editors: Balasubramanian Shanmugam/Steven Patt

Typeset by Stallion Press
Email: enquiries@stallionpress.com

Printed in Singapore

# Preface

As I stated in *Difficult Engineering Concepts Better Explained: Statics and Applications*, I would like to follow the example of the legendary physicist, Dr. Richard P. Feynman, by explaining engineering concepts in such a way that

*Perhaps they (students) will have fun thinking them through — or going on to develop some of the ideas further.*

In this follow-up book on dynamics, I asked myself these questions: Can I present undergraduate Engineering Dynamics in a unique or different way? Can I explain difficult concepts better or more excitingly? Can I instill students with enough motivation to think deeper and to develop the ideas further? After all, learning is not simply memorizing how to solve standardized problems. In view of many, many excellent textbooks on Dynamics, some at their 15th edition or higher, my goal is apparently a tall order.

Although far from the stratosphere of Feynman's intellectual capabilities, after nearly 30 years in higher education as an engineering professor, I have developed a few bags of tricks that I feel are worth sharing. Statics and Dynamics present substantial challenges for Mechanical Engineering students because they form the foundation for later more advanced subjects in material strength, fluid mechanics, manufacturing, and engineering design. With standardized test problems and the all-out attacks of smartphones and Internet search capabilities, the teaching of these two fundamental subjects, as well as many other subjects, has become more like short-term memory exercises. Students do not "have fun thinking them through" but simply recite the solutions they have practiced earlier, often just one or two nights before the test. Many times, I have noticed that students' brains would cease to function if they are confronted with problems they have never practiced before. If allowed to get help, students would "Google" for answers or simply venture into some statements based on their "gut feeling" without proper scientific analysis based on basic laws in physics. In other words, they would speculate and hope they would get lucky and be awarded some partial credits for offering speculations. We do not design airplanes or cars by speculations; if I may paraphrase Einstein's words, "God does not play dice with the universe."

We, as educators in higher education, must not forget the importance of motivating students to think through difficult problems, while having fun doing so. It takes major effort to think, but one who does so will be rewarded generously.

This book will follow the same style of presentation of the first book of mine in Statics and utilize the vigorous force analysis method presented there to conduct analysis, design, and problem solving for engineering problems involving dynamics. I will also focus on how the fundamental techniques in dynamics can be applied to many practical applications in machine components, vehicles, and manufacturing, in addition to some interesting phenomena in daily life.

Finally, in this book, all the graphics, if not drawn by myself personally, are obtained from "Bing image search" using Microsoft Word and only limited to those from "Creative Commons". All these graphics are provided with the links where they are from, so as to observe their copyrights.

# About the Author

 **Jay Tu** graduated from National Taiwan University in 1981 and received his Ph.D. degree in Mechanical Engineering from the University of Michigan, Ann Arbor, in 1991. He is currently a full professor at North Carolina State University (NCSU). Before joining NCSU in 2003, he was an assistant professor and a tenured associate professor at Purdue University from 1992 to 2003.

Dr. Tu works closely with industry to conduct high-impact research to enhance the precision, reliability, and productivity of modern manufacturing systems through fundamental research in modeling, testing, and design. His specific research areas include high-speed machining, high-speed spindles, and laser material processing, all of which involve complex optical, electrical, and mechanical systems. Dr. Tu has infused his research interests into his teaching in both his graduate and undergraduate courses, with a strong emphasis on enhancing the thinking capabilities of students. In his classes, he presents numerous real-world examples and stories to show how fundamental laws can be used to solve practical engineering problems and to explain common events in music, sports, and daily life.

# Contents

# Chapter 1

# Introduction: Principles of Dynamics

Principles of Dynamics is a subset of Mechanics that deals with bodies in motion under the action of forces. The subject of Dynamics is completely captured by *Newton's Second Law*, $\sum \vec{F} = m\vec{a}$. To study Dynamics, we must be able to handle correct force analysis. In the book, *"Difficult Engineering Concepts Better Explained: Statics and Applications,"*[1] a vigorous method of force analysis, named the ABCC method for constructing correct free-body diagrams, is presented. We will present the key elements of the ABCC method in Chapter 2.

The right side of *Newton's Second Law* involves two important elements. One is the system, represented by $m$, and the other is the motion of the system, represented by $\vec{a}$. The motion analysis of different systems, particles and rigid bodies, will be discussed in Chapters 3 and 4, for both 2D and 3D motions, respectively.

The applications of *Newton's Second Law* to different systems will be presented in Chapter 5. In Chapter 6, we reformulate *Newton's Second Law* again to a different form, which leads to the concept of momentum. The concept of momentum is important in explaining problems related to impacts and collisions. Different forms of energy are discussed in Chapter 7. The concepts of energy and work are derived by converting the vector equation of *Newton's Second Law* into a scalar equation. In Chapters 8 and 9, we discuss the

---

[1]Tu, J.F. *Difficult Engineering Concepts Better Explained: Statics and Applications*. World Scientific, 2020. ISBN 978911213786.

1

dynamics analysis for many interesting real world problems related to automotive vehicles and other useful engineering applications in manufacturing. In Chapter 10, we extend the analysis into dynamics of non-rigid body systems. Finally, in Chapter 11, we provide detailed solutions to 32 difficult problems in Dynamics.

As stated, the main objective of studying Dynamics is to conduct correct force and motion analyses for mechanical or integrated systems. To this end, we, as engineers, must approach the analysis quantitatively, with proper knowledge about precision and uncertainties. We will discuss key concepts and practices to ensure confidence in the calculations.

Finally, in Chapter 1, we will suggest a general problem-solving protocol to promote a systematic and creative thought process. Again, the idea is that we can do a little thinking to solve new problems, not limited to the problems we have studied before.

## 1.1  Space and Time

Time keeping is critical in describing motion, which is related to the right side of *Newton's Second Law*. Space and time (or spacetime) are fundamental metrics we need for solving problems. To quantify a space, typically, we will define a reference coordinate. Typically, in Mechanics, we deal with a 1D, 2D, or 3D space, defined commonly by a suitable coordinate such as the $O-x-y-z$ Cartesian coordinate, shown in Figure 1.1. Each axis of the $O-x-y-z$ coordinate is typically fixed in direction in Statics, but not necessarily so in dynamics.

We need to define metrics along each axis. In other words, how do we define length, for example, the length of one meter in the SI system? In the old days, there was a "National Prototype Metre Bar No. 27, made in 1889 by the international Bureau of Weights and Measures (BIPM) and it was given to the United States to serve as the standard for defining all units of length in the US from 1893 to 1960."[2] "The bars were to be made of a special alloy, 90% platinum and 10% iridium, which is significantly harder than pure platinum,

---

[2]https://en.wikipedia.org/wiki/History_of_the_metre#International_prototype_metre.

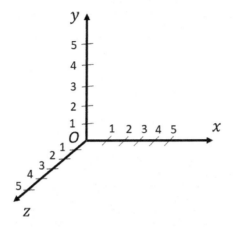

Figure 1.1: A Cartesian coordinate with metrics defined along each axis.

and have a special X-shaped cross section (a 'Tresca section,' named after French engineer Henri Tresca) to minimize the effects of torsional strain during length comparisons."[3]

Today, the standard of length is no longer defined by a meter bar, but by time, based on Einstein's theory of special relativity, which assumes that the speed of light in vacuum is constant. The official definition is[4]

> *The metre, symbol m, is the SI unit of length. It is defined by taking the fixed numerical value of the speed of light in vacuum c to be 299792458 when expressed in the unit m/s, where the second is defined in terms of the caesium frequency $\Delta\nu Cs$.*

Therefore, length is based on time. The question now is how do we achieve precise time-keeping? According to NIST,[5] "since 1967, the International System of Units (SI) has defined a second as the period equal to 9,192,631,770 cycles of the radiation, which corresponds to the transition between two energy levels of the ground state of the

---

[3]Nelson, R.A. Foundations of the international system of units (SI). *The Physics Teacher* **19**(9) (1981) 596–613.

[4]BIPM *SI Brochure*, 9th edition, 2019, p. 131.

[5]https://www.nist.gov/pml/time-and-frequency-division/timekeeping-and-clocks-faqs.

Cesium-133 atom. This definition makes the cesium oscillator (sometimes referred to generically as an atomic clock) the primary standard for time and frequency measurements." In other words, we rely on counting to establish time and thus the standard of length.

## 1.2  Concept of Precision

Being quantitative is one important requirement for engineers. We need to handle numbers properly.

In Section 1.1, we talked about using counting for time-keeping, precisely, at least down to $1/9,192,631,770$ seconds, or 0.10873 nanoseconds, based on the definition of 1 second.

As we count, we could make mistakes or simply reach different counts because the event has changed. In other words, there are errors and uncertainties. Errors are due to mistakes. The uncertainties are due to factors out of our control, no matter how carefully we try to avoid mistakes. The subject of precision deals with errors and uncertainties. There are three main concepts in precision, which are accuracy, repeatability, and resolution.[6]

For example, if we use a ruler to measure the length of a rod, the accuracy is the deviation of the measurement from the true value of the length. The problem is that we usually do not know the true value. When we take measurements with a ruler, the ruler has its limit of resolution, which is 1 mm for an office ruler. We can get a 0.01 mm resolution using a caliper.

With a simple ruler, we align one end of the rod to the zero mark of the ruler and check where the other end ends. If the other end lands between the marks of 11 and 12 mm, we can only guess that the true length is, judging where it lands between the marks, as 11.4 mm, for example. In fact, we are not sure if it is 11.3, 11.4, or 11.5 mm, and different people might guess it differently. As a result, we have an uncertainty, which is at least 0.2 mm, in this case. This uncertainty affects how repeatable the measurement could be.

---

[6]Slocum, A. *Precision Machine Design.* Society of Manufacturing Engineers, Dearborn, Michigan, 1992. ISBN-13: 978-0872634923.

One practical way of achieving a higher precision is to make multiple measurements. For example, if we want to estimate the value of an antique piece, we can ask many experts to estimate its value. Because we cannot trust one expert more than others, the best practice is to calculate the average of all the estimates. The average, thus, is the best estimate to incorporate all the inputs from all the experts. To determine the accuracy, we should use the best estimate (the average) and calculate its deviation from the true value (if we know it). The variation among the estimates from different experts represents repeatability. The resolution is related to the smallest money unit used in the estimates for this case.

### 1.2.1 *Significance of figures*

When we make a measurement or present a number, we often do not explicitly state its precision in terms of accuracy, repeatability, and resolution. Instead, we use the concept of significant figures to indicate the precision of a measurement. This is the subject of arithmetic precision. "A significant figure is any one of the digits 1 to 9; zero is a significant figure except when it is used to fix the decimal point or to fill the places of unknown or discarded digits."[7] "Thus in the number 0.000532, the significant figures are 5, 3, and 2, while in the number 2,076, all the digits, including the zero, are significant. For a number such as 2,300, the zeros may or may not be significant. To convey which figures are significant, we write this as $2.3 \times 10^3$ if two significant figures are intended, $2.30 \times 10^3$ if three, $2.300 \times 10^3$ if four, and so forth (see footnote 7)."

We often have to compute a variable based on several measurements with different significant figures. When we are doing hand calculation, it is preferred to round the number with more significant figures to be the same as the one with less significant figures. There are practical reasons for this rounding practice such as: (1) the final results cannot have a higher number of significant figures; and (2)

---

[7]Doebelin, E.O. (1990). *Measurement Systems Application and Design.* McGraw-Hill Publishing Company, New York, NY. pp. 58-67, 19990 ISBN 0-07-017338-9.

handling a long string of digits can incur more mistakes. However, in today's computing practice using computers, the second concern is no longer an issue. We should still round the number of the final result to recognize its limited significant figures and to enable easy reading.

A widely used rounding practice is as follows (see footnote 7):

*To round a number to n significant figures, discard all digits to the right of the nth place. If the discarded number is less than one-half a unit in the nth place, leave the nth digit unchanged. If the discarded number is greater than one-half a unit in the nth place, increase the nth digit by 1. If the discarded number is exactly one-half a unit in the nth place, leave the nth digit unchanged if it is an even number and add 1 to it if it is odd.*

Therefore, if we are given a number, 12.35, we should know that the actual number could range from 12.346 to 12.355, or $12.35^{+0.005}_{-0.004}$. The uncertainty range could be larger than 0.009 because most people may not follow the rounding rule consistently.

In some applications, we need to emphasize the number of decimal places (the number of significant figures following the decimal point). For example, in your bank statement, the balance of your checking account is rounded to two places after the decimal point. This is of course for practical reasons because we do not have a coin smaller than one cent.

In general, when one is performing addition and/or subtraction, it is better to round the result to the same number of decimal places as the number with the lowest number of decimal places. When the computation involves multiplication and/or division, it is better to round the result to be the same as the one with the lowest significant figures.

In this book, armed with modern computing power, all numbers of final results will be rounded to two decimal places unless a higher precision is required. If a number is very small, such as 0.0017345, it would be rounded to 0.00173, to contain three significant figures, unless a higher precision is required.

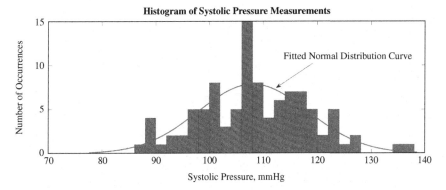

Figure 1.2:    Histogram of 100 measurements of systolic pressure.

## 1.2.2   *Statistical nature of uncertainty*

In the above discussion, we specified the range of uncertainty of a measurement using the plus and minus expression. If a number is expressed as $12.35^{+0.005}_{-0.004}$, we only specify that all the measurements would be between 12.346 and 12.355. To understand this uncertainty more, we should also define the probability distribution within the range of uncertainty. Essentially, there are no fundamental laws to help us decide how often a number would appear within the range of uncertainty.

To discuss issues related to the probability distribution, a basic discussion on statistics is in order. If we measure the blood pressure of a person several times, it is highly unlikely we will get the same reading. Let us say that we take the measurement 100 times, consecutively (hypothetically, not advised to do so), and obtained the measurements of the systolic pressure as shown in Figure 1.2 (this figure is denoted as a histogram).

The measurements are grouped into 26 bins and each bin covers 2 mmHg. From Figure 1.2, we found that, for example, there are 15 readings between 106 and 108 mmHg, only 2 readings over 130 mmHg, 10 readings between 120 and 130 mmHg, and 5 readings below 90 mmHg. Therefore, is the blood pressure of the person normal? There are 83 readings within the normal range between 90 and 120 mmHg, but 12 times when the pressure is in the pre-high blood pressure range. What is causing all these variations? The actual blood pressure could fluctuate, and the measurement device could have measurement errors. The question is still "Is the blood pressure of this person normal?" The same scenario can be related to the strength of a structure or the quality of a part after we obtain multiple measurements.

One common statistical practice is to assume that the probability distribution of the reading is a normal distribution. We can fit a normal distribution curve to the histogram, as shown in Figure 1.2. This assumption is typically valid due to the Central Limit Theorem when we have a large number of samples. Often, we simply assume the normal distribution for convenience. There are, of course, many exceptions. Based on the fitted normal distribution, we assume that if we repeat the measurement an almost infinite number of times, the histogram will approach a normal distribution curve (not necessarily the same as the fitted one in Figure 1.2). Based on the fitted curve, we estimate the average of the reading, $m$, is 108.0 mmHg, and the standard deviation, $\sigma$, is 10.2 mmHg. From the normal distribution curve, we know that 68.27% of the readings will fall in the range of $108.0 \pm 10.2$ mmHg (97.8 − 118.2 mmHg), 95.45% in the range of $108.0 \pm 2 \times 10.2$ mmHg (87.6 − 128.4 mmHg), and 99.73% in the range of $108.0 \pm 3 \times 10.2$ mmHg (77.4 − 138.6 mmHg).

From these values, we are pretty confident that the person is not suffering from high blood pressure over 138.6 mmHg because there is only 0.14% chance for it to be so.

The average of 108.0 mmHg provides us with the best guess of the true blood pressure, while we are 99.73% confident that the true blood pressure is between 77.4 and 138.6 mmHg, which represents the repeatability or the range of the uncertainty. Unfortunately, we can never know what the true blood pressure is; we can only give our best guess and indicate how confident we are with our guess.

When we have additional information about a measurement, the implied uncertainty of using the significant figures and number of decimal places is no longer adequate. We should represent the measurement in the format of plus and minus, such as $108.0 \pm 30.6$ mmHg. If not specifically stated, we will assume that this is defined with 99.73% confidence. In many cases, a 95.45% confidence is sufficient; therefore, the measurement should be expressed as $108.0 \pm 20.4$ mmHg with 95.45% or simply 95% confidence. The confidence should be explicitly stated if it is not 99.73%.

### 1.2.3 *Computation with uncertainties*

We will use an excellent example and discussion from Doeblin (1990) (see footnote 7) to demonstrate how to handle uncertainties in computation. Let a variable $y = f(x_1, x_2, x_3, \ldots, x_n)$, where $x_i$'s denote the measured variables with the uncertainties $\Delta x_1, \Delta x_2, \Delta x_3, \ldots, \Delta x_n$. What will be the uncertainty $\Delta y$?

Using the Taylor series, we expand function $f$ as

$$y = y_0 + \Delta y = f(x_1 + \Delta x_1, x_2 + \Delta x_2, x_3 + \Delta x_3, \ldots, x_n + \Delta x_n)$$

$$= f(x_{10}, x_{20}, x_{30}, \ldots, x_{n0}) + \Delta x_1 \frac{\partial f}{\partial x_1} + \Delta x_2 \frac{\partial f}{\partial x_2} + \Delta x_3 \frac{\partial f}{\partial x_3}$$

$$+ \cdots + \Delta x_n \frac{\partial f}{\partial x_n} + \text{H.O.T.} \tag{1.1}$$

where all the partial derivatives are to be evaluated at the known values of $x_{i0}$'s and the values of $\Delta x_i$'s could be positive or negative. The variable $y_0$ is the nominal value of $f$ with respect to $x_{i0}$'s. We can ignore the Higher Order Terms (H.O.T.) if the values of $\Delta x_i$'s are small.

To determine $\Delta y$, we usually consider the worst-case and the best-case scenarios. The real case most likely falls somewhere in between.

For the worst-case scenario, we define the range of $\Delta x_i$ as $R\Delta x_i$; therefore, $-\frac{1}{2}R\Delta x_i < \Delta x_i < \frac{1}{2}R\Delta x_i$ and the range of $\Delta y$ becomes,

$$R\Delta y = \left( \left| R\Delta x_1 \frac{\partial f}{\partial x_1} \right| + \left| R\Delta x_2 \frac{\partial f}{\partial x_2} \right| + \left| R\Delta x_3 \frac{\partial f}{\partial x_3} \right| \right.$$

$$\left. + \cdots + \left| R\Delta x_n \frac{\partial f}{\partial x_n} \right| \right) \tag{1.2}$$

We should express $y$ as

$$y = y_0 \pm \frac{1}{2} R\Delta y \tag{1.3}$$

If we consider $\Delta x_i$'s as random variables with normal distributions, then we can compute the standard deviation of $\Delta y$ as

$$\sigma_{\Delta y} = \sqrt{ \left( \sigma_{\Delta x_1} \frac{\partial f}{\partial x_1} \right)^2 + \left( \sigma_{\Delta x_2} \frac{\partial f}{\partial x_2} \right)^2 + \left( \sigma_{\Delta x_3} \frac{\partial f}{\partial x_3} \right)^2 + \cdots + \left( \sigma_{\Delta x_n} \frac{\partial f}{\partial x_n} \right)^2 }$$

$$\tag{1.4}$$

From Equation (1.4), if we present the uncertainty of $y$ with 99.73% confidence, we have

$$y = y_o + \Delta y = y_o \pm 3\sigma_{\Delta y} \tag{1.5}$$

Equation (1.5) represents the best-case scenario.

What if the uncertainties are not in normal distribution? In that case, there is no easy analytical way to determine $\Delta y$. However, with today's computing power, we can use the Monte Carlo simulation to determine the probability distribution of $\Delta y$ and plot a histogram similar to Figure 1.2.

Let us consider an example from Doeblin (1990) regarding the measurement obtained from a dynamometer. The output power of a dynamometer can be written as

$$P = \frac{2\pi RF\left(\frac{L}{12}\right)}{550t} = \frac{2\pi}{550 \times 12} \frac{RFL}{t} \tag{1.6}$$

where $P$ is the power in hp, $R$ is the revolution of the shaft during the time period $t$, $F$ is the measured force in lbf at the end of the torque arm, $L$ is the length of the torque arm in inches, and $t$ is time in second, s.

For a specific run, if the data are

$$F = 10.12 \pm 0.040 \, \text{lbf} \tag{1.7}$$

$$R = 1202 \pm 1.0 \, \text{rev} \tag{1.8}$$

$$L = 15.63 \pm 0.050 \, \text{in} \tag{1.9}$$

$$t = 60.0 \pm 0.50 \, \text{s} \tag{1.10}$$

the uncertainties of these measurements are determined from the sensors' calibration. Note that all the uncertainty terms are expressed with one extra decimal place than the nominal value.

The nominal value of $P$ is computed based on Equation (1.1). We then have

$$P_0 = \frac{2\pi}{550 \times 12} \frac{1202.0 \times 10.12 \times 15.63}{60.0} = 3.01668 \cong 3.02 \tag{1.11}$$

We round the above value to two decimal places.

Now, we compute the partial derivatives in Equation (1.1) to three significant figures as

$$\frac{\partial f}{\partial x_1}\bigg|_0 = \frac{\partial P}{\partial F}\bigg|_0 = \frac{2\pi}{550 \times 12} \frac{LR}{t}\bigg|_0 = \frac{2\pi}{550 \times 12} \frac{15.63 \times 1202}{60.0}$$
$$= 0.298\,(\text{hp/lbf}) \qquad (1.12)$$

$$\frac{\partial f}{\partial x_2}\bigg|_0 = \frac{\partial P}{\partial R}\bigg|_0 = \frac{2\pi}{550 \times 12} \frac{FL}{t}\bigg|_0 = 0.00251\,(\text{hp/rev}) \qquad (1.13)$$

$$\frac{\partial f}{\partial x_3}\bigg|_0 = \frac{\partial P}{\partial L}\bigg|_0 = \frac{2\pi}{550 \times 12} \frac{FR}{t}\bigg|_0 = 0.193\,(\text{hp/in}) \qquad (1.14)$$

$$\frac{\partial f}{\partial x_4}\bigg|_0 = \frac{\partial P}{\partial t}\bigg|_0 = \frac{-2\pi}{550 \times 12} \frac{FLR}{t^2}\bigg|_0 = -0.0500\,(\text{hp/s}) \qquad (1.15)$$

For the worst-case scenario, the ranges of uncertainties are based on Equations (1.7)–(1.10)

$$R\Delta F = 0.080\,\text{lbf} \qquad (1.16)$$
$$R\Delta R = 2.0\,\text{rev} \qquad (1.17)$$
$$R\Delta L = 0.100\,\text{in} \qquad (1.18)$$
$$R\Delta t = 1.00\,\text{s} \qquad (1.19)$$

From Equation (1.2), we have

$$R\Delta y = \left( \left| R\Delta F \frac{\partial P}{\partial F} \right| + \left| R\Delta R \frac{\partial P}{\partial R} \right| + \left| R\Delta L \frac{\partial P}{\partial L} \right| + \left| R\Delta t \frac{\partial P}{\partial t} \right| \right)$$
$$= 0.098\,\text{hp} \qquad (1.20)$$

From Equation (1.3), we have

$$P = P_0 \pm \frac{1}{2} R\Delta y = 3.02 \pm 0.049\,(\text{hp}) \cong 3.02 \pm 0.05\,(\text{hp}) \qquad (1.21)$$

For the best-case scenario, we first assume that the expressions of Equations (1.7)–(1.10) are based on 99.73% confidence level of normal distributions. Therefore, we have

$$\sigma_{\Delta F} = \frac{1}{3}(0.040)\,\text{lbf} \tag{1.22}$$

$$\sigma_{\Delta R} = \frac{1}{3}(1.0)\,\text{rev} \tag{1.23}$$

$$\sigma_{\Delta L} = \frac{1}{3}(0.050)\,\text{in} \tag{1.24}$$

$$\sigma_{\Delta t} = \frac{1}{3}(0.50)\,\text{s} \tag{1.25}$$

From Equation (1.4), the standard deviation of $\Delta P$ is

$$\sigma_{\Delta P} = \sqrt{\left(\sigma_{\Delta F}\frac{\partial P}{\partial F}\right)^2 + \left(\sigma_{\Delta R}\frac{\partial P}{\partial R}\right)^2 + \left(\sigma_{\Delta L}\frac{\partial P}{\partial L}\right)^2 + \left(\sigma_{\Delta t}\frac{\partial P}{\partial t}\right)^2}$$

$$= 0.00977\,\text{hp} \tag{1.26}$$

Finally,

$$P = P_0 \pm 3\sigma_{\Delta P} = 3.017 \pm 3\,(0.00977) \cong 3.017 \pm 0.029\,(\text{hp}) \tag{1.27}$$

In other words, we are confident (99.73% sure) that the actual value of $P$ will lie between 2.988 and 3.046 hp. With a lower confidence of 95.45%, the true value of $P$ will lie between 2.997 and 3.036 hp. There is only a 4.28% chance that the true value lies between 3.036 and 3.046 hp or between 2.988 and 2.997 hp. We use three decimal places for the best-case scenario.

### 1.2.4  *Precision consideration in differential quantities*

As shown in Equation (1.1), the H.O.T. can be neglected if the values of $\Delta x_i$'s are small. This is the case when we discuss the differential terms. For example, the Taylor expansion of $\sin\theta$ is

$$\sin\theta = \theta - \frac{\theta^3}{3!} + \frac{\theta^5}{5!} - \cdots \tag{1.28}$$

For a differential angle $d\theta$, it becomes

$$\sin d\theta = d\theta - \frac{d\theta^3}{3!} + \frac{d\theta^5}{5!} - \cdots \cong d\theta \tag{1.29}$$

when the H.O.T are neglected. What is the error if we neglect the H.O.T.? In this case, the error ratio, $\varepsilon$, defined as the error to the true value, is

$$\varepsilon = \frac{|\sin d\theta - d\theta|}{\sin d\theta} = \frac{\frac{d\theta^3}{3!} - \frac{d\theta^5}{5!} + \frac{d\theta^7}{7!} - \cdots}{\sin d\theta} < \frac{\frac{d\theta^3}{3!}}{\sin d\theta} \cong \frac{\frac{d\theta^3}{3!}}{d\theta} = \frac{d\theta^2}{3!}$$

(1.30)

which is infinitesimal. Similarly, we found that $\cos d\theta \cong 1$ and $\tan d\theta \cong d\theta$.

In engineering analysis, there will be cases where integration will need to be done over different geometric shapes to determine the surface area, volume, and moments of inertia. Often, we have to decide on the inclusion or exclusion of differential terms. We will present a few cases for illustration.

As shown in Figure 1.3, we want to determine the volume of a straight cone with a height $h$ and a base radius $r$. We can slice the cone into infinitesimally thin disks and add up the volume of all the disks. The top surface of the thin disk in Figure 1.3(b) has a radius $x$ and the bottom surface has a radius $x + dx$. Let us say that we do not know how to determine the differential volume, $dV$, of the thin disk of (b), but we know it will be between those two cylinders with a radius $x$ and $x + dx$, respectively, as shown in Figure 1.3(c). Therefore, we have

$$\pi x^2 dy < dV < \pi(x + dx)^2 dy \qquad (1.31)$$

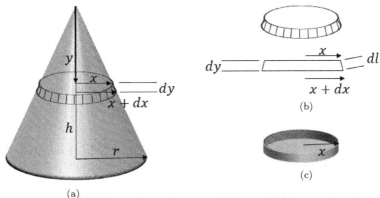

Figure 1.3: (a) Volume calculation of a straight cone; (b) an infinitesimal conical layer; and (c) an infinitesimal cylindrical layer.

If we decide to set $dV = \pi x^2\, dy$, the error ratio, $\varepsilon$, will be

$$\varepsilon < \frac{\pi(x+dx)^2\, dy - \pi x^2\, dy}{\pi x^2\, dy} = \frac{2\, dx}{x^2} + \frac{dx^2}{x^2} \quad (1.32)$$

which is still negligible.

Now, integrating $dV$ over $y$, we have the well-known result

$$\int_0^V dV = \int_0^h \pi x^2\, dy = \int_0^h \pi \frac{r^2}{h^2} y^2\, dy = \frac{\pi}{3} r^2 h \quad (1.33)$$

For the same straight cone, if we would like to determine the exterior cone surface, can we use the exterior surface of the thin cylindrical disk and carry out similar integration?

If we do so, then the exterior cone surface becomes

$$\hat{S} = \int_0^{\hat{S}} d\hat{S} = \int_0^h 2\pi x\, dy = \int_0^h 2\pi \frac{r}{h} y\, dy = \pi r h \quad (1.34)$$

which is incorrect. So what went wrong?

The actual exterior surface of the thin disk in Figure 1.3(b) should be

$$dS = 2\pi x\, dl = 2\pi x\sqrt{dx^2 + dy^2} = 2\pi \frac{r}{h}\sqrt{\frac{r^2 + h^2}{h^2}} y\, dy \quad (1.35)$$

The error ratio between $dS$ and $d\hat{S}$ is

$$\varepsilon = \frac{dS - d\hat{S}}{dS} = \frac{\sqrt{r^2 + h^2} - h}{\sqrt{r^2 + h^2}} = \frac{l - h}{l} \quad (1.36)$$

where $l = \sqrt{r^2 + h^2}$, which is the length along the cone exterior surface. As shown in Equation (1.36), the error ratio is finite, not negligible. As a result, the error will accumulate, leading to mistakes. The correct answer is

$$S = \int_0^S dS = \int_0^h 2\pi x\, dl = \int_0^h 2\pi \frac{r}{h}\sqrt{\frac{r^2 + h^2}{h^2}} y\, dy = \pi r l \quad (1.37)$$

Between $\hat{S}$ and $S$, we have the same error ratio as in Equation (1.36). A similar consideration should be applied when calculating the exterior surface of a paraboloid disk, as discussed in Problem 11.19 of Chapter 11 in the book of Difficult Engineering Concepts Better Explained (see footnote 1).

## 1.3 Problem-Solving Protocol in Dynamics

To solve a problem in Statics, we should follow a set of systematic steps as follows:

(a) Problem statements;
(b) Force analysis;
(c) Motion analysis;
(d) Governing equations;
(e) Solving equations;
(f) Answer verifications;
(g) Extensions.

In the problem statements, one should make proper sketches and define proper reference coordinates. List all the given conditions and the unknowns to be found.

For force analysis, we will review the systematic method, denoted the ABCC method, in Chapter 2 so that the force analysis can be conducted correctly every time.

Chapters 3 and 4 discuss systematic ways of motion analysis. As regards governing equations, we have several choices based on *Newton's Second Law* and its different variations (Chapters 5–7).

As regards solving equations, after we have established relevant governing equations, we should always count the number of the unknowns associated with the governing equations. If the number of unknowns is greater than the number of the governing equations, we are not able to solve for all the unknowns. When this occurs, we should carry out the following steps to overcome this problem. First, we should check if there is an equation involving only one unknown. If so, this particular unknown can be solved. As a result, the number of unknowns is reduced by one, but the number of equations is also reduced by one. We should also check if there is a subset of the governing equations involving the same number of unknowns. If so, we can solve for those unknowns. Once we exhaust all the equations that can be solved, we still need to find additional equations.

To find additional equations, we should examine the unknowns and consider if they have specific relationships among them so that we can establish additional equations.

If we cannot find additional equations, then we have to make assumptions to define the values of some unknowns. For example, in the case of a smooth surface the friction can be assumed to be zero.

Finally, as engineers, we should know that we can conduct measurements using sensors to determine the values of some unknowns. Once we reduce the number of unknowns to be the same as that of the governing equations, we can proceed to solve for all the unknowns.

In solving equations, we can use computing aids for help. Today, there are equation solvers useful for solving complicated equations, numerically or symbolically. We provide a few programming examples in this book.

After we have solved the equations and obtained the values of unknowns, we should conduct answer verifications. Do they look reasonable? We could have made some mistakes along the way. Do they violate the assumptions? Do they violate basic laws?

Finally, we should think about the implications of the results for extensions.

## 1.4  Concluding Remarks

Practice makes perfect. It applies to both honing the problem-solving skill and deepening the understanding of the concept. In Chapter 11, we provided detailed solutions to 32 difficult problems based on Appendix A of the classic Dynamics textbook by Meriam.[8] These problems, along with the examples presented in Chapters 1–10, are useful to learn how to think properly to solve difficult Dynamics problems through proper understanding of difficult concepts in Dynamics.

---

[8]Meriam, J.L. *Dynamics*, 2nd Edition. John Wiley & Sons, Inc., Hoboken, NJ, 1975.

# Chapter 2

# How to Conduct Force Analysis Correctly "Almost" Every Time?

The entire subject of Statics and Dynamics basically deals with the effects or consequences of forces. The concept of forces was first more scientifically defined based on their effect on the motion of an object by *Newton's Second Law*,

$$\sum \vec{F} = m\vec{a} \tag{2.1}$$

The left-hand side of this equation, $\sum \vec{F}$, represents the summation of many forces acting on an object, while the right-hand side, $m\vec{a}$, is related to the effects of the force on the object, where $m$ is the mass of the object and $\vec{a}$ is related to the motion of the object, or more precisely, the acceleration of the object. This famous, apparently simple, equation actually contains three drastically different subjects of physics, and each of them could warrant a lifelong study to uncover their fundamental roots. However, for the purpose of basic engineering, we will only address them from the point of solving engineering problems.

Let us also agree for the time being that *Newton's Second Law* is correct. Basically, we have three unknowns involved in this equation. We must know two of them in order to determine the third unknown based on Equation (2.1). If we are interested in determining the motion of the object, we must independently determine the quantities of the left side, the forces, and that of the mass, in order to determine the motion.

In Dynamics, we must study all three unknowns in Equation (2.1). Motion analysis will be discussed in Chapter 3 for particles and in Chapter 4 for rigid bodies. In this chapter, we provide a review of the ABCC method for achieving correct force analysis "almost" every single time. We should strive for eliminating the word "almost" from the last statement. For more advanced materials in force analysis, please refer to footnote 1 cited in Chapter 1.

## 2.1   Contact and Non-Contact Transmitted Forces

Let us agree first that we will address some specific "influences" which cause some specific effects on objects as forces. By observing these effects, we can define different types of forces. In Physics, there are forces involved within the molecular and atomic structures, which exert influences on how the atoms bond together, how the electrons orbit around a nucleus, etc. These forces are beyond the scope of basic engineering subjects regarding Statics and Dynamics. For basic engineering, it is useful to put forces into two different categories based on how a force is applied to an object. If a force must be exerted on an object by putting it in contact or bonding with another object, the force is considered as a contact-transmitted force; otherwise, it is a non-contact transmitted force.

The non-contact transmitted forces are fundamental forces due to physical fields, such as gravitational field, electric field, and magnetic field. Among them, the gravitational field is the most important to mechanical systems. When a satellite is orbiting in the space around the Earth, it is apparently not in direct contact with any other object, but a gravitational force can still act on it by the Earth; therefore, gravitational forces are the non-contact transmitted forces.

The contact transmitted forces are more obvious to us. If you bang your head against a wall, you feel a "force" at the point where your head is in contact with the wall. Because you can remove your head off the wall readily, the "contact" condition between the head and the wall is denoted as "simple *contact*," or just "*contact*." On the other hand, if you pull your right hand index finger lightly with your left hand, you can feel the force exerted by your left hand on the index finger through the simple contact. Because your index finger is not pulled off your right palm, you also feel another force which keeps your right index finger from coming off. Technically, the finger and the palm are not just in contact, but joined together via muscle and tendon. They cannot be readily separated. In other words, they are bonded together, not just in a simple contact. We will call this more complicated contact condition as "constraint". It becomes clear that contact transmitted forces can only be transmitted through simple contacts or constraints.

One important contact transmitted force is the frictional force. If you rub your face with your hand, you feel this force because your hand is in contact with the face. When you are driving, the air particles are in contact with the car, exerting a contact force, called the wind drag. Because air is not visible, we often forget that we are in contact with air all the time. Frictional forces are very important, and many engineering solutions are devised to overcome them. However, in most standardized problems of textbooks in Statics and Dynamics, air resistance and friction are often neglected without proper justification. One could be mistaken that we live in a giant vacuum and a frictionless world from those assumptions.

## 2.2    Mathematical Representation of Forces

When math joins hands with physics, it opens up new ways to understand physics. Therefore, we should define useful math tools to describe forces.

Judging the effect of a force, it is observed that a force is directional, with a magnitude or size, and acting on a specific point of an object, as shown in Figure 2.1. Therefore, mathematically, a force can be represented by a vector. Typically, we use $\vec{F}$ to represent a force. Here we use a vector sign over a letter to represent a vector, instead of using a bold-face letter. This is done because when we solve problems by hand, we write it as $\vec{F}$, not $\boldsymbol{F}$. It is very hard to distinguish $F$ from $\boldsymbol{F}$ in handwriting. Therefore, in this book, we opted to use $\vec{F}$ instead of $\boldsymbol{F}$. Mathematically, the direction of vector $\vec{F}$ can be determined based on a reference frame, such as the $O$–$x$–$y$–$z$ or the $O'$–$x'$–$y'$–$z'$ coordinate in Figure 2.1.

It should be noted that one can choose any reference frame, $O$–$x$–$y$–$z$ or $O'$–$x'$–$y'$–$z'$, to describe a force, but the size and the direction of the force do not change because of the choice of the

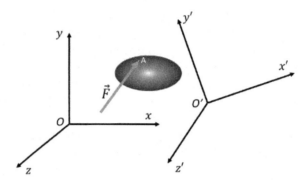

Figure 2.1:   A force vector $\vec{F}$ defined by different reference frames.

reference frame. As shown in Figure 2.1,

$$\vec{F} = a\vec{i} + b\vec{j} + c\vec{k} = a'\vec{i'} + b'\vec{j'} + c'\vec{k'} \tag{2.2}$$

where $\vec{i}, \vec{j}, \vec{k}$ and $\vec{i'}, \vec{j'}, \vec{k'}$ are unit vectors of the reference frames, $O\text{–}x\text{–}y\text{–}z$ and $O'\text{–}x'\text{–}y'\text{–}z'$, respectively. The directions and sizes of these unit vectors may or may not be the same. Typically, we only have to choose one convenient reference frame for force analysis. For the moment, there is no concern regarding the properties of the reference frame, such as if the origin of the reference frame is moving or not, or if the unit vectors could be changing in their directions. In fact, the force observation should be independent of the reference frame, and we only use the unit vectors for representing directions.

For common conventions, when designating a Cartesian reference frame, we need to follow the right-hand rule. If we place four fingers of the right hand to point to the positive direction of the $x$-axis, and then fold the fingers naturally toward the palm to align with the positive direction of the $y$-axis, the thumb then points to the positive direction of the $z$-axis. We cannot arbitrarily assign the axes of a reference frame. For example, $y$ and $z$ cannot be switched in Figure 2.1.

Once we have selected a reference frame to define a force as a vector, we need to specify the component of the force along each axis. The best way to express these components is by the angles between the force vector and the axes. There are several ways of doing this. For a 2D case, as shown in Figure 2.2(a), we can define an angle, $\theta$, between $\vec{F}$ and $\vec{i}$, which is the unit direction vector of the $x$-axis.

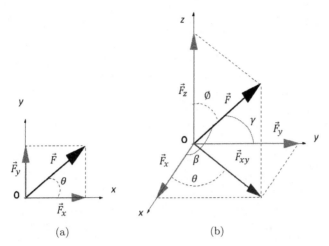

Figure 2.2: A force can be broken down into its components: (a) a 2D case; and (b) a 3D case.

The magnitude of $\vec{F}$ is $|\vec{F}|$, and graphically, it is the length of the vector arrow. The vector expression of $\vec{F}$ becomes

$$\vec{F} = |\vec{F}|(\cos\theta\vec{i} + \sin\theta\vec{j}) = |\vec{F}|(\cos\theta\vec{i} + \cos(90^\circ - \theta)\vec{j}) \qquad (2.3)$$

where $(90^\circ - \theta)$ is the angle between $\vec{F}$ and $\vec{j}$.

The second expression of Equation (2.3) provides a convenient form when we extend it to a 3D case, as shown in Figure 2.2(b). The direction vector is simply the cosine of the angle of $\vec{F}$ with respect to each axis; therefore,

$$\vec{F} = |\vec{F}|(\cos\beta\vec{i} + \cos\gamma\vec{j} + \cos\phi\vec{k}) \qquad (2.4)$$

where $\beta$ is the angle between $\vec{F}$ and $\vec{i}$, $\gamma$ is the angle between $\vec{F}$ and $\vec{j}$, and $\phi$ is the angle between $\vec{F}$ and $\vec{k}$.

The vector, $\cos\beta\vec{i} + \cos\gamma\vec{j} + \cos\phi\vec{k}$, is denoted as direction cosine.

Although Equation (2.4) is simple, the angles, $\beta, \gamma$, and $\phi$ are not easily measured in practice. These angles must be measured on the basis of the individual plane formed by $\vec{F}$ and the corresponding axis (see Figure 2.2(b)). These planes are different for different $\vec{F}$. Practically, it is difficult to mount a sensor that can be used to measure angles on different planes.

Angle $\theta$ represents the angle between $\vec{F}_{xy}$ and the $x$ axis. $\vec{F}_{xy}$ is the projection of $\vec{F}$ on the $x$–$y$ plane. We can install a second angle sensor, perpendicular to $\vec{F}_{xy}$, to measure the angle between $\vec{F}_{xy}$ and $\vec{F}$, on the plane defined by $\vec{F}_{xy}$ and $\vec{F}$. Technically, this can be done without much difficulty. This angle is $90° - \phi$, while $\phi$ is the angle between $\vec{F}$ and the $z$-axis, as defined before.

Now, we have a different expression for $\vec{F}$ based on $\theta$ and $\phi$,

$$\vec{F} = |\vec{F}|\,(\sin\phi\cos\theta\vec{i} + \sin\phi\sin\theta\vec{j} + \cos\phi\vec{k}) \qquad (2.5)$$

Equation (2.5) is more practical, requiring only two sensors, but its expression is more complex, while Equation (2.4) is just the opposite. Typically, Equation (2.5) is preferred.

Finally, how do we determine $|\vec{F}|$? If the force is expressed as Equation (2.2), then

$$|\vec{F}| = \sqrt{a^2 + b^2 + c^2} = \sqrt{a'^2 + b'^2 + c'^2} \qquad (2.6)$$

The expression in Equations (2.4) or (2.5) is very convenient because $|\vec{F}|$ represents the size, while $(\sin\phi\cos\theta\vec{i} + \sin\phi\sin\theta\vec{j} + \cos\phi\vec{k})$ is the direction unit vector for $\vec{F}$ because its magnitude is 1 and its direction is the same as $\vec{F}$.

We can further generalize the force vector as

$$\vec{F} = |\vec{F}|\vec{e}_F \qquad (2.7)$$

where $\vec{e}_F = (\sin\phi\cos\theta\vec{i} + \sin\phi\sin\theta\vec{j} + \cos\phi\vec{k}) = (\cos\beta\vec{i} + \cos\gamma\vec{j} + \cos\phi\vec{k})$, representing the directional unit vector of $\vec{F}$. Most importantly, if we know $\vec{F}$, $\vec{e}_F$ can be easily calculated as

$$\vec{e}_F = \frac{\vec{F}}{|\vec{F}|} \qquad (2.8)$$

## 2.3 Combining the Effects of Forces

The vector mathematical operations include vector addition, subtraction, both dot and cross multiplications, etc. Once forces are represented by vectors, many basic vector operations can be applied to forces to determine $\sum \vec{F}$. However, there are a few additional rules to be considered for the physics of forces.

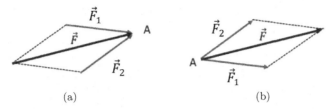

(a)                                    (b)

Figure 2.3:   Two forces acting on the same point can be combined as vectors into one single force: (a) with forces pointing to point A; and (b) with forces originating from point A. Both ways are correct.

The physics of forces requires that we consider where the force is applied, denoted as the point of application, and the line of action, along which the force is heading, passing through the point of application. The property of the object on which the force is applied should be considered as well.

In general, in order to combine two forces, both forces must be acting on the same point of application. As shown in Figure 2.3, two forces, $\vec{F_1}$ and $\vec{F_2}$, are acting on the same point, A. We have the choice of representing the force acting on point A as pointing at point A (Figure 2.3(a)) or starting from point A (Figure 2.3(b)). We simply need to choose one way and be consistent with it. We can then combine these two forces to form a new force, $\vec{F}$, using the vector addition. Graphically, the addition is carried out as shown in Figure 2.3. Mathematically, it is simpler. We simply add up each component.

$$\vec{F} = \vec{F_1} + \vec{F_2} = (a_1\vec{i} + b_1\vec{j} + c_1\vec{k}) + (a_2\vec{i} + b_2\vec{j} + c_2\vec{k})$$

$$= (a_1 + a_2)\vec{i} + (b_1 + b_2)\vec{j} + (c_1 + c_2)\vec{k} \qquad (2.9)$$

The reason that two forces not applied at the same point could not be combined can be explained with practical experience. As shown in Figure 2.4, if we push our left and right cheeks in with fingers, the force applied by right index finger cannot be combined with the one by the left finger. This is because each force creates its own local depression, and so the two cannot be combined. For many engineering applications, these local effects may not be significant if the object is very rigid. In an extreme case, if we assume that the object is so rigid that there is no local effect whatsoever, then we can call this kind of object a rigid body. Therefore, a rigid body never deforms, no matter how many or how big the forces are applied to it.

Figure 2.4: Forces applied to the human face cannot be combined as vectors.

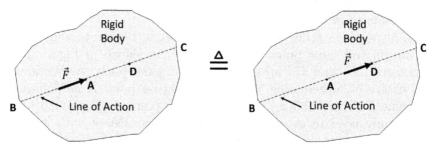

Figure 2.5: A force can move along the line of action on a rigid body without changing its effect on the rigid body.

This assumption of a rigid body is valid in many engineering applications when the deformation is negligible due to the strength of the material used for the object. For example, a steel nail is considered rigid when you hammer it into a piece of wood, while the wood is not a rigid body.

Once an object can be assumed as a rigid body, then a force can be moved along its line of action without changing its effect on the rigid body, which is explained in Figure 2.5.

The line of action of a force starts and ends on the rigid body. For Figure 2.6, the line of action of $\vec{F}$ starts at point B and ends at point C. The same force can be applied at point A or D, both on the same line of action, and the effects on the rigid body are the same, as shown in Figure 2.5. With this reasoning, we have a bit of a silly situation as shown in Figure 2.6.

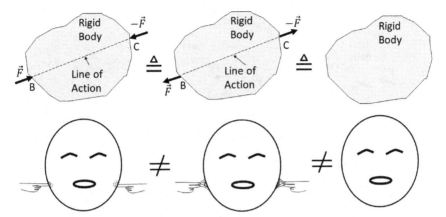

Figure 2.6:  A pair of forces with equal size and opposite directions can be canceled on the rigid body, but not on a non-rigid body.

As can be seen from Figure 2.6, a pair of compressive forces has the same effect on a rigid body as a pair of tensile forces, as well as no forces, because these pairs of forces can be canceled if they are the same size after moving them along the line of action. Apparently, this is not the case on a non-rigid human face. The cheeks could be pulled out, snapped in, or no changes with different force conditions. This silly example reminds us of the limit of the rigid body assumption.

The assumption of rigid body is important. If it is a reasonable assumption, we can now combine forces even though they are not applied at the same point of application. Any two forces, not in parallel, now can be moved along their respective lines of action to the point where these two lines of action intersect, if they do have an intersection point on the rigid body. For a pair of forces, with equal size, but in opposite directions, and not on the same line of action, we apparently cannot combine them, as shown in Figure 2.7. The combined action of such a force pair causes a rotation of the rigid body. This force pair is denoted as a couple or a force couple. This rotation action is described by the product of the size of the forces and the distance between the two lines of action. As in the case of Figure 2.7, this force couple, which forms a moment, denoted as $\vec{\tau}$, is in clockwise rotation. Based on the right-hand rule, the clockwise rotation points to the direction of $-\vec{k}$.

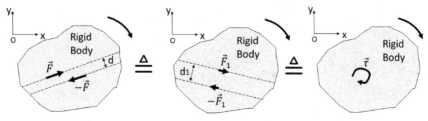

Figure 2.7:  A pair of forces with equal size and opposite directions can be canceled on a rigid body if they are on the two parallel lines of action. The combined action is a rotating effect.

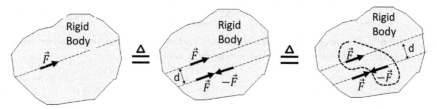

Figure 2.8:  A force can be moved to a new parallel line of action, but it would result in an accompanying torque.

It turns out that different force pairs can form the same moment as $\vec{\tau}$, as shown in Figure 2.7 and Equation (2.10).

$$\vec{\tau} = (|\vec{F}| \cdot d)(-\vec{k}) = (|\vec{F}_1| \cdot d_1)(-\vec{k}) \qquad (2.10)$$

This moment $\vec{\tau}$ does not have a point of application on a rigid body. We can apply a moment anywhere on the rigid body, and its rotating action is the same. In other words, $\vec{\tau}$ is a free vector and it can be anywhere on the rigid body as long as the direction and size are kept the same. Of course, this free vector property does not exist for non-rigid bodies. For example, if you twist your shirt at different spots, you would create different wrinkles.

Finally, we realize that we can also move a force out of its line of action, but there will be a consequence of a force couple. This result is explained in Figure 2.8. To move a force to a new line of action, we first create a null pair of forces: two forces on the same point with equal size but in opposite directions. The magnitude of the null force pair is the same as that of the original force. This null pair essentially is zero; therefore, adding this null pair does not change the loading condition of the rigid body. If we let one force of the null pair form a force couple with the original force vector, the remaining force of the null pair becomes the original force but on a new line of action. The force couple (the moment) is free to be anywhere on the rigid body. The new force line of action and the moment have the same effect on the rigid body as the original force on the original line of action.

Based on the results of Figure 2.8, we can move any force acting on a rigid body so that all of them act at the same point. Once they are at the same point, they can be combined as vectors. All the resulting torques can also be added together as vectors. Finally, we conclude that on a 2D rigid body, all the forces can be combined to become a total force and a total moment on a specific line of action as shown in Figure 2.9. We can then move this total force to a different line of action so that the resulting moment can cancel the total moment. The end result is that all forces on a 2D object can be combined into one total force acting on a specific line of action if the rigid body is large enough. A special case of this result is when an object is subjected to gravitational forces on every particle of the rigid body. All these forces can be combined into a total force acting on the mass center of the object. The mass center concept will be explained in later chapters.

On a 3D rigid body, the case is a bit more complicated because the resulting force and moment may not be perpendicular as in the case of a 2D rigid body. However, as shown in Figure 2.10, we can always split the moment into two components: one aligned with the total force, $\vec{\tau}_W$, and the other perpendicular to the total force, $\vec{\tau}_M$. The perpendicular component can be nulled by moving the total force to a different line of action, similar to Figure 2.9; however, the moment component aligned with total force, $\vec{\tau}_W$, cannot be nulled. The aligned moment, $\vec{\tau}_W$, is called a wrench because it is similar to using a screwdriver to tighten or loosen a screw.

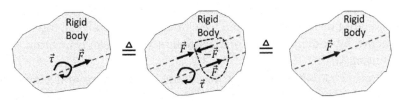

Figure 2.9: All forces can be combined into one force and one moment, and further simplified to just one force along a specific line of action on a 2D rigid body.

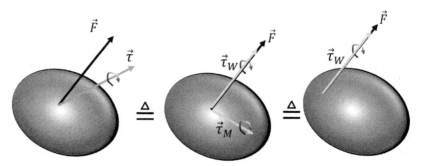

Figure 2.10:   On a 3D rigid body, all forces can be combined into one force and one moment, and further simplified to just one force with a wrench along a specific line of action.

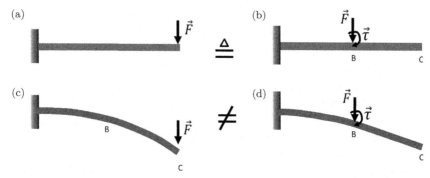

Figure 2.11:   For rigid bodies as in (a) and (b), we can move a force accordingly. However, for a non-rigid body, even if the force conditions are equivalent, the resulting deformation of the non-rigid body is different. Note that the section of BC of beam; and (c) is curved, while the same section of beam (d) is straight.

Combining all the forces and moments acting on a rigid body to a total force and a wrench appears to simplify the force representation, but this has very little practical value. One main reason is that most objects are not rigid bodies and we cannot combine all the forces and moments by moving forces and moments to the same position and combining them.

Figure 2.11 illustrates why we cannot move forces and moments on a non-rigid body as in a rigid body. For a rigid beam, as in Figures 2.11(a) and 2.11(b), we can move the force accordingly.

However, for a non-rigid beam, as in Figures 2.11(c) and 2.11(d), the beam will deform differently even if the force and moment conditions are the same. In particular, in Figure 2.11(c), section BC of the beam is curved, while in Figure 2.11(d), it is straight.

This actually brings up a fundamental question regarding if we can combine forces and moments in general. For the time being, it is worthwhile to pay attention to the rigid body assumption so that we do not have the wrong impression that we can move a force at will.

### 2.4    How to Calculate Moments

The method illustrated by Figure 2.8 to calculate a moment is not very convenient. There are simpler ways if we use vector math. We shall start with 2D cases and proceed to 3D cases.

As shown in Figure 2.12, the resulting moment can be calculated with the following equation:

$$\vec{\tau}_A = \vec{r}_{B/A} \times \vec{F} \tag{2.11}$$

The notations used here are important and should be defined clearly. $\vec{\tau}_A$ is defined as the resulting moment if we move $\vec{F}$ from its original position B to point A. Because we do not want to actually move the force to determine $\vec{\tau}_A$, we can only state that $\vec{\tau}_A$ is the moment of $\vec{F}$ with respect to point A. $\vec{r}_{B/A}$ is a position vector of

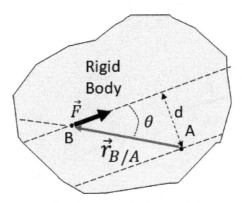

Figure 2.12:    The vector analysis for calculating the moment of a force with respect to a point.

point B with respect to point A. Note that B/A implies B with respect to A. Graphically, we draw an arrow from point A to point B for $\vec{r}_{B/A}$. To determine a position vector, the easiest way is to use the position coordinates of the points. Therefore,

$$\vec{r}_{B/A} = \vec{r}_{B/O} - \vec{r}_{A/O} \tag{2.12}$$

For a 2D case, $\vec{\tau}_A$ of Figure 2.12 has a direction, based on the right-hand rule, as pointing into the paper, while the size of $\vec{\tau}_A$ is

$$|\vec{\tau}_A| = |\vec{r}_{B/A}|\,|\vec{F}|\sin(\pi - \theta) = |\vec{r}_{B/A}|\,|\vec{F}|\sin(\theta) = d\,|\vec{F}| \tag{2.13}$$

The result of Equation (2.13) is the same for any force on the same line passing through point B. Therefore, for 2D cases, we can use the right-hand rule to determine the direction of the resulting moment.

However, for 3D cases, the right-hand rule is very hard to use. It would be much easier if we simply use the vector multiplication of Equation (2.11) to determine the moment.

With respect to Figure 2.13, for example, if a force, $\vec{F} = 4\vec{i} - 2\vec{j} - 3\vec{k}$, acts at point B, the moment $\vec{\tau}_A$ due to $\vec{F}$ with respect to point A is

$$\vec{r}_{B/A} \times \vec{F} = \begin{vmatrix} \vec{i} & \vec{j} & \vec{k} \\ -2 & 4 & 5 \\ 4 & -2 & -3 \end{vmatrix} = -2\vec{i} + 14\vec{j} - 12\vec{k} \tag{2.14}$$

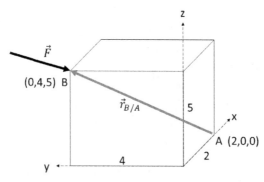

Figure 2.13: The vector analysis for calculating the moment of a force with respect to a point for a 3D case.

If $\vec{F}$ is not given, but the size is given with its direction expressed graphically, we can conveniently find two points along the force vector direction and determine a vector along this force direction. The unit force vector direction can then be determined similar to Equation (2.8). For example, if the force vector with a size 5 (we will discuss the unit in the next chapter) passes through point C with a coordinate, (1, 3, 6), and point D, (3, −2, 1), and is pointing from C to D, the unit direction vector of $\vec{F}$ is

$$\vec{e}_F = (\vec{r}_{D/C})/|\vec{r}_{D/C}| = \frac{(2\vec{i} - 5\vec{j} - 5\vec{k})}{\sqrt{2^2 + (-5)^2 + (-5)^2}} = \frac{(2\vec{i} - 5\vec{j} - 5\vec{k})}{\sqrt{54}}$$

$$(2.15)$$

Therefore,

$$\vec{F} = |\vec{F}|\vec{e}_F = \frac{5(2\vec{i} - 5\vec{j} - 5\vec{k})}{\sqrt{54}} \tag{2.16}$$

## 2.5   The Use of Dot Multiplication

The direction vector, $\vec{e}_F$, is very useful. We can use dot multiplication to determine the component of a force vector along any direction. For example, for a force vector, $\vec{F} = a\vec{i} + b\vec{j} + c\vec{k}$, the component of $\vec{F}$ in the $\vec{i}$ direction can be determined as follows:

$$a = \vec{F} \cdot \vec{i} = (a\vec{i} + b\vec{j} + c\vec{k}) \cdot \vec{i} \tag{2.17}$$

Similarly, we can find out the components in $\vec{j}$ and $\vec{k}$ directions. Let us revisit Figure 2.10 to show how we can split the total moment vector, $\vec{\tau}$, into two components: one aligned with the total force, denoted as wrench ($\vec{\tau}_W$) and the other perpendicular to the total force ($\vec{\tau}_M$).

We can easily determine the direction vector, $\vec{e}_F$, of $\vec{F}$ using Equation (2.8). Once $\vec{e}_F$ is determined, $\vec{\tau}_W$ can be determined easily as

$$\vec{\tau}_W = (\vec{\tau} \cdot \vec{e}_F)\vec{e}_F \tag{2.18}$$

and

$$\vec{\tau}_M = \vec{\tau} - \vec{\tau}_W \tag{2.19}$$

## 2.6  How to Conduct Force Analysis Correctly Most of the Time?

The title of this section is a bit conservative. As engineers, we should strive for correct force analysis all the time, not just most of the time. A simple force analysis test is given in the first lecture for the engineering courses taught by the author as a survey for force analysis skills. The correct rate of force analysis at best has been 10%, be it in the sophomore, senior, or graduate courses. Most of the force analyses appear more like random guesses than trained engineering analyses. Why so? All these students have completed Engineering Statics before coming to my classes. The reason that I found is related to how they conduct the force analysis. Mostly, it was done in an arbitrary way. Therefore, the success of their force analysis is based on luck. We cannot design engineering products by chance. We must seek 100% success, aside from occasional slips due to unforeseen mistakes. We should not count on genius or unique talent to achieve this success either. Therefore, what is the best way to achieve correct force analysis results? The answer is that we must be thorough and systematic. To be thorough, we have to be exhaustive, seeking out all possibilities. By doing force analysis systematically, we can avoid mistakes due to carelessness and omissions. Force analysis is not a creative process, but proper exercise of rigorous engineering disciplines.

How can we be thorough? We must consider all the possible forces that could act on the object. We first group all possible forces acting on an object into four types: *Applied Forces, Body Forces, Contact Forces*, and *Constraint Forces*. We will present a systematic way to account for all these four different force types to avoid any omissions. Remember, if you miss a force, the entire force analysis is wrong.

We will discuss these four force types further in the following sections. Before we do so, we need to discuss the concept of the free-body diagram, which is the most common method for force analysis. A free-body idea is about isolating a portion of a system for which we want to conduct a thorough force analysis.

## 2.7 Free-Body Diagram

We will not address exactly how this term, "free-body," was proposed. Instead, we can interpret it literally that we want to "free" a body from the rest of the world. However, we must do this without changing the force condition that the object is subject to. In our engineering world, we hardly have a body which is free, isolated, or as an ideal point with an infinitesimal size. On the other hand, we cannot afford to do a force analysis for the entire world all the time. Therefore, we should only consider a specific "body" part of an object to conduct the force analysis. To do so, we will "free" or isolate this "body" of interest.

## 2.8 Draw a Boundary Around the Free-Body

To isolate this "body" of interest, we must identify it first. The best way to identify it is to draw a boundary around it. In this way, we can clearly indicate what this "free-body" is, what are included within this "free-body," and how it is connected to the rest of the world. For example, if we are interested in designing a better backpack for reducing lower back pain (Figure 2.14(a)), the first intuition would be to isolate the backpack and consider the forces acting on the backpack as shown in Figure 2.14(b). However, the objective is to design a backpack to reduce the lower back pain. Therefore, it is probably better to isolate a free-body which includes the upper body and the backpack, as shown in Figure 2.14(c). The boundary line cutting through the lower back carries the most significance because this part of the boundary is where the contact forces or constraint forces will reside. We will elaborate this further in the next sections. Lower back pain must be related to the forces acting at the lower back, mainly on the spinal vertebrae. We can extend the boundary to the knees if we are interested in the burden to the knees. The decision of choosing a free-body by drawing a boundary around it is not trivial. One would become more efficient and choose a "better" free-body when he or she becomes more experienced.

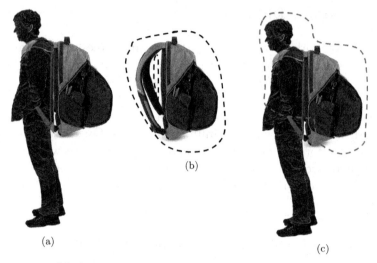

Figure 2.14:  (a) A person with a backpack; (b) FBD for the backpack; and (c) FBD for the upper body and the backpack.

*Modified from the sources*:  https://freepngimg.com/thumb/man/3-man-png-image.png;  http://media.gadgetsin.com/2015/06/incase_reform_action_camera_backpack_3.jpg.

## 2.9  The ABCC Method for Force Accounting

Once we draw a boundary to isolate a free-body for analysis, the next step naturally is to consider all possible forces that could act on the free-body. As we discussed before, there are only four different types of forces: *Applied Forces, Body Forces, Contact Forces*, and *Constraint Forces*. Taking the first letter of these four types of forces, we have ABCC. Each of these four letters represents a force type. If we go through this list of ABCC, we will exhaust all possible forces, and avoid missing any forces. These four letters are easy to remember, as easy as ABCC.

Most mistakes students would make in the free-body diagram analysis is due to omitting forces. They simply mark down some forces here and there, but they rarely are confident that they have included all possible forces. By going through this list of ABCC, such a mistake can be avoided.

Now let us elaborate on these four force types further.

**Applied forces (A):** Applied forces are those forces we want to apply to the free-body. We do not have to specify how an applied force is applied (contact or non-contact), but we need to specify where, which direction, and how much the force is. For homework problems, a statement such as, "There is a force of 500 N applied to the top center of the block, vertically down," indicates that an applied force should be included. An applied force could also be a loading condition for which we would like to verify if it could cause damages. In conclusion, applied forces are those forces we need to consider but which are not specifically related to body forces, contact forces, and constraint forces.

**Body forces (B):** Body forces are non-contact forces due to gravitational, electric, magnetic fields, etc. Body forces act on a body as distributed forces. For a rigid body, the magnitude of the gravitational body force is $mg$, its direction is along the gravitational acceleration, $\vec{g}$, and it acts at the mass center.

**Contact forces (C):** Contact forces, represented by the first letter **C**, are the first type of contact transmitted forces. They occur at the contact points where the free-body is in touch with the rest of the world. A contact is not permanent, and the free-body can be readily separated at the contact point. For example, when your feet are on the floor, there are contact areas between your shoes and the ground. You can readily lift your feet up and lose the contact. When we put our hands together, palm against palm, we create a contact. The contact is only created when two hands are pressed against each other, but we lose the contact when we pull them apart. This is an important observation because the force perpendicular (or normal) to the contact surface is always compressive. Whenever there is a contact, there are contact forces. If we draw a boundary through a contact, we must replace the contact by corresponding contact forces at the boundary. A ball that rests on a surface has a contact point, but it is not permanently bonded to the surface. When we only have a contact point, we have concentrated contact forces. If the contact is a line or an area, we have distributed contact forces. However, similar to body forces, it is easier to replace the distributed forces as concentrated forces at specific locations. Sometimes, it is important to determine the distribution of the contact forces. In such cases, the contact forces should be represented by the profile of the distributed forces.

**Constraint forces (C):** Constraint forces, represented by the second letter **C**, are the second type of contact transmitted forces. Constraint forces are different from the first type of contact transmitted forces. Constraint forces occur when we try to isolate a free-body by "cutting" it off the rest of the world. For example, if we want to make a hand a free-body, then we have to "cut" the hand off at the wrist. Our hand and the arm are not just in contact. We can pull our hand hard, but the hand won't lose the "contact" with the arm. This is an important distinction of a constraint from a contact. Furthermore, when we isolate a free-body through a constraint point, line, or area, we might need to replace the constraint not only with forces but also with moments. This difference is further illustrated in Figure 2.15.

As shown in Figure 2.15, when a water bottle is placed upon a surface, there is a contact between the bottom of the water bottle and the surface. When a force is applied to topple the water bottle, the contact can only offer a limited counter-moment to resist the toppling, as shown in Figure 2.15(a). On the other hand, if the water bottle is glued to the surface, the constraint by the glue can offer a strong counter-moment to resist the toppling. We do not have to label the counter-moment explicitly if we specify an asymmetric distributed force across the contact or constraint area.

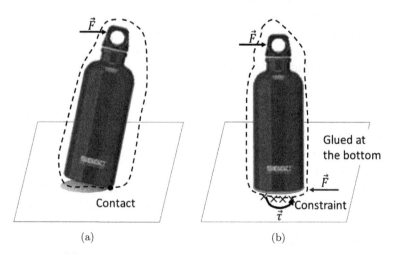

Figure 2.15: (a) A water bottle in contact with a surface; and (b) a water bottle glued to a surface.

*Source*: https://marchantscience.wikispaces.com/envisrh.

However, we should not conclude that a contact can never offer a counter-moment. For example, for the same water bottle in Figure 2.15, if we spin the water bottler around the vertical axis, the friction at the bottom of the water bottle can still resist the spin. This counter-moment is limited in the toppling case, because the water bottle can lose the contact and the normal contact force can only be compressive.

Finally, the forces and moments related to contacts and constraints are all located at the boundary. Therefore, we cannot emphasize enough how important it is to draw a boundary as the first step for the free-body diagram analysis.

## 2.10   The ABCC Method in Three Steps

We propose a systematic and thorough procedure for the free-body diagram analysis and denote it as the ABCC method. Let us use the case in Figure 2.16 to demonstrate how to use the ABCC method. A force, $\vec{R}$, is applied to a homogeneous and rigid block A with mass $M$, resting on a surface, tied to a rope. After the force $\vec{R}$ is applied, we notice that the rope is tightened, but the block stays motionless, which means that an equilibrium is achieved.

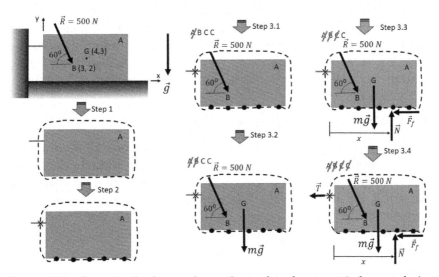

Figure 2.16:   Steps involved to conduct a thorough and systematic force analysis.

## Step 1: Draw the boundary to isolate the free-body

The choice of boundary is not trivial as discussed before. For this case, if we are interested in verifying if the rope is strong enough to keep the block from moving, we should draw the boundary through the rope. We then draw around the entire block to form a complete loop, or a closed boundary, as shown in Step 1 of Figure 2.16.

## Step 2: Go around the boundary to mark all the contacts and constraints

Once the boundary is drawn, we should start at any point of the boundary and go around the boundary to mark all the contacts and constraints. Use a heavy dot to represent a contact point, or several heavy dots for a contact line or area, as shown in Step 2 of Figure 2.16. Use a cross to represent a constraint. In this case, the constraint is at the rope where the boundary cuts through.

## Step 3: Account for all the forces simply as ABCC

Now, we are ready to account for all possible forces by following through the ABCC list. Write the four letters, ABCC, on the side of the free-body.

## Step 3.1: A for applied forces

Ask yourself, "Do we have any applied forces?" In this case, the answer is "yes" because there is an applied force, $\vec{R}$, specified in the problem statement. Draw it and mark it correctly onto the free-body diagram, as shown in Step 3.1 of Figure 2.16. Cross out the letter A from the ABCC list to indicate that all applied forces have been accounted for.

## Step 3.2: B for body forces

Now, ask yourself, "Do we need to consider body forces?" The answer is "yes" in this case because the block is in a gravitational field. We can identify the mass center and mark a concentrated gravitational force, $m\vec{g}$. Typically, it is better to draw the gravitational force from the mass center G and downward, as shown in Step 3.2 of Figure 2.16. Cross out the letter B to indicate that the body forces have been accounted for.

**Step 3.3: C for contact forces**

Now, ask yourself, "Do we have any **contact** points?" Look at the boundary to check if there are any heavy dots (which we drew in Step 2). In this case, we have many dots to form a contact line under the block. Instead of marking distributed forces across the contact line, we can just mark a concentrated normal force, $\vec{N}$, and define an application point, which is $x$ distance away from the left edge. The normal contact force can only be compressive; therefore, the force must be pointing toward the boundary. Next, we mark the concentrated frictional force, $\vec{F}_f$. We only need to draw the frictional force in the most probable direction. If the assumed direction is wrong, the force value will be negative once we solve the force equations. Sometimes, it is critical that we draw the frictional force direction correctly. In such cases, we need to judge the tendency of the movement. Cross out the first letter C to indicate that all the contact forces have been accounted for, as shown in Step 3.3 of Figure 2.16.

**Step 3.4: C for constraint forces**

Finally, ask yourself, "Do we have any **constraint** points?" Look at the boundary again to check for cross marks. In this case, we have a constraint at the rope. The rope can only sustain tensile force; therefore, the constraint force, $\vec{T}$, should be drawn away from the boundary. Cross out the second letter C to indicate that all the constraint forces have been accounted for, as shown in Step 3.4 in Figure 2.16. Again, tensile forces must be pointing away from the boundary, while the compressive forces, pointing toward the boundary.

*Mission accomplished!*

Once we have gone through Steps 1–3, we would have accounted for all possible forces, thoroughly and systematically. We can be confident that we have constructed a proper free-body diagram for force analysis. The next stage is to examine if we can determine all the forces identified by the free-body diagram. Finally, we do not need to label the forces with the vector signs as long as we mark the direction of the force clearly.

## 2.11 Force Analysis at Equilibrium

If we choose an arbitrary point and combine all the forces, the total resulting force $\sum \vec{F}$ and moment, $\sum \vec{\tau}$, must be zero if block A is at equilibrium. Therefore,

$$\sum \vec{F} = 0 \qquad (2.20)$$

$$\sum \vec{\tau} = 0 \qquad (2.21)$$

We denote these equations as the force equilibrium equations. For a 2D problem, these represent three scalar equations. The equilibrium equations with respect to mass center G become

$$R\cos(60°) - T - F_f = 0 \qquad (2.22)$$

$$-R\sin(60°) - mg + N = 0 \qquad (2.23)$$

$$\vec{r}_{B/G} \times \vec{R} + 2T\,\vec{k} - (x - 4)N\,\vec{k} - 3\,F_f\vec{k} = 0 \qquad (2.24)$$

where $\vec{r}_{B/G} = \vec{r}_B - \vec{r}_G = -1\vec{i} - 1\vec{j}$ and $\vec{R} = 500(-\cos 60°\vec{i} - \sin 60°\vec{j})$.

For Equations (2.22)–(2.24), we have four unknowns $T, N, F_f$, and $x_1$, but only three equations. As a result, we cannot solve for all the unknowns. This is not uncommon. Despite the fact that we cannot readily find out these unknowns, block A still reaches an equilibrium. As discussed in Chapter 1, we can resolve this issue in several ways. First, we can assume that the friction is zero. By assuming the friction as zero, we are determining the tension of the rope as the worst-case scenario. Without friction, the rope needs to carry all the loading due to $\vec{R}$, according to Equation (2.22). Therefore, a rope size chosen by this assumption will be able to handle the worst and maximal loading condition. If the objective is not about choosing a proper size for the rope, but precise force calculations, we can install sensors to measure some of the forces. For example, if we measure how much the rope is stretched, we can determine $T$. The unknowns are reduced to three, $N, F_f$, and $x_1$. They can now be determined by solving the force equilibrium equations. Another solution is that we can remove the rope. However, we will be counting on the friction to achieve equilibrium. Finally, we can make the friction between the block and the surface very small via polishing the surface and adding lubricants.

As a result, the frictional force, $F_f$, can be assumed to be zero. We can then solve for $T, N$, and $x_1$. This is how the worst-case scenario described above is satisfied. It should be noted, yet again, that we should not automatically assume that the friction is negligible.

For 3D objects, the equilibrium force equations will have six equations, which means that we can have up to six unknowns.

## 2.12   Backpack Example Revisited

Let us revisit the example of Figure 2.14 and apply the ABCC method. Because the objective is to design a better backpack for reducing lower back strain, we choose the boundary as shown in Step 1 of Figure 2.17. The entire backpack is inside the boundary. In Step 2, we go around the boundary and mark all the contact and constraint points. The contact points where the boundary passes through are at the contact between the backpack and the butt of the person. The constraint points are at the boundary where it cuts through the lower back. Remember that these contact and constraint points are only on the boundary. Therefore, we do not consider the

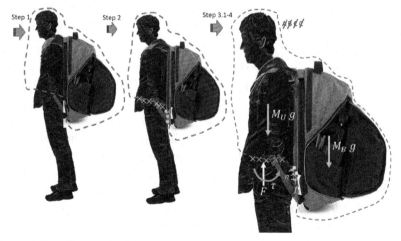

Figure 2.17:   Free-body diagram construction using the ABCC method for the backpack analysis.

*Modified from the sources*: https://freepngimg.com/thumb/man/3-man-png-image.png; http://media.gadgetsin.com/2015/06/incase_reform_action_camera_backpack_3.jpg.

contact between the back and the backpack, or the contact between the top strap and the shoulder. These contact points are inside the boundary. In Step 3, we go through the ABCC list.

**Step 3.1:** We do not have any Applied forces. Cross the letter A out.

**Step 3.2:** For Body forces, we have two masses to consider. One is the upper body of the person. We mark $M_U g$ for the gravitational force due to the upper body mass, $M_U$. We draw a downward arrow from approximately where the mass center of the upper body is. The other mass is the backpack. We label it as $M_B g$ and mark it accordingly at the mass center of the backpack. Cross the letter B out.

**Step 3.3:** For Contact forces, we note that we have some contact points (dots). We mark the normal forces, $n$, and frictional forces, $f$, and draw the arrows accordingly. Cross the first letter C out.

**Step 3.4:** For Constraint forces, we note that we have constraint points (crosses). We mark a concentrated force, $F$, and a moment, $\tau$, at the constraint points across the lower back. Cross out the second letter C.

### *Again, Mission Accomplished! Correct and complete!*

By examining this free-body diagram in Figure 2.17, our goal is to reduce the force, $F$, and the moment, $\tau$, which cause lower back strain. Without solving the force equilibrium equations, we can quickly see that if we reduce our upper body weight (i.e., $M_U g$), or carry less inside the backpack (i.e., $M_B g$), it will help. Also, a thinner backpack to keep $M_B g$ closer horizontally to the lower back will reduce the moment $\tau$. Finally, we also notice that the contact forces between the backpack and the butt are actually counteracting the gravitational forces of the upper body and the backpack. This observation leads to a possible backpack design to shift some weight of the backpack to below the lower back so that the force and the moment to the lower back are less. Finally, if we can have a design so that there is a force pushing up at the bottom of the backpack (for example someone is holding it), the lower back strain will be greatly reduced. This might lead to a backpack design with some sticks tightened to the lower legs of the person so that the backpack weight is partially carried by the lower legs directly without going through the lower back. Of course, we can also mount a stick with wheels to the ground to carry some weight of the backpack.

Before we conclude this section, it should be noted that we did not introduce a systematic way to determine the forces due to contact and constraint conditions. We did not elaborate enough on the equilibrium condition either. All these issues will be addressed in the next section to determine the contact and constraint forces systematically for common mechanical devices.

Finally, according to the teaching experience of the author, students often discount the importance of not skipping any steps of the ABCC method. Once taught the ABCC method, a student might think that it is so easy and he or she does not need to go through the entire sequence of steps anymore. For example, they will not draw the boundary. Without the boundary, the contact and constraint points are no longer properly defined. Often, they will not actually go through the ABCC list. According to the personal experience of the author, if the author does not follow through every step of the ABCC method, mistakes could still happen. It is, therefore, a puzzle to the author why most students, immediately after being taught the ABCC method, feel so comfortable in not following it.

## 2.13   How Do We Set a Free-Body Free?

In fact, a free-body is not free. A free-body is still subjected to many forces and moments, but it is now free of physical contacts and constraints with the rest of the world. In other words, when we construct a free-body diagram, we are replacing the contacts and constraints by equivalent forces and moments. It would be helpful if we have a database to show how to carry out this replacement for different contacts and constraints often found in practical mechanical systems.

In standard textbooks of statics, there are tables or charts to show how to carry out this replacement. However, these charts are not organized on the basis of how forces are transmitted through contacts and constraints. As a result, they are not very easy to follow and are difficult to implement correctly.

In the following sections, we will provide detailed discussions and charts, organized according to the ABCC method, i.e., contact conditions first, and then constraint conditions, going from 1D and 2D to 3D situations. Practical examples are also provided so that readers can identify contacts and constraints correctly for practical mechanical systems.

## 2.14 Common Contact Conditions

As defined before, contact points are not permanent bonds. A free-body can be readily separated from the neighboring objects at these contact points. When two objects are in contact, the contact could be a point, a line, or an area. The line could be straight or curved, while the contact area could be a flat or curved surface.

### 2.14.1 *Point contacts*

**Point contacts between 3D objects:** As shown in Figure 2.18, a point contact is formed between a sphere and a flat surface (Figure 2.18(a)), between two spheres (Figure 2.18(b)), between two

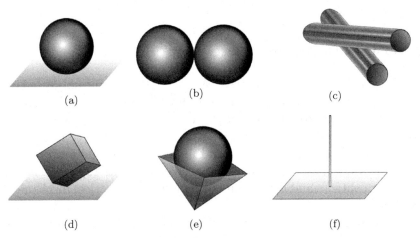

Figure 2.18: 3D objects with point contacts: (a) between a ball and a flat surface; (b) between two balls; (c) between two skewed cylinders; (d) between the corner of a cube and a flat surface; (e) between a ball and the inner surface of a cone; and (f) between a needle and a flat surface.

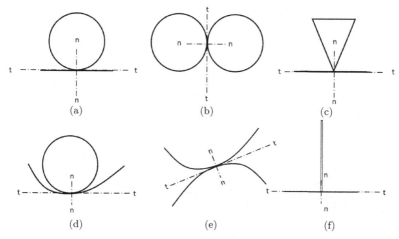

Figure 2.19:  2D objects with point contacts: (a) between a disk and a straight line; (b) between two disks; (c) between a triangle tip and a straight line; (d) between a disk and a curved line; (e) between two curved lines; and (f) between a needle and a straight line.

misaligned cylinders (Figure 2.18(c)), between a sharp corner of a cube and a flat surface (Figure 2.18(d)), and between a needle and a flat surface (Figure 2.18(f)). These contact conditions can be extended from flat surfaces to curved surfaces and from spheres to non-spherical surfaces. Multiple and countable contact points, such as four contact points between a sphere and an internal cone (Figure 2.18(e)), can also be formed.

**Point contacts between 2D objects:** As shown in Figure 2.19, a point contact is formed between a thin disk and a flat surface (Figure 2.19(a)), between two thin disks (Figure 2.19(b)), between the sharp corner of a thin plate and a flat surface (Figure 2.19(c)), between a thin disk and a curved surface (Figure 2.19(d)), between two curved 2D surfaces (Figure 2.19(e)), and between a needle and a flat surface (Figure 2.19(f)).

In Figure 2.19, we also define the tangential line (t–t) and the normal line (n–n) at the contact point. At a 2D contact point, the normal contact force will be at the contact point along the normal line, while the frictional force will be along the tangential line (t–t) in either direction depending on the tendency of the movement. For the 3D objects shown in Figure 2.18, we define two tangential lines and one normal line at the contact points (see Figures 2.20(b1)–2.20(b3)).

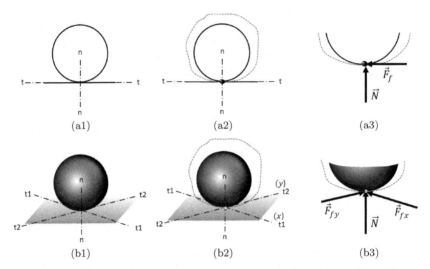

Figure 2.20: How to replace a contact point to equivalent forces: (a) a 2D case; and (b) a 3D case.

### 2.14.2 *Equivalent force condition at the point contact*

Based on the point contact condition, if we remove the neighboring object in contact with the free-body, we need to replace this neighboring object with an equivalent normal force and frictional forces. This is shown in Figure 2.20.

In Figures 2.20(a3) and 2.20(b3), the normal force, $\vec{N}$, is always compressive and it should be pointing toward the boundary; the frictional force can, however, be in any direction along the tangential line. For a 3D point contact, we can assume a pair of tangential lines perpendicular to each other as an $x$–$y$ coordinate. As shown in Figure 2.20(b1), the tangential line pair (t1–t1 and t2–t2) can be rotated to any orientation. For 3D contact points, we have two frictional forces and one compressive normal force, as shown in Figure 2.20(b3).

### 2.14.3 *Line contacts*

As shown in Figure 2.21, a line contact is formed between a cylinder and a flat surface (Figure 2.21(a)), between two cylinders (Figure 2.21(b)), between a cone and a flat surface (Figure 2.21(c)),

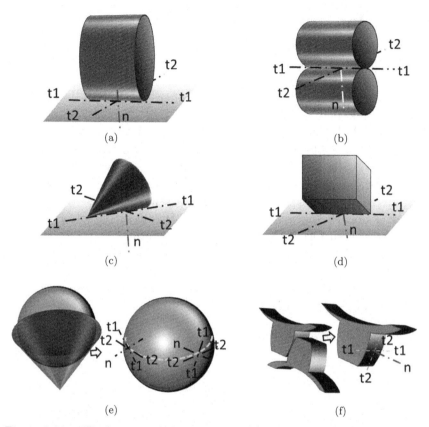

Figure 2.21: 3D objects with line contacts: (a) between a cylinder and a flat surface; (b) between two aligned cylinders; (c) between a cone and a flat surface; (d) between an edge of a cube and a flat surface; (e) between a ball and the inner surface of a cone; and (f) between two engaged gears.

between an edge of a cube and a flat surface (Figure 2.21(d)), between the internal surface of a cone and a sphere (Figure 2.21(e)), and between two gear teeth (Figure 2.21(f)). The contact lines can be straight lines (Figures 2.21(a)–2.21(d) and 2.21(f)), a circular line (Figure 2.21(e)), or a curve. Similarly, we can define the tangential lines and normal line for the line contact. One of the tangential lines will be along the contact line, while the other will be perpendicular to it. The normal line is then defined accordingly. For a circular contact line, the second tangential line and the normal line are different at different contact points on the contact circle (Figure 2.21(e)).

### 2.14.4 *Area contacts*

As shown in Figure 2.22, an area contact is formed between a cube and a flat surface (Figure 2.22(a)), between a cylinder and a ring (Figure 2.22(b)), between a cone and a tapered ring (Figure 2.22(c)), between a sphere and an internal sphere in a ball joint (Figure 2.22(d)), between a diamond and a ring (Figure 2.22(e)), and between a bolt and a nut (Figure 2.22(f)). The contact area can be a straight surface (Figure 2.22(a)), a cylindrical surface (Figure 2.22(b)), a cone surface (Figure 2.22(c)), a spherical surface (Figure 2.22(d)), facets of a diamond cone (Figure 2.22(e)), or a threaded surface (Figure 2.22(f)). For Figure 2.22(a), we can define the tangential lines and the normal line related to the area contact. For other area contact surfaces, the tangential and normal lines are defined locally at different points within the contact surface.

### 2.14.5 *Free-body representation for line and area contacts*

Similar to the point contact condition, if we remove the neighboring object in contact with the free-body, we need to replace this

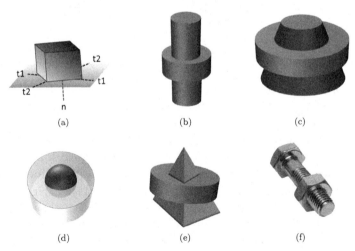

Figure 2.22: Area contact between different 3D objects: (a) between a cube and a flat surface; (b) between a cylinder inside a larger cylinder; (c) between a cone inside an inner cone; (d) between a ball and its housing; (e) between a pyramid and its covering; and (f) between the bolt and the nut.

*Source*: (f) http://pngimg.com/download/3001.

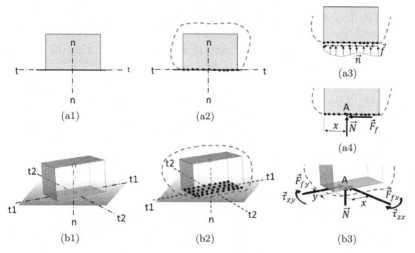

Figure 2.23:   Equivalent contact forces for line and area contacts: (a) a 2D case; and (b) a 3D case.

neighboring object with equivalent normal and frictional forces for the line or the area contact. This is shown in Figure 2.23. However, because a contact line or a contact area can have many (uncountable) contact points, we have distributed frictional forces and normal forces.

If we choose to draw the entire distributed force profiles as shown in Figure 2.23(a3), we will replace the line contact with the frictional distributed force, $\vec{f}$, and the distributed normal force, $\vec{n}$. However, typically, it is very difficult to obtain the correct profile of these distributed forces. For some idealized cases, certain models can be used for the distributed force profile prediction, such as the Hertzian contact theory for contacts between elastic bodies, or sensors can be used to measure the force distribution. As discussed before, we can convert a distributed force into a concentrated force acting on a specific point. For a line contact, we will then replace it with a concentrated normal force, $\vec{N}$, and a concentrated frictional force, $\vec{F}_f$. The frictional force, $\vec{F}_f$, is along the tangential line, while the normal force, $\vec{N}$, needs to be placed at a specific point, defined by a distance, $x$, from the left edge of the contact line, shown as point A in Figure 2.23(a4). It should be noted that the location of the point of application of $\vec{N}$ is very important and must not be ignored. As a result, for a line contact in a 2D case, we have three unknowns, $\vec{F}_f$,

$\vec{N}$, and $x$. For a line contact in a 3D case, we will have three contact forces and two location variables, for a total of five unknowns.

Similarly, for an area contact, if we convert all the distributed forces to concentrated forces, as shown in Figure 2.23(b3), we have three concentrated forces, two location variables, and two additional moments. The two location variables, $x$ and $y$, are related to the point of application of the normal force, $\vec{N}$, shown as point A in Figure 2.23(b3). The line of action of the frictional force, $\vec{F}_{fx}$, however, does not necessarily pass through point A. We can move $\vec{F}_{fx}$ to point A, resulting in a moment, around axis $z$, denoted as $\vec{\tau}_{zx}$ since it is related to $\vec{F}_{fx}$. Similarly, we have z, $\vec{F}_{fy}$, and $\vec{\tau}_{zy}$. Thus, there are seven unknowns. Those two moments in the z direction can be combined as one unknown to reduce the total unknowns to six. The number of unknowns presents a difficult problem. In many practical applications, however, the force distribution of the contact area may be symmetric. If so, we can quickly determine the location of point A at the geometric center of the contact area and there will be no moments $\vec{\tau}_{zx}$ and $\vec{\tau}_{zy}$. As a result, there will be only three unknowns, $\vec{N}$, $\vec{F}_{fx}$, and $\vec{F}_{fy}$.

Of course, there are many cases when the moments $\vec{\tau}_{zy}$ and $\vec{\tau}_{zx}$ can be significant. One example is the frictional force between the tire and the road when one makes a turn. There will be a counter-moment, $\vec{\tau}_z$, to partially turn the tire back when one lets go of the steering wheel.

### 2.14.6 *Common parts with point, line, and area contacts*

A few common mechanical devices with point, line, or area contacts are shown in Figure 2.24. For ball bearings (Figure 2.24(a)), the contact between the ball and the bearing's inner (outer) ring is typically considered as a point contact, or more specifically a 2D point contact because the ball can only move along the circumferential direction. In the case of roller bearings (Figure 2.24(b)), the contact between the roller and the inner (outer) ring is a line contact. The contact line could be straight or curved. Similarly, this line contact is a 2D line contact. Typically, we can assume symmetric contact forces for the analysis of roller bearings. For an HSK tool holder shown in Figure 2.24(c), there are two contact areas, one related to the cone

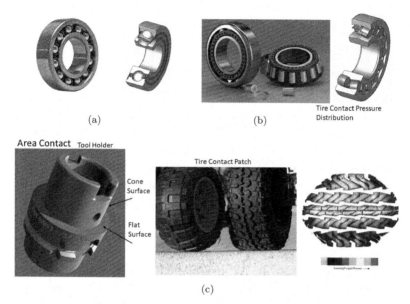

(a)        (b)

(c)

Figure 2.24: Practical devices for different contact conditions: (a) point contact, ball bearings; (b) line contact, roller bearings; and (c) area contact, tool holder and tires.

*Sources*: (a) http://www.911uk.com/viewtopic.php?p=1141885, https://commons.wikimedia.org/wiki/File:Angular-contact-ball-bearing_single-row_din628_type-b_120.png; (b) https://www.thingiverse.com/thing:781635, https://commons.wikimedia.org/wiki/File:Spherical-roller-bearing_double-row_din635-t2_120.png; and (c) https://commons.wikimedia.org/wiki/File:DIN_69893_HSK_A63_3drender_1.png, https://www.flickr.com/photos/42988571@N08/7524617342, https://commons.wikimedia.org/wiki/File:Tirefootprint.jpg

and one related to the flat surface of the ring. Conventional 7/24 tool holders may have the cone surface contact only. Note that once the surface contact is locked in by locking devices, such as bolts, nuts, and clamps, the contact becomes a constraint, which we will discuss in the following sections. Another important contact area example is the contact patch between a tire and the road. The force distributions on a tire contact patch are very complicated and have been the focus of intensive research by tire companies so as to enable them to optimize the tire performance based on traction, wear, comfort, stiffness, and energy consumption. The moments due to the area contact in Figure 2.23(b3) are the phenomena we experience on making a turn while driving. At the end of the turn, if you let go of the steering wheel, the steering wheel will rotate back partially by itself due to the presence of the moments.

### 2.14.7 *Practical consideration of point and line contacts*

In practice, there are no true point and line contacts. A point contact is likely a small circular or oval area contact, while a line contact is a narrow rectangular area contact. For example, if we are interested in the motion analysis of a car, we will conveniently assume the contact between a tire and the road as a point contact. However, if we are interested in assessing the tire wear, the contact will be an area contact with highly complicated distributed force profiles. We make these assumptions based on the objective and the precision requirements of the analysis.

If we consider point, line, and area contacts as area contacts, the size of the contact area is the smallest for the "point" contact and the largest for the area contact. For the same force, the contact stress (force/area) is higher for a smaller contact area. The material failure is typically proportional to the magnitude of the contact stress. As a result, the loading capacity, or the amount of the force that a contact area can carry, is the smallest for a point contact and the largest for an area contact. This is very true, as can be seen by the fact that we know how easily a sharp needle can penetrate through the skin with a small force while an unsharpened pencil cannot despite being applied with a much larger force.

On the other hand, a point contact or a line contact is involved with fewer unknowns; therefore, it is easier for analysis and more predictable for the loading condition.

As engineers, we balance between these considerations to choose different contact conditions for our design and analysis.

## 2.15 Common Constraint Conditions

As defined earlier, constraint points are permanent bonds unlike contact points. A free-body cannot be readily separated from the neighboring objects at these constraint points. There are many ways two objects can be "bonded" together through constraint points. Those important constraints, which are discussed here, include the following: (1) cable and wire constraint; (2) pin and hinge constraint; (3) ball joint constraint; (4) non-spherical surface constraint; (5) pined–pined rod constraint; (6) sliding collar constraint; and (7) total constraint. We will discuss the features of these constraints and determine their equivalent forces and moments.

### 2.15.1 *Wire/cable/rope constraint*

One of the simplest constraints is achieved by using a wire, a cable, or a rope. Wires (or cables or ropes) are flexible. They can only be subjected to tensile force because they buckle or bend when compressive forces are applied. For engineering applications, the mass of a wire is typically much less than that of the object to be constrained by the wire. We can reasonably assume that a piece of wire is massless. As shown in Figure 2.25(a), a plumb bob is attached to a rope and hung vertically for a vertical reference, commonly used in construction. The general case of a wire constraint is shown in Figure 2.25(b). If we draw a boundary across the wire to isolate a free-body (Figure 2.25(c)), we should mark an "×" where the boundary intersects with the wire. We can also isolate a piece of the wire as a free-body as shown in Figure 2.25(d). We realize that the two forces acting on the wire free-body must be in tension, thus pointing away from the boundary and they must be of an equal size and in opposite directions, aligned with the stretched wire, to reach an equilibrium because the wire is assumed to be massless. As a result, based on *Newton's Third Law*, the constraint force at the point where the "×" is marked should be the same as the tensile force as shown in Figure 2.25(c). The wire constraint is the simplest constraint. One should note that the constraint force at the wire constraint must be tensile, pointing away from the boundary, and aligned with the stretched wire. For this constraint, the only unknown is the magnitude of the force because the direction is determined by the orientation of the wire.

### 2.15.2 *Pin/hinge constraints and their extensions*

A pin or a hinge constraint is used to join two objects so that they can be joined at the pin location while being capable of rotating freely with respect to each other, as shown in Figure 2.26(a). These two objects will be rotating on the same plane (or on parallel planes).

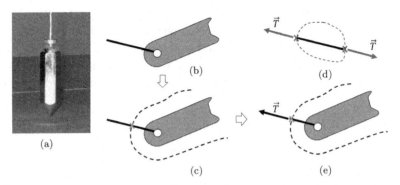

Figure 2.25: (a) A plumb bob; and (b)–(e) the construction of the corresponding FBD.
*Source*: (a) https://asacredrebel.wordpress.com/2012/07/15/gods-plumb-line/.

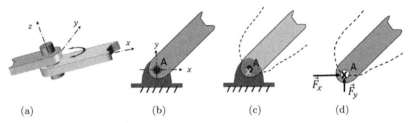

Figure 2.26: (a) A pin linking two members; (b) a member fixed by a pin; and (c–d) construction of the corresponding FBD.

Therefore, a pin constraint is considered a 2D constraint, and we can represent it as shown in Figure 2.26(b), where a linkage is pined to a foundation at point A. The linkage can rotate freely around point A, but point A will be stationary. The pin constraint provides constraint forces to keep point A from moving but provides no constraint moment to restrict the rotation of the linkage around point A. If we want to remove this pin constraint, we should draw a boundary across the pin, as shown in Figure 2.26(c), and mark an "×" at the center of the pin. The equivalent constraint forces will be two forces $\vec{F}_x$ and $\vec{F}_y$ as shown in Figure 2.26(d).

These two forces can be combined as one resultant force pointing to a specific direction. The two unknowns become the magnitude and the direction of the combined resultant force. In general, it is easier and more convenient to use $\vec{F}_x$ and $\vec{F}_y$ when we derive the force equilibrium equations. Note that $\vec{F}_x$ and $\vec{F}_y$ are plotted to the

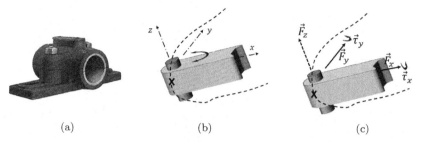

(a)                          (b)                          (c)

Figure 2.27:   (a) A deep flange as a 3D pin; (b) motion of a 3D pin; and (c) FBD of a 3D pin.
*Source*: (a) https://upload.wikimedia.org/wikipedia/commons/thumb/c/c0/ Pillow-block-bearing.jpg/1200px-Pillow-block-bearing.jpg

right and upward, respectively. However, the actual directions can be reversed. In other words, the combined resultant force can be compressive or tensile (pointing toward or away from the boundary). We simply assume the most probable directions. If the directions are wrong, the magnitudes of the forces will be negative. As a 2D constraint, the pin constraint is the second simplest constraint, and no moment is present.

In practice, all objects are 3D objects. In a practical pin constraint as shown in Figure 2.26(a), the linkages can only rotate on an $x-y$ plane. Therefore, a pin constraint, when viewed as a 3D object, can offer moment constraints around both $x$ and $y$ directions as shown in Figure 2.27, in particular for a deep flange as shown in Figure 2.27(a). A 3D pin can also be made to resist the linear motion along the $z$-axis so that a constraint force $\vec{F}_z$, plus two moments, $\vec{\tau}_x$ and $\vec{\tau}_y$, exist.

However, when designing a mechanical system, we should avoid using the moment constraint capability of a 3D pin for carrying heavy loading. In other words, it is better to design a system with several 2D pins to carry moments. This is the typical case of using hinges to secure a door, as shown in Figure 2.28. Hinges, as shown in Figure 2.28, only allow rotation around the $z$-axis, and they do not allow rotation around the $x$ and $y$ axes. However, we should not rely on one hinge to secure a door to keep it from rotating around the $x$ and $y$ axes. The moment resisting capabilities of a single hinge are too low (Figure 2.28(a)). To achieve high load-carrying capacity to resist $\vec{\tau}_x$ and $\vec{\tau}_y$, we use multiple hinges at different locations. As shown in Figure 2.28(b), three hinges are used for each door. Each hinge will only carry forces in $x$ and $y$ directions, as well as the weight of the

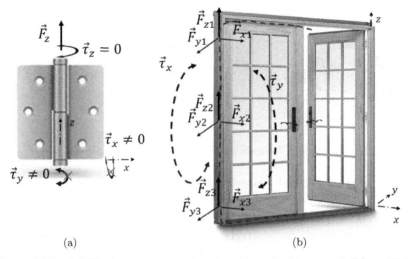

(a)                                          (b)

Figure 2.28: (a) The movements and constraints of a hinge; and (b) multiple hinges used to secure a door.

*Sources*: (a) https://www.publicdomainpictures.net/en/view-image.php?image-9211&picture=hinge&jazyk=JP; and (b) https://www.flickr.com/photoswind ows/6237870220/in/photostream.

door in the $z$ direction. The forces in the $x$ directions of these hinges combine to form a moment to counter $\vec{\tau}_y$, while the forces in the $y$ directions form a moment to counter $\vec{\tau}_x$, as shown in Figure 2.28(b). Just like 2D pins, we only have to replace each hinge with two forces as shown in Figure 2.26(d).

Another example of using several 2D pin constraints is the universal joint. As shown in Figure 2.29(b), a shaft is attached to a rod with two pin joints to allow rotation around the $x$-axis (line AB), while a second shaft is attached to a second rod to allow rotation around the $y$-axis (line CD). These two rods are then joined together as a rigid cross, as shown in Figures 2.29(a) and 2.29(b). A universal joint can be considered as an assembly of four pin constraints on a 2D plane. As can be seen from Figures 2.29(b) and 2.29(c), there are four pins, A, B, C, and D. Each pin will provide two constraint forces as shown in Figure 2.29(c) for eight forces, which then form two force couples for those forces in the $x-y$ plane. The combination of these forces and force couples can be summarized into three constraint forces and one moment, as shown in Figure 2.29(d). Because the cross is rigid, these two rods are not allowed to rotate with respect to each other.

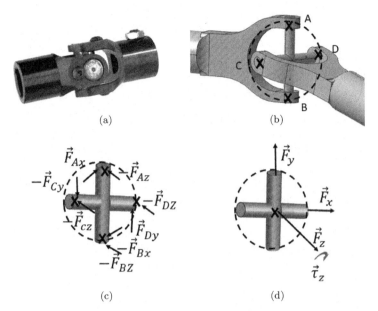

(a)

(b)

(c)

(d)

Figure 2.29:    (a) A universal joint; and (b–d) the construction of the corresponding FBDs.

*Sources*:    (a)    https://mechanics.stackexchange.com/questions/24550/clicking-noise-in-opel-astra;    and    (b)    http://catiatutorial.blogspot.com/2009/11/modeling-universal-joint-in-catia-v5.html .

As a result, a universal joint can provide constraint forces in all three directions but only a moment constraint around the $z$-direction, as shown in Figure 2.29(d). The $z$-axis is defined as normal to the plane defined by the cross.

A universal joint allows a rotation to be transmitted from one shaft to another shaft that is not aligned in a straight line with the first shaft. Two universal joints are often used for the main drive shaft of a rear-wheel drive car when the engine crank shaft is higher than the shaft of the differential, as shown in Figure 2.30.

### 2.15.3    *Ball joint constraint*

A ball constraint is used to join two objects so that they can be joined at the ball joint location, allowing for free rotation with respect to each other around any axis, as shown in Figure 2.31(a). A ball joint constraint is considered a 3D constraint.

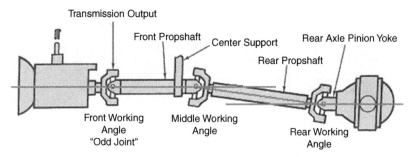

Figure 2.30: Two universal joints are used to connect two different axes of drive shafts.
*Source*: https://mechanics.stackexchange.com/questions/6800/1-piece-versus-2-piece-drive-shaft.

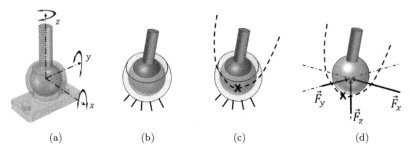

Figure 2.31: (a) A ball joint; and (b–d) the construction of the corresponding FBDs.
*Source*: (a) https://www.thingiverse.com/thing:939326.

The linkage connected to the ball joint can rotate freely, but the center of the ball will be stationary. Therefore, the ball joint constraint provides constraint forces in three directions to keep the ball center from moving but provides no constraint moment to restrict the rotation. If we want to remove this ball constraint, we should draw a boundary across the ball center, as shown in Figure 2.31(c), and mark an "×" at the center of the ball. The equivalent constraint forces will be three forces, $\vec{F}_x$, $\vec{F}_y$, and $\vec{F}_z$, as three unknowns, as shown in Figure 2.31(d).

Ball joint constraints are used in many important applications, such as in the suspension and steering systems of automobiles (Figures 2.32(a) and 2.32(b)), in metrology for ballbar measurement (Figure 2.32(c)), and in artificial knee replacement (Figure 2.32(d)). For the automobile applications, the ball joints are subjected to huge

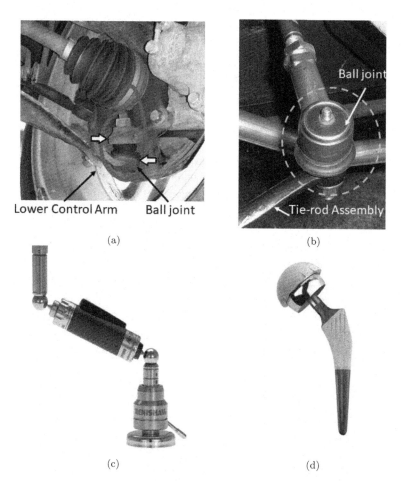

Figure 2.32:   Ball joints used for different applications: (a) lower control arm;
(b) tie-rod assembly; (c) ball bar measurement tool; and (d) an artificial hip.
*Sources*: (a) https://mechanics.stackexchange.com/questions/9884/macpherson-
strut-lower-control-arm-ball-joint-play-and-suspension-geometry;(b)https://www.
flickr.com/photos/mwanasimba/16966732482/; (c) http://ims-eng.net/products/
renishaw/; and (d) https://www.geripal.org/2013/01/metal-on-metal-hip-replace
ments-tragic.html.

loading. The fit between the ball and its socket must be tight without play in order to maintain accurate positioning. Typically, grease
is applied to reduce the wear of the ball. Rubber boots are used to
keep the grease from being contaminated by dirt. When a rubber

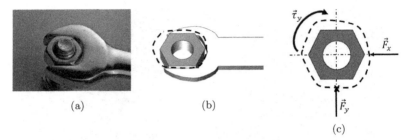

(a)                              (b)

(c)

Figure 2.33:   Non-spherical constraint examples: (a) a wrench tightening a nut; and (b–c) the construction of the corresponding FBDs.
*Source*:   (a)   https://imbratisare.blogspot.com/2013/02/el-fmi-sigue-apretando-las-tuercas-los.html.

boot breaks due to aging, dirt can get into the interface between the ball and the socket, causing severe wear. Excessive ball joint wear in the car's suspension system can cause large play and excessive tire wear. Worst of all, it could progress to catastrophic failures when the ball separates from the socket, leading to major car accidents. Ball joints for automotive applications are not failure-proof designs; therefore, drivers should pay close attention to their conditions.

### 2.15.4  *Non-spherical surface constraints*

Figure 2.33 shows the joining between a nut and a wrench as an example of non-spherical constraint. This constraint contains an extra moment, $\vec{\tau}_z$, in addition to two forces $\vec{F}_x$ and $\vec{F}_y$ (Figure 2.33(c)). However, the two forces are not essential. For the hand tool example, one might think this is a contact condition because the wrench can be easily removed from the nut. However, we should consider it as a constraint.

Non-spherical constraints are typically used for transmitting torques. More examples are shown in Figure 2.34. The example in Figure 2.34(a) is a valve handle that fits into a square stem for turning the valve or a faucet open or close. In Figure 2.34(b), it is a shaft coupling using splines. In Figure 2.34(c), it is the square fitting for rachet wrenches.

The ability of a non-spherical contact surface to sustain both forces and a torque has many other applications. One popular application is the trailer hitch shown in Figure 2.35. With a lock pin and

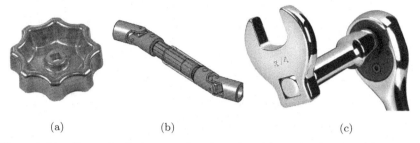

(a)      (b)      (c)

Figure 2.34:  Non-spherical constraint examples: (a) a valve handle; (b) a spline coupling; and (c) a ratchet wrench and a crowfoot tool.

*Sources*: (a) https://diy.stackexchange.com/questions/125237/what-is-the-miss ing-part-that-would-fix-this-outdoor-spigot; (b) https://commons.wikimedia.org/ wiki/File:Cardan-joint_spline-shaft_3D_transparent_animated.gif; and (c) https:// diy.stackexchange.com/questions/103422/remove-fastener-nut-for-kitchen-fau cetrusted.

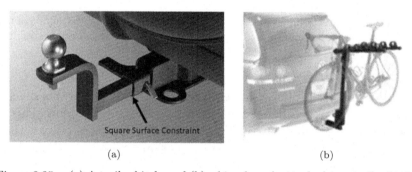

(a)      (b)

Figure 2.35:  (a) A trailer hitch; and (b) a bicycle rack attached to a trailer hitch.

*Sources*: (a) https://bicycles.stackexchange.com/questions/8958/bike-rack-for-hat chback-vehicle; and (b) http://www.cyclecityusa.com/index.php?title=Accesories.

a square contact surface, this trailer hitch is capable of sustaining forces and moments in all three directions. This is called a full constraint, and it will be discussed later.

## 2.15.5  *Pined–pined rod constraints*

There is a special case of the pined constraint, which was discussed earlier. As shown in Figures 2.36(a) and 2.36(b), a pined–pined rod is defined as a 2D element constrained by two pins on each end and no external forces on the element except at the pins. The corresponding free-body diagram is shown in Figure 2.36(c). Normally, we should

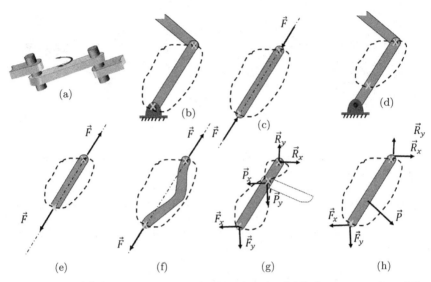

Figure 2.36: (a) A member pined at both ends only; (b–f) the construction of the corresponding FBDs; and (g–h) a member with additional forces or constraints.

label two force components on each pin constraint, but in this special case, we only need to label one force along the rod direction. This is because the element is at equilibrium, and the only possibility is that the resultant force at one pin must cancel that at the other to result in this equilibrium. As a result, if we draw a straight line between these two pins, the resultant force at one pin must align with this straight line as shown in Figures 2.36(c) and 2.36(f). We can draw the boundary across the middle of the beam, and the force at the constraint of the beam (not a pin) will still be aligned with the rod, as shown in Figures 2.36(d) and 2.36(e). Therefore, a 2D element with only two pined constraints, without any other forces, is similar to a rope constraint, except that the force could be either tensile or compressive. A 2D element with the pined–pined rod constraint is very useful because we already know the force direction at the pined constraint. Therefore, we only have one unknown — the magnitude of the force.

It is very important that there are only two pined constraints and no external forces acting on the element other than on the pins. If these conditions are not met, as shown in Figures 2.36(g) and 2.36(h), then we do not have a pined–pined rod constraint. In Figure 2.36(g), there is a third pined element in the middle, i.e., more than two pins. Therefore, we need to replace each pined constraint with two unknown forces. In Figure 2.36(h), there is an external force, $\vec{P}$, acting on the element. As a result, we need to replace each constraint with two unknown forces.

In reality, it is very difficult to satisfy the condition that requires no external forces acting on the element in locations other than those of the two end pins. The gravitational force will always be acting on the mass center of the element. However, if the constraint forces are expected to be substantially larger than the weight of the element, we can ignore the gravitational force.

There are many important applications of the pined–pined rod constraint as shown in Figure 2.37. In Figure 2.37(a), a bridge was built with many beams, called trusses, which are riveted only at the end. It is important to know that these trusses are not weightless, but we can assume that the weight is split in two equal parts to act on each end pin. Typically, the force sustained by the truss is much higher than its weight to justify this assumption. Figure 2.37(b) shows a multiple-axis suspension design for automobiles. Each axis is formed by a 2D element with the pined–pined rod constraint. In this way, we know each force is applied (compressive or tensile) along each axis. For a 5-axis suspension design, only one degree of freedom is allowed (up and down motion). Figure 2.37(c) shows a shock absorber (circled element), used in the rear suspension of a bicycle. This pined–pined rod is no longer rigid but allows movement along the longitudinal direction. The movement is resisted by viscous friction inside the rod to dissipate vibration energy. Figure 2.37(d) depicts a hydraulic cylinder (circled element) used in a crane hoist. The hydraulic pressure inside the cylinder extends along its length and can withstand forces along the longitudinal direction.

### 2.15.6 *Sliding collar constraints*

If we allow a pin to move along a slot, we have the sliding collar constraint. It can also be a collar over a rod as shown in Figure 2.38(a).

A sliding collar constraint is different from a sliding contact because the normal force between the sliding pin and the slot can be either compressive or tensile.

In fact, the pin is in contact with the slot at two points, but only one of the contact points will be engaged and subjected to compressive normal forces. We do not have to identify which contact

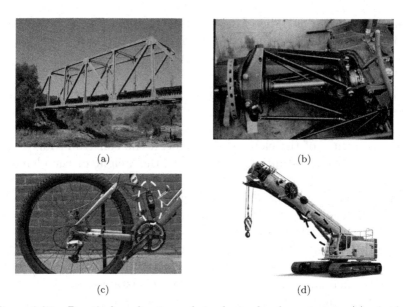

(a)  (b)

(c)  (d)

Figure 2.37:    Practical applications of pined–pined rod constraints: (a) a bridge; (b) multiple-axis rear suspension system; (c) shocks of a bicycle; and (d) hydraulic actuator of a crane hoist.
*Sources*:   (a)   http://www.bridgeofweek.com/2009;    (b)   https://www.deviant art.com/f1-history/art/Lotus-97T-Rear-Suspension-Belgium-1985-384304250; (c) https://en.wikipedia.org/wiki/File:Mountain_Bike_Suspension.jpg; and (d) http://pngimg.com/download/20218.

point is in contact. We only have to label a normal force and allow this force to be in either of the directions. The frictional force is usually small compared with the normal force, in particular if lubrication is applied. Therefore, we only identify one normal force in a sliding collar constraint, as shown in Figures 2.38(c) and 2.38(d).

Examples of elements with a sliding collar constraint are shown in Figure 2.39, which includes a power window regulator used in cars (Figure 2.39(a)), guide rails used in a linear stage of a servo control positioning system (Figure 2.39(b)), a rail for monorail trains in cities (Figure 2.39(c)), and the overhead beam used for a sliding crane host (Figure 2.39(d)).

### 2.15.7   *Total constraint*

The last constraint type is called total or full constraint. When an element is attached to another element with a full constraint, there

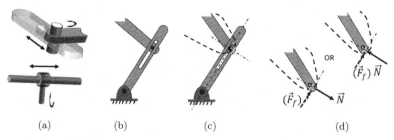

(a)          (b)          (c)          (d)

Figure 2.38: (a) Sliding collar examples; (b) 2D representation; and (c–e) the construction of the corresponding FBDs.

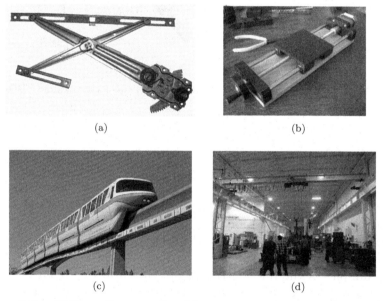

(a)                              (b)

(c)                              (d)

Figure 2.39: Sliding collar examples: (a) an automatic window regulator; (b) a sliding stage; (c) a monorail train; and (d) an overhead crane hoist.

*Sources*: (a) https://robotics.stackexchange.com/questions/1935/how-to-open-a-sliding-window; (b) https://www.thingiverse.com/thing:898212; (c) https://en.wikipedia.org/wiki/Walt_Disney_World_Monorail_System; and (d) https://www.flickr.com/photos/obrieninstallations/7703847972.

will be no relative movement between these two elements. A full constraint means that all degrees of freedom have been eliminated between these two elements. As shown in Figure 2.40(a), a rod is buried inside a plate such that the rod and the plate are bonded together as one element. Similarly, a post can be bolted to a plate

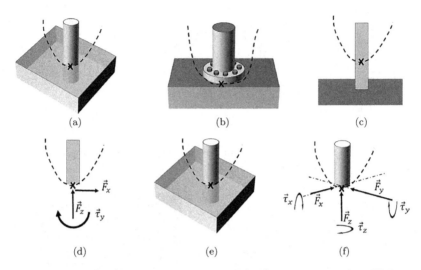

Figure 2.40:  (a–c) Total constraints; and (d–f) the corresponding FBDs.

to achieve the same constraint (Figure 2.40(b)). In a 2D case, the full constraint should be replaced with two equivalent forces and a moment as shown in Figures 2.40(c) and 2.40(d). Note that the boundary can cut across anywhere on the rod, which would result in different forces and moments. Similarly, in a 3D case, the full constraint should be replaced with three equivalent forces and three equivalent moments.

Examples of the full constraints are shown in Figure 2.41, which includes a post bolted to a plate (Figure 2.41(a)), the bridge foundation buried firmly into the river bed (Figure 2.41(b)), a workpiece held tightly by a three-jaw chuck and a centering drill bid held by a drill chuck (Figure 2.41(c)), and a few square tubes welded together (Figure 2.41(d)).

## 2.16  Contact and Constraint Table

Table 2.1 summarizes all the contact and constraint cases discussed in Section 6.2. Students should refer to this when they construct free-body diagrams.

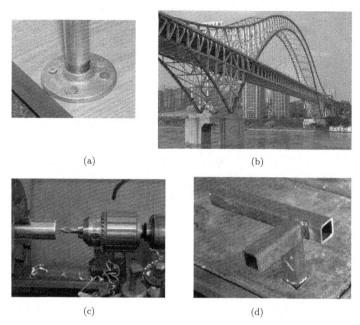

(a)  (b)

(c)  (d)

Figure 2.41:   The total constraint examples: (a) bolted post; (b) bridge foundations; (c) workpiece and tool held by chucks; and (d) welded tubes.

*Sources*: (a) http://www.philobiblon.com/nigropress/; (b) https://en.wikipedia .org/wiki/Chaotianmen_Bridge; (c) https://www.flickr.com/photos/designand technologydepartment/5162652710; and (d) http://newprotest.org/categories.pl? forging.

## 2.17   Concluding Remarks

Table 2.1 provides a convenient reference for constructing free-body diagrams. In the contact cases, the normal forces are always compressive, while in the constraint cases, the force and moment can be in either direction. At equilibrium, there are three forces/moments governing equations in a 2D case, while there are six forces/moments governing equations for a 3D case. Therefore, when constructing free-body diagrams, it is important to keep track of unknowns when drawing a boundary to "free" a body. In a 2D case, the unknowns are limited to three, while in a 3D case the unknowns number six. In some special cases, we can add additional governing equations based on geometry or other conditions. However, it is better to observe the number of unknowns when constructing a free-body diagram.

Table 2.1:  Free-body diagram reference table.

| Type/Application | Free Body | Equivalent Forces/Moments, 2D | Equivalent Forces/Moments, 3D | Remarks |
|---|---|---|---|---|
| Point contact | | $\vec{N}$, $\vec{F}_f$ | $\vec{N}$, $\vec{F}_{fx}$, $\vec{F}_{fy}$ | The normal force is always compressive. Friction can be neglected if the surface is smooth or lubricated. |
| Line contact | | $\vec{N}$, $\vec{F}_f$; $A$, $x$, $\vec{N}$, $\vec{F}_f$ | — | The equivalent forces can be represented as distributed or concentrated. As a concentrated force, the application point of the normal force should be identified. |
| Area contact | | — | $\vec{F}_{fy}$, $\vec{\tau}_{xy}$, $A$, $x$, $\vec{N}$, $\vec{F}_{fx}$, $\vec{\tau}_{zx}$ | The application point of the normal force should be identified and the two horizontal forces could be moved to the same point with resulting moments. |
| Wire constraint | | $\vec{T}$ | — | The force is always tensile, pointing away from the boundary. |

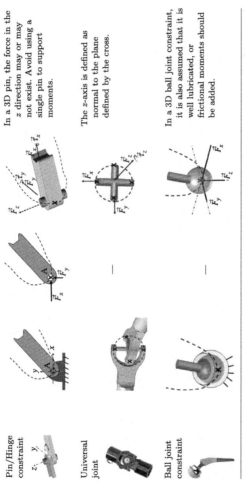

**Pin/Hinge constraint** — In a 3D pin, the force in the z direction may or may not exist. Avoid using a single pin to support moments.

**Universal joint** — The z-axis is defined as normal to the plane defined by the cross.

**Ball joint constraint** — In a 3D ball joint constraint, it is also assumed that it is well lubricated, or frictional moments should be added.

(*Continued*)

Table 2.1:  (*Continued*)

| Type/Application | Free Body | Equivalent Forces/Moments, 2D | Equivalent Forces/Moments, 3D | Remarks |
|---|---|---|---|---|
| Non-spherical constraint | | | — | The main purpose of a non-spherical constraint is for applying a torque. |
| Pined–pined rod constraint | | | — | The force is aligned with the straight line passing through both pins. |
| Sliding collar constraint | | | — | Frictional force can be neglected if well lubricated. The normal force could be in either direction. |
| Total constraint | | | | The total constraint will have three unknowns in the 2D case and six unknowns in the 3D case. |

Chapter 3

# Particle Kinematics — Motion Analysis for Particles

As we discussed before, the entire subject of dynamics is based on *Newton's Second Law*, $\sum \vec{F} = m\vec{a}$. In Chapter 2, we discussed how to conduct correct force analysis, which is related to the left-hand side of the Law. In this chapter, we will discuss the right-hand side of *Newton's Second Law*, which is related to the system, represented by $m$, and the motion, represented by $\vec{a}$. The motion analysis is denoted as Kinematics. For the system, we will start with one particle and proceed to multiple particles. A special multiple-particle system is denoted as the rigid body. We will discuss the motion of one rigid body and then multiple-rigid-body systems as in machine assemblies in the next chapter. For particle kinematics discussed in this chapter, we will proceed from 1D motion to 2D motion, and then 3D motion.

## 3.1 Which Motion Observation is Correct?

Looking at *Newton's Second Law*, we must point out that the force analysis, as discussed in Chapter 2, is totally independent of the motion analysis. However, we also know from our experience that different observers might observe the same motion differently. For example, for the same motion of a car on a highway, its speed is different for a police in a parked car, compared with another police in a car in chase. The speed of the chasing police car skews the observation of the car speed. Note that it is not the velocity involved in

*Newton's Second Law*, but the acceleration. Therefore, we should choose an observer that will not skew the acceleration observation. There are two ways that the acceleration observation could be skewed. First, if the observer is accelerating, the observer could perceive an acceleration of an object even if the object is not moving. Second, the observer must not be spinning. If a person spins, he or she will see the world as spinning, though in fact, it is not.

We should now define what the observation of a motion is. As shown in Figure 3.1, the observer is $O$ and the object to be observed is B. The observer needs some references and metrics in order to define the motion of B properly. This is done typically by using a Cartesian coordinate, $x-y-z$, as shown in Figure 3.1. The motion of B observed by $O$ should include its position, velocity, and acceleration, expressed as $\vec{r}_{B/O}$, $\vec{v}_{B/O}$, and $\vec{a}_{B/O}$, respectively. Note that the notation, $B/O$, is important and it reads as $B$ with respect to $O$ or $B$ observed by $O$.

Now we need to consider the motion of point $O$ and the references used for observing the motion of B. It is obvious that the position of $O$ will affect the measurement of $\vec{r}_{B/O}$. The measured velocity $\vec{v}_{B/O}$ will be affected by the velocity of $O$, while the measured acceleration $\vec{a}_{B/O}$ will be affected by the acceleration of $O$. Considering *Newton's Second Law* is related to the force and the acceleration, we apparently need point $O$ to have no acceleration.

As we know, motion variables, $\vec{r}_{B/O}$, $\vec{v}_{B/O}$, and $\vec{a}_{B/O}$, are all vectors with magnitudes and directions. The direction is defined with

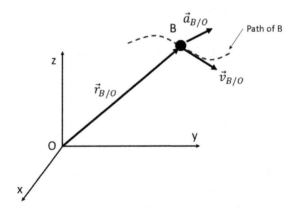

Figure 3.1:   A Cartesian coordinate used by observer $O$ to observe the motion of object B.

respect to the reference frame, $x-y-z$. If these axes are not pointing to fixed directions, then the vectors of $\vec{r}_{B/O}$, $\vec{v}_{B/O}$, and $\vec{a}_{B/O}$ will change even if B is not moving.

From the above two observations, it becomes important that we define the properties of the observer and $x-y-z$ before we can even discuss *Newton's Second Law*. In fact, *Newton's Second Law* is valid only if $O-x-y-z$ has the property that point $O$ is not accelerating and each axis of $x-y-z$ is pointing to a fixed direction without changing. A reference frame with this property is denoted as an Inertial Reference Frame (IRF).

Unfortunately, the definition of the IRF brings forth more questions. Who will observe $O-x-y-z$ to define its property? For example, is a spinning gyro pointing to a fixed direction as commonly believed? This is actually a very profound question and some think that there is no definitive answer to it yet.[1]

We will revisit this question in more depth later. For now, let's accept that, for the purpose of engineering, we can define a proper IRF. If point $O$ has no acceleration, we can show that point $O$ must be moving in a straight line at a constant speed. In other words, the velocity of point $O$ is constant. We should note the difference between the speed and the velocity. The speed is a scalar, representing the magnitude of the velocity. For each axis of $x-y-z$ to point at a fixed direction, then the reference frame of $x-y-z$ must not be rotating. If one is not using an IRF to observe the motion of an object, *Newton's Second Law* is not valid to link the motion of the object and the forces acting on it.

For now, let's just accept that we have the convenience of defining an IRF for most of our engineering problems.

Based on an IRF, $O-x-y-z$, as shown in Figure 3.1, *Newton's Second Law* can be applied for object $B$ as

$$\sum \vec{F} = m\,\vec{a}_{B/O} \qquad (3.1)$$

where $m$ is the mass of object $B$, $\sum \vec{F}$ is the total force acting on $B$, and $\vec{a}_{B/O}$ is the acceleration of $B$.

Equation (3.1) can be used to predict $\vec{a}_{B/O}$ if we know $m$ and $\sum \vec{F}$. We can also determine $\sum \vec{F}$ if we know $m$ and measure $\vec{a}_{B/O}$

---

[1]Rothman T. The Forgotten Mystery of Inertia, *American Scientist* **105**(6), (2017) 344. DOI:10.1511/2017.105.6.344.

with respect to an IRF. Finally, we can calculate $m$ if we know $\sum \vec{F}$ and $\vec{a}_{B/O}$ independently. For different fields of applications, *Newton's Second Law* is used differently.

## 3.2 Units of Forces

Based on *Newton's Second Law*, we can define the unit of force. If we define a force based on International System of Units (SI Units), the mass is defined in kilogram (kg), the time in second (s), and the length in meter (m). The length unit is also called the metric system. The acceleration takes on the unit of $m/s^2$.

The unit for the force becomes $kg\,m/s^2$. However, this unit is too cumbersome to write. Therefore, the SI Units conveniently define

$$1\,N = (1\,kg)\left(1\frac{m}{s^2}\right) = 1\left(kg\frac{m}{s^2}\right) \tag{3.2}$$

where N stands for the *newton* as a new unit for the force. Equation (3.2) is conveniently defined and can be described as: if we observe that an object with $1\,kg$ of mass is undergoing $1\,m/s^2$ of acceleration, then the amount of the net force applied to the object is $1\,N$. The same object if placed near the surface of the Earth will be subjected to a gravitational pull by the Earth that is

$$F_w = mg = (1\,kg)(9.81\,m/s^2) = 9.81\,kg\,m/s^2 = 9.81\,N \tag{3.3}$$

Equation (3.2) can be rewritten as

$$1 = \frac{N}{(kg\,m/s^2)} = \frac{(kg\,m/s^2)}{N} \tag{3.4}$$

Equation (3.4) is denoted as an unity equation. It is very useful for unit conversion because we can multiply a unity equation to any equation without altering it. For example, if we want to change the unit of $F_w$ of Equation (3.3) from $kg\,m/s^2$ to N, we can proceed as follows:

$$F_w = 9.81\,kg\,m/s^2$$

$$= 9.81\,kg\,m/s^2 \times 1 = 9.81\,kg\,m/s^2 \times \frac{N}{(kg\,m/s^2)}$$

$$= 9.81\,N \tag{3.5}$$

In fact, we can define many different forms of Equation (3.4) to convert units.

Other commonly used units, particularly in the US, are based on the foot (ft), the pound mass (lbm), and the second (s). Quite often, these units are referred to as the English units in the US, but they are more customarily used in the US than in the UK, in particular, for the length. If a person walks into a hardware store in the US to buy hand tools, the clerk might ask: "Do you want standard or metric?" The "standard" here means the length system in inches and feet. The standing joke is that this "standard" is not the standard used in the world. Moreover, it should not to be referred to as the English unit either, because, in the UK, meters are used. Therefore, the metric system should be referred to as the standard, while the "*standard*" should be referred to as "American." The question that should be rephrased as: "Do you want standard or American?"

Another confusion with this unit system is related to the word "pound." In this book, we should be very precise in indicating which "pound" we are referring to because the same word is used to indicate either mass or force in most standard textbooks in Statics and Dynamics. This causes a lot of confusion, and it can result in major mistakes in numerical values during calculation. Therefore, in this book, when the word "*pound*" is used for mass, we always refer to it as "*pound mass*", expressed as lbm. When the word "pound" is used as a unit for the force, we will call it "*pound force*," expressed as lbf. Now we can define one *pound force*, 1 lbf, as follows:

$$1\,\text{lbf} = (1\,\text{lbm})(32.2\,\text{ft/s}^2) = 32.2\,\text{lbm}\,\text{ft/s}^2 \qquad (3.6)$$

If we try to explain this definition based on Equation (3.6), then one *pound force* (lbf) is equal to the gravitational force the Earth applies to an object with one pound mass placed near the Earth's surface. However, because the value of $g$ varies at different locations on the Earth, the value of 9.81 m/s$^2$ or 32.2 ft/s$^2$ is for the gravitational acceleration at the sea level and at a latitude of 45°. At the North Pole, the value of $g$ is higher, while at the equator, it is lower.

Of course, we are not restricted to that particular spot on the Earth to use the unit of lbf. We need to standardize the definition of lbf strictly based on the numerical values of Equation (3.6). One should notice that this definition is not very convenient, unlike the SI unit system. In addition, if we do not distinguish lbf from lbm, it could become very confusing when one tries to use this definition correctly. To emulate the convenience of the SI system, a new definition was proposed to convert Equation (3.6) to

$$1 \, \text{lbf} = (1 \, \text{slug})(1 \, \text{ft/s}^2) \tag{3.7}$$

and

$$1 \, \text{slug} = 32.2 \, \text{lbm} \tag{3.8}$$

Instead of one equation, we now have two equations to define the pound force, lbf. A new mass unit as the *slug* is defined by Equation (3.8). In fact, it is easier to stay with Equation (3.6); hence, the unit of slug is not commonly used. Some people do prefer using the unit *slug*.

It is easier to remember Equation (3.6) by stating "One pound-force is the gravitational pull acting at an object with one pound mass placed on (a special spot of) the Earth."

Based on Equation (3.6), we can also establish a unity equation similar to Equation (3.4),

$$1 = \frac{\text{lbf}}{(32.2 \, \text{lbm ft/s}^2)} = \frac{(32.2 \, \text{lbm ft/s}^2)}{\text{lbf}} \tag{3.9}$$

Equation (3.9) is useful to ensure correct unit and calculation when the "American" system of units is used. For example, if an object with one *pound mass* undergoes 1 ft/s² acceleration, then the net force applied to the object is

$$F_1 = (1 \, \text{lbm}) \left( \frac{1 \, \text{ft}}{\text{s}^2} \right)$$

$$= 1 \left( \text{lbm} \frac{\text{ft}}{\text{s}^2} \right) \times 1 = 1 \left( \text{lbm} \frac{\text{ft}}{\text{s}^2} \right) \times \frac{\text{lbf}}{\left( 32.2 \, \text{lbm} \frac{\text{ft}}{\text{s}^2} \right)} = \frac{1}{32.2} \, \text{lbf} \tag{3.10}$$

The common mistake for the value of $F_1$ is 1 lbf, which is off by 3,220%. Major disaster can happen due to such a silly but easy mistake.

Equations (3.4) and (3.9) are useful for converting the force unit from N to lbf, or vice versa.

Before doing so, we need to define more unity equations.

$$1 = \frac{\text{ft}}{0.3048 \text{ m}} = \frac{0.3048 \text{ m}}{\text{ft}} = \frac{\text{ft}}{12 \text{ in}} = \frac{\text{in}}{25.4 \text{ mm}} = \frac{\text{m}}{1,000 \text{ mm}}$$

$$= \frac{\text{mile}}{5,280 \text{ ft}} = \frac{0.4536 \text{ kg}}{\text{lbm}} \tag{3.11}$$

The conversion between lbf and N can be determined as

$$1 \text{ lbf} = 32.2 \text{ lbm} \frac{\text{ft}}{\text{s}^2} = 32.2 \text{ lbm} \frac{\text{ft}}{\text{s}^2} \times \frac{0.4536 \text{ kg}}{\text{lbm}} \times \frac{0.3048 \text{ m}}{\text{ft}}$$

$$= 32.2 \times 0.4536 \times 0.3048 \text{ kg} \frac{\text{m}}{\text{s}^2} = 4.451 \text{ kg} \frac{\text{m}}{\text{s}^2} = 4.451 \text{ N}$$

$$\tag{3.12}$$

However, this result is different from the standard conversion number as 1 lbf = 4.448 N because of the rounded-off numbers used in Equation (3.11). To get 1 lbf = 4.448 N, we have to use 1 lbf = 32.174 lbm ft/s$^2$ and 1 lbm = 0.45359243 kg, which yields 1 lbf = 4.44821549 N.

By using Equation (3.11) for conversions, we could have 0.0674% error, which may not seem huge but could sometimes be critical. Therefore, it might be useful to establish one more unity equation for convenience

$$1 = \frac{\text{lbf}}{4.448 \text{ N}} = \frac{4.448 \text{ N}}{\text{lbf}} \tag{3.13}$$

Finally, as an exercise, let us calculate how much force is needed to make a typical full size car with 3,500 lbm to achieve 0.5 g acceleration.

$$F_w = (3,500 \text{ lbm})(0.5)\left(32.2 \frac{\text{ft}}{\text{s}^2}\right) \tag{3.14}$$

$$= 5,6350 \left(\text{lbm} \frac{\text{ft}}{\text{s}^2}\right) = 56,350.0 \left(\text{lbm} \frac{\text{ft}}{\text{s}^2}\right) \frac{\text{lbf}}{\left(32.2 \text{lbm} \frac{\text{ft}}{\text{s}^2}\right)} \tag{3.15}$$

$$= 1,750.0 \text{ lbf} = 1,750.0 \text{ lbf} \times \frac{4.448 \text{ N}}{\text{lbf}} = 7,784.0 \text{ N} \tag{3.16}$$

We have one last remark on the force unit. Quite often, when referring to the body weight, we will use kg as the unit. This is incorrect from the engineering point of view because weight is a force, not a mass. Perhaps we should say "What is your body mass?" or "Your body weight is 637.65 newtons (65 kg)." It could sound strange but correct. Similarly, we should say "His weight is 150 pound-force." or "His body mass is 150 pound-mass."

## 3.3   General Motion Analysis

In order to describe the motion of a particle precisely, we must define its position, $\vec{r}_{B/O}$, velocity, $\vec{v}_{B/O}$, and acceleration, $\vec{a}_{B/O}$, with respect to an IRF, as shown in Figure 3.1. These variables are related as

$$\vec{v} = \frac{d\vec{r}}{dt} = \lim_{\Delta t \to 0} \frac{\Delta \vec{r}}{\Delta t} \tag{3.17}$$

$$\vec{a} = \frac{d\vec{v}}{dt} = \lim_{\Delta t \to 0} \frac{\Delta \vec{v}}{\Delta t} \tag{3.18}$$

Equations (3.17) and (3.18) can be expressed graphically as shown in Figure 3.2. When $\Delta t \to 0$, $\Delta \vec{r}$ becomes tangential to the path position at $t_1$; therefore, the velocity at $t_1$ is tangential to the path. Similarly, the velocity vector at $t_2$ is tangential to the new position on the path. The difference in the velocity vectors is related to the acceleration, as shown in Figure 3.2. However, as shown in Figure 3.1, the direction of $\vec{a}$ is generally unknown and could not be determined

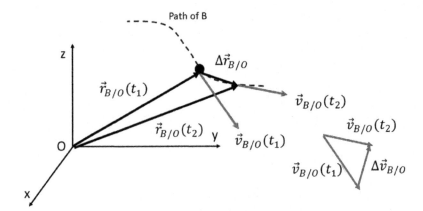

Figure 3.2:   Graphical representations of motion variables

by the path alone. We should recognize that the velocity vector is always tangential to the path.

If $O-x-y-z$ is an IRF, Equations (3.17) and (3.18) can be further simplified as

$$\vec{v} = \frac{d\vec{r}}{dt} = \frac{d(x\vec{i} + y\vec{j} + z\vec{k})}{dt}$$

$$= \frac{dx}{dt}\vec{i} + \frac{d\vec{i}}{dt}x + \frac{dy}{dt}\vec{j} + \frac{d\vec{j}}{dt}y + \frac{dz}{dt}\vec{k} + \frac{d\vec{k}}{dt}z$$

$$= \frac{dx}{dt}\vec{i} + \frac{dy}{dt}\vec{j} + \frac{dz}{dt}\vec{k} = \dot{x}\vec{i} + \dot{y}\vec{j} + \dot{z}\vec{k}$$

$$= v_x\vec{i} + v_y\vec{j} + v_z\vec{k} \tag{3.19}$$

$$\vec{a} = \frac{d\vec{v}}{dt} = \frac{d(v_x\vec{i} + v_y\vec{j} + v_z\vec{k})}{dt}$$

$$= \frac{dv_x}{dt}\vec{i} + \frac{d\vec{i}}{dt}v_x + \frac{dv_y}{dt}\vec{j} + \frac{d\vec{j}}{dt}v_y + \frac{dv_z}{dt}\vec{k} + \frac{d\vec{k}}{dt}v_z$$

$$= \frac{dv_x}{dt}\vec{i} + \frac{dv_y}{dt}\vec{j} + \frac{dv_z}{dt}\vec{k} = \ddot{x}\vec{i} + \ddot{y}\vec{j} + \ddot{z}\vec{k}$$

$$= a_x\vec{i} + a_y\vec{j} + a_z\vec{k} \tag{3.20}$$

Note that because of the use of the inertial frame, the unit vectors along each axis, $\vec{i}$, $\vec{j}$, and $\vec{k}$, are constants with their directions fixed; therefore, their time derivatives are zero.

Equations (3.19) and (3.20) are easy to implement and we can just measure one of the three variables and obtain the other two either by taking derivatives or by integration. We will discuss sensors for measuring motion variables later in this chapter.

## 3.4 One-Dimensional Motion — Straight Line Motion

In this section, we consider the simplest type of motion in which a particle travels along a straight line. Without loss of generality, we can define that this straight line aligns with the $x$-axis, as shown in Figure 3.3(a). Along this straight line, moving to the right is defined as a positive direction, while to the left is negative. As a result, we can represent the motion variables as scalars, $\vec{r} \rightarrow x$, $\vec{v} \rightarrow v$,

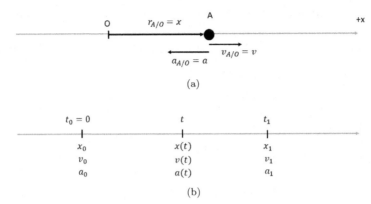

(a)

(b)

Figure 3.3: (a) One-dimensional motion of particle A; and (b) the time line depicting the motion variables.

and $\vec{a} \to a$, with the positive and negative signs used for directions. In Figure 3.3(a), we also define the origin of an IRF at point $O$. Because we are only considering one particle and one IRF, we will skip the subscript notations. For the example in Figure 3.3(a), particle A has a positive position, a positive velocity, and a negative acceleration. We can also define a time line to bring the time variable to the motion variables, as shown in Figure 3.3(b). At every instant of time, the three motion variables are specified and they are typically time-dependent; i.e., $x(t)$, $v(t)$, and $a(t)$ are functions of time.

With the motion variables defined as scalars, Equations (3.17) and (3.18) can be simplified as

$$v = \frac{dx}{dt} = \lim_{\Delta t \to 0} \frac{\Delta x}{\Delta t} \tag{3.21}$$

$$a = \frac{dv}{dt} = \lim_{\Delta t \to 0} \frac{\Delta v}{\Delta t} \tag{3.22}$$

We can also construct a few variations of Equations (3.21) and (3.22) as

$$dt = \frac{dx}{v} \tag{3.23}$$

$$dt = \frac{dv}{a} \tag{3.24}$$

Combining Equations (3.23) and (3.24), we have

$$a \, dx = v \, dv \tag{3.25}$$

Equation (3.25) is very useful in some applications because it does not contain the time variable.

Equation (3.21) can also be reconstructed in the integration form,

$$\int_{t_0}^{t} v(t)\, dt = \int_{x_0}^{x} dx = x(t) - x_0 \qquad (3.26)$$

In order to carry out the integration for the left of Equation (3.26), we must know the velocity as a function of time. From Equation (3.26), we obtain the position as function of time, $x(t)$, if we know the initial position, $x_0$.

Accordingly, we can construct the integration versions of Equations (3.22)–(3.25) as

$$\int_{t_0}^{t} a(t)\, dt = \int_{v_0}^{v} dv = v(t) - v_0 \qquad (3.27)$$

$$t - t_0 = \int_{t_0}^{t} dt = \int_{x_0}^{x} \frac{dx}{v(x)} \qquad (3.28)$$

$$t - t_0 = \int_{t_0}^{t} dt = \int_{v_0}^{v} \frac{dv}{a(v)} \qquad (3.29)$$

$$\int_{x_0}^{x} a(x)\, dx = \int_{v_0}^{v} v\, dv = \frac{1}{2}(v^2 - v_0^2) \qquad (3.30)$$

Note that in Equation (3.27), we must know the acceleration as a function of time. Similarly, In Equation (3.28), the velocity must be known as a function of position; in Equation (3.29), acceleration as a function of velocity; and in Equation (3.30), acceleration as a function of position.

Equations (3.23)–(3.30) are useful, but they do not need to be memorized. Instead, they should be derived from Equations (3.21) and (3.22). We can also make a table of these equations for references when we use them to solve problems.

### 3.4.1 *Sensors for measuring motion variables*

To describe the motion of a particle with respect to an IRF, we need to define the three motion variables, as discussed before. Typically, we also need to have the information of time to render these variables as functions of time. To save cost, we can only measure one motion variable and the time. The other two motion variables can then be calculated using two equations out of Equations (3.21)–(3.30) through taking derivatives or integrations.

A brief discussion of motion sensors is in order. We will not discuss the measurement of time even though timing is a difficult issue in physics. We are confident that the timing is very easy and accurate for typical engineering applications.

First, let's discuss the position measurement. For machines, the most common position sensor is called encoder. There are encoders for the linear displacement measurement, denoted as linear encoders or linear scales (see Figure 3.4(a)). The internal structure of a linear encoder is shown in Figure 3.4(b). A light beam passes through a piece of film with dark strips. On the other side, detectors are mounted to pick up the light signals. When the light source is moving

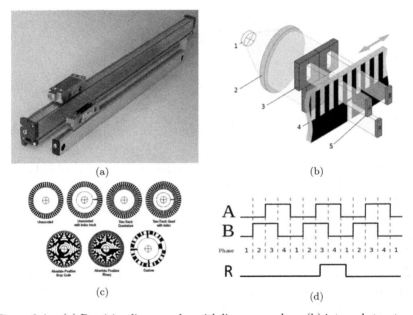

Figure 3.4: (a) Precision linear scales with linear encoders; (b) internal structure of a linear encoder with three signals, A, B and R; (c) different designs of shaft encoders; and (d) the signal outputs as pulse trains of A, B, and R.

*Sources*: (a) https://c1.staticflickr.com/7/6092/6333918787_9c010bd1f2_b.jpg; (b) https://upload.wikimedia.org/wikipedia/commons/thumb/e/e8/Linear_Scale_Scheme.svg/308px-Linear_Scale_Scheme.svg.png; (c) http://i.stack.imgur.com/H6pKj.gif; and (d) https://upload.wikimedia.org/wikipedia/commons/thumb/6/68/Quadrature_Diagram.svg/360px-Quadrature_Diagram.svg.png.

along the film, series of pulses, as shown in Figure 3.4(d), will appear. The appearance of each pulse indicates that the object has moved a fixed distance equal to the distance between two neighboring stripes on the film. This is the resolution of the encoder. Typically, two of such pulse signals, denoted as $A$ and $B$, are arranged to be 90° apart. Depending on which signal is leading, we can determine the movement direction. The signals of $A$ and $B$ can also be combined to provide a higher resolution (four times higher) for the displacement. A third signal, denoted as $R$ or $Z$, is used to provide a reference position. The precision position is defined by the rising or falling edges of the pulses. In Figure 3.4(c), the same encoder structure can be arranged to a circle, forming a shaft encoder for angle measurement. In fact, shaft encoders are cheaper to make. Linear displacement can be measured by combining a shaft encoder and a ball screw, which converts a rotational motion to a linear motion. Such position measurements are commonly seen in machine tools with digital readouts to precisely measure the position of the cutting tool. There are other position sensors, but we will not discuss this topic further.

To obtain the velocity and acceleration signals from the position measurement, we simply utilize Equations (3.21) and (3.22) by taking derivatives of the position signal with respect to time. However, we should always consider the noise problems in taking multiple derivatives. The noise problem is more pronounced for a slow motion because the signal to noise ratio is lower.

As for the velocity sensor, it turns out that the rotational speed is the easiest to measure. By reversing a DC motor, we have a tachometer and its output voltage is proportional to the rotational speed, as shown in Figure 3.5. When used as a motor, the torque of the motor is proportional to the current passing through the coil. A DC motor typically has a tachometer attached co-axially for providing the measurement of the rotational speed. Such a device is commonly used as the speedometer of older cars. We can easily convert the rotational speed of a tire to the speed of the car if we know the tire diameter. Note that the speed display will be incorrect if we use a different size of tires.

Another commonly used velocity sensor is the radar gun, based on electromagnetic waves or laser beams for the speed measurement, as shown in Figure 3.6. A radar gun is useful for measuring the speed of an object at a distance, but the resolution and accuracy are not as good as the encoder.

(a)

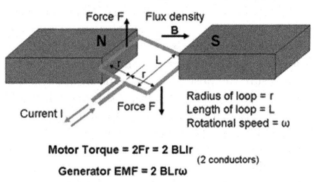

(b)

Figure 3.5: (a) A DC motor with a tachometer attached coaxially at the rear end; and (b) the mechanism shared by the dc motor and the tachometer.

*Sources*: (a) https://i.stack.imgur.com/Jz439.png; and (b) https://i.stack.imgur.com/Jz439.png.

To obtain the acceleration, we can take a derivative from the velocity signal (Equation (3.22)). By using Equation (3.26), we can obtain the position information.

The invention of the laser beam opened the door for precise and large scale motion measurement. Such a device is called the laser interferometer, as shown in Figure 3.7. We will not discuss the details of laser interferometers, but readers are encouraged to learn more about them.

Finally, it turns out that acceleration can be measured quite handily by a miniaturized device, called the accelerometer. In

(a) (b)

Figure 3.6: (a) A radar gun using electromagnetic waves for vehicle speed measurement; and (b) a radar gun using a laser beam for the speed measurement.
*Sources*: (a) https://upload.wikimedia.org/wikipedia/commons/thumb/b/b3/Radarvelocidade20022007.jpg/1200px-Radarvelocidade20022007.jpg; and (b) http:// www.delonixradar.com.au/new-south-wales/images/index.2.jpg.

Figure 3.8(a), a mechanical accelerometer is attached to a beam for modal analysis. An accelerometer can also be made into micro-scale by using the semiconductor technology, as in micro-electromechanical systems (MEMS) for multi-directional measurements (Figure 3.8(b)). MEMS-based accelerometers can be produced at low cost and are widely deployed in modern cars for vibration and impact measurements, as shown in Figure 3.8(c).

To obtain the information of the position and the velocity from the acceleration measurement, we simply take integrations using Equations (3.26) and (3.27). In general, the noise problem is less severe in integration compared with taking derivatives. In Figure 3.9, an accelerometer is used to measure the acceleration of a piano key when it is pressed downward. The corresponding velocity and position signals are then obtained by taking integration. This was a study conducted by the author to study the sound quality of a piano and the piano playing skills.

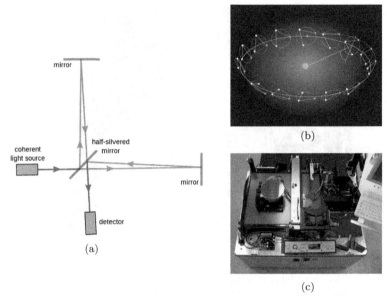

Figure 3.7: (a) The layout of a basic laser interferometer; (b) laser interferometers used for astrophysics research for measuring large-scale distances and positioning; and (c) laser interferometers used in semiconductor manufacturing with ultra-precision positioning and control.

*Sources*: (a) https://upload.wikimedia.org/wikipedia/commons/thumb/e/e7/ Interferometer.svg/1200px-Interferometer.svg.png; (b) https://upload.wikimedia. org/wikipedia/commons/f/f4/LISA-orbit.jpg; and (c) https://upload.wikimedia. org/wikipedia/commons/5/5b/Wafer_prober_service_configuration.jpg.

### 3.4.2  *Real-World examples of one-dimensional motion*

Let's consider practical examples of the one-dimensional (1D) motion. We will consider examples of constant speed, constant acceleration, acceleration as a function of speed, acceleration as a function of position, and velocity as a function of position.

#### 3.4.2.1  *Constant speed motion*

Newton's Laws of Motion state: "Every object persists in its state of rest or uniform motion in a straight line unless it is compelled to change that state by forces impressed on it."

Therefore, according to Professor Newton, if an object is at rest or moving along a straight line at a constant speed, the net force acting on the object is zero. If the object is at rest, we have the force

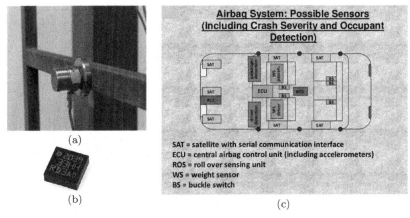

Figure 3.8: (a) An accelerometer attached to a structure for modal analysis; (b) a semiconductor accelerometer capable of measuring accelerations in three directions; and (c) accelerometers and other sensors deployed in a car for safety control and the activation of airbags.

*Source*: (a) https://upload.wikimedia.org/wikipedia/commons/thumb/a/af/ST_Triple-Axis-Accelerometer_LIS331HH_09755-01.jpg; (b) https://cdn.jvejournals.com/articles/17588/xml/img4.jpg; and (c) https://www.slideshare.net/abhisheksutrave/embedded-systems-in-automobile.

equilibrium as discussed in Statics. For the case with a constant velocity, the object will move along one fixed direction at a constant speed. Therefore, from Equations (3.21) and (3.22), we have

$$v = \frac{dr}{dt} = v_0 \tag{3.31}$$

$$a = \frac{dv}{dt} = 0 \tag{3.32}$$

Introducing Equation (3.31) to Equation (3.26), we have

$$\int_{t_0}^{t} v(t)\, dt = \int_{t_0}^{t} v_0\, dt = v_0(t - t_0) = x(t) - x_0 \tag{3.33}$$

From above, we obtain the position as a function of time, $x(t)$. In the real world, we have many constant velocity cases. If we activate the cruise control of our car to travel at a constant speed along a straight road, the driving force of the car and the resistance to the car will be the same. When a spaceship travels in outer space, away from any planets and stars, it will travel along a straight line at a constant speed if it does not fire its engine.

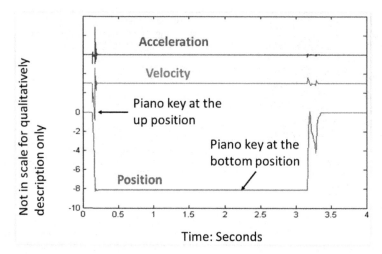

Figure 3.9: The measurement of the acceleration of a piano key when it was pressed down rapidly. The corresponding velocity and position signals were derived by taking integration. This is a slide from a lecture by Dr. Jay Tu, Department of Mechanical and Aerospace Engineering, NCSU and Dr. Olga Kleiankina, Music Department, NCSU Presented for Raleigh Piano Teachers Association. *Source*: https://www.youtube.com/watch?v=zPf7TURAd5M&t=1561s.

### 3.4.2.2 *Constant acceleration motion*

If we ignore the air resistance, a free falling object is a typical case of one-dimensional motion of constant acceleration.

### Example 3.1: Tennis Ball Vertical Movement

If a tennis ball is tossed up vertically, ignoring air resistance and the spinning of the ball, at an initial speed of $v_0$, how high will it reach and what is the speed when it falls back to the same vertical position?

**Solution:** We will follow the rigorous steps outlined in Chapter 1 to solve this problem.

**Problem Statement:** We will draw some sketches to represent the problem graphically as in Figure 3.10.

We define an IRF with the $y$-axis and show that object A starting from the origin at point $O$, reaching the top position, and returning

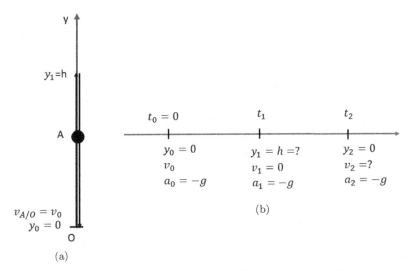

Figure 3.10: (a) A tennis ball tossed vertically upward; and (b) the time line representation.

to point $O$. We also draw a time line similar to the one in Figure 3.3. We define $t_0 = 0$ at the initial position with the initial speed, $v_0$. At $t_1$, object A reaches its highest position at $y_1 = h$, which is unknown when the speed is zero. Finally, at $t_2$, it returns to the original position with a downward speed, $v_2$, also an unknown. Throughout this entire process, the acceleration is $-g$.

**Force Analysis:** Skipped.

**Motion Analysis:** We make sure that we define a proper IRF as shown. The upward direction is defined as positive. The motion type of object A is a straight line motion with a constant acceleration. The acceleration of $-g$ indicates that the acceleration is pointing downward.

**Governing Equations:** Because this is a one-dimensional motion, we have Equations (3.21)–(3.30) as our governing equations. We don't know immediately which equation we should use, but we can simply go down the list and find suitable ones. First, we need to consider motion variables across different times; therefore, we should focus on Equations (3.26)–(3.30).

**Solving Equations:** Let's go down the equation list to decide which one or ones to use. We cannot use Equation (3.26) because we don't have the information of the velocity as a function of time. Similarly, we cannot use Equation (3.28) because we don't have the velocity as a function of position. Because $a = -g$, a constant, we can use Equations (3.27), (3.29), and (3.30). However, Equations (3.27) and (3.29) cannot be used because they contain an extra unknown of the time variable.

Based on Equation (3.30), we have

$$\int_0^h a(y)\, dy = (-g)(h-0) = \frac{1}{2}(0^2 - v_0^2) \tag{3.34}$$

The maximum height is found to be $h = v_0^2/(2g)$. Applying Equation (3.30) again for the falling process, we have

$$\int_h^0 a(y)\, dy = (-g)(0-h) = \frac{1}{2}(v_2^2 - 0^2) \tag{3.35}$$

The velocity $v_2$ is found to be $\pm\sqrt{2gh}$. As Object A is moving downward, the correct answer is $v_2 = -\sqrt{2gh}$. The magnitude of $v_2$ (which is called speed) is the same as that of $v_0$.

This result is actually interesting and important in tennis. When a person is ready to hit an overhead for a ball dropping downward, if the ball is dropping from a high position, the speed will be faster. Therefore, a player would have a shorter reaction time to hit a proper overhead. It would be better to let the ball bounce off the ground first.

We can now apply Equation (3.29) to find out $t_1$ and $t_2$. We easily obtain $t_1 = \frac{v_0}{g}$ and $t_2 = \frac{2v_0}{g}$. The time needed to go to the top position is the same as that to go down from the top to the original position.

**Answer Verification/Extensions:** At this point we should consider the assumptions we made. We have ignored the effect of the air resistance and the spinning of the ball. We should consider how valid these two assumptions are.

As will be discussed later, the air resistance will be significant if the speed is high. How high is too high for our assumptions? It depends on the application and the precision requirement. We should

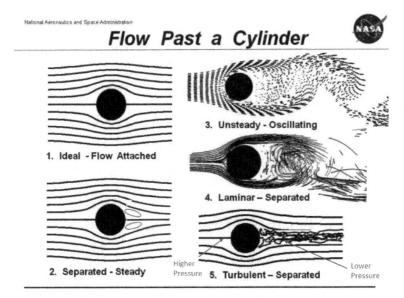

Figure 3.11: The flow patterns when a fluid medium flows past a cylinder at different speeds.
*Source*: https://www1.grc.nasa.gov/beginners-guide-to-aeronautics/drag-of-a-sphere/.

understand that, by ignoring the air resistance, we obtain an upper bound result. Object A probably won't reach the height, $h$, as we predicted if the air resistance is considered. Similarly, $v_2$ could be less than $v_0$. Finally, the spinning of the ball will make the ball move side ways, no longer maintaining its vertical trajectory; thus, it is no longer a one-dimensional motion.

### 3.4.2.3  *Acceleration as a function of speed*

The wind drag, or wind resistance, to an object traveling through air at high speeds is a function of the traveling speed.

Figure 3.11 is a diagram from NASA to show why there is wind drag. We will focus on the #5 case for a cylinder traveling through a fluid medium at high speeds. Without getting into details of fluid mechanics, we note that in front of the cylinder, the pressure is higher, while the pressure is lower at the back. This pressure difference creates the wind drag. At low speeds, it would be more like case #1 of Figure 3.11 without wind drag. From experiments, it is found that the wind drag is often proportional to the square of the speed,

$$F_D = C_D v^2 = \hat{C}_D A v^2 = F_D(v) \tag{3.36}$$

where $C_D$ is a general wind drag coefficient with an unit of $\mathrm{Ns^2/m^2}$, related to the object's geometry, the fluid medium properties, etc, while $\hat{C}_D$ is also commonly used with an unit of $\mathrm{Ns^2/m^4}$ and $A$ is the equivalent cross-section area of the subject. Now, we have a case of acceleration as a function of speed.

Let's rework Example 3.1 to include the wind drag effect.

### Example 3.2: Tennis Ball Dropping Speed with Wind Drag

What is the speed of the ball some time after it is dropped from at rest? The wind drag coefficient $C_D$ is found to be $0.3 \frac{\mathrm{s^2 N}}{\mathrm{m^2}}$.

**Solution:**

**Problem Statement:** Skipped.

**Force Analysis:** Skipped.

**Motion Analysis:** We have the same 1D motion as in Example 3.1. The acceleration is now $a = g - 0.3\, v^2/m$ due to both the gravity and the wind drag. The mass of the ball is $10\,\mathrm{kg}$. We define downward direction as positive. The expression of $a$ actually should be obtained from *Newton's Second Law*, but let's accept it for now.

**Governing Equations:** Again, we should focus on Equations (3.26)–(3.30).

**Solving Equations:** With the same reasoning before, we realize that the only equation we can use is Equation (3.29), which will give us the speed as a function of time. We will do that first.

Based on Equation (3.29), we have

$$t = \int_0^v \frac{dv'}{a(v')} = \int_0^v \frac{dv'}{9.81 - 0.03v'^2} \tag{3.37}$$

The integration of the equation above is not trivial. Therefore, we look up the integration table to complete it. We obtain

$$t = 1.84321 \left(0.5 \ \ln|0.05530v + 1| - 0.5 \ \ln|0.05530v - 1|\right) \tag{3.38}$$

Applying the same equation for the case without the wind drag, we obtain

$$t = \frac{v}{9.81} \qquad (3.39)$$

Let's compare the above two results at different time instants. Based on Equation (3.39), at $t = 2$s, we have $v = 19.62$ m/s. Let's verify how long it would take for the same ball to achieve the same speed. Introducing $v = 19.62$ m/s to Equation (3.38), we obtain $t = 2.949$ s. Apparently, the wind drag slows down the ball because it takes longer to reach the same speed with the wind drag. For reference, the wind drag coefficient of modern cars is about $\hat{C}_D = 0.3 \, \text{Ns}^2/\text{m}^4$; therefore, the wind will slow down the car when we are driving.

### 3.4.2.4 *Acceleration as a function of Position*

There are many examples of the acceleration as a function of the position. *Newton's gravitational law* describes the gravitational force between two objects with masses as

$$F = G\frac{m_1 \, m_2}{r^2} \qquad (3.40)$$

Therefore, the acceleration due to the gravitational force is a function of position. We will discuss *Newton's gravitational law* in greater detail in later chapters.

Let's consider the magnetic force. When two magnets are placed at $r$ distance apart, the attractive or repulsive force between them can be expressed as

$$F = \frac{K}{r^2} \qquad (3.41)$$

where $K$ is related to the magnitudes of magnetic charge on magnetic poles and the permeability of the intervening medium. Similar to the gravitational force, the acceleration due to the magnetic force is a function of position.

Let's consider one more example. For a linear spring, Hooke's law states,

$$F = Kx \qquad (3.42)$$

where $x$ is the deformation of the spring. If we attach an object to a spring and define the position of the spring at its neutral length as $x_0 = 0$, then the force the object is subjected to is described by Equation (3.42) as a function of position. Therefore, the acceleration due to the spring force is a function of position.

## Example 3.3: Spring Compression when Hit by a Ball

What is the maximum compression of the spring when it is hit by a ball of mass $m = 10\,\text{kg}$ dropped from a height of 1 m. The neutral length of the spring is 0.1 m. The spring constant is $k = 100{,}000\,\text{N/m}$. In addition, at which height will the object achieve the highest speed? Note that the acceleration of the object before hitting the spring is $a_1 = -g$, and after hitting the spring, it is $a_2 = -g + \frac{k}{m}(h_0 - y)$, where $y$ is the vertical distance of the ball from the ground.

**Solution:**

**Problem Statement:** The graphical representation of the problem is shown in Figure 3.12.

**Force Analysis:** Skipped.

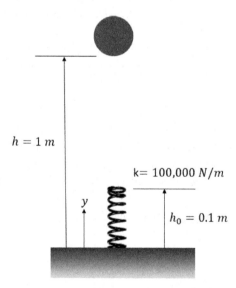

Figure 3.12: An example of acceleration as a function of position.

**Motion Analysis:** This is a 1D, straight line motion. The acceleration during $h_0 < y < h$ is $a_1 = -g$, while $a_2 = -g + \frac{k}{m}(h_0 - y)$ when $0 < y < h_0$.

**Governing Equations:** Again, for this one-dimensional motion, we consider Equations (3.21)–(3.30). Because the acceleration is a function of position, it is obvious that we should consider Equation (3.30).

**Solving Equations:**
Based on Equation (3.30), for $h_0 < y < h$ , we have

$$\int_h^{h_0} a(y)\, dy = (-g)(h_0 - h) = \frac{1}{2}(0^2 - v_1^2) \tag{3.43}$$

From this result, the speed of the object $v_1$ when it hits the top of the spring is found to be 0.42 m/s. To determine the maximum compression of the spring, we can use Equation (3.30) again,

$$\int_h^y a(y)\, dy = \int_h^{h_0} a_1(y)\, dy + \int_{h_0}^y a_2(y)\, dy = \frac{1}{2}(0^2 - v_2^2) \tag{3.44}$$

Letting $v_2 = 0$ and solving the equation above, we found $y = 0.057$ m. Therefore, the maximum compression of the spring is 0.043 m.

To answer the question that at which height the speed of the object is at maximum, we can take the result of Equation (3.44) and derive an expression for the speed as

$$v = \sqrt{-\frac{k}{m}y^2 - 2\left(g - \frac{kh_0}{m}\right)y + 2gh - \frac{kh_0^2}{m}} \tag{3.45}$$

For the maximal speed condition, we can take one derivative of Equation (3.45) and let it be zero. Solving the resulting equation, we have $y = 0.09990$ m, which is when the spring is compressed by 0.000981 m. The maximum speed is 4.203 m/s, slightly higher than the one right before the ball hits the spring.

Another way of solving this problem is to recognize that the highest speed is when the acceleration is zero. If we have the force information, we will see this happens when the spring force and the gravitational force cancel each other; thus $ky = mg$. From there, we get the same result.

### 3.4.2.5   *Velocity as a function of position*

In a potential flow field, the fluid velocity is a function of the position. We will use a simple potential flow example for illustration.

### Example 3.4: Potential Flow

In a point source potential flow, the fluid emerges from a single point and flows away along the radial direction. As it flows away, its speed in the radial direction becomes slower and can be defined as

$$v = \frac{Q}{2\pi r} \tag{3.46}$$

where $Q$ is the strength of the point source and $r$ is the radial distance from the source.

As shown in Figure 3.13, we let a little ball, $A$, come out of the point source and flow away. Determine the time it takes for the ball to reach the position when $r = d$. Also determine acceleration as a function of $r$.

**Solution:**

**Problem Statement:** Skipped.

**Force Analysis:** Skipped.

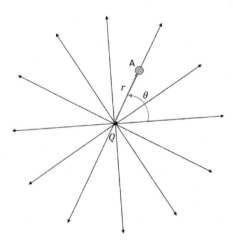

Figure 3.13:   An example of velocity as a function of position.

**Motion Analysis:** This is a 1D motion, where ball A flows along a straight line along a radial direction as shown in Figure 3.13. The speed of ball A is defined by Equation 3.46, as function of the radial position.

**Governing Equations:** Again, for this one-dimensional motion, we should consider Equations (3.21)–(3.30). Because the velocity is a function of position, it is obvious that we should consider Equation (3.28).

**Solving Equations:** Based on Equation (3.28), for $0 \leq r \leq d$, we have

$$t = \int_0^d \frac{dr}{v(r)} = \int_0^d \frac{2\pi r dr}{Q} = \frac{\pi d^2}{Q} \tag{3.47}$$

To determine the acceleration as a function of position, we need to use Equation (3.25). Why? Because this equation is the only one in the derivative form not related to time. We, however, should reconstruct Equation (3.25) and obtain

$$a(r) = v(r) \frac{dv(r)}{dr} = \frac{-Q^2}{4\pi^2} \frac{1}{r^3} \tag{3.48}$$

The negative sign indicates that the acceleration is negative along the radial distance; therefore, ball A slows down as it drifts further away. Note that the acceleration is proportional to $\frac{-1}{r^3}$, while the velocity is proportional to $\frac{1}{r}$.

One common mistake is to assume $a(r) = \frac{dv}{dr}$, which does not exist based on Equations (3.17)–(3.30). We cannot simply invent a new equation by intuition.

### 3.4.3 *Non-time-based motion planning*

Equation (3.48) actually has important applications. One of them is the non-time-based motion planning. For example, the train speed entering and leaving a train station is based on the train distance to the station platform, not time.

When we only consider the speed and the traveling distance in a one way motion, it can be considered as a 1D motion even if the actual travel path is not straight.

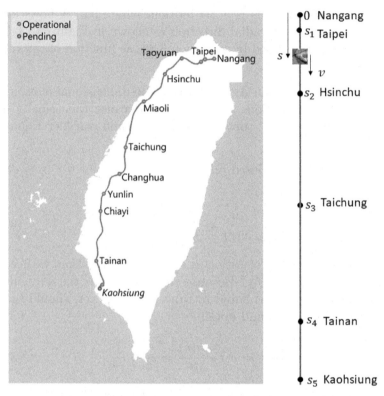

Figure 3.14: Winding high speed train rail route in Taiwan as a straight line for analysis.

*Source*: https://upload.wikimedia.org/wikipedia/commons/thumb/c/c8/Taiwan_High_Speed_Rail.svg/723px-Taiwan_High_Speed_Rail.svg.png

The high speed train system in Taiwan is a good example for illustration. As shown in Figure 3.14, the high speed train rail in Taiwan runs from north to south in a winding curve. We can define the speed and acceleration of the high speed train as a function of the traveling distance from the first station (Nangang) as if it is traveling along a straight line Figure 3.14.

Let's do a simple motion planning for an express train from Taipei to Kaohsiung, with one stop at Taichung. Leaving Taipei, the speed of the train will accelerate to its top speed of 300 km/h after traveling for about 15 km and cruise at this top speed. It will start to decelerate 10 km from Taichung station and come to a complete stop at the station for boarding. It will then accelerate with the same pattern to

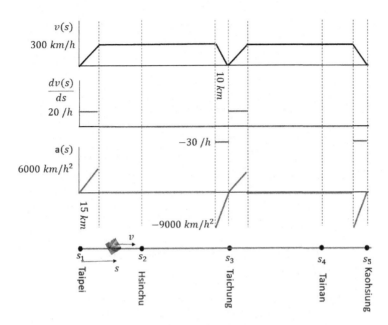

Figure 3.15: A simple motion planing for high speed trains.

its top speed. Finally, it will decelerate 10 km away from Kaohsiung before arriving at the destination. This motion is highly simplified for illustration purposes. The motion planer needs to calculate the acceleration as a function of distance so that the train control can be programmed accordingly. We plot the speed as a function of distance in the top chart of Figure 3.15.

To determine the acceleration as a function of distance, we need to use Equation (3.48). We will first calculate $\frac{dv(s)}{s}$ graphically, as shown in the middle chart of Figure 3.15. To obtain the bottom chart of $a(s)$, we multiply the top chart with the middle chart. You can do a point-wise multiplication and then connect the dots. For example, at 15 km away from Taipei, $\frac{dv(s)}{ds} = 20/h$, while $v(s) = 300$ km/h. As a result, the value of the acceleration can be calculated as 6000 km/h$^2$. Immediately after this, the acceleration is zero. Then, the acceleration becomes $-9,000$ km/h$^2$, through braking. We obtain the bottom chart of Figure 3.15.

It is interesting to note that the acceleration in the beginning of an acceleration cycle is small and the acceleration becomes the highest when it is close to the top cruising speed. On the other hand, the braking is the hardest at the beginning of the deceleration cycle.

The actual motion planing of high speed trains will likely have smooth speed curves than the one in the top chart. Therefore, the acceleration control will be smoother for a more comfortable ride.

## 3.5   Two-and Three-Dimensional Motions

In the previous section, we derived all the equations which are useful for straight line motions. The same set of equations can describe a non-straight line motion if we only consider the traveling distance and speed of a one-way motion as in the high speed train example. Can we apply the same equations to a general 2D or even 3D motions?

### 3.5.1   *2D and 3D motion descriptions using the Cartesian coordinate*

Based on Equations (3.19) and (3.20), we know Equations (3.17) and (3.18) can be extended from 1D motion to 2D and 3D motions if we treat the motion related to each axis independently. We then combined them via the vector addition to describe the total motion.

$$\vec{r} = x\vec{i} + y\vec{j} + z\vec{k} \tag{3.49}$$

$$\vec{v} = \dot{x}\vec{i} + \dot{y}\vec{j} + \dot{z}\vec{k} \tag{3.50}$$

$$\vec{a} = \ddot{x}\vec{i} + \ddot{y}\vec{j} + \ddot{z}\vec{k} \tag{3.51}$$

We will use a projectile motion for illustration.

### Example 3.5: Projectile Motion Analysis

As shown in Figure 3.16, a superman is trapped in a cart flying off a cliff. The speed of the cart is $10\,\mathrm{m/s}$ when it just leaves the cliff. Ignoring the air resistance (will consider it later), where should a smart shark position itself so that it would have the first bite of the superman? What is the speed of the cart when it hits the water? Finally, how much time does the smart shark have in order to get to the right position?

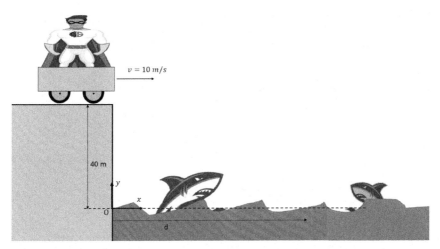

Figure 3.16: A projectile motion example for 2D motion analysis. https://upload.wikimedia.org/wikipedia/en/thumb/a/a5/Hawaii_Pacific_Sharks_logo.svg/1200px-Hawaii_Pacific_Sharks_logo.svg.png; https://p0.pikist.com/photos/223/243/superman-designing-human.jpg

**Solution:**

**Problem Statement:** We define a reference frame $O-x-y$ as shown in Figure 3.16. We need to determine the horizontal distance $d$. The height of the cliff relative to the surface of the water is 40 m.

**Force Analysis:** Skipped.

**Motion Analysis:** This is a 2D motion. For a projectile motion, we can prove it via *Newton's Second Law* that the motion in the horizontal direction has zero acceleration, while the vertical motion is at a constant acceleration. We can use the same time line technique shown in Figure 3.3, but use it twice, as shown in Figure 3.17.

**Governing Equations:** We only need to choose appropriate equations from Equations (3.21)–(3.30) for both horizontal and vertical motions separately.

**Solving Equations:** For the vertical motion, we can first use Equation (3.30) to find out the vertical speed when the cart hits the water.

$$\int_0^{v_{y1}} v_y\, dv_y = \int_{40}^0 -g\, dy \qquad (3.52)$$

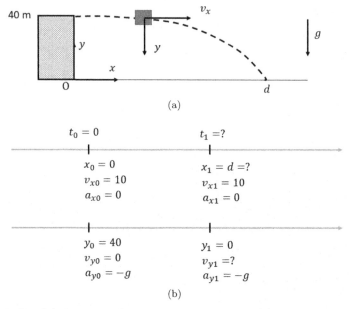

(a)

(b)

Figure 3.17: (a) the projectile motion trajectory; and (b) the double time lines depicting the motion variables.

We quickly find that $v_{y1} = -28.01 \, \text{m/s}$. The negative sign indicates that it is pointing downward. We now use Equation (3.27) to find out the time it hits the water. We obtain $t_1 = 2.856 \, \text{s}$.

We subsequently calculate the horizontal distance using $t_1$ to obtain $d = v_x \, t_1 = 28.56 \, \text{m}$.

The total speed when the cart hits the water is $v_1 = \sqrt{(v_{x1})^2 + (v_{y1})^2} = 48.83 \, \text{m/s}$.

In fact, sharks are a lot smarter than we thought. The horizontal distance predicted above is actually wrong because the air resistance likely needs to be considered due to the strong gust on the coast. Sharks can figure in all these factors better than us. Animals actually have mathematical formulas hard wired to their brains to perform many amazing tasks easily.[2]

Let's consider the air resistance and see how the results will be affected. A more thorough analysis related to the air resistance will

---

[2]Develin, K. *The Math Instinct: Why You're a Mathematical Genius (Along with Lobsters, Birds, Cats, and Dogs)*, Thunder's Mouth Press, 2005. ISBN-13: 978-1560256724.

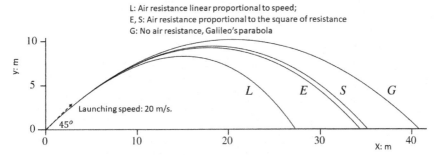

Figure 3.18:   Different projectile motion trajectories based on different air resistance models from Hackborn (2008).
*Note*: The launching speed should be 20 m/s for this simulation.

be presented in later chapters. Here, let's just examine the effects of air resistance. A great discussion of air resistance, based on different air drag models and different numerical solutions, can be found in Hackborn (2008).[3]

We cite one of the results in the Hackborn paper to compare the trajectories of a projectile under different air resistance models. As shown in Figure 3.18, an object launched at 20 m/s at 45° can reach a height of 10.19, 9.40, 9.15, 8.35 m, respectively, for the trajectories with no air resistance (case G), with air resistance proportional to square of speed (case E and S), and with air resistance linearly proportional to the object speed (case L). The horizontal traveling distance is found to be 41.55, 37.13, 36.24, and 28.82 m, respectively, for those trajectories.

Now, we should ask the question: which air resistance model is most accurate? The answer depends on the actual speed. Let's cite some experimental results to conclude this discussion.

A study for rocket, artillery, and mortar trajectories was reported by Ramezani and Rothe (2017).[4] Figure 3.19 shows the rocket flight trajectories by simulation and by radar measurement. The simulation

---

[3]Hackborn, W. W. *Projectile Motion: Resistance is Fertile*, The Mathematical Association of America **115** (2008) 813–819.

[4]Ramezani, A. and Rothe, H. Simulation-based early prediction of rocket, artillery, and mortar trajectories and real-time optimization for counter-RAM systems, *Mathematical Problems in Engineering* **8157319** (2017). https://doi.org/10.1155/2017/8157319.

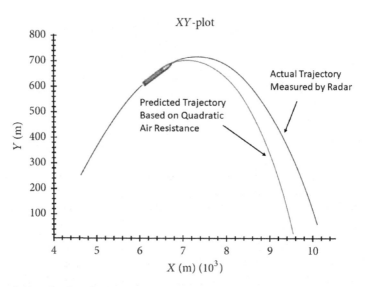

Figure 3.19: Rocket flight trajectories predicted by simulation using quadratic air resistance model and by radar measurement reported by Ramezani and Rothe (2017) (see footnote 4).

model is based on that the air resistance is proportional to the square of the traveling speed. The horizontal landing position is predicted to be at about 9.6 km, while the actual position is 10.2 km, off by 600 m, or about 6%. The rocket speed is in the range of 500 m/s. If the air resistance is not considered, the error would be huge.

Let's consider another experimental result in a smaller scale and at lower speeds. As reported by Shrivastava *et al.* (2015),[5] the horizontal projectile trajectory of a small ball launched by a de-soldering pump matches the prediction of the model assuming no air resistance, as shown in Figure 3.20. The trajectory of this horizontal projectile motion is similar to Figure 3.17. We can derive a relationship between the horizontal traveling distance ($x$), the launching speed ($u$), and the height ($h$) as

$$x = \sqrt{\frac{2}{g}} \sqrt{h}\, u \qquad (3.53)$$

[5]Shrivastava, A., Ramesh, K.P., and Raghavendra, H.N. *An Experiment on Projectile Motion*, 2015. https://www.researchgate.net/publication/324888726.

We take the logarithm for both sides and obtain the following expressions,

$$\log(x) = \log\left(\sqrt{\frac{2}{g}}u\right) + 0.5\log(h) \qquad (3.54)$$

and

$$\log(x) = \log\left(\sqrt{\frac{2h}{g}}\right) + \log(u) \qquad (3.55)$$

If we plot $\log(h)$ versus $\log(x)$ and $\log(u)$ versus $\log(x)$, we should get the relationships in straight lines. Figure 3.20 shows the linear regression lines of the experimental results. The theoretical slope for Figure 3.20(a), according to Equation (3.54), is 0.5, while for Figure 3.20(b), it should be 1.0. The experimental results are 0.47 and 0.95, respectively. The errors are 6% and 5%, respectively, similar to the complicated model used in Figure 3.19.

We conclude that the air resistance should be considered based on the speed, object size, the flight of time, etc. To complicate the issue further, if the object is not considered as a particle, we need to consider the spin of the object and the air lift, which is the force by the air perpendicular to the motion direction. In tennis, a system, called Hawk-Eye,[6] is used to make line calls by computer simulation combined with measurements by several high speed video cameras. The accuracy is claimed to be within 3.6 mm, which is actually not insignificant and many championships were won or lost by the simulation results of this system. During matches, the simulation could zoom in to the ball landing spot, showing hairline separation between the ball and the line , thus, calling a ball in or out. This tiny hairline gap could be far smaller than 3.6 mm of its accuracy. Therefore, the call by the Hawk-eye system could be meaningless but we simply accept it without questioning.

A newer system can now show video images of the moment when the ball hits the ground, not just by computer simulation. This greatly enhances the confidence of this technology.

---

[6]https://en.wikipedia.org/wiki/Hawk-Eye.

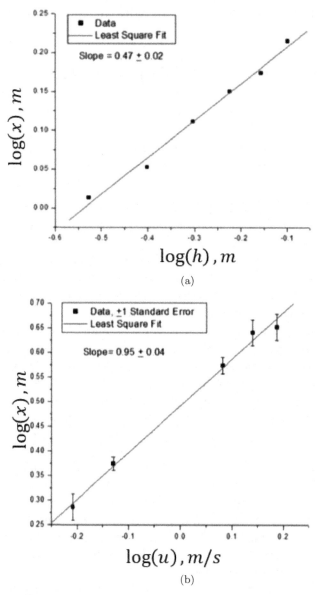

(a)

(b)

Figure 3.20: Experimental results of the horizontal projectile motion of a small ball launched by a de-soldering pump, from Shrivastava *et al.* (2015): (a) range vs height, log scale; and (b) range vs speed, log scale.

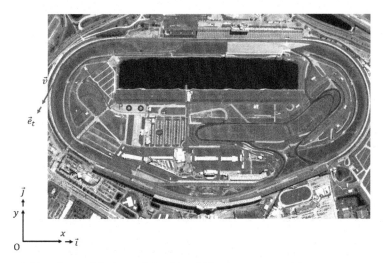

Figure 3.21: An aerial view of the famed Daytona race track.
*Source*: http://1.bp.blogspot.com/-OZrpbtULkJM/T_MmYHCDgXI/AAAAAA
AAEFs/RncLdbx1hhw/s1600/GoogleEarthDaytonaInternationalSpee.jpg

## 3.6 Tangential and Normal Description of a General Motion

Although the description using the $x-y-z$ reference frame can describe basically any 2D or 3D motions, there are more convenient ways to describe some specific motion types.

For example, in car racing, such as NASCAR, the driver, the crew chief, and the spotters are not too concerned with the $x-y$ position of the car except when it is close to the finish line. During the race, they are more concerned with the speed and the acceleration. They do not need to know the directions of the velocity and the acceleration. Because the velocity is always tangential to the trajectory, we can define the velocity vector as

$$\vec{v} = v\vec{e}_t \tag{3.56}$$

where $v$ is the speed, or the tangential speed, and $\vec{e}_t$ is a unit directional vector along the tangential direction. As shown in Figure 3.21, car A is moving with a velocity $\vec{v}$ and its direction is described by a unit vector $\vec{e}_t$, similar to that $\vec{i}$ defines the positive $x$ direction.

Equation (3.56) is simple in appearance but actually quite complicated because the unit vector, $\vec{e}_t$, is not pointing to the same direction

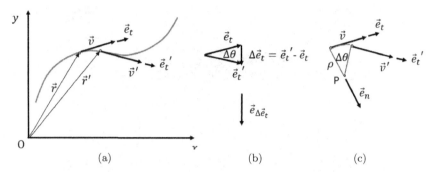

Figure 3.22: (a) The definition and derivation of the tangential and normal directions; (b) the unit directional vector change; and (c) the local center and radius when the time interval is very small.

as in the case of $\vec{i}$. When car A moves to a new position of the race track, the direction of $\vec{e}_t$ becomes different.

To determine the acceleration, we can take a derivative for Equation (3.56),

$$\vec{a} = \dot{\vec{v}} = \dot{v}\vec{e}_t + v\dot{\vec{e}}_t \tag{3.57}$$

We could not proceed further with Equation (3.57) because we do not know how to take a derivative of $\vec{e}_t$. We will show how to determine the derivative of $\vec{e}_t$ graphically. From the definition of the derivative, we have

$$\dot{\vec{e}}_t = \lim_{\Delta t \to 0} \frac{\vec{e}_t' - \vec{e}_t}{\Delta t} \tag{3.58}$$

Based on Equation (3.58), we need to determine the term, $\vec{e}_t' - \vec{e}_t$.

As shown in Figure 3.22, an object moves along its trajectory. At $t$, its velocity points at $\vec{e}_t$, and at $t + \Delta t$, it points to $\vec{e}_t'$. Graphically, we can express the term, $\vec{e}_t' - \vec{e}_t$, as shown in Figure 3.22(b), labeled as $\Delta \vec{e}_t$.

We simply need to figure the size of $\Delta \vec{e}_t$, which is $|\Delta \vec{e}_t|$, and the direction of $\Delta \vec{e}_t$, represented as a unit vector, $\vec{e}_{\Delta \vec{e}_t}$. As shown in Figure 3.22(b), $|\Delta \vec{e}_t|$ is the length of the arrow of $\Delta \vec{e}_t$, while its direction is represented by the arrow of $\vec{e}_{\Delta \vec{e}_t}$.

As $\Delta t$ becomes smaller and smaller, the new position gets very close to the original position. The trajectory between the two positions can be approximated by an arc with respect to a local center $P$.

The radius of the arc is $\rho$, as shown in Figure 3.22(c). As $\Delta t$ becomes smaller and smaller, the length of the arc can be approximated as $v\Delta t$, which is the travel distance during $\Delta t$. Correspondingly, the little angle, $\Delta\theta$, becomes

$$\Delta\theta = \frac{v\Delta t}{\rho} \tag{3.59}$$

Note that this little angle $\Delta\theta$ is the same angle between $\vec{e}_t'$ and $\vec{e}_t$. From Figure 3.22(b), the value of $|\Delta\vec{e}_t|$ is

$$|\Delta\vec{e}_t| = |\vec{e}_t|\Delta\theta = \Delta\theta \tag{3.60}$$

We have solved the first puzzle regarding the size of $\Delta\vec{e}_t$. Now, we need to figure out its direction. We will use imagination to figure this out. As $\Delta t$ becomes smaller and smaller, the direction of $\Delta\vec{e}_t$, i.e., the unit vector of $\vec{e}_{\Delta\vec{e}_t}$, becomes perpendicular to $\vec{e}_t$. As shown in Figure 3.22(c), it becomes the direction represented by the arrow of $\vec{e}_n$, which points toward the local center $P$. In other words, $\vec{e}_t$ and $\vec{e}_n$ are perpendicular. We denote $\vec{e}_n$ as the unit vector of the normal direction, as compared with $\vec{e}_t$ as the unit vector of the tangential direction. Note that $\vec{e}_n$ should always point to the local center $P$. We can now complete the derivation of Equation (3.58).

$$\dot{\vec{e}}_t = \lim_{\Delta t \to 0} \frac{\vec{e}_t' - \vec{e}_t}{\Delta t} = \lim_{\Delta t \to 0} \frac{\Delta\vec{e}_t}{\Delta t} = \left(\lim_{\Delta t \to 0} \frac{\Delta\theta}{\Delta t}\right)\vec{e}_n$$

$$= \left(\lim_{\Delta t \to 0} \frac{v\Delta t}{\rho}\frac{1}{\Delta t}\right)\vec{e}_n = \frac{v}{\rho}\vec{e}_n \tag{3.61}$$

We should feel a sense of achievement that we completed the derivation to determine $\dot{\vec{e}}_t$, but there is still one problem. What is the value of $\rho$? It will have to be given, measured, or derived. If we know the trajectory of the object as $y = f(x)$, there is a formula to determine the value of $\rho$ as

$$\rho = \frac{\left[1 + (dy/dx)^2\right]^{3/2}}{|d^2y/dx^2|} \tag{3.62}$$

If we are to design a race track or a highway, we can assign $\rho$ to a desirable value. What is a desirable value of $\rho$? To answer this question, we need to finish the derivation of the acceleration in Equation (3.57).

$$\vec{a} = \dot{\vec{v}} = \dot{v}\,\vec{e}_t + v\,\dot{\vec{e}}_t = \dot{v}\,\vec{e}_t + \frac{v^2}{\rho}\vec{e}_n \qquad (3.63)$$

From Equation (3.63), we know if a race car is running around the curve track, as shown in Figure 3.21, there are two accelerations involved. One is along the direction of $\vec{v}$, i.e., $\vec{e}_t$, with a value of $\dot{v}$. The other acceleration is pointing toward local center $P$, i.e., the $\vec{e}_n$ direction, with a value of $\frac{v^2}{\rho}$.

We denote $\vec{a}_t = \dot{v}\vec{e}_t$ as the tangential acceleration and $\vec{a}_n = \frac{v^2}{\rho}\vec{e}_n$ as the normal acceleration.

Now, we can understand the conversation between the crew chief and the driver such as

*You are going too fast on the turn!*

You want to go fast in a race, but why would it become too fast during a race when running around the curve? The answer is the normal acceleration term. It relies on the grip of the tire to provide the required normal acceleration. If the race track is slanted, gravity can provide additional help. When the driver goes too fast during the turn, the car may not have the needed traction for making the turn. The result is that the car would slam into the wall, causing spectacular crashes.

The same argument clarifies why we must slow down before exiting a highway. This is even more important when it rains because the grip of the tires is a lot less. When it rains, NASCAR always cancels the race because it becomes too dangerous.

Finally, the two unit vectors $\vec{e}_t$ and $\vec{e}_n$ define a plane. If a motion is 3D, we need to define a third direction perpendicular to the first plane. Based on the right-hand rule, by sweeping from $\vec{e}_t$ to $\vec{e}_n$, the thumb will point to the third direction, which is denoted as $\vec{e}_b$.

A final note on the $t-n-b$ description of motion. We should be aware that because all the unit directional vectors, $\vec{e}_t$, $\vec{e}_n$, and $\vec{e}_b$, are not fixed in their directions, an observer using such descriptions is NOT an inertial observer. To make such observation inertial, we must define these unit vectors with respect to an IRF, such as $x-y-z$.

In other words, we should add the following information to Equations (3.56) and (3.63).

$$\vec{e_t} = a_1\vec{i} + a_2\vec{j} + a_3\vec{k} \qquad (3.64)$$

$$\vec{e_n} = b_1\vec{i} + b_2\vec{j} + b_3\vec{k} \qquad (3.65)$$

$$\vec{e_b} = c_1\vec{i} + c_2\vec{j} + c_3\vec{k} \qquad (3.66)$$

The coefficients in the above equations, $a_i$, $b_i$, and $c_i$, $i = 1, 2, or\ 3$, need to determined for the instant of the time. We often skip Equations (3.64)–(3.66) when using the $t-n-b$ description, but we should be aware of it.

Finally, a stationary observer using the $t-n-b$ reference frame immediately next to the moving object will only observe

$$\hat{v} = v\vec{e_t} \qquad (3.67)$$

and

$$\hat{a} \neq \dot{v}\,\vec{e_t} + \frac{v^2}{\rho}\vec{e_n} \qquad (3.68)$$

Equation (3.68) is an inequality because, in general, we do not know what acceleration a non-inertial observer will perceive. The object is also moving away from this stationary observation with the $t-n-b$ frame, which could result in extra acceleration terms. The acceleration perceived by a non-inertial observer will be discussed in the sections related to the relative motion analysis in Chapters 3 and 4. In other words, the motion observed by a non-inertial observer is not complete or could be skewed due to the rotation of the reference frame.

In conclusion, Equations (3.56) and (3.63) are correct for an inertial observer together with the unit vectors defined in Equations (3.64)–(3.66), but this is only valid for that particular instant of time. The distinction between an inertial observer and a non-inertial observer is very important. We will discuss this topic in greater depth in later sections of this chapter.

## 3.7 Radial and Angular Description of Motion

Let's consider another situation when the motion description is inconvenient for the $x-y-z$ frame. When a balloon slips off a kid's hand and flies away, how do we track its position, velocity, and acceleration? As shown in Figure 3.23(a), it would be very difficult to measure its $x-y-z$ positions with giant rulers or digital scales. A better sensor will be a device similar to a radar or one of those shown in Figures 3.6 and 3.7.

Let's consider a 2D motion problem first. As shown in Figure 3.23(a), to determine the position vector $\vec{r}$ of the object, the radar will fire a wave aiming at the far away object. When the wave is bounced back from the object, the distance $|\vec{r}|$ between the radar and the object can be determined. To know the direction of $\vec{r}$, we only need to know the aiming angle of the radar. We will define a directional unit vector $\vec{e}_r$, denoted as the radial direction unit vector, to represent the direction of $\vec{r}$. The position vector can now be defined as

$$\vec{r} = r\vec{e}_r \qquad (3.69)$$

Just like $\vec{e}_t$ and $\vec{e}_n$ in the $t-n$ description, $\vec{e}_r$ is not fixed in direction. We can define $\vec{e}_r$ based on the information of $\theta$,

$$\vec{e}_r = \cos\theta\vec{i} + \sin\theta\vec{j} \qquad (3.70)$$

Equations (3.69) and (3.70) complete an inertial observation of the position.

To find out the velocity and the acceleration, we only need to take derivatives as before.

$$\vec{v} = \dot{\vec{r}} = \dot{r}\vec{e}_r + r\dot{\vec{e}}_r \qquad (3.71)$$

Again, we need to take a derivative of a directional unit vector.

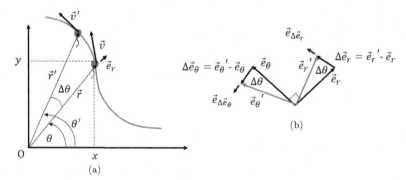

(a)

(b)

Figure 3.23: (a) The definition and derivation of the radial and transverse directions; and (b) the unit directional vector changes.

We can use the same graphical approach to determine it, as shown in Figure 3.23(b). We will define a second directional unit vector, $\vec{e}_\theta$. This direction is called the transverse direction and it is perpendicular to $\vec{e}_r$. We should use the right-hand rule to sweep from the direction of $\vec{e}_r$ to $\vec{e}_\theta$, as shown in Figure 3.23(b). We also need to define $\vec{e}_\theta$ with respect to $O-x-y$.

$$\vec{e}_\theta = -\sin\theta\,\vec{i} + \cos\theta\,\vec{j} \tag{3.72}$$

When the balloon moves from $\vec{r}$ to $\vec{r}'$, $\vec{e}_r$ changes to $\vec{e}_r'$ and $\vec{e}_\theta$ to $\vec{e}_\theta'$. The angle change is $\Delta\theta$, as shown in Figure 3.23(b). Just like Equation (3.61), we obtain

$$\dot{\vec{e}}_r = \lim_{\Delta t \to 0} \frac{\vec{e}_r' - \vec{e}_r}{\Delta t} = \lim_{\Delta t \to 0} \frac{\Delta \vec{e}_r}{\Delta t}$$

$$= \left( \lim_{\Delta t \to 0} \frac{\Delta\theta}{\Delta t} \right) \vec{e}_\theta = \dot{\theta}\vec{e}_\theta \tag{3.73}$$

When $\Delta t \to 0$, $\vec{e}_{\Delta\vec{e}_r}$ becomes $\vec{e}_\theta$.

We can also find out the derivative of $\vec{e}_\theta$ as

$$\dot{\vec{e}}_\theta = \lim_{\Delta t \to 0} \frac{\vec{e}_\theta' - \vec{e}_\theta}{\Delta t} = \lim_{\Delta t \to 0} \frac{\Delta \vec{e}_\theta}{\Delta t}$$

$$= \left( \lim_{\Delta t \to 0} \frac{\Delta\theta}{\Delta t} \right) (-\vec{e}_r) = -\dot{\theta}\vec{e}_r \tag{3.74}$$

When $\Delta t \to 0$, $\vec{e}_{\Delta\vec{e}_\theta}$ becomes $-\vec{e}_r$. Note the negative sign in Equation (3.74).

Equation (3.71) can now be completed as

$$\vec{v} = \dot{\vec{r}} = \dot{r}\vec{e}_r + r\dot{\vec{e}}_r = \dot{r}\vec{e}_r + r\dot{\theta}\,\vec{e}_\theta \tag{3.75}$$

The acceleration becomes

$$\vec{a} = \dot{\vec{v}} = \ddot{r}\vec{e}_r + \dot{r}\dot{\vec{e}}_r + \dot{r}\dot{\theta}\vec{e}_\theta + r\ddot{\theta}\vec{e}_\theta + r\dot{\theta}\dot{\vec{e}}_\theta \tag{3.76}$$

Introducing Equations (3.73) and (3.74) into (3.76), we obtain

$$\vec{a} = (\ddot{r} - r\dot{\theta}^2)\vec{e}_r + (r\ddot{\theta} + 2\dot{r}\dot{\theta})\vec{e}_\theta \tag{3.77}$$

$\dot{\theta}$ is called the angular velocity and $\ddot{\theta}$ is the angular acceleration. We often use $\omega = \dot{\theta}$ and $\alpha = \dot{\omega} = \ddot{\theta}$.

To extend the $r - \theta$ description to 3D, we can add the $z\vec{k}$ to Equation (3.69). This $r-\theta-z$ description is called the cylindrical coordinate. A 3D motion description without using the $z$ axis is more complicated and we will discuss it later.

Equation (3.77) is rather complex and hard to remember. In particular, the term $2\dot{r}\dot{\theta}$ is quite surprising, which is denoted as the Coriolis acceleration. The Coriolis acceleration is very important and will be elaborated later in this chapter.

Again, we need to point out that the $r - \theta$ description is NOT an IRF because the unit directional vectors, $\vec{e}_r$ and $\vec{e}_\theta$, are not fixed in their directions. We need Equations (3.70) and (3.72) to make Equations (3.69), (3.75), and (3.77) as inertial observations.

A stationary observer using the $r - \theta$ reference frame will only observe

$$\hat{r} = r\vec{e}_r \tag{3.78}$$

$$\hat{v} = \dot{r}\vec{e}_r \tag{3.79}$$

$$\hat{a} = \ddot{r}\vec{e}_r \tag{3.80}$$

This is because, to this non-inertial observer, $\vec{e}_r$ is fixed and the motion appears to be a straight line motion along the radial direction; therefore, its derivative should be zero.

## 3.8   Perfect Circular Motion

Let's consider a special motion when an object is traveling around a perfect circular path, as shown in Figure 3.24.

We can relate all three descriptions of 2D motion in a perfect circular motion. From Figure 3.24, we have

$$x = r\cos\theta \tag{3.81}$$

$$y = r\sin\theta \tag{3.82}$$

$$\vec{e}_t = \vec{e}_\theta \tag{3.83}$$

$$\vec{e}_n = -\vec{e}_r \tag{3.84}$$

Based on Equations (3.82) and (3.84), and comparing Equations (3.51), (3.56), and (3.75), we have

$$\vec{v} = \dot{\vec{r}} = \dot{x}\vec{i} + \dot{y}\vec{j} = \dot{r}\vec{e}_r + r\dot{\theta}\vec{e}_\theta = v\vec{e}_t \tag{3.85}$$

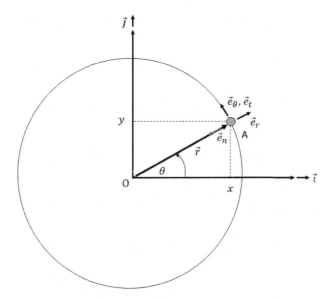

Figure 3.24: A perfect circular motion when different descriptions of motion are related.

Because $\dot{r} = 0$ in a perfect circular motion, we have $v = r\dot{\theta}$, $\dot{x} = -v\sin\theta$, and $\dot{y} = v\cos\theta$.

Comparing Equations (3.51), (3.63), and (3.77), we have

$$\vec{a} = \ddot{x}\vec{i} + \ddot{y}\vec{j} = \dot{v}\vec{e}_t + \frac{v^2}{r}\vec{e}_n = (\ddot{r} - r\dot{\theta}^2)\vec{e}_r + (r\ddot{\theta} + 2\dot{r}\dot{\theta})\vec{e}_\theta \qquad (3.86)$$

Similarly, due to $r$ being constant, we have $\ddot{r} = \dot{r} = 0$. We obtain $\dot{v} = r\ddot{\theta}$, $\frac{v^2}{r} = r\dot{\theta}^2$, $\ddot{x} = -r\dot{\theta}^2\cos\theta - r\ddot{\theta}\sin\theta$, and $\ddot{y} = -r\dot{\theta}^2\sin\theta + r\ddot{\theta}\cos\theta$.

The acceleration, $r\dot{\theta}^2$, is often referred to as the centripetal acceleration (not centrifugal acceleration).[7]

The straight line motion and the perfect circular motion are very special because we have complete equations to define them. These two motions are denoted as SIMPLE motions in this book. In fact, we should know that our human brain can only handle these two SIMPLE motions.

---

[7]Centripetal acceleration points to the center, which is real, while centrifugal acceleration points outward and is a fictitious acceleration. We will elaborate on this in later chapters.

## 3.9 Relative Motion Analysis #1: Different Observers with the Same Reference Frame

In this section, we will discuss how we reconcile the observations of different observers using the same reference frames. We then proceed to the situation when different reference frames are used in the next section.

As shown in Figure 3.25, two observes $O$ and $B$ are observing the motion of object $A$. In our notations, we must specify very clearly who is doing the observation. The position vector of $A$ observed by $O$ is expressed $\vec{r}_{A/O}$, which should be read as the position of $A$ with respect to $O$. Similarly, we define $\vec{r}_{A/B}$ and $\vec{r}_{B/O}$, as the position of $A$ with respect to $B$ and the position of $B$ with respect to $O$, respectively. From Figure 3.25, graphically, the relationship of these position vectors are defined as

$$\vec{r}_{A/O} = \vec{r}_{B/O} + \vec{r}_{A/B} \tag{3.87}$$

Note that Equation (3.87) relates two position observations using the same reference frame. We did not specify if observer $O$ is an inertial observer. It does not have to be so. However, because there are no other observers or observation frames, both observer $O$ and

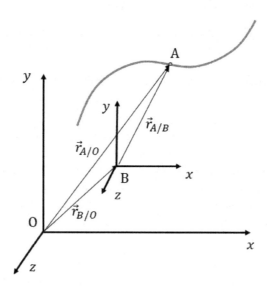

Figure 3.25: Relative position vectors with respect to different observers using the same reference frame.

$B$ believe that their reference frame $x-y-z$ is fixed in direction. Without loss of generality, we can simply consider $x-y-z$ as a fixed reference frame.

To relate the velocity and acceleration observations, we can simply take derivatives of Equation (3.87) to obtain

$$\vec{v}_{A/O} = \vec{v}_{B/O} + \vec{v}_{A/B} \tag{3.88}$$

and

$$\vec{a}_{A/O} = \vec{a}_{B/O} + \vec{a}_{A/B} \tag{3.89}$$

Equations (3.87)–(3.89) are really simple and can be readily applied to many different situations.

## Example 3.6: Relative Motion Analysis between Two Objects in Straight Line Motions

As shown in Figure 3.26(a), jet fighter $B$ is taking off at 250 km/h with respect to the carrier, while jet fighter $A$ is also taking off at the same speed along a different runway, which is 30° to the runway

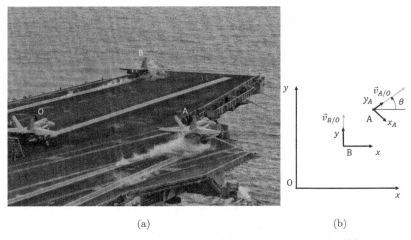

(a)                                    (b)

Figure 3.26:   Relative motion analysis for jets taking off a carrier: (a) the actual image of the carrier; and (b) the motions in different reference frames.
*Source*: https://upload.wikimedia.org/wikipedia/commons/8/8e/US_Navy_0811 24-N-3659B-305_F-A-18C_Hornets_launch_from_the_Nimitz-class_aircraft_carrier USS_Ronald_Reagan_(CVN_76).jpg.

of jet fighter $B$. What is the velocity of jet fighter $A$ observed by the pilot of jet fighter $B$?

**Solution:**

**Problem Statement:** We first plot the jet fighters and the reference frame as shown in Figure 3.26(b). Observer $O$ is the pilot of the parked jet fighter $O$. Both pilots $O$ and $B$ use the same reference frame. According to them, their forward direction is the $y$-axis, while their right-hand side is the $x$-axis. we need to determine $\vec{v}_{A/B}$.

**Force Analysis:** Skipped.

**Motion Analysis:** We define the reference frame as shown in Figure 3.26(b). However, pilot $O$, who is motionless with respect to the carrier, is not necessarily motionless with respect to, for example, an observer on the shore. We do not define the motion of the carrier in this example and we don't need to, as explained earlier.

**Governing Equations:** First, we have $\vec{v}_{A/O} = 250 \cos 30° \, \vec{i} + 250 \sin 30° \, \vec{j}$ (km/h) and $\vec{v}_{B/O} = 250 \, \vec{j}$ (km/h). From Equation (3.88), we have

$$\vec{v}_{A/O} = 250 \cos 30° \, \vec{i} + 250 \sin 30° \, \vec{j} = \vec{v}_{B/O} + \vec{v}_{A/B} = 250 \, \vec{j} + \vec{v}_{A/B}$$

$$(3.90)$$

**Solving Equations:** There is only one unknown in the equation above. We obtain $\vec{v}_{A/B} = 250 \cos 30° \, \vec{i} + 250(\sin 30° - 1) \, \vec{j}$ (km/h).

How about $\vec{v}_{B/A}$? If we follow the same logic about the reference frame. Pilot $A$ has a forward direction which is defined as axis $y_A$ and the corresponding right-hand side is $x_A$. They are different from $x - y$ used by pilot $O$. Does it mean that we do not have a relative velocity equation? We should realize that the reference frame $x_A - y_A$ is also fixed in directions. We have the following relationship and it does not change for the time duration of this problem.

$$\vec{i}_A = \sin 30° \, \vec{i} - \cos 30° \, \vec{j} \qquad (3.91)$$

$$\vec{j}_A = \cos 30° \, \vec{i} + \sin 30° \, \vec{j} \qquad (3.92)$$

In other words, pilot $A$ can be considered as using the same reference frame as pilot $O$. As a result, we have

$$\vec{v}_{B/O} = 250 \, \vec{j} = \vec{v}_{A/O} + \vec{v}_{B/A} = 250 \cos 30° \, \vec{i} + 250 \sin 30° \, \vec{j} + \vec{v}_{B/A}$$

$$(3.93)$$

Solving the above equation, we obtain $\vec{v}_{B/A} = -250\,\cos 30°\,\vec{i} - 250(\sin 30° - 1)\,\vec{j}$ (km/h). It turns out that

$$\vec{v}_{A/B} = -\vec{v}_{B/A} \tag{3.94}$$

However, we must be cautious about Equation (3.94) because it is only valid if $A$ and $B$ use the same reference frame.

Let's consider an example for which $\vec{v}_{A/B} \neq -\vec{v}_{B/A}$.

### Example 3.7: Relative Motion Analysis for Objects not in Straight Lines

As shown in Figure 3.27, car $B$ is moving at 100 km/h and slows down at 100 km/h$^2$, while car $A$ is traveling through a ramp to merge into the same highway. Car $A$ is at 50 km/h and is speeding up at 1,000 km/h$^2$. What are the velocity and acceleration of car $A$ observed by the driver of car $B$?

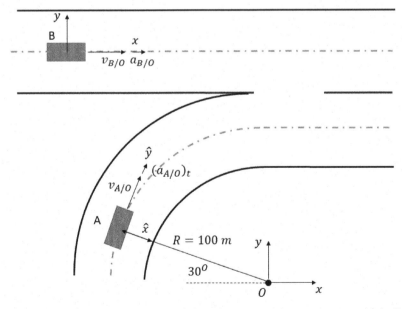

Figure 3.27:   The relative motion analysis between two cars, one in straight line motion, while the other one in a circular motion.

## Solution:

**Problem Statement:** As shown in Figure 3.27, there are three observers, driver $A$, driver $B$, and a person on the ground at point $O$. Observer $O$ and driver $B$ are using the same $x-y$ reference frame, while driver $A$ is using a reference frame $\hat{x}-\hat{y}$, which is not fixed in direction. It will change as car $A$ moves along the ramp.

**Force Analysis:** Skipped.

**Motion Analysis:** Skipped.

**Governing Equations:** To determine $\vec{v}_{A/B}$ and $\vec{a}_{A/B}$, we can invoke Equations (3.88) and (3.89) as before because driver $B$ and observer $O$ use the same reference frame.

**Solving Equations:** We quickly find out $\vec{v}_{A/B} = (50 \ \sin 30° - 100)\vec{i} + 50 \ \cos 30°\vec{j} = -75\vec{i} + 43.3\vec{j}$ (km/h). To determine $\vec{a}_{A/B}$, we need to determine $\vec{a}_{A/O}$ first. Because car $A$ is in a circular motion, there are tangential and normal acceleration components. The tangential acceleration is given as 1000 km/h². The normal component is determined as $(v_{A/O})^2/R = 25000$ km/h². Therefore,

$$\vec{a}_{A/O} = 25000\hat{i} + 1000\hat{j} = \vec{a}_{B/O} + \vec{a}_{A/B} = -100 \ \vec{i} + \vec{a}_{A/B} \quad (3.95)$$

where $\hat{i}$ and $\hat{j}$ are unit directional vectors of $\hat{x}$ and $\hat{y}$, respectively.

We need to convert $\hat{i}$ and $\hat{j}$ to $\vec{i}$ and $\vec{j}$. Based on Figure 3.27, we have

$$\hat{i} = \cos 30°\vec{i} - \sin 30°\vec{j} \quad (3.96)$$

$$\hat{j} = \sin 30°\vec{i} + \cos 30°\vec{j} \quad (3.97)$$

Introducing Equations (3.96) and (3.97) into (3.95), we obtain $\vec{a}_{A/B} = 22250.6\vec{i} - 11634.0\vec{j}$ (km/h²).

How about $\hat{v}_{B/A}$ and $\hat{a}_{B/A}$? Driver $A$ uses a reference frame, $\hat{x}-\hat{y}$, whose direction is changing. With respect to driver $A$, we have to express the relative motion in $\hat{v}$ and $\hat{a}$ instead of $\vec{v}$ and $\vec{a}$. For now, we do not know the relative motions observed by $A$ or $B$ are different. We will try to answer this question and the question related to Equation (3.68) in the next section.

## 3.10 Relative Motion Analysis #2: Different Observers with Different Reference Frames

We discuss $\hat{v}$ and $\hat{a}$, the motion observed by a rotating reference frame, in this section. What is seen by the driver of car A of Figure 3.27? To answer this question, we need to define a reference frame which rotates. When something rotates, its orientation changes, or we can state that it has an angular displacement, as opposed to the position displacement vectors of Equation (3.87).

### 3.10.1 *Angular displacement*

Let's refer to Figure 3.24 again. When object $A$ moves along a circular path, its position vector is $\vec{r}_{A/O}$. We can define an angular position by the angle, $\theta$. Can the angular position be represented as a vector? If so, which direction does it represent?

If we use the right hand rule as shown in Figure 3.28 with the four fingers following the rotation, then the thumb is pointing to a direction, which represents the direction of the angular displacement.

Based on the right-hand rule of Figure 3.28, we will express the angular displacement, $\theta$, as a vector,

$$\vec{\Theta} = \theta \vec{k} \tag{3.98}$$

Therefore, the rotation of plane $x-y$ has a direction in the $z$ axis. Note that the counterclockwise rotation is considered as positive when looking down at the $x-y$ plane from the $z$ axis. Similarly, we can define the rotation of the $y-z$ plane as

$$\vec{\Phi} = \phi \vec{i} \tag{3.99}$$

and that of the $z-x$ plane as

$$\vec{\Psi} = \psi \vec{j} \tag{3.100}$$

Note that, for the above equation to define the angular displacement around the $y$ axis, we need to sweep the four fingers from the $z$ axis to the $x$ axis, so that the thumb point, to the $y$ axis.

Next question: Do we have a 3D angular displacement vector? In other words, similar to a position vector, can we express a 3D angular displacement vector by combining Equations (3.98)–(3.100)?

Figure 3.28: The right hand rule to represent the direction of a rotation.
*Source*: https://upload.wikimedia.org/wikipedia/commons/thumb/b/b9/Right-hand_screw_rule.svg/1200px-Right-hand_screw_rule.svg.png.

If we define the 3D angular displacement as $\vec{\Gamma}$, do the following different expressions of $\vec{\Gamma}$ equal to each other?

$$\vec{\Gamma} \overset{?}{=} \theta\vec{k} + \phi\vec{i} + \psi\vec{j} \overset{?}{=} \theta\vec{k} + \psi\vec{j} + \phi\vec{i} \overset{?}{=} \psi\vec{j} + \phi\vec{i} + \theta\vec{k} \qquad (3.101)$$

By adding up the angular displacements as a vector, we imply that the sequence of adding up the angular displacement vectors should not matter, just like in the case of the position vector.

Let's see if Equation (3.101) is valid. As shown in Figure 3.29, let's examine how vector $\overrightarrow{OA}$ will rotate following two angular displacements. First, if the angular displacement sequence is $\theta = 90°$ and then $\phi = 90°$, vector $\overrightarrow{OA}$ will move as shown in Figure 3.29(b) from $\overrightarrow{OA}$ to $\overrightarrow{OA_1}$, and finally to $\overrightarrow{OA_2}$, pointing to the positive $z$ axis.

On the other hand. if we follow the sequence of $\phi = 90°$ and then $\theta = 90°$, vector $\overrightarrow{OA}$ will rotate as shown in Figure 3.29(c). It ends up pointing to the direction of the $y$ axis. The interesting thing is that if $\phi = 90°$ is applied first, there is no change of vector $\overrightarrow{OA}$; i.e., $\overrightarrow{OA} = \overrightarrow{OA_1}$.

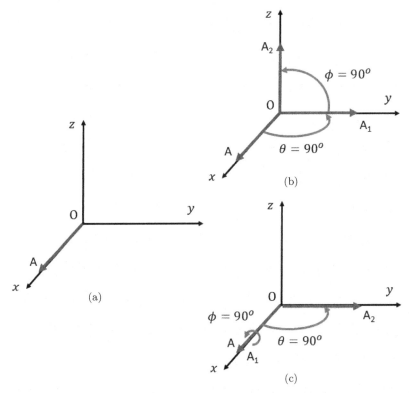

Figure 3.29: Different sequences of angular displacements could lead to different results. (a) Vector $\vec{OA}$ before angular displacement; (b) vector $\vec{OA}$ with $\theta = 90°$, then $\phi = 90°$; and (c) vector $\vec{OA}$ with $\phi = 90°$, then $\theta = 90°$.

From Figure 3.29, it becomes clear that the angular displacement vector $\vec{\Gamma}$ does not exist, unlike the position vector $\vec{r}$. Therefore, Equation (3.101) is not valid.

Let's address the scenario that an angular movement does not seem to result in any changes, as in the case of the first angular rotation of Figure 3.29(c). Because we are conducting 3D angular movements, we should look at a 3D object instead of a vector, which is a line.

As shown in Figure 3.30, we examine a 3D structure in the form of a reference frame. We have the reference frame $x-y-z$ and a rigid frame $\hat{x}-\hat{y}-\hat{z}$. The rigid frame $\hat{x}-\hat{y}-\hat{z}$ initially is aligned with $x-y-z$ as shown in Figure 3.30(a). We can go through similar angular displacements as in Figures 3.29(b) and 3.29(c).

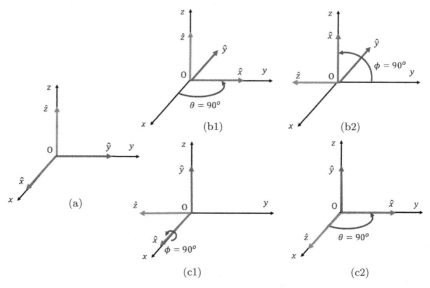

Figure 3.30: Different sequences of angular displacements for a 3D frame which could lead to different results. (a) Frame $\hat{x}-\hat{y}-\hat{z}$ before angular displacement; (b1–b2) $\hat{x}-\hat{y}-\hat{z}$ with $\theta = 90°$, then $\phi = 90°$; and (c1–c2) $\hat{x}-\hat{y}-\hat{z}$ with $\phi = 90°$, then $\theta = 90°$.

From Figures 3.29 and 3.30, we reach the same conclusion that angular displacements in different sequences lead to different results.

If we use different reference frames to observe the motion of an object, we will need to define the relationship between different reference frames. However, as shown in Figure 3.30, clearly, there is no unique relationship between different reference frames.

We understand that it is very difficult to visualize 3D rotations. We will introduce a new mathematical tool which can process rotations much more easily.

### 3.10.2 *Moving coordinate concept (a.k.a. coordinate transformation)*

A great tool to establish the relationship between different reference frames is commonly known as the coordinate transformation. However, the discussion for the coordinate transformation is often confusing. In this section, we will present the coordinate transformation in a different way to make it easier to understand. This different perspective is denoted as the Moving Coordinate Concept.[8]

To explain the moving coordinate concept, we will define some new notations. First, we will only consider rotations. Second, we will designate the first reference frame as $X-Y-Z$, in capitals. Third, a little moving coordinate, denoted as $x-y-z$ is introduced. Finally, the rotating reference frame to be related to $X-Y-Z$ is denoted as $\hat{x}-\hat{y}-\hat{z}$. All of these reference frames have the same origin at point $O$ because we only consider rotations here. The same technique can also address reference frames with different origins, but, without loss of generality, but we will only consider the same origin. Figures 3.31(a)–3.31(d) illustrate how these reference frames are related.

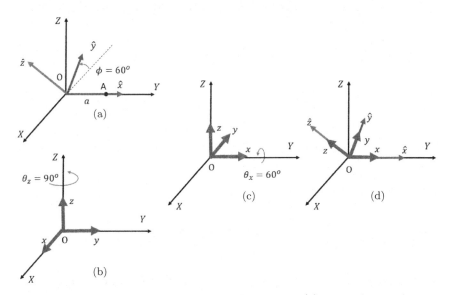

Figure 3.31: Moving coordinate concept illustration: (a) Two reference frames, $X-Y-Z$ and $\hat{x}-\hat{y}-\hat{z}$ are to be related; (b) a little moving coordinate, $x-y-z$, is used to match $\hat{x}-\hat{y}-\hat{z}$. First step is to rotate around $z$ by 90°; (c) the little coordinate becomes the one shown here. A 60° rotation with respect to the new $x$ is then performed; and (d) the little coordinate $x-y-z$ now matches $\hat{x}-\hat{y}-\hat{z}$.

[8]Tu, J.F., Bossmanns, B., and Hung, S.C.C. Modeling and error analysis for assessing spindle radial error motions. *Precision Engineering* **21**(2–3) (1997) 90–101. https://doi.org/10.1016/S0141-6359(97)00065-2.

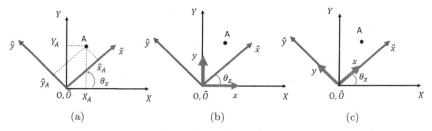

Figure 3.32: Coordinate transformation and moving coordinate concept: (a) two reference frames $X-Y-Z$ and $\hat{x}-\hat{y}-\hat{z}$ are used to describe the motion of object $A$; (b) a little moving coordinate, $x-y-z$, is first aligned with $X-Y-Z$; and (c) after rotating $\theta_z$, the little moving coordinate is now aligned with $\hat{x}-\hat{y}-\hat{z}$.

We will also use simpler expressions for the rotating angles. If a rotation is with respect to the $x$ axis, the rotating angle is denoted as $\theta_x$. Similarly, we define $\theta_y$ and $\theta_z$ for the rotations around the $y$ and $z$ axes, respectively. All the rotations are carried out based on the little coordinate $x-y-z$ at its current orientation.

Figure 3.32 illustrates how the moving coordinate concept and coordinate transformation can be achieved. As shown in Figure 3.32(a), we want to observe the motion of object $A$. According to reference frame $X-Y-Z$, the position of Object $A$ is $\vec{r}_{A/O} = (X_A, Y_A, Z_A)$. However, based on the reference frame $x-y-z$, $\hat{r}_{A/\hat{O}} = (\hat{x}_A, \hat{y}_A, \hat{z}_A)$. How are these expressions related? From Figure 3.32(a), we know

$$X_A = \hat{x}_A \cos\theta_z - \hat{y}_A \sin\theta_z \tag{3.102}$$

$$Y_A = \hat{x}_A \sin\theta_z + \hat{y}_A \cos\theta_z \tag{3.103}$$

$$Z_A = \hat{z}_A \tag{3.104}$$

If we organize Equations (3.102)–(3.104) in a matrix from, we arrive at

$$
\begin{bmatrix} X_A \\ Y_A \\ Z_A \\ 1 \end{bmatrix} =
\begin{bmatrix}
\cos\theta_z & -\sin\theta_z & 0 & 0 \\
\sin\theta_z & \cos\theta_z & 0 & 0 \\
0 & 0 & 1 & 0 \\
0 & 0 & 0 & 1
\end{bmatrix}
\begin{bmatrix} \hat{x}_A \\ \hat{y}_A \\ \hat{z}_A \\ 1 \end{bmatrix}
\tag{3.105}
$$

In Equation (3.105), a fourth equation is introduced but it is trivial as $1 = 1$. This fourth equation is not trivial if reference frame $X-Y-Z$ and $\hat{x}-\hat{y}-\hat{z}$ are not using the same metric for length, which means that unit length in $\hat{x}-\hat{y}-\hat{z}$ is different from that in $X-Y-Z$. We will not discuss the metric transformation here, but we will use the expanded form as shown in Equation (3.105).

Equation (3.105) should be read as: the position vector of $A$ is transformed from the expression of $[\hat{x}_A\,\hat{y}_A\,\hat{z}_A\,1]'$ by a $4\times 4$ matrix to the expression of $[X_A\,Y_A\,Z_A\,1]'$. The symbol, $'$, represents an operator, denoted as transpose, which converts a row vector into a column vector and vice versa.

Applying the moving coordinate concept as shown in Figure 3.32(b) and 3.32(c), this transformation and the resulting transformation matrix are due to the rotation of the little coordinate, $x-y-z$ by $\theta_z$. We will label the transformation matrix in Equation (3.105) as

$$T_{tz}(\theta_z) = \begin{bmatrix} \cos\theta_z & -\sin\theta_z & 0 & 0 \\ \sin\theta_z & \cos\theta_z & 0 & 0 \\ 0 & 0 & 1 & 0 \\ 0 & 0 & 0 & 1 \end{bmatrix} \tag{3.106}$$

The notation of $T_{tz}(\theta_z)$ is very easy to follow. The letter $T$ means transformation. The subscript $tz$ means that it is rotating around the $z$ axis of the little moving coordinate and the independent variable of $T_{tz}$ is $\theta_z$. We can introduce any value of $\theta_z$ into Equation (3.106).

Similarly, we can define two additional transformation matrices for the rotations with respect to the $x$ and $y$ axes, respectively.

$$T_{tx}(\theta_x) = \begin{bmatrix} 1 & 0 & 0 & 0 \\ 0 & \cos\theta_x & -\sin\theta_x & 0 \\ 0 & \sin\theta_x & \cos\theta_x & 0 \\ 0 & 0 & 0 & 1 \end{bmatrix} \tag{3.107}$$

$$T_{ty}(\theta_y) = \begin{bmatrix} \cos\theta_y & 0 & \sin\theta_y & 0 \\ 0 & 1 & 0 & 0 \\ -\sin\theta_y & 0 & \cos\theta_y & 0 \\ 0 & 0 & 0 & 1 \end{bmatrix} \tag{3.108}$$

If we move the little moving coordinate through a sequence of rotations, we simply record the sequence by multiplying the transformation matrices in the correct order. For example, if we want to determine the transformation matrix between reference frames $X-Y-Z$ and $\hat{x}-\hat{y}-\hat{z}$ of Figure 3.31, we can simply record the rotation sequences of the little moving coordinate from $X-Y-Z$ to $\hat{x}-\hat{y}-\hat{z}$. Write the transformation matrix of the first rotation on the left and then multiply the next rotation from the right, as follows:

$$\prod T_{ti} = T_{tz}(90°) \cdot T_{tx}(60°). \tag{3.109}$$

The actual transformation equation becomes

$$\begin{bmatrix} X_A \\ Y_A \\ Z_A \\ 1 \end{bmatrix} = T_{tz}(90°) \cdot T_{tx}(60°) \cdot \begin{bmatrix} \hat{x}_A \\ \hat{y}_A \\ \hat{z}_A \\ 1 \end{bmatrix} \tag{3.110}$$

We read Equation (3.110) in this way: The coordinate of object $A$ with respect to $X-Y-Z$ is equal to it if we first rotate a little moving coordinate, $x-y-z$, initially aligned with $X-Y-Z$, with respect to the $z$ axis by $\theta_z = 90°$. After that, the little $x$ axis is aligned with the $Y$ axis. However, after the rotation, we do not need to recognize how the new orientation of $x-y-z$ is related to $X-Y-Z$. We only focus on the little coordinate $x-y-z$ at its current state. We then rotate with respect to the new $x$ axis of the little coordinate by $\theta_x = 60°$. After that, we observe that the position of object $A$ with respect to the reference frame, which is defined as $\hat{x}-\hat{y}-\hat{z}$. $[X_A\ Y_A\ Z_A\ 1]'$ and $[\hat{x}_A\ \hat{y}_A\ \hat{z}_A\ 1]'$, are now related by Equation (3.110).

Let's verify if Equation (3.110) is correct. According to Figure 3.31, object $A$ has a position as $[0, a, 0, 1]$ with respect to $X-Y-Z$ and $[a, 0, 0, 1]$ with respect to $\hat{x}-\hat{y}-\hat{z}$. Therefore,

$$T_{tz}(90°) \cdot T_{tx}(60°) \cdot \begin{bmatrix} a \\ 0 \\ 0 \\ 1 \end{bmatrix} = T_{tz}(90°) \cdot \begin{bmatrix} 1 & 0 & 0 & 0 \\ 0 & \cos 60° & -\sin 60° & 0 \\ 0 & \sin 60° & \cos 60° & 0 \\ 0 & 0 & 0 & 1 \end{bmatrix} \begin{bmatrix} a \\ 0 \\ 0 \\ 1 \end{bmatrix}$$

$$= T_{tz}(90°) \cdot \begin{bmatrix} a \\ 0 \\ 0 \\ 1 \end{bmatrix} = \begin{bmatrix} \cos 90° & -\sin 90° & 0 & 0 \\ \sin 90° & \cos 90° & 0 & 0 \\ 0 & 0 & 1 & 0 \\ 0 & 0 & 0 & 1 \end{bmatrix} \cdot \begin{bmatrix} a \\ 0 \\ 0 \\ 1 \end{bmatrix} = \begin{bmatrix} 0 \\ a \\ 0 \\ 1 \end{bmatrix}$$

$$\tag{3.111}$$

The order of applying the transformation matrices is very important because the end result could be different as discussed before.

With the definition of transformation matrices and the little moving coordinate concept, let's consider a very small angle rotation. Therefore, let's consider infinitesimal angles of rotations, $\delta\theta_x$, $\delta\theta_y$, and $\delta\theta_z$. Because these angles are so small, $\sin(\delta\theta) \simeq \delta\theta$ and $\cos(\delta\theta) \simeq 1$. The transformation matrices of Equations (3.105)–(3.108) for infinitesimal rotations are

$$T_{tz}(\delta\theta_z) = \begin{bmatrix} 1 & -\delta\theta_z & 0 & 0 \\ \delta\theta_z & 1 & 0 & 0 \\ 0 & 0 & 1 & 0 \\ 0 & 0 & 0 & 1 \end{bmatrix} \tag{3.112}$$

$$T_{tx}(\delta\theta_x) = \begin{bmatrix} 1 & 0 & 0 & 0 \\ 0 & 1 & -\delta\theta_x & 0 \\ 0 & \delta\theta_x & 1 & 0 \\ 0 & 0 & 0 & 1 \end{bmatrix} \tag{3.113}$$

$$T_{ty}(\delta\theta_y) = \begin{bmatrix} 1 & 0 & \delta\theta_y & 0 \\ 0 & 1 & 0 & 0 \\ -\delta\theta_y & 0 & 1 & 0 \\ 0 & 0 & 0 & 1 \end{bmatrix} \tag{3.114}$$

Let's apply a sequence of infinitesimal rotations with respect to all three axes of the little moving coordinate. We have

$$\begin{bmatrix} X_A \\ Y_A \\ Z_A \\ 1 \end{bmatrix} = T_{tz}(\delta\theta_z) \cdot T_{tx}(\delta\theta_x) \cdot T_{ty}(\delta\theta_y) \cdot \begin{bmatrix} \hat{x}_A \\ \hat{y}_A \\ \hat{z}_A \\ 1 \end{bmatrix}$$

$$= \begin{bmatrix} \hat{x}_A - \delta\theta_z\hat{y} + \delta\theta_y\hat{z} \\ \delta\theta_y\hat{x}_A + \hat{y} - \delta\theta_x\hat{z} \\ -\delta\theta_y\hat{x}_A + \delta\theta_x\hat{y} + \hat{z} \\ 1 \end{bmatrix} \tag{3.115}$$

If we try out different rotation sequences, such as $T_{tx}(\delta\theta_x) \cdot T_{ty}(\delta\theta_y) \cdot T_{tz}(\delta\theta_z)$ or $T_{tz}(\delta\theta_z) \cdot T_{ty}(\delta\theta_y) \cdot T_{tx}(\delta\theta_x)$, we reach the same result of Equation (3.115). In other words, when the angles of rotation are infinitesimal, the order of rotations is no longer important. This result is of utmost importance. Remember that we do not have Equation (3.101) with finite rotation angles, but with infinitesimal angles, we now have

$$\delta\vec{\Gamma} = \delta\theta_x\vec{i} + \delta\theta_y\vec{j} + \delta\theta_z\vec{k} = \delta\theta_y\vec{j} + \delta\theta_x\vec{i} + \delta\theta_z\vec{k} = \delta\theta_z\vec{k} + \delta\theta_y\vec{j} + \delta\theta_x\vec{i} \tag{3.116}$$

In other words, infinitesimal angle rotations can be expressed as a vector. If we further examine the resultant transformation matrix of $T_{tx}(\delta\theta_x) \cdot T_{ty}(\delta\theta_y) \cdot T_{tz}(\delta\theta_z)$, we obtain

$$T_{txyz}(\delta\theta_x, \delta\theta_y, \delta\theta_z) = T_{tx}(\delta\theta_x) \cdot T_{ty}(\delta\theta_y) \cdot T_{tz}(\delta\theta_z)$$

$$= \begin{bmatrix} 1 & -\delta\theta_z & \delta\theta_y & 0 \\ \delta\theta_z & 1 & -\delta\theta_x & 0 \\ -\delta\theta_y & \delta\theta_x & 1 & 0 \\ 0 & 0 & 0 & 1 \end{bmatrix} \tag{3.117}$$

$T_{txyz}(\delta\theta_x, \delta\theta_y, \delta\theta_z)$ again confirms that the order of the infinitesimal rotations makes no difference.

Accordingly, if we divide $\delta\vec{\Gamma}$ with $\delta t$ and let $\delta t$ approach zero, we will have an angular velocity vector of a reference frame

$$\vec{\Omega} = \lim_{\delta t \to 0} \frac{\delta\vec{\Gamma}}{\delta t} = \frac{\delta\theta_x}{\delta t}\vec{i} + \frac{\delta\theta_y}{\delta t}\vec{j} + \frac{\delta\theta_z}{\delta t}\vec{k} = \Omega_x\vec{i} + \Omega_y\vec{j} + \Omega_z\vec{k} \tag{3.118}$$

where $\Omega_x$, $\Omega_y$, and $\Omega_z$ are the angular velocity components of frame $x - y - z$ with respect to frame $X - Y - Z$.

### 3.10.3 *Angular velocity and acceleration vectors of a rotating reference frame*

We can take the time derivative for the transformation matrices of Equations (3.112)–(3.114) and obtain $T_{tx}(\Omega_x)$, $T_{ty}(\Omega_y)$, and $T_{tz}(\Omega_y)$, respectively. These angular velocity transformation matrices have the same properties as the infinitesimal angle matrices. However, we normally do not use the transformation matrices in the angular velocity form.

If we have three reference frames, for which $x_0-y_0-z_0$ is fixed, $\hat{x}_1-\hat{y}_1-\hat{z}_1$ has an angular velocity $\vec{\Omega}_1$ with respect to $x_0-y_0-z_0$, and $\hat{x}_2-\hat{y}_2-\hat{z}_2$ has an angular velocity $\vec{\Omega}_2$ with respect to $\hat{x}_1-\hat{y}_1-\hat{z}_1$. What is the relative angular velocity of $\hat{x}_2-\hat{y}_2-\hat{z}_2$ with respect to $x_0-y_0-z_0$? Based on Equation (3.117), we can simply add up the relative angular velocities. This is equivalent to conducting a sequence of coordinate transformation. We obtain

$$\vec{\Omega}_{2/0} = \vec{\Omega}_{1/0} + \vec{\Omega}_{2/1} \tag{3.119}$$

Equation (3.119) is called the additivity of the angular velocity, similar to the additivity of the position vector (Equation (3.87)). Note that we do not have the additivity of the finite angular displacements. We do have the additivity of the infinitesimal angular displacement. Equation (3.119) is highly important and plays a critical role to determine the relative motions observed by reference frames with different frame rotations.

Similarly, we can take one more derivative to obtain the relationship of angular acceleration vectors. We will use $\dot{\Omega}$ as the angular acceleration for a reference frame.

$$\dot{\vec{\Omega}}_{2/0} = \dot{\vec{\Omega}}_{1/0} + \dot{\vec{\Omega}}_{2/1} \tag{3.120}$$

Note that the total time derivative of $\dot{\vec{\Omega}}_{2/1}$ is more complicated if $\hat{x}_1-\hat{y}_1-\hat{z}_1$ is rotating. The total time derivative of a vector representing by a rotating reference is presented in the next section.

Equation (3.119) is important because it offers a convenient way to determine the relation of one reference frame to another. In other words, we can always break down a complicated rotational motion of a reference frame into its sub-components and then add them up.

### 3.10.4   *Relative motion analysis with rotating reference frames*

As shown in Figure 3.33, we would like to observe the motion of particle $A$ with respect to an IRF frame, $O-X-Y-Z$. However, the motion of A is too complicated and we could not write down explicit expressions to describe $\vec{v}_{A/O}$ and $\vec{a}_{A/O}$. On the other hand, we noticed that if we use a rotating reference frame, $\hat{B}-\hat{x}-\hat{y}-\hat{z}$, we can write down $\hat{v}_{A/\hat{B}}$ and $\hat{a}_{A/\hat{B}}$ quite easily because the motion of $A$ now appears to be a simple motion (straight line or perfect circular motion). Situations like this are quite common. For example, if we

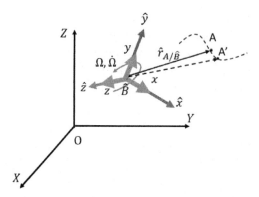

Figure 3.33: Relative motion analysis between the observations of particle A by $O-X-Y-Z$ and by $\hat{B}-\hat{x}-\hat{y}-\hat{z}$.

are on the ground, it would be difficult to describe the motion of a person jumping up and down in a bus which is circling around a race track. However, for an observer inside the bus, the motion of the person jumping up and down is quite simple, only up and down, i.e., a straight line motion.

How can we relate $(\vec{v}_{A/O}, \vec{a}_{A/O})$ to $(\hat{v}_{A/\hat{B}}, \hat{a}_{A/\hat{B}})$? To answer this question, we will introduce a little coordinate, $B-x-y-z$. At this instant of time, $\hat{B}$ and $B$ are the same point. Observer $B$ is using a fixed reference frame $x-y-z$ which aligns with $\hat{x}-\hat{y}-\hat{z}$ at this instant of time. In the next instant of time, $\hat{x}-\hat{y}-\hat{z}$ and $x-y-z$ will be different by an infinitesimal angular displacement, while point $A$ moves to point $A'$, as shown in Figure 3.33. The reference frame, $x-y-z$, is fixed and it could be defined by conducting a sequence of finite angular displacements from $X-Y-Z$. With this new reference frame, we only need to find out how $(\vec{v}_{A/B}, \vec{a}_{A/B})$ and $(\hat{v}_{A/\hat{B}}, \hat{a}_{A/\hat{B}})$ are related. After that, we can easily relate $(\vec{v}_{A/B}, \vec{a}_{A/B})$ to $(\vec{v}_{A/O}, \vec{a}_{A/O})$ using Equations (3.88) and (3.89).

We can use Equation (3.115) to relate the position vectors of point $A$ observed by $\hat{x}-\hat{y}-\hat{z}$ and $x-y-z$.

$$
\begin{bmatrix} x \\ y \\ z \\ 1 \end{bmatrix}_{A/B} = T_{tx}(\theta_x) \cdot T_{ty}(\theta_y) \cdot T_{tz}(\theta_z) \cdot \begin{bmatrix} \hat{x} \\ \hat{y} \\ \hat{z} \\ 1 \end{bmatrix}_{A/\hat{B}} \tag{3.121}
$$

where $\theta_x$, $\theta_y$, and $\theta_z$ are zero at $t = t_0$ and become $\delta\theta_x$, $\delta\theta_y$, and $\delta\theta_z$ at the next instant, $t = t_0 + \delta t$.

To determine the relationship between $\vec{v}_{A/B}$ and $\hat{v}_{A/\hat{B}}$, we can take a derivative of Equation (3.121) to obtain

$$
\frac{d}{dt}
\begin{bmatrix} x \\ y \\ z \\ 1 \end{bmatrix}_{A/B}
=
\frac{d}{dt}
\left[
T_{tx}(\theta_x) \cdot T_{ty}(\theta_y) \cdot T_{tz}(\theta_z) \cdot
\begin{bmatrix} \hat{x} \\ \hat{y} \\ \hat{z} \\ 1 \end{bmatrix}_{A/\hat{B}}
\right]
\tag{3.122}
$$

which leads to

$$
\vec{v}_{A/B} =
\begin{bmatrix} \dot{x} \\ \dot{y} \\ \dot{z} \\ 1 \end{bmatrix}_{A/B}
=
\frac{d}{dt}\left[T_{tx}(\theta_x) \cdot T_{ty}(\theta_y) \cdot T_{tz}(\theta_z)\right] \cdot
\begin{bmatrix} \hat{x} \\ \hat{y} \\ \hat{z} \\ 1 \end{bmatrix}_{A/\hat{B}}
$$

$$
+ \left[T_{tx}(\theta_x) \cdot T_{ty}(\theta_y) \cdot T_{tz}(\theta_z)\right]_{t=t_0} \cdot \frac{d}{dt}\left(
\begin{bmatrix} \hat{x} \\ \hat{y} \\ \hat{z} \\ 1 \end{bmatrix}_{A/\hat{B}}
\right)
\tag{3.123}
$$

Based on Equation (3.117), we have

$$
\frac{d}{dt}\left[T_{tx}(\theta_x) \cdot T_{ty}(\theta_y) \cdot T_{tz}(\theta_z)\right] =
\begin{bmatrix}
0 & \dfrac{-d\theta_z}{dt} & \dfrac{d\theta_y}{dt} & 0 \\
\dfrac{d\theta_z}{dt} & 0 & \dfrac{-d\theta_x}{dt} & 0 \\
\dfrac{-d\theta_y}{dt} & \dfrac{d\theta_x}{dt} & 0 & 0 \\
0 & 0 & 0 & 1
\end{bmatrix}
$$

$$
=
\begin{bmatrix}
0 & -\Omega_z & \Omega_y & 0 \\
\Omega_z & 0 & -\Omega_x & 0 \\
-\Omega_y & \Omega_x & 0 & 0 \\
0 & 0 & 0 & 1
\end{bmatrix}
\tag{3.124}
$$

and

$$[T_{tx}(\theta_x) \cdot T_{ty}(\theta_y) \cdot T_{tz}(\theta_z)]_{t=t_0} = \begin{bmatrix} 1 & 0 & 0 & 0 \\ 0 & 1 & 0 & 0 \\ 0 & 0 & 1 & 0 \\ 0 & 0 & 0 & 1 \end{bmatrix} \qquad (3.125)$$

Introducing Equations (3.124) and (3.125) into Equation (3.123), we arrive at

$$\vec{v}_{A/B} = \begin{bmatrix} \dot{x} \\ \dot{y} \\ \dot{z} \end{bmatrix}_{A/B} = \vec{\Omega} \times \begin{bmatrix} \hat{x} \\ \hat{y} \\ \hat{z} \end{bmatrix}_{A/\hat{B}} + \begin{bmatrix} \dot{\hat{x}} \\ \dot{\hat{y}} \\ \dot{\hat{z}} \end{bmatrix}_{A/\hat{B}}$$

$$= \vec{\Omega} \times \hat{r}_{A/\hat{B}} + \hat{v}_{A/\hat{B}} \qquad (3.126)$$

The equality between $\vec{\Omega} \times \hat{r}_{A/\hat{B}}$ and the first term on the right-hand side of Equation (3.123) can be shown by multiplying both out.

Note that $\vec{\Omega} = \Omega_x \vec{i} + \Omega_y \vec{j} + \Omega_z \vec{k}$, which is the angular velocity vector of frame $\hat{x}-\hat{y}-\hat{z}$ with respect to $x-y-z$. Therefore, the unit directional vectors are $\vec{i}$, $\vec{j}$, and $\vec{k}$. On the other hand, $\hat{i}$, $\hat{j}$, and $\hat{k}$ are used for $\hat{r}_{A/\hat{B}}$ and $\hat{v}_{A/\hat{B}}$ in Equation (3.126). How do we conduct the cross multiplication for $\vec{\Omega} \times \hat{r}_{A/\hat{B}}$ when different unit vectors are used? Fortunately, as we defined it, at $t = t_0$, $\hat{i} = \vec{i}$, $\hat{j} = \vec{j}$, and $\hat{k} = \vec{k}$. However, we should know that $d\hat{i}/dt \neq 0$, while $d\vec{i}/dt = 0$, etc., because $\hat{x}-\hat{y}-\hat{z}$ is rotating while $x-y-z$ is fixed in its directions.

We can also release the restriction that $x-y-z$ and $\hat{x}-\hat{y}-\hat{z}$ are aligned at the current instant of time. If we keep $x-y-z$ the same as $X-Y-Z$, Equation (3.126) would be the same and we only need to define the transformation between the unit vectors. In other words, we have the choice, for example, to choose to let $\vec{i} = \vec{I}$ or $\vec{i} = \hat{i}$.

To determine $\vec{v}_{A/O}$, we simply use Equation (3.88) with some adjustments in order to convert $\vec{i}$, $\vec{j}$, and $\vec{k}$ to $\vec{I}$, $\vec{J}$, and $\vec{K}$ of $X-Y-Z$.

The complete relation velocity equation is

$$\vec{v}_{A/O} = \vec{v}_{B/O} + \vec{v}_{A/B} = \vec{v}_{B/O} + \vec{\Omega} \times \hat{r}_{A/\hat{B}} + \hat{v}_{A/\hat{B}} \qquad (3.127)$$

The term $\vec{\Omega} \times \hat{r}_{A/\hat{B}}$ in Equation (3.127) is a consequence due to the rotation of $\hat{x}-\hat{y}-\hat{z}$.

To simplify the notation of Equation (3.127), we will not distinguish between $B$ and $\hat{B}$, but rely on $\hat{r}$ and $\hat{v}$ to indicate if a rotating reference frame is used for the observation. We now have a simpler form as

$$\vec{v}_{A/O} = \vec{v}_{B/O} + \vec{\Omega} \times \hat{r}_{A/B} + \hat{v}_{A/B} \qquad (3.128)$$

We can read Equation (3.128) as

*The velocity of A with respect to O can be obtained by adding the velocity of A observed by the non-inertial observer B and the velocity of B observed by O, plus the correction term $\vec{\Omega} \times \hat{r}_{A/B}$.*

Let's revisit Example 3.7 and answer the question "what are $\hat{v}_{B/A}$ and $\hat{a}_{B/A}$?"

## Example 3.8: Relative Velocity Analysis with Rotating Frames

As shown in Figure 3.34, car $B$ moves at $100 \, \text{km/h}$ and slows down at $100 \, \text{km/h}^2$, while car $A$ travels through a ramp to merge into the same highway. Car $A$ is at $50 \, \text{km/h}$ and speeds up at $1{,}000 \, \text{km/h}^2$. What are the velocity and acceleration of car $B$ observed by the driver of car $A$?

**Solution:**

**Problem Statement:** We are to determine $\hat{v}_{B/A}$ and $\hat{a}_{B/A}$.

**Force Analysis:** Skipped.

**Motion Analysis:** As shown in Figure 3.34, we redefine the little non-rotating coordinate $x-y$ to be aligned with $\hat{x}-\hat{y}$ at this instant of time. The IRF at $O$ is now renamed to be $X-Y$. From geometry, we found that $\hat{i} = \vec{i} = \cos 30° \vec{I} - \sin 30° \vec{J}$, $\hat{j} = \vec{j} = \sin 30° \vec{I} + \cos 30° \vec{J}$, and $\hat{k} = \vec{k} = \vec{K}$. We also found that $\vec{I} = \cos 30° \vec{i} + \sin 30° \vec{j}$, $\vec{J} = -\sin 30° \vec{i} + \cos 30° \vec{j}$.

**Governing Equations:** First, we need to determine the rotation of $\hat{x}-\hat{y}$. From Figure 3.34, we know as car $A$ moves along the curved

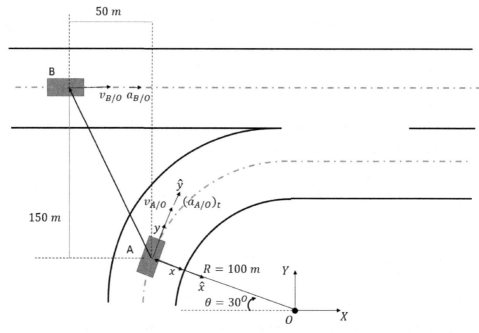

Figure 3.34: Relative motion analysis for determining $\vec{v}_{B/A}$ and $\vec{a}_{B/A}$, where observer $A$ uses a rotating reference frame.

road, the car body rotates, rending $\hat{x}-\hat{y}$ to rotate at the same rate. Therefore,

$$\vec{\Omega}_A = \frac{v_{A/O}}{R}(-\vec{k}) = \frac{50 \cdot 1000}{100 \cdot 60 \cdot 60}(-\vec{k}) = -0.139\,\vec{k}\ (\text{rad/s})$$

$$\dot{\vec{\Omega}}_A = \frac{(a_{A/O})_t}{R}\,(\vec{k}) = -0.000772\,\vec{k}\ (\text{rad/s}^2) \tag{3.129}$$

Another way to determine $\Omega_A$ and $\dot{\Omega}_A$ is to recognize that the rotating frame is related to the angle $\theta$. Therefore, $\Omega_A = \dot{\theta}$ and $\dot{\Omega}_A = \ddot{\theta}$. $\dot{\Omega}_A$ will be used later for relative acceleration analysis.

The relative velocity equation can now be defined as

$$\vec{v}_{B/O} = \frac{100 \cdot 1,000}{3,600}\,\vec{I} = \vec{v}_{A/O} + \vec{\Omega} \times \hat{r}_{B/A} + \hat{v}_{B/A}$$

$$= \frac{50 \cdot 1,000}{3,600}\,\hat{j} + (-0.139\,\vec{k}) \times (-50\vec{I} + 150\vec{J}) + \hat{v}_{B/A} \tag{3.130}$$

Note that in the above equation, we express $\hat{r}_{B/A}$ in terms of $\vec{I}$ and $\vec{J}$ for convenience.

Solving Equation (3.130) and converting $\vec{I}$ and $\vec{J}$ into $\vec{i}$ and $\vec{j}$, we obtain $\hat{v}_{B/A} = 34.1\hat{i} - 59.1\hat{j}$(km/h) or $\hat{v}_{B/A} = 0.0\vec{I} - 68.3\vec{J}$(km/h). Compared with the result of Example 3.7, in which $\vec{v}_{A/B} = -75\vec{I} + 43.3\vec{J}$ (km/h), apparently $\hat{v}_{B/A} \neq -\vec{v}_{A/B}$. Driver $A$ notices some speed components not seen by driver $B$ due to the rotating frame used by driver $A$.

For the relative acceleration analysis, we can take a derivative of Equation (3.128), which yields

$$\vec{a}_{A/O} = \frac{d}{dt}(\vec{v}_{A/O}) = \frac{d}{dt}[\vec{v}_{B/O} + \vec{\Omega} \times \hat{r}_{A/B} + \hat{v}_{A/B}]$$

$$= \vec{a}_{B/O} + \frac{d}{dt}[\vec{\Omega} \times \hat{r}_{A/B}] + \frac{d}{dt}[\hat{v}_{A/B}] \qquad (3.131)$$

Before we complete the derivation of Equation (3.131), let's generalize Equation (3.126) as

$$\frac{d}{dt}(\vec{c}) = \vec{\Omega} \times \vec{c} + \frac{\partial}{\partial t}\vec{c} \qquad (3.132)$$

where $\vec{c}$ is any vector expressed in $\hat{i}$, $\hat{j}$, and $\hat{k}$ with a frame rotation $\vec{\Omega}$. $\frac{d}{dt}(\vec{c})$ is the correct total time derivative of $\vec{c}$ observed by an IRF, recognizing $\hat{i}$, $\hat{j}$, and $\hat{k}$ are not fixed in directions.

On the other hand, the time derivative of $\vec{c}$ assuming $\hat{i}$, $\hat{j}$, and $\hat{k}$ are fixed is $\frac{\partial}{\partial t}\vec{c}$.

Let's verify Equation (3.132) using the $r - \theta$ description. The velocity of an object with respect to an IRF but expressed in $\vec{e}_r$ and $\vec{e}_\theta$ is

$$\vec{v} = \dot{r}\vec{e}_r + r\dot{\theta}\vec{e}_\theta \qquad (3.133)$$

Applying Equation (3.132) for $\vec{v}$ expressed in $\vec{e}_r$ and $\vec{e}_\theta$, we have

$$\vec{a} = \frac{d}{dt}(\vec{v}) = \vec{\Omega} \times \vec{v} + \frac{\partial}{\partial t}\vec{v} \qquad (3.134)$$

With $\vec{\Omega} = \dot{\theta}\vec{k}$, Equation (3.134) becomes

$$\vec{a} = \dot{\vec{v}} = (\dot{\theta}\vec{k}) \times (\dot{r}\vec{e}_r + r\dot{\theta}\vec{e}_\theta) + \frac{\partial}{\partial t}(\dot{r})\vec{e}_r + \frac{\partial}{\partial t}(r\dot{\theta})\vec{e}_\theta$$

$$= \dot{r}\dot{\theta}\vec{e}_\theta - r\dot{\theta}^2\vec{e}_r + \ddot{r}\vec{e}_r + r\ddot{\theta}\vec{e}_\theta + \dot{r}\dot{\theta}\vec{e}_\theta$$

$$= (\ddot{r} - r\dot{\theta}^2)\vec{e}_r + (r\ddot{\theta} + 2\dot{r}\dot{\theta})\vec{e}_\theta \qquad (3.135)$$

We obtain the same result of Equation (3.80).

Again, we must pay attention that $\frac{d}{dt}\vec{c}$ is the total time derivative of $\vec{c}$ with respect to the fixed frame, recognizing that $\vec{c}$ is expressed in unit directional vectors which are not fixed. On the other hand, $\frac{\partial}{\partial t}\vec{c}$ is the time derivative assuming the unit directional vectors are fixed. In other words,

$$\frac{\partial}{\partial t}\vec{c} = \dot{c}_x\hat{i} + \dot{c}_y\hat{j} + \dot{c}_z\hat{k} \neq \frac{d}{dt}(\vec{c}) \tag{3.136}$$

Let's apply Equation (3.132) for the unit vectors we defined for the $r-\theta$ and the $t-n$ rotating reference frames. Both of these frames are found to have a frame rotation as $\vec{\Omega} = \dot{\theta}\vec{k}$. For the $t-n$ system, we also have $\dot{\theta} = v/\rho$.

$$\dot{\vec{e}}_r = \vec{\Omega} \times \vec{e}_r + \frac{\partial}{\partial t}\vec{e}_r = \dot{\theta}\vec{e}_\theta \tag{3.137}$$

where $\frac{\partial}{\partial t}\vec{e}_r = 0$ because $\vec{e}_r$ is considered as fixed. In addition, we have $\vec{k} \times \vec{e}_r = \vec{e}_\theta$. Similarly, we obtain

$$\dot{\vec{e}}_\theta = \vec{\Omega} \times \vec{e}_\theta + \frac{\partial}{\partial t}\dot{\vec{e}}_\theta = -\dot{\theta}\vec{e}_r \tag{3.138}$$

$$\dot{\vec{e}}_t = \vec{\Omega} \times \vec{e}_t + \frac{\partial}{\partial t}\vec{e}_t = \dot{\theta}\vec{e}_n = \frac{v}{\rho}\vec{e}_n \tag{3.139}$$

Note that the frame rotation of the $t-n$ Reference frame is the opposite that of the $r-\theta$ Reference frame.

We obtain the same results as before, but the derivation is much easier.

By applying Equation (3.132) for $\vec{\Omega} \times \hat{r}_{A/B}$, we have

$$\frac{d}{dt}(\vec{\Omega} \times \hat{r}_{A/B}) = \vec{\Omega} \times (\vec{\Omega} \times \hat{r}_{A/B}) + \frac{\partial}{\partial t}[\vec{\Omega} \times \hat{r}_{A/B}]$$

$$= \vec{\Omega} \times (\vec{\Omega} \times \hat{r}_{A/B}) + \dot{\vec{\Omega}} \times \hat{r}_{A/B} + \vec{\Omega} \times \hat{v}_{A/B} \tag{3.140}$$

$$\frac{d}{dt}[\hat{v}_{A/B}] = \vec{\Omega} \times \hat{v}_{A/B} + \frac{\partial}{\partial t}\hat{v}_{A/B}$$

$$= \vec{\Omega} \times \hat{v}_{A/B} + \hat{a}_{A/B} \tag{3.141}$$

Introducing Equations (3.140) and (3.141) to Equation (3.131), we finally obtain the relative acceleration equation,

$$\vec{a}_{A/O} = \vec{a}_{B/O}$$

$$+\vec{\Omega} \times (\vec{\Omega} \times \hat{r}_{A/B}) + \dot{\vec{\Omega}} \times \hat{r}_{A/B} + 2\vec{\Omega} \times \hat{v}_{A/B}$$

$$+\hat{a}_{A/B} \tag{3.142}$$

The middle three terms of Equation (3.142) on the right-hand side are the consequences due to the frame rotation of $\hat{x}$–$\hat{y}$–$\hat{z}$. The term $2\vec{\Omega} \times \hat{v}_{A/B}$ is famously known as the Coriolis acceleration. We can read Equation (3.142) as

*The acceleration of A with respect to O can be obtained by adding the acceleration of A observed by the non-inertial observer B and the acceleration of B observed by O, plus three correction terms, $\vec{\Omega} \times (\vec{\Omega} \times \hat{r}_{A/B})$, $\dot{\vec{\Omega}} \times \hat{r}_{A/B}$, and $2\vec{\Omega} \times \hat{v}_{A/B}$.*

Let's revisit the $r - \theta$ description of Section (3.7). As shown in Figure 3.35, similar to Figure 3.23, we define three sets of reference frames. The rotating reference frame, $\hat{x}$–$\hat{y}$, has the unit directional vectors defined as $\hat{i} = \vec{e}_r$ and $\hat{j} = \vec{e}_\theta$, respectively. The observer $B$ with the rotating frame $\hat{x}$–$\hat{y}$ or non-rotating frame $x$–$y$ is at the same position as the origin of $X$–$Y$ at all times.

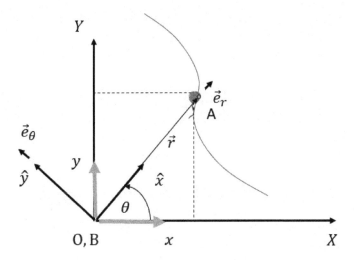

Figure 3.35: Relative acceleration analysis related to the $r - \theta$ description.

Based on Equation (3.142), we have

$$\vec{a}_{A/O} = \vec{a}_{B/O} + \vec{\Omega} \times (\vec{\Omega} \times \hat{r}_{A/B}) + \dot{\vec{\Omega}} \times \hat{r}_{A/B}$$
$$+ 2\vec{\Omega} \times \hat{v}_{A/B} + \hat{a}_{A/B}$$
$$= \vec{\Omega} \times (\vec{\Omega} \times \hat{r}_{A/B}) + \dot{\vec{\Omega}} \times \hat{r}_{A/B} + 2\vec{\Omega} \times \hat{v}_{A/B} + \ddot{r}\vec{e}_r \quad (3.143)$$

In Equation (3.143), we find that $\hat{a}_{A/B} = \ddot{r}\vec{e}_r$ because with respect to $\hat{x}-\hat{y}$, balloon $A$ is always on the $\hat{x}$ axis. To observer $B$ with the rotating reference frame, the only acceleration observed is the acceleration of the radial position, $r$. Similarly, the only velocity observed by $B$ with the rotating frame will be $\hat{v}_{A/B} = \dot{r}\vec{e}_r$. Because observer $B$ is at point $O$ at all times, $\vec{a}_{B/O} = 0$. We can now determine the middle three terms of Equation (3.143) as

$$\vec{\Omega} \times (\vec{\Omega} \times \hat{r}_{A/B}) = \dot{\theta}\vec{k} \times (\dot{\theta}\vec{k} \times (r\vec{e}_r)) = -r\dot{\theta}^2\vec{e}_r \quad (3.144)$$

$$\dot{\vec{\Omega}} \times \hat{r}_{A/B} = (\ddot{\theta}\vec{k}) \times (r\vec{e}_r) = r\ddot{\theta}\ \vec{e}_\theta \quad (3.145)$$

$$2\vec{\Omega} \times \hat{v}_{A/B} = 2(\dot{\theta}\vec{k}) \times (\dot{r}\vec{e}_r) = 2\dot{r}\dot{\theta}\ \vec{e}_\theta \quad (3.146)$$

Combining Equations (3.143)–(3.146), we reach the same expression of Equation (3.80) again.

Regarding the tangential–normal description, the motion observed by the rotational frame is more complicated, not necessarily as a straight-line motion. This is because we do not have the displacement information in the tangential–normal description.

### 3.10.5 *How to select the second observer with a rotating reference frame?*

When we try to observe a motion, we cannot express it explicitly unless the motion is simple, i.e., either in straight line or in perfectly circular motion.

Therefore, one important guideline in selecting a second observer with a specific rotating reference frame is to make a complicated motion appear as a simple motion to the second observer. To borrow a great slogan:

*Make a Complicated Motion Simple Again!*

In some cases, the motion might be so complicated that we need a third or more observers. The key is to make the motion from one observer to the next appear SIMPLE!

We can now complete the acceleration analysis of Example 3.8.

## Example 3.9: Relative Acceleration Analysis with a Rotating Frame

With respect to Figure (3.34), the acceleration of $B$ to $O$ based on observer $A$ using a rotating frame can be expressed as

$$\vec{a}_{B/O} = \vec{a}_{A/O} + \vec{\Omega} \times (\vec{\Omega} \times \hat{r}_{B/A}) + \dot{\vec{\Omega}} \times \hat{r}_{B/A} + 2\,\vec{\Omega} \times \hat{v}_{B/A} + \hat{a}_{B/A} \quad (3.147)$$

where

$$\vec{a}_{B/O} = -100\,\vec{I}\,(\text{km/h}^2) = -0.00772\,\vec{I}\,(\text{m/s}^2)$$

$$\vec{a}_{A/O} = \frac{(v_{A/O})^2}{R}\,\hat{i} + (a_{A/O})_t\,\hat{j}$$

$$= 25000\hat{i} + 1,000\hat{j}\,(\text{km/h}^2) = 1.93\hat{i} + 0.0772\hat{j}\,(\text{m/s}^2)$$

$$\vec{\Omega} \times (\vec{\Omega} \times \hat{r}_{B/A}) = (-0.139\vec{k}) \times [(-0.139\vec{k}) \times (-50\vec{I} + 150\vec{J})]$$

$$= 0.966\vec{I} - 2.898\vec{J}\,(\text{m/s}^2)$$

$$\dot{\vec{\Omega}} \times \hat{r}_{B/A} = (-0.000772\vec{k}) \times (-50\vec{I} + 150\vec{J})$$

$$= 0.116\vec{I} + 0.0386\vec{J}\,(\text{m/s}^2)$$

$$2\,\vec{\Omega} \times \hat{v}_{B/A} = 2(-0.139\vec{k}(\text{rad/s})) \times (34.1\hat{i} - 59.1\hat{j}(\text{km/h}))$$

$$= -4.56\,\hat{i} - 2.63\,\hat{j}\,(\text{m/s}^2)$$

Finally, we obtain $\hat{a}_{B/A} = 0.256\hat{i} + 4.49\hat{j}\,(\text{m/s}^2)$.

The result from Example 3.7 is $\vec{a}_{A/B} = 1.717\vec{I} - 0.898\vec{J}(\text{m/s}^2)$. We need to convert $\vec{I}$ and $\vec{J}$ to $\hat{i}$ and $\hat{j}$ for comparison.

After conversion, we obtain $\vec{a}_{A/B} = 1.930\hat{i} + 0.0810\hat{j}\,(\text{m/s}^2)$. Apparently, $|\vec{a}_{A/B}| \neq |\hat{a}_{B/A}|$ because observer $A$ uses a rotating reference frame.

From this example, we understand that the extra acceleration terms observed by $A$ are related to $-\vec{a}_{A/O}$, $-\vec{\Omega} \times (\vec{\Omega} \times \hat{r}_{B/A})$, $-\dot{\vec{\Omega}} \times \hat{r}_{B/A}$, and $-2\,\vec{\Omega} \times \hat{v}_{B/A}$ because

$$\hat{a}_{B/A} = \vec{a}_{B/O} - \vec{a}_{A/O} - \vec{\Omega} \times (\vec{\Omega} \times \hat{r}_{B/A}) - \dot{\vec{\Omega}} \times \hat{r}_{B/A}$$

$$-2\vec{\Omega} \times \hat{v}_{B/A} \qquad (3.148)$$

The result of Equation (3.148) explains the expression of Equation (3.68) because often we do not know what a non-inertial observer will perceive. In the next section, we present many strange phenomena experienced due to the use of rotating reference frames.

### 3.10.6 *Examples of strange phenomena due to the use of a rotating reference frame*

In this section, we present several examples of strange phenomena when an observer is using a rotating reference frame, caused by, in particular, the negative effect of Coriolis's acceleration, $-2\,\vec{\Omega} \times \hat{v}$.

### Example 3.10: Can a Ball Move in a Curved Line?

We are not talking about the curve ball in the baseball pitching, which is mainly due to the spinning action of the ball interacting with the air. As shown in Figure 3.36, a ball is thrown straight across a rotating platform. What will person $P$ standing at the center of the

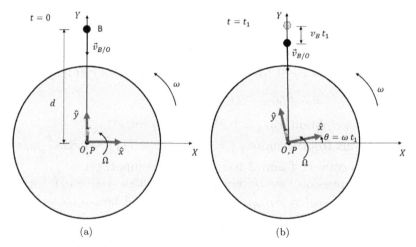

Figure 3.36: A ball traveling in a straight line may appear to be curved: (a) when the ball is thrown at $t = 0$; and (b) at $t = t_1$, the ball is in a closer position to the person standing at the center of the platform, rotating with the platform.

platform and rotating with the platform see? The platform rotates at a constant rotating speed, $\omega$.

**Solution:**

**Problem Statement:** We want to determine $\hat{v}_{B/P}$ and $\hat{a}_{B/P}$ which are what person $P$ will observe.

**Force Analysis:** Skipped.

**Motion Analysis:**

At $t = 0$, the ball was released with a speed, $\vec{v}_{B/O} = -v_B \vec{J}$ (Figure 3.36(a)). For the inertial observer, $O-X-Y$, this ball will travel through the platform in a straight line to the center and hit the person. At $t_1$, as shown in Figure 3.36(b), the ball has moved to a new position along the straight line path, closer to the person. The platform would have rotated by an angle $\theta = \omega \cdot t_1$.

The person at the center of the platform will be using a rotating reference frame, $P-\hat{x}-\hat{y}$, with a frame rotating speed, $\vec{\Omega} = \omega \vec{k}$ and and $\dot{\vec{\Omega}} = 0$, identified by $O-X-Y$. Initially, $P-\hat{x}-\hat{y}$ and $O-X-Y$ are aligned. They would be deviated by the angle $\theta = \omega t_1$ at $t_1$.

Based on Figure 3.36, the relationship between directional unit vectors can be found as

$$\vec{I} = \cos \omega t \, \hat{i} - \sin \omega t \, \hat{j}; \quad \vec{J} = \sin \omega t \, \hat{i} + \cos \omega t \, \hat{j} \tag{3.149}$$

Based on the relative motion analysis, we have

$$\vec{v}_{B/O} = \vec{v}_{P/O} + \vec{\Omega} \times \hat{r}_{B/P} + \hat{v}_{B/P} \tag{3.150}$$

$$\vec{a}_{B/O} = \vec{a}_{P/O} + \vec{\Omega} \times (\vec{\Omega} \times \hat{r}_{B/P}) + \dot{\vec{\Omega}} \times \hat{r}_{B/P}$$
$$+ 2 \vec{\Omega} \times \hat{v}_{B/P} + \hat{a}_{B/P} \tag{3.151}$$

The velocity and acceleration of $P$ observed by the IRF are both zero. The velocity and acceleration of the ball, observed by $P$, are

$$\hat{v}_{B/P} = \vec{v}_{B/O} - \vec{\Omega} \times \hat{r}_{B/P} \tag{3.152}$$

$$\hat{a}_{B/P} = -\vec{\Omega} \times (\vec{\Omega} \times \hat{r}_{B/P}) - 2 \vec{\Omega} \times \hat{v}_{B/P} \tag{3.153}$$

Introducing Equation (3.149) into Equations (3.152) and (3.153), we obtain

$$\hat{v}_{B/P} = -v_B \sin \omega t \, \hat{i} - v_B \cos \omega t \, \hat{j} - \vec{\Omega} \times \hat{r}_{B/P}$$

$$= (-v_B \sin \omega t + \omega(d - v_B t) \cos \omega t) \, \hat{i}$$
$$+ (-v_B \cos \omega t - \omega(d - v_B t) \sin \omega t) \, \hat{j} \tag{3.154}$$

where

$$\hat{r}_{B/P} = \vec{r}_{B/O} = (d - v_B t)\vec{J}$$

$$= (d - v_B t)(sin\omega t \ \hat{i} + cos\omega t \ \hat{j}) \tag{3.155}$$

$$-\vec{\Omega} \times (\vec{\Omega} \times \hat{r}_{B/P}) = -\omega^2 (d - v_B t)(\sin \omega t \ \hat{i} + \cos \omega \ \hat{j}) \tag{3.156}$$

$$-2 \ \vec{\Omega} \times \hat{v}_{B/P} = -2\omega(v_B \cos \omega t + \omega(d - v_B t)\sin \omega t) \ \hat{i}$$

$$-2\omega(-v_B \sin \omega t + \omega(d - v_B t)\cos \omega t) \ \hat{j} \tag{3.157}$$

We can plot Equation (3.155) to illustrate the trajectory of the ball perceived by $P$.

Let's consider a case with $d = 25$ m, $v = 100$ m/s, and $\omega = 1$ rad/s. The speed of $100$ m/s is similar to the bullet speed of a firearm muzzle and is about $1/3$ of the speed of sound. The rotational speed of $1$ rad/s is about $9.5$ rpm. The second hand of a wrist watch rotates at $1$ rpm. Figure 3.37 shows the trajectory of the ball, observed by $P$ using the rotating reference frame. Instead of observing the ball flying straight toward observer $P$, $P$ will actually see the ball move along a curved path. The "observed" speed, the Coriolis acceleration, and the "centripetal" acceleration of the ball at different time steps are also listed in Figure 3.37. The time of travel until the ball hits $P$ is the

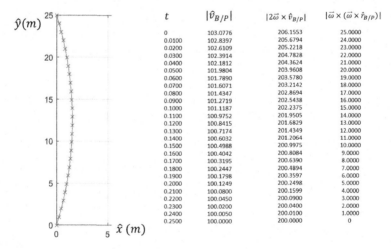

| $t$ | $|\hat{v}_{B/P}|$ | $|2\vec{\omega} \times \hat{v}_{B/P}|$ | $|\vec{\omega} \times (\vec{\omega} \times \hat{r}_{B/P})|$ |
|---|---|---|---|
| 0 | 103.0776 | 206.1553 | 25.0000 |
| 0.0100 | 102.8397 | 205.6794 | 24.0000 |
| 0.0200 | 102.6109 | 205.2218 | 23.0000 |
| 0.0300 | 102.3914 | 204.7828 | 22.0000 |
| 0.0400 | 102.1812 | 204.3624 | 21.0000 |
| 0.0500 | 101.9804 | 203.9608 | 20.0000 |
| 0.0600 | 101.7890 | 203.5780 | 19.0000 |
| 0.0700 | 101.6071 | 203.2142 | 18.0000 |
| 0.0800 | 101.4347 | 202.8694 | 17.0000 |
| 0.0900 | 101.2719 | 202.5438 | 16.0000 |
| 0.1000 | 101.1187 | 202.2375 | 15.0000 |
| 0.1100 | 100.9752 | 201.9505 | 14.0000 |
| 0.1200 | 100.8415 | 201.6829 | 13.0000 |
| 0.1300 | 100.7174 | 201.4349 | 12.0000 |
| 0.1400 | 100.6032 | 201.2064 | 11.0000 |
| 0.1500 | 100.4988 | 200.9975 | 10.0000 |
| 0.1600 | 100.4042 | 200.8084 | 9.0000 |
| 0.1700 | 100.3195 | 200.6390 | 8.0000 |
| 0.1800 | 100.2447 | 200.4894 | 7.0000 |
| 0.1900 | 100.1798 | 200.3597 | 6.0000 |
| 0.2000 | 100.1249 | 200.2498 | 5.0000 |
| 0.2100 | 100.0800 | 200.1599 | 4.0000 |
| 0.2200 | 100.0450 | 200.0900 | 3.0000 |
| 0.2300 | 100.0200 | 200.0400 | 2.0000 |
| 0.2400 | 100.0050 | 200.0100 | 1.0000 |
| 0.2500 | 100.0000 | 200.0000 | 0 |

Figure 3.37: The traveling path of the ball of Figure (3.36) and the values of the speed and accelerations at different time steps.

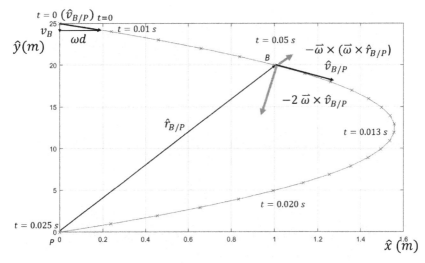

Figure 3.38:    The traveling path of the ball of Figure (3.37) at a different scale. The velocity and acceleration vectors are shown at different points along the path.

same at $0.25\,\text{s}$. Therefore, the ball must travel faster along the longer curved path in order to make it at the same time. The highest speed is $103.08\,\text{m/s}$ compared with $100\,\text{m/s}$. The extra speed is due to $-\vec{\omega}\times\hat{r}_{B/P}$. When the ball hits $P$, $\hat{r}_{B/P}=0$, the observed speed becomes the same at $100\,\text{m/s}$. Similarly, the "centripetal" acceleration is the highest when the ball was just released and becomes zero when the ball hits $P$.

However, the Coriolis acceleration, $-2\vec{\Omega}\times\hat{v}_{B/P}$, never reduces to zero. It is the highest when the observed velocity $\hat{v}_{B/P}$ is the highest.

We re-plot Figure 3.37 in a different scale to exaggerate the curvature of the traveling path for better graphical explanation of the observed velocity and acceleration.

As shown in Figure 3.38, at $t=0$, the observed velocity is $v_B\hat{i}+\omega d\hat{j}=-100\hat{j}+25\hat{i}\,\text{m/s}$. Because of the second term, $25\hat{i}\,\text{m/s}$, $P$ would think that the ball is not thrown at him. However, the ball makes a turn and comes back to hit him. The magnitude and direction of the velocity are changed by the acceleration. There are two fictitious accelerations observed by $P$, but there is no acceleration of the ball observed by an inertial observer. At $t=0.005\,\text{s}$, the negative "contripetal" acceleration, $-\vec{\Omega}\times(\vec{\Omega}\times\hat{r}_{B/P})$, points at the same direction as $\hat{r}_{B/P}$. This acceleration will change the velocity of the

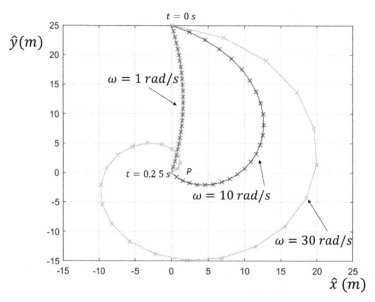

Figure 3.39: Ball traveling paths observed by a non-inertial observer at different frame rotating speeds.

ball further to the right, away from $P$. However, the Coriolis acceleration, $-2\vec{\Omega} \times \hat{v}_{B/P}$, is perpendicular to the ball velocity and has a higher magnitude than the "centripetal" acceleration. As a result, the ball eventually is bent back to hit $P$. The Coriolis acceleration bends the ball path and its effects are more pronounced if $\omega$ is larger or $\hat{v}_{B/P}$ is higher. In Figure 3.39, we plot the travel paths of the ball at $\omega = 1, 10$, and $30 \, \text{rad/s}$.

For observer $P$, it would be very confusing as to why the ball will come back to hit him. At a higher $\omega$, the ball could move past him but still turns back to hit him. The event of the ball hitting $P$ happens for both the inertial and non-inertial observers, but how it happens is quite different.

The ball turning phenomena discussed above can be demonstrated. In the Ashton Graybiel Spatial Orientation Laboratory at Brandeis University, there is an Artificial Gravity Facility: otherwise known as the rotating room.[9] Inside the rotating room, people do not realize that they are rotating. When they throw a ball from one

---

[9]https://www.youtube.com/watch?v=bJ_seXo-Enc&t=5s.

person to another, the ball, instead of going straight to the target, curves to the side. These people inside the rotating room use a rotating reference frame without knowing it.

The readers are encouraged to watch a video listed in the footnote to witness this seemingly impossible phenomenon.

In a short story by Ernest Hemingway, *"The Short Happy Life of Francis Macomber,"* Hemingway wrote the following passage about a hunting trip in Africa of Mr. and Mrs. Macomber:

> *..., and Mrs. Macomber, in the car, had shot at the buffalo with the 6.5 Mannlicher as it seemed about to gore Macomber and had hit her husband about two inches up and a little to one side of the base of his skull.*

Presumably, Mrs. Macomber aimed at the buffalo ("Mrs. Macomber, in the car, had shot at the buffalo with the 6.5 Mannlicher"), but the bullet hit her husband instead. If we do not consider any ill intention of Mrs. Macomber as the story suggests, what could be a scientific explanation other than Ms. Macomber is a bad shooter? For Mr. Macomber, he was happy to see the bullet not aimed at him, but he was only happy for a little while. The bullet made a turn to hit him. What could cause it to turn? Could it be caused by the Coriolis acceleration? It would be fun to see a defense lawyer suggesting this.

These questions, however, lead to an important question: Are we, humans, living on earth, inertial observers?

As shown in Figure 3.40(a), we might think that we are inertial observers because the directions, east, west, north, and south seemingly are fixed. However, from space, the Earth is rotating as shown in Figure 3.40(b). As the Earth turns, these "fixed" directions are changing. Therefore, we, humans are using a rotating reference frame, and the frame rotation speed is

$$\Omega = \frac{2\pi}{day} = 0.000694 \text{ rpm} = 0.0000727 \text{ rad/s} \qquad (3.158)$$

The frame rotating speed is quite slow. However, if we observe an object traveling at high speeds, the Coriolis acceleration could be detectable. We will discuss a few examples of strange phenomena we experience on earth because we are not inertial observers.

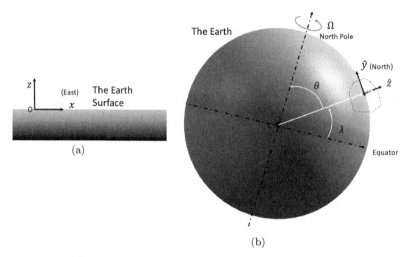

Figure 3.40: (a) Humans on Earth consider themselves as non-inertial observers with a non-rotating reference frame $x-y-z$; and (b) due to the Earth, rotation, the reference frame is actually rotating and should be labeled as $\hat{x}-\hat{y}-\hat{z}$.

## Example 3.11: Foucault Clock: Why does the Giant Pendulum Precess?

One great attraction in a science museum is the demonstration of the Foucault clock, which is a giant pendulum, swinging back and forth, but also slowly precessing, first demonstrated by French physicist Léon Foucault in 1851 for proving that the Earth is rotating. Why is the giant pendulum of the Foucault clock precessing, not simply going back and forth as in the case of a small pendulum.

As shown in Figure 3.41(b), if the Foucault clock is located at a latitude of $\lambda$, the angle between the local vertical axis, $\hat{z}$, and the axis of rotation of the Earth is $\theta = 90° - \lambda$. In the local reference frame, we define $\hat{x}$ as pointing to the east and $\hat{y}$ to the north.

A classical treatment[10] of the Foucault pendulum shows the precession trajectories of the pendulum projected onto the $\hat{x}-\hat{y}$ plane (Figure 3.41(b)). Note that the simulation of Figure 3.41(b) is for a much higher $\Omega$ to demonstrate the precessing more clearly. An

---

[10]Boulanger, N. and Buisseret, F. The formulations of classical mechanics with Foucault's pendulum, *Physics* **2**(2020) 531–540. Doi:10.3390/physics2040030.

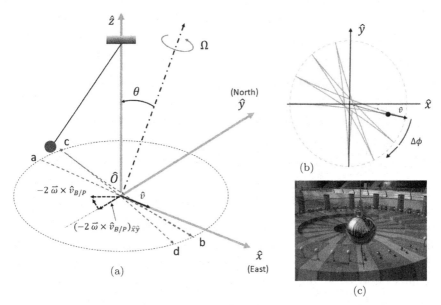

Figure 3.41: (a) The reference frame of a Foucault pendulum; (b) the projected processing paths of the Foucault pendulum; and (c) an actual apparatus of the Foucault pendulum.

*Source:* (b) https://www.mdpi.com/2624-8174/2/4/30/htm; and (c) https://up load.wikimedia.org/wikipedia/commons/thumb/7/71/Foucault_pendulum_clo seup.jpg/800px-Foucault_pendulum_closeup.jpg.

actual Foucault pendulum apparatus is shown in Figure 3.41(c) and the precessing rate is very slow.

If we conduct a force analysis, we can show that there are only two forces, the tension of the pendulum arm and the gravity, without a force to cause the pendulum to precess. Therefore, how come the pendulum does not move along a straight path? Or is it actually moving along a straight path, but our observation is skewed due to the Earth's rotation?

In Figure 3.41(a), a few pendulum paths projected on the $\hat{x}-\hat{y}$ plane are plotted for a Foucault pendulum in the north hemisphere. From point $a$, the pendulum will move to point $b$, then to $c$, and to $d$. Comparing points $b$ and $d$, as well as $a$ and $c$, we observe that the pendulum is precessing in a clockwise manner, as shown in Figure 3.41(b).

We will not attempt to derive the precessing rate of the Foucault pendulum here and readers can read the article cited in the footnote for the derivation. We will only consider this problem from the view point of motion analysis. The reference frame, $\hat{x}-\hat{y}-\hat{z}$, which we use to observe the Foucault pendulum, rotates with the Earth. Therefore, its frame rotation is the same as the Earth. Depending on the latitude of the location, the axis of rotation of the Earth is at an angle $\theta$ from the local vertical axis $\hat{z}$. This Earth axis of rotation lies on the plane of $\hat{y}-\hat{z}$. The angular velocity of the Earth can be expressed in the local reference frame as

$$\vec{\Omega} = \Omega \sin\theta \hat{j} + \Omega \cos\theta \hat{k} \qquad (3.159)$$

As discussed before, our observation using a rotating reference frame will contain a negative Coriolis acceleration, $-2\,\vec{\Omega} \times \hat{v}$. As shown in Figure 3.41(a), if the pendulum happens to pass through point $\hat{O}$ and is heading to the east, the velocity vector observed by us is $\hat{v} = v\hat{i}$. The Coriolis acceleration is

$$-2\vec{\Omega} \times \hat{v} = -2(\Omega \sin\theta \hat{j} + \Omega \cos\theta \hat{k}) \times (v\hat{i}) = -2v\Omega \cos\theta \hat{j} + 2v\Omega \sin\theta \hat{k} \qquad (3.160)$$

In Equation (3.160), the component of the negative Coriolis acceleration in the $\hat{k}$ direction, is restricted by the pendulum arm, not to be observed, but the component in $\hat{j}$ direction, $-2v\Omega \cos\theta$, pointing to the south, is observable (Figure 3.41(a)). This acceleration component will always be perpendicular to $\hat{v}$. As a result, the pendulum will not stay on the $\hat{x}$ axis but deviates to point $b$. If we imagine that we are riding with the pendulum and facing forward, the pendulum will always steers to the right side. Therefore, from point $b$, we face the opposite direction, the pendulum steers to our right, which now points to the positive $\hat{j}$ direction. The pendulum then ends up at point $c$. Similarly, from point $c$, it would end at point $d$. The precessing rate of the Foucault pendulum is well known as $\dot{\phi} = \Omega \cos\theta$, which is slower than the Earth's rotation unless we are at the North Pole. In the Northern Hemisphere, we perceive the Earth's rotation as counterclockwise, and the Foucault pendulum precesses in the clockwise direction. However, in the Southern Hemisphere, the Earth's rotation is perceived as clockwise, which results in a counterclockwise precessing of the Foucault pendulum.

## Example 3.12: Hurricane, an Illusion?

Before we discuss how hurricanes and the Coriolis acceleration are related, it would be interesting to learn how the theory of the Coriolis acceleration had been developed. It is generally recognized that the French physicist, Gustave-Gaspard Coriolis, described this acceleration, now named after him, in the most general way in his article, "Sur les Équations du Mouvement Relatif des Systèmes de Corps," published in 1835.[11]

This article is not easy to read for modern readers, but a discussion of it by Alexandre Moatti is excellent to help understand the history of this theory.[12] Moatti's article is very interesting and the readers are encouraged to get a copy and read it.

According to the Moatti article, when Foucault did his pendulum experiment in 1851, he did not cite Coriolis's work. The article observes that

*Scientific bodies have only gradually come to accept the concept of Coriolis in all its generality, taking it out of its initial field of application and disregarding Coriolis's demonstrative approach. After having been shown again kinematically and extracted from the theory of machines, Coriolis's theorem will explain numerous phenomena, such as Foucault's pendulum or the erosion of waterways.*

It is Charles-Eugéne Delaunay (1816–1872), an Ecole polytechnique student, astronomer, and mathematician, who cited the work of Coriolis to explain the Foucault pendulum and the "movement of waterways, which tends to bring the water to the right side of their bed," thus eroding the right side of the waterway. Delaunay is credited with the naming of the Coriolis acceleration.

---

[11]Gaspard-Gustave de Coriolis, Sur les Équations du Mouvement Relatif des Systémes de Corps," *Journal de l'Ecole Royale Polytechnique*, **15**, 144–154.

[12]Alexandre Moatti, "Coriolis: The Birth of a Force," Chief Mining Engineer, Doctor in History of Science Associate Researcher at the University of Paris-VII, http://www.bibnum.education.fr/sites/default/files/analyse-coriolis-force-en.pdf.

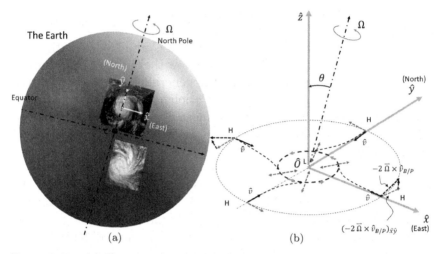

Figure 3.42: (a) Hurricanes in the Northern and Southern Hemispheres rotate differently; and (b) the Coriolis acceleration in the formation of hurricanes in the Northern Hemisphere.

*Source*: (a) North Hemisphere hurricane image is from images:katrina-08-28-2005.jpg (1920–1200) (swling.com) and (b) South Hemisphere hurricane image is from https://www.nasa.gov/images/content/515330main_20110207_Yasi-MODIS_full.jpg.

The article published by Anders Persson in 1998 provides a detailed account of the Coriolis effect in meteorology including jet streams.[13]

We will focus on the Coriolis acceleration and the formation of hurricanes here. As shown in Figure 3.42(a), in the Northern Hemisphere, hurricanes rotate in a counterclockwise manner, and in a clockwise manner in the Southern Hemisphere. Why so? As shown in Figure 3.42(b), heat and evaporation could form a low pressure spot (marked as $L$) over the ocean. The air in the surrounding cooler area is therefore at a higher pressure (marked as $H$). The pressure gradient draws air particles from the high pressure region toward the low pressure region.

---

[13]Persson, A. How do we understand the Coriolis force? *Bulletin of the American Meteorological Society*, **79**(1998), https://doi.org/10.1175/1520-0477(1998) 079¡1373:HDWUTC¿2.0.CO;2.

After the air particles gain speed toward the center ($L$), the Coriolis acceleration becomes significant. As shown in Figure 3.42(b), the horizontal component of the negative Coriolis acceleration, $-2\,\vec{\Omega} \times \hat{v}$, pushes the air particles to the right if you are riding with the air particles. This applies to all air particles from all directions. Remember that the Coriolis acceleration is always perpendicular to the velocity, it does not increase the speed of the air particles, but it changes its direction. The speed gets higher and higher due to the pressure gradient as the air particles get closer to the center. However, the velocity of the air particles no longer points at the center due to the Coriolis acceleration. It can happen that the air particles now have a velocity tangential to the circular loop shown in Figure 3.42(b). Once the air particles enter the circular loop, to us, non-inertial observers, it has the centripetal acceleration toward the center and the Coriolis acceleration away from the center. The centripetal acceleration is much bigger because it is proportional to $v^2$, while the Coriolis acceleration is proportional to $v$. Therefore, the net acceleration is toward the center, maintained by the pressure gradient, to keep the air particle in the circular path.

### A Hurricane is born!

The wind speed of the hurricanes can increase or decrease depending on the pressure gradient. Let's do a calculation to identify the size of different acceleration terms of an actual hurricane. The satellite image of the hurricane in Figure 3.42(a) in the Northern Hemisphere is actually the famous Katrina hurricane, which hit New Orleans, LA, USA in 2005 as a category 3 storm, causing 1,800 deaths and 125 billion US dollars in damage. At its peak strength, it was a category 5 hurricane with the speed reaching 280 km/h.

Before Katrina hit New Orleans, its speed reached over 200 km/h which typically happens at the eye wall. The hurricane eye radius is estimated to be 15 km. Let's use these two numbers for calculation, assuming that the radius of the loop in Figure 3.42(b) is 15 km and the tangential speed is 200 km/h. If we trace one air particle, $P$, in the loop and at the hurricane eye, $\hat{O}$, observe this air particle in a circular path, we will see the acceleration of the air particle as $\hat{a}_{P/\hat{O}} = \hat{\omega}_1 \times (\hat{\omega}_1 \times \hat{r}_{P/\hat{O}})$. Because we use a rotating reference frame, this acceleration is not the actual acceleration $\vec{a}_{P/O}$ observed by an

IRF. $\hat{a}_{P/\hat{O}}$ and $\vec{a}_{P/O}$ are related by the relative acceleration equation as follows:

$$\hat{a}_{P/\hat{O}} = \hat{\omega}_1 \times (\hat{\omega}_1 \times \hat{r}_{P/\hat{O}}) = \vec{a}_{P/O} - \vec{\Omega} \times (\vec{\Omega} \times \hat{r}_{P/\hat{O}}) - 2\,\vec{\Omega} \times \hat{v}_{P/\hat{O}} \quad (3.161)$$

where $|\hat{\omega}_1|$ is the angular speed of the hurricane observed by us and $\vec{a}_{P/O} = \frac{\vec{F}_P}{m_P}$ is the acceleration of an air particle calculated from *Newton's Second Law* due to the pressure difference, for which $\vec{F}_P$ is pressure force vector and $m_P$ is the air particle mass.

In the above equation, the term, $-\vec{\Omega} \times (\vec{\Omega} \times \hat{r}_{P/\hat{O}})$ is very small because $\Omega$ is very small and can be neglected. As a result, the centripetal acceleration we observed can be calculated as

$$|\hat{a}_{P/\hat{O}}| = |\hat{\omega}_1 \times (\hat{\omega}_1 \times \hat{r}_{P/\hat{O}})|$$

$$= \frac{200{,}000 \text{ m}/3600 \text{ s}}{15{,}000 \text{ m}} \cdot \frac{200{,}000 \text{ m}/36{,}00 \text{ s}}{15{,}000 \text{ m}} \cdot 15{,}000 \text{ m}$$

$$= 0.205 \text{ m/s}^2 \quad (3.162)$$

The Coriolis acceleration is calculated to be

$$|-2\,\vec{\Omega} \times \hat{v}_{P/\hat{O}}| = 2 \cdot 0.0000727 \text{ rad/s}$$

$$\cdot \frac{200{,}000 \text{ m}}{3{,}600 \text{ s}} \cdot \cos 60° = 0.004 \text{ m/s}^2 \quad (3.163)$$

where the latitude of New Orleans is $30°$.

From the above calculation, we found that the acceleration of the air particle is

$$\vec{a}_{P/O} = (0.004 + 0.205) = 0.209 (\text{m/s}^2) \quad (3.164)$$

The pressure gradient needed to achieve such acceleration can be determined as

$$\frac{\partial P}{\partial r} = \rho \vec{a}_{P/O} = 1.003 \text{ kg/m}^3 \cdot 0.209 \text{ m/s}^2 = 2.10 \text{ mbar/km} \quad (3.165)$$

where the air density is $1.003 \text{ kg/m}^3$.

The eye pressure of Katrina was recorded as $920\,\text{mbar}$ and the atmosphere pressure typically is $1013\,\text{mbar}$. The net pressure difference is $93\,\text{mbar}$. The largest pressure gradient of a hurricane occurs

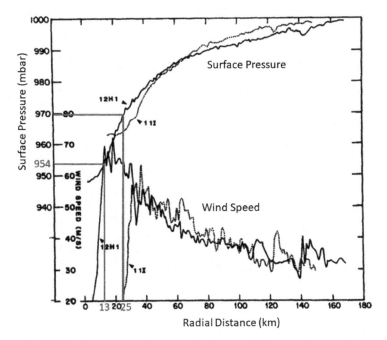

Figure 3.43: Surface pressure and wind speed measurements of hurricane Frederic in 1982.
*Source*: Powell, M., The transition of the hurricane Frederic boundary-layer wind field from the open gulf of Mexico to landfall, *Mon. Weather Rev.*, **10**(1982) 1912–1932.

near the eye wall, where the wind speed is the highest. Let's compare this calculation with the actual pressure and speed measurement of hurricane Frederic of 1982, shown in Figure 3.43.[14]

From Figure 3.43, the eye wall of hurricane Frederic ranged from 13 to 25 km radius from the center, with an average wind speed 230 km/h (64 m/s). The surface pressure dropped from 970 to 954 mbar. The average pressure gradient is 1.33 mbar/km, which is in the same order of magnitude as the result of Equation (3.165).

Finally, we should notice that the Coriolis acceleration is only 2% of the observed centripetal acceleration. Therefore, the role of the Coriolis acceleration is to change the direction of the air traveling

---

[14]Powell, M. The transition of the hurricane Frederic boundary-layer wind field from the open Gulf of Mexico to landfall, *Mon. Weather Rev.* **10**(1982) 1912–1932.

velocity to cause the rotation, not the rotational speed itself. It makes us wonder if hurricanes are illusions or for real? The answer is: "Of course, it is real." The rotational wind speed of the hurricanes is not due to the Coriolis acceleration, but due to the pressure gradient. The Coriolis acceleration bends the air particle path so that it enters into the eye wall in the tangential direction and circles around the eye. If an alien traveling in an UFO in a straight line takes a picture of the hurricane, will the alien see a hurricane, similar to the images taken by a satellite? Note that the satellite is not an inertial observer because it is orbiting around the Earth. Based on Equation (3.160), the answer should be "Yes!" because $\vec{a}_{P/O}$ essentially is equal to $\hat{a}_{P/\hat{O}}$ based on the fact that the terms $-\vec{\Omega} \times (\vec{\Omega} \times \hat{r}_{P/\hat{O}})$ and $-2\vec{\Omega} \times \hat{v}_{P/\hat{O}}$ are both small. An alien using an inertial reference should still see a hurricane because the acceleration of $\vec{a}_{P/O}$ essentially is a centripetal acceleration. NASA's Cassini spacecraft flew by Saturn in 2006 and took a picture of the swirling hurricane-like vortex at Saturn's south pole, and the images show clearly the rotation.[15] Spacecraft Cassini was not using the same rotating frame of Saturn when it took those images. The rotation is real, not an illusion. Therefore, the hurricane is real, not an illusion.

This answer, however, invites more questions. From our analysis, the rotational motion of the hurricane was caused by an illusion of a curved path due to the Coriolis acceleration, similar to the Foucault clock (Example 3.11) or the curved ball path (Example 3.10). How can a hurricane start as an illusion and turn into something that is real?

Let's review Examples 3.10 and 3.11 more closely to answer this question. Both of these phenomena observed by us using the Earth rotation frame are illusions because an inertial observer (for example, an alien on a spacecraft moving in straight line at a constant speed) will see a ball going along a straight line and a pendulum moving back and forth without precessing. As we discussed earlier, an alien will see a real hurricane, not much different from what we see.

For the curving ball and Foucault pendulum, there are either no forces (curving ball) or only the gravitation and pendulum arm forces (the Foucault pendulum) involved. However, in the hurricane case,

---

[15]https://www.nasa.gov/mission_pages/cassini/multimedia/pia08332.html.

we considered one air particle and provided the initial analysis to infer that the velocity should be from the high pressure to the low pressure area, along a straight line. However,when the air particle starts to turn side ways, it would encounter other air particles, with resulting force conditions much different from the initial conditions. To explain this observation, consider Example 3.10, if the ball, moving along in a straight line, hits an object on the rotating platform before hitting $P$, the ball will no longer stay in the straight line trajectory. When the ball bounces off the object, it might settle into a real rotational motion because it was hit by a rotating object. This is similar to when you throw a ball into a roulette in a casino, the ball, once it hits the roulette, will exhibit a rotational motion as well.

In conclusion, we can state that the Coriolis acceleration kick-starts the condition of forming hurricanes and after the process has started, whether a hurricane can form or not depends on many factors. Once the hurricane is formed, the Coriolis acceleration is no longer significant, as indicated by the calculation of Equations (3.162) and (3.163).

## Example 3.13: Deviation to the East of a Vertically Dropped Heavy Object

As discussed in the article by Moatti,[12] another phenomenon due to the Coriolis acceleration was actually observed before the Foucault experiment in 1851. In 1833, Ferdinand Reich, a German chemist and physicist, discovered that when a heavy object was dropped down a mine shaft in Freiberg (Saxony), of a depth of 158 m, he had measured on average, after 106 tests, a deviation of 28.3 mm to the east. Why is there such a deviation to the east instead of the object falling vertically down?

First, let's do some calculation based on those data obtained by Reich in his experiments. Assuming no air resistance, the time and the speed of an object released from at rest are 5.68 s and 55.68 m/s, respectively, after dropping vertically 158 m based on Equations (3.29) and (3.30). With a deviation of 28.3 mm to the east, the average eastward speed is only 0.0050 m/s. The eastward speed is very small, compared with the vertical falling speed. Consequently, the eastward acceleration is very small as well.

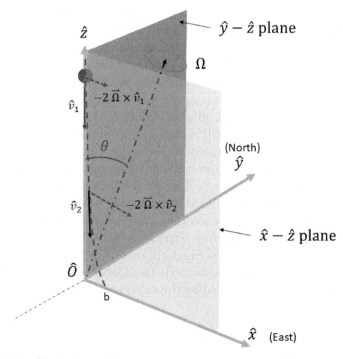

Figure 3.44: Deviation to the east of a heavy object dropped vertically, discovered by Riech in 1833.

Let's do a calculation of the Coriolis acceleration. As shown in Figure 3.44, we will consider the vertical component of the velocity only when calculating the Coriolis acceleration because the velocity components in other two directions are very small based on the calculations above.

$$-2\vec{\Omega} \times \hat{v}_{P/\hat{O}} = -2\Omega(\cos\theta\hat{k} + \sin\theta\hat{j}) \times (-gt\hat{k}) = 2\Omega\sin\theta gt\hat{i}$$

$$= 2 \cdot 0.0000727 \cdot \sin 39° \cdot 9.81 \cdot t\,\hat{i} = 0.000898\,t\,\hat{i}(\text{m/s}^2) \quad (3.166)$$

where the latitude of Saxony, Germany, is 51°.

From Equation (3.166), the Coriolis acceleration is in the $\hat{i}$ direction, which is aligned with the east, and is a linear function of time $t$. We can integrate the Coriolis acceleration to obtain the eastward speed and displacement as

$$v_E = \int_0^t \left| -2\,\vec{\Omega} \times \hat{v}_{P/\hat{O}} \right| dt$$

$$= \int_0^t 0.000898\ t\ dt = 0.000449\ t^2 \mathrm{(m/s)}$$

$$s_E = \int_0^t v_E\ dt = 0.0000150\ t^3 \mathrm{(m/s^2)} \tag{3.167}$$

With $t = 5.68$ s, the eastward deviation is found to be 27.4 mm, which is only 3.2% lower than the experimental average value from Reich. Note that in our calculation, we do not consider the air resistance, which could increase the time of the fall, as discussed in Example 3.2. We do not have the details about the shape of the heavy objects used in Reich's experiment. Therefore, we cannot determine an appropriate wind drag coefficient to consider the air resistance.

From the above discussion, we can conclude quite convincingly that the eastward deviation is due to the Coriolis acceleration.

An exact solution of the eastward deviation by considering every velocity component during the fall can be found in Classical Mechanics[16]

$$x = \frac{g\sin\theta}{2\Omega}\left(t - \frac{\sin 2\Omega t}{2\Omega}\right) \tag{3.168}$$

$$y = -g\sin\theta\cos\theta\left(\frac{t^2}{2} - \frac{1-\cos 2\Omega t}{4\Omega^2}\right) \tag{3.169}$$

$$z = h - \frac{g}{2}t^2 + g\sin^2\theta\left(\frac{t^2}{2} - \frac{1-\cos 2\Omega t}{4\Omega^2}\right) \tag{3.170}$$

where $h$ is the fall height in meter. We can let $z = h$ and solve for $t$. The value of the fall time, $t$, is then used to determine the eastward deviation ($x$)) and the southward deviation ($-y$). According to Equation (3.169), there is deviation to the south too, although its, very small.

Before we conclude this chapter, we need to be careful that we do not attribute everything which rotates on Earth to the Coriolis

---

[16]Grener, W. *Classical Mechanics, Equation (2.10)*, Chapter 2, Springer-Verlag, Berlin, Heidelberg, 2010. DOI:10.1007/978-3-642-03434-3_2.

effect. For example, when we flush a toilet, we notice that the water is swirling. Is it due to the Coriolis acceleration? A calculation will conclude that there is not enough time and speed of the water flowing down from a toilet tank to achieve that much rotational speed. The rotation of the water flow is mainly due to the design of the toilet. However, if we design a toilet, do we want to design it to comply with the nature, despite an extremely minute effect, or against the nature? Most toilets in the Northern Hemisphere display a counterclockwise swirling pattern, while those in the Southern Hemisphere, a clockwise swirling pattern. However, occasionally, you might find a clockwise swirling toilet in the Northern Hemisphere.

## 3.11 Concluding Remarks

In this chapter, we started with the definition of an IRF and explained its importance. We then proceeded to derive the equations which define the general motion of a particle (Section 3.3). The governing equations for a particle in a straight line motion are presented in Section 3.4. In Section 3.5, the motion analysis is expanded to 2D and 3D motions. Two convenient ways of describing 2D and 3D motions are presented in Sections 3.6 and 3.7, denoted as the $t-n$ and the $r-\theta$ descriptions. In Section 3.8, we derived the equations to define the velocity and acceleration of a perfect circular motion. The position, velocity, and acceleration of a perfect circular motion or a straight line motion are very well defined. We denote these two motions as "Simple Motions" in this book. In Section 3.9, the relative motion analysis between different reference frames are discussed. Initially, the discussion is limited to reference frames without rotations. In Section 3.10, the relative motion analysis is expanded to reference frames with rotations. The guideline to select different reference frames for motion analysis is coined as "Make a Complicated Motion Simple Again!"

Many real-world examples are provided, in particular for the Coriolis acceleration due to the Earth's rotation. Throughout this chapter, the emphasis is on the practical aspects of the theories, not merely presenting idealized conditions which match the standard equations.

In the next chapter, the motion analysis will be extended from one particle to rigid bodies.

Chapter 4

# Motion Analysis for Rigid Bodies

From the discussions in Chapters 2 and 3, we have obtained a proper understanding of motion analysis for particles. In this chapter, we expand the analysis to rigid bodies which contain infinite number of particles bonded together in a specific way. The rigid body assumption is discussed in Chapter 2, which leads to collective common behavior of all particles within a rigid body.

## 4.1 Rigid Body Revisited

As discussed in Section 2.3, a rigid body is defined as an object which does not deform no matter how many or how big the forces applied to it are. In this section, we revisit the definition of the rigid body and derive the common kinematic properties of a rigid body.

As shown in Figure 4.1, if we choose any three points, $A$, $B$, and $C$, on a rigid body and draw a triangle to link them, due to the rigid body assumption we know that this $A-B-C$ triangle will not change its shape and size. Point $B$, for example, can move relative to point $A$, but point $B$ must keep the same distance from point $A$. Therefore, point $B$ can only rotate around point $A$. The same conclusion can be made for point $C$ with respect to point $A$. The question is whether or not the rotations of $C$ and $B$ with respect to $A$ are the same. Because the angle of $BAC$, $\phi$, as shown in Figure 4.1 should stay the same, the sweeping angle, $\theta_B$ must be equal to $\theta_C$. In other words,

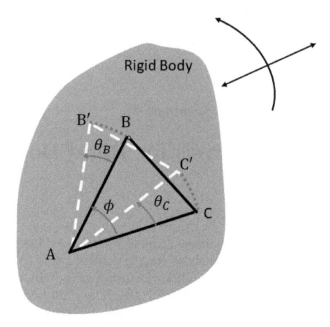

Figure 4.1: A rigid body in a generation motion with three points identified which form a triangle. This triangle will remain the same size and shape.

the rotations of $B$ and $C$ with respect to $A$ are the same. Because we pick these three points arbitrarily, the same conclusion can be applied to the entire rigid body. With this result, we come to a very important property of the rigid body:

*Rotation is a universal property of a rigid body*

In other words, there will be only one angular velocity, $\vec{\omega}$, and one angular acceleration, $\vec{\alpha}$, associated with a rigid body at any given time. Between any two points on the rigid body, one will appear to rotate around the other with the same $\vec{\omega}$ and $\vec{\alpha}$. We can express this property as

$$\vec{v}_{B/A} = \vec{\omega} \times \vec{r}_{B/A} \tag{4.1}$$

$$\vec{v}_{A/B} = \vec{\omega} \times \vec{r}_{A/B} \tag{4.2}$$

and

$$\vec{a}_{B/A} = \vec{\alpha} \times \vec{r}_{B/A} + \vec{\omega} \times (\vec{\omega} \times \vec{r}_{B/A}) \tag{4.3}$$

$$\vec{a}_{A/B} = \vec{\alpha} \times \vec{r}_{A/B} + \vec{\omega} \times (\vec{\omega} \times \vec{r}_{A/B}) \tag{4.4}$$

## 4.2 Motion Analysis of a Rigid Body

With Equations (4.1)–(4.4), we can try to describe the motion of every point of a rigid body. As shown in Figure 4.2, if we know the motion of point $A$ of a rigid body and we know the rotational properties, $\vec{\omega}$ and $\vec{\alpha}$, of the rigid body, then based on the relative motion analysis of Section 3.9, we have

$$\vec{v}_{i/O} = \vec{v}_{A/O} + \vec{v}_{i/A} = \vec{v}_{A/O} + \vec{\omega} \times \vec{r}_{i/A} \tag{4.5}$$

$$\vec{a}_{i/O} = \vec{a}_{A/O} + \vec{a}_{i/A} = \vec{a}_{A/O} + \vec{\alpha} \times \vec{r}_{i/A} + \vec{\omega} \times (\vec{\omega} \times \vec{r}_{i/A}) \tag{4.6}$$

where point $i$ is any point on the rigid body.

Equations (4.5) and (4.6) pretty much are all we need to describe the motion of a rigid body. These governing equations can be applied to many special cases and new equations can be derived from them.

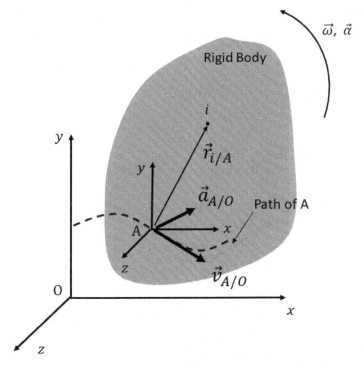

Figure 4.2: If we know the motion of point $A$ of a rigid body and we know its rotational properties, $\vec{\omega}$ and $\vec{\alpha}$, then we can define the motion of any point on the rigid body.

Based on Equations (4.5) and (4.6), we can also conclude that if we know the motion of two points on the rigid body, then we know the motion of every point on the rigid body. Why? Because knowing the velocities and accelerations of two points, we can use Equations (4.5) and (4.6) to calculate $\vec{\omega}$ and $\vec{\alpha}$ of the rigid body. After that, using Equations (4.5) and (4.6) again, we can determine the motion of every point on the rigid body.

### 4.2.1 Case #1: Motion of a rigid body with one point fixed (2D and 3D cases)

We will start with 2D cases. As shown in Figure 4.3, a rigid arm is pinned at point $O$ and rotates with an angular velocity $\omega$, and an angular acceleration $\alpha$. Because point $O$ is pinned at the base, it is stationary; as a result, $\vec{v}_O = 0$ and $\vec{a}_O = 0$. For a 2D rigid body shown in Figure 4.3, both $\vec{\omega}$ and $\vec{\alpha}$ are in the $\vec{k}$ direction, which makes the computation rather easy.

The velocity and acceleration of point $A$ can be easily determined as

$$\vec{v}_{A/O} = 0 + \vec{\omega} \times \vec{r}_{A/O} \tag{4.7}$$

$$\vec{a}_{A/O} = 0 + \vec{\alpha} \times \vec{r}_{A/O} + \vec{\omega} \times (\vec{\omega} \times \vec{r}_{A/O}) \tag{4.8}$$

For example, if $\theta = 60°$, $\vec{\omega} = 2\vec{k}$ and $\vec{\alpha} = -3\vec{k}$, and the distance from $A$ to $O$ is 10, then we easily find that $\vec{r}_{A/O} = 10\cos 60°\vec{i}$

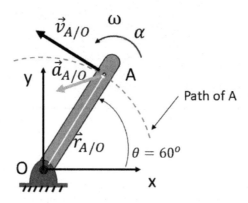

Figure 4.3:   The motion analysis of a 2D rigid body with a stationary point.

$+ 10 \sin 60° \vec{j}$. Putting these numbers into Equations (4.7) and (4.8), we find that

$$\vec{v}_{A/O} = 0 + 2\vec{k} \times (10 \cos 60° \vec{i} + 10 \sin 60° \vec{j}) = -17.32\vec{i} + 10\vec{j} \quad (4.9)$$

$$\vec{a}_{A/O} = 0 + (-3\vec{k}) \times (10 \cos 60° \vec{i} + 10 \sin 60° \vec{j})$$

$$+ 2\vec{k} \times (2\vec{k} \times (10 \cos 60° \vec{i} + 10 \sin 60° \vec{j}))$$

$$= 5.98\vec{i} - 49.64\vec{j} \quad (4.10)$$

It is recommended that math is used to complete the calculation, not through the right-hand rule to determine the directions of the velocity and acceleration vectors to avoid mistakes.

With one point stationary, all other points on the rigid body rotate around that point. There is no need to remember Equations (4.7) and (4.8).

For 3D rotations, the analysis is more complicated. As shown in Figure 4.4, an axisymmetric rigid body, $B$, rotates around a fixed point. The following analysis is not limited to axisymmetric bodies, but it is used here for easier illustration.

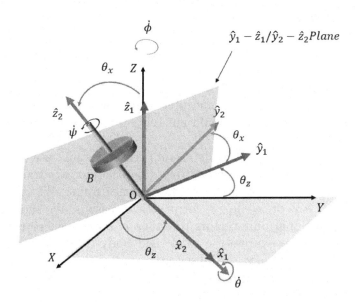

Figure 4.4: The motion analysis of a 3D rigid body with a stationary point.

The rotation of the rigid body in Figure 4.4 is too complicated to be observed by the inertial reference frame, $X-Y-Z$. As discussed in Chapter 3, we need to choose a new reference frame so that the rigid body motion appears to be SIMPLE! If we attach $\hat{x}_2-\hat{y}_2-\hat{z}_2$, as shown in Figure 4.4, to move with the rigid body but not spin with it, the rigid body appears to be in a simple spin around the $\hat{z}_2$ axis. This spinning motion is then expressed as

$$\hat{\omega}_{B/2} = \dot{\psi}\hat{k}_2 \tag{4.11}$$

How about the rotation of $\hat{x}_2-\hat{y}_2-\hat{z}_2$? Using $\hat{x}_1-\hat{y}_1-\hat{z}_1$, we observe that $\hat{x}_2-\hat{y}_2-\hat{z}_2$ has a frame rotation of $\hat{\omega}_1$ as

$$\hat{\omega}_{2/1} = \dot{\theta}\hat{i}_1 \tag{4.12}$$

Finally, the rotation of $\hat{x}_1 - \hat{y}_1 - \hat{z}_1$ is seen by $X-Y-Z$ as

$$\vec{\omega}_{1/0} = \dot{\phi}\vec{K} \tag{4.13}$$

Using the additivity law of angular velocity presented in Section 3.10.3 in Chapter 3, we obtain the angular velocity of the rigid body in Figure 4.4 as

$$\vec{\omega} = \hat{\omega}_{B/2} + \hat{\omega}_{2/1} + \vec{\omega}_{1/0} = \dot{\psi}\hat{k}_2 + \dot{\theta}\hat{i}_1 + \dot{\phi}\vec{K} \tag{4.14}$$

where $\dot{\psi}$ is often referred to as spin, $\dot{\theta}$ as nutation, and $\dot{\phi}$ as precession. However, the nutation and precession could look alike in some situations. Therefore, we should not rely on these terms to describe the 3D rotations.

Equation (4.14) is not really useful because we mix-use too many different directional unit vectors from different reference frames. It is better to use the unit vectors of one reference frame. We can either choose $X-Y-Z$ or $\hat{x}_2-\hat{y}_2-\hat{z}_2$ for practical reasons. We should convert all directional unit vectors to either $\vec{I}$, $\vec{J}$, and $\vec{K}$, or to $\hat{i}_2$, $\hat{j}_2$, and $\hat{k}_2$. When we express the total rigid body rotation in terms of $\hat{i}_2-\hat{j}_2-\hat{k}_2$, we still have an observation corresponding to an inertial reference, simply expressed in $\hat{i}_2-\hat{j}_2-\hat{k}_2$ for convenience. We can imagine $\vec{\omega}$ is observed by an inertial reference frame $x-y-z$, which overlaps with $\hat{x}_2-\hat{y}_2-\hat{z}_2$ at this instant of time.

To relate different directional unit vectors, we can use the moving coordinate concept discussed in Section 3.10.2. To convert everything to $\hat{x}_2 - \hat{y}_2 - \hat{z}_2$, we need the following transformation relationships:

$$\begin{bmatrix} X \\ Y \\ Z \\ 1 \end{bmatrix} = T_{tz}(\theta_z) \cdot T_{tx}(\theta_x) \cdot \begin{bmatrix} \hat{x}_2 \\ \hat{y}_2 \\ \hat{z}_2 \\ 1 \end{bmatrix} \tag{4.15}$$

$$\begin{bmatrix} \hat{x}_1 \\ \hat{y}_1 \\ \hat{z}_1 \\ 1 \end{bmatrix} = T_{tx}(\theta_x) \cdot \begin{bmatrix} \hat{x}_2 \\ \hat{y}_2 \\ \hat{z}_2 \\ 1 \end{bmatrix} \tag{4.16}$$

Working out Equations (4.15) and (4.16) and replacing all the variables by unit vectors, we have

$$\hat{i}_1 = \hat{i}_2 \tag{4.17}$$

$$\hat{j}_1 = \cos\theta_x \hat{j}_2 - \sin\theta_x \hat{k}_2 \tag{4.18}$$

$$\hat{k}_1 = \sin\theta_x \hat{j}_2 + \cos\theta_x \hat{k}_2 \tag{4.19}$$

$$\vec{I} = \cos\theta_z \hat{i}_2 - \sin\theta_z \ \cos\theta_x \ \hat{j}_2 + \sin\theta_z \ \sin\theta_x \ \hat{k}_2 \tag{4.20}$$

$$\vec{J} = \sin\theta_z \hat{i}_2 + \cos\theta_z \ \cos\theta_x \hat{j}_2 - \cos\theta_z \ \sin\theta_x \hat{k}_2 \tag{4.21}$$

$$\vec{K} = \sin\theta_x \hat{j}_2 + \cos\theta_x \hat{k}_2 \tag{4.22}$$

We only need $\hat{i}_1 = \hat{i}_2$ and $\vec{K} = \sin\theta_x \hat{j}_2 + \cos\theta_x \hat{k}_2$ for the conversion:

$$\begin{aligned} \vec{\omega} &= \dot{\psi}\hat{k}_2 + \dot{\theta}\hat{i}_2 + \dot{\phi}(\sin\theta_x \hat{j}_2 + \cos\theta_x \hat{k}_2) \\ &= \dot{\theta}\hat{i}_2 + \dot{\phi}\sin\theta_x \hat{j}_2 + (\dot{\phi} \ \cos\theta_x + \dot{\psi})\hat{k}_2 \\ &= \hat{\omega}_x \hat{i}_2 + \hat{\omega}_y \hat{j}_2 + \hat{\omega}_z \hat{k}_2 \end{aligned} \tag{4.23}$$

Equation (4.23) is the well-known Euler expression. The Euler expression is based on an inertial reference overlapping the rotating reference frame, $\hat{i}_2$, $\hat{j}_2$, and $\hat{k}_2$, at this instant of time.

If we want to express a general 3D rotation with the same inertial observer, we need to convert $\vec{\omega}$ to $X-Y-Z$. We can use the following transformation relationships:

$$
\begin{bmatrix} \hat{x}_2 \\ \hat{y}_2 \\ \hat{z}_2 \\ 1 \end{bmatrix} = T_{tx}(-\theta_x) \cdot T_{tz}(-\theta_z) \cdot \begin{bmatrix} X \\ Y \\ Z \\ 1 \end{bmatrix} \tag{4.24}
$$

$$
\begin{bmatrix} \hat{x}_1 \\ \hat{y}_1 \\ \hat{z}_1 \\ 1 \end{bmatrix} = T_{tz}(-\theta_z) \cdot \begin{bmatrix} X \\ Y \\ Z \\ 1 \end{bmatrix} \tag{4.25}
$$

Based on Equations (4.24) and (4.25), we obtain

$$\hat{i}_1 = \cos\theta_z \vec{I} + \sin\theta_z \vec{J} \tag{4.26}$$

$$\hat{j}_1 = -\sin\theta_z \vec{I} + \cos\theta_z \vec{J} \tag{4.27}$$

$$\hat{k}_1 = \vec{K} \tag{4.28}$$

$$\hat{i}_2 = \cos\theta_z \vec{I} + \sin\theta_z \vec{J} \tag{4.29}$$

$$\hat{j}_2 = -\cos\theta_x \, \sin\theta_z \vec{I} + \cos\theta_x \, \cos\theta_z \vec{J} + \sin\theta_x \vec{K} \tag{4.30}$$

$$\hat{k}_2 = \sin\theta_x \, \sin\theta_z \vec{I} - \sin\theta_x \, \cos\theta_z \vec{J} + \cos\theta_x \vec{K} \tag{4.31}$$

$$
\begin{aligned}
\vec{\omega} &= \dot{\psi}(\sin\theta_x \, \sin\theta_z \vec{I} - \sin\theta_x \, \cos\theta_z \vec{J} + \cos\theta_x \vec{K}) \\
&\quad + \dot{\theta}(\cos\theta_z \vec{I} + \sin\theta_z \vec{J}) + \dot{\phi}\vec{K} \\
&= (\dot{\psi}\sin\theta_x \, \sin\theta_z + \dot{\theta}\cos\theta_z)\vec{I} \\
&\quad + (-\dot{\psi}\sin\theta_x \, \cos\theta_z + \dot{\theta}\sin\theta_z)\vec{J} \\
&\quad + (\dot{\psi}\cos\theta_x + \dot{\phi})\vec{K} \\
&= \omega_X \vec{I} + \omega_Y \vec{J} + \omega_Z \vec{K}
\end{aligned} \tag{4.32}
$$

The expression of Equation (4.32) is much more complicated, but it is more useful if one wants to produce an animation of the 3D rotation.

For example, if $\dot{\theta} = 0$, $\vec{\omega} = \dot{\phi}\sin\theta_x \hat{j}_2 + (\dot{\phi} \, \cos\theta_x + \dot{\psi})\hat{k}_2$; i.e., there are only rotations around $\hat{y}_2$ and $\hat{z}_2$ axes, a seemly 2D rotation.

However, the actual rotation, viewed by an inertial observer, will be $\vec{\omega} = (\dot{\psi}\sin\theta_x \ \sin\theta_z)\vec{I} + (-\dot{\psi}\sin\theta_x \ \cos\theta_z)\vec{J} + (\dot{\psi}\cos\theta_x + \dot{\phi})\vec{K}$, a full 3D rotation and the rotational speeds are not constant.

If we want to derive the angular acceleration, taking a derivative of Equation (4.32) is more straightforward, but the expression would still be quite complicated. In a simpler form, we can express it as

$$\vec{\alpha} = \dot{\omega}_X\vec{I} + \dot{\omega}_Y\vec{J} + \dot{\omega}_Z\vec{K} \qquad (4.33)$$

If we use Euler's expression of Equation (4.23), the derivative is not just a partial time derivative because the unit directional vectors are not fixed. However, the final expression is actually simpler. Using Equation (3.132) of Chapter 3, we have

$$\vec{\alpha} = \dot{\hat{\omega}}_x\hat{i}_2 + \dot{\hat{\omega}}_y\hat{j}_2 + \dot{\hat{\omega}}_z\hat{k}_2 + \vec{\Omega}_{2/0} \times (\hat{\omega}_x\hat{i}_2 + \hat{\omega}_y\hat{j}_2 + \hat{\omega}_z\hat{k}_2) \qquad (4.34)$$

where $\Omega_{2/0} = \hat{\omega}_{2/1} + \vec{\omega}_{1/0} = \dot{\theta}\hat{i}_2 + \dot{\phi}\sin\theta_x\hat{j}_2 + \dot{\phi}\ \cos\theta_x\hat{k}_2$.

Although Euler's expression is simpler, there is a catch. We still need to define the unit vectors of $\hat{i}_2$, $\hat{j}_2$, and $\hat{k}_2$ with respect to $X-Y-Z$. Therefore, Equations (4.29)–(4.31) should be linked with Equations (4.23) and (4.34) for the complete inertial description. This is similar to the $t-n$ and $r-\theta$ descriptions in Sections 3.6 and 3.7 that we need to define those unit directional vectors with respect to an inertial reference frame.

Once we obtain the angular velocity and acceleration of a 3D rigid body rotating around a fix point, the velocity and the acceleration of any point on the rigid body can be determined by Equations (4.7) and (4.8).

### 4.2.2 Case #2: Motion of a rigid body with one point at zero velocity at an instant (2D and 3D cases)

As shown in Figure 4.5, a perfect circular disk with a radius $R$ rolls to the right over a flat surface. It is in a special type of motion as it is rolling without slipping over the flat surface. This rolling without slipping condition dictates the contact point $A$ and its neighboring point on the flat surface, $A'$, must have the same velocity. We will define this rolling without slipping condition more rigorously later. Because the flat ground is not moving, $\vec{v}_{A/O} = 0$. However, we should be aware that the acceleration of point $A$ is not zero, to be shown

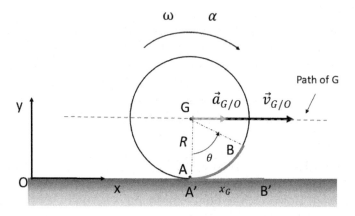

Figure 4.5: The motion of a circular disk which is rolling without slipping.

later. This zero velocity condition of point $A$ only applies at this instant of time. At the next time instant, a new pair of contact points will appear.

As shown in Figure 4.5, with the rolling without slipping condition, when point $B$ is in contact with the ground, the horizontal distance between $A'$ and $B'$ will be equal to the arc length between point $A$ and $B$ along the circumference of the disk. Because the center $G$ is always vertical up from the contact point, $G$ will travel through the same distance. Note that point $G$ is in a straight line motion. We then obtain the velocity of point $G$ as

$$\vec{v}_{G/O} = \frac{dx_G}{dt}\,\vec{i} = \frac{dR\theta}{dt}\,\vec{i} = R\dot{\theta}\,\vec{i} = R\omega\,\vec{i} \qquad (4.35)$$

Taking one more derivative of Equation (4.35), we obtain the acceleration of $G$,

$$\vec{a}_{G/O} = \frac{d\vec{v}_G}{dt} = R\ddot{\theta}\,\vec{i} = R\alpha\,\vec{i} \qquad (4.36)$$

With the motion of the center, $G$, completely defined, we can now determine the motion of any point on the disk. Let's verify the motion of point $A$.

$$\vec{v}_{A/O} = \vec{v}_{G/O} + \vec{\omega} \times \vec{r}_{A/G} = 0 \qquad (4.37)$$

$$\vec{a}_{A/O} = \vec{a}_{G/O} + \vec{\alpha} \times \vec{r}_{A/G} + \vec{\omega} \times (\vec{\omega} \times \vec{r}_{A/G}) = R\,\omega^2\vec{j} \qquad (4.38)$$

where $\vec{r}_{A/G} = -R\vec{j}$, $\vec{\omega} = -\omega\vec{k}$, and $\vec{\alpha} = -\alpha\vec{k}$.

Equations (4.37) and (4.38) confirm that point $A$ has a zero velocity, but has an acceleration pointing to point $G$. The rigid disk of Figure 4.5 has only one degree of freedom. If we know $\vec{\omega}$ and $\vec{\alpha}$, we know the complete motion of the disk. Similarly, if we know the velocity and acceleration of point $G$, we know the entire motion of the disk.

If the rolling without slipping condition does not hold, then Equations (4.35)–(4.38) are no longer valid. We will need both the motion of $G$ and the rotation of the disk to define its entire motion.

Let's extend the disk rolling example to ball rolling. As shown in Figure 4.6, a ball, with a radius $R$, has an angular velocity with components in all three axes. At this instant of time, the ball is in contact with the surface at point $O$ and is rolling without slipping. Therefore, as in the rolling disk example, the velocity at the contact point is zero. We can determine the velocity of every point of the ball if the angular velocity is known. The velocity of the center of the ball, $G$, becomes

$$\vec{v}_{G/O} = \vec{\omega} \times \vec{r}_{G/O} = (\omega_x \vec{i} + \omega_y \vec{j} + \omega_z \vec{k}) \times (R\vec{k})$$

$$= R\omega_y \vec{i} - R\omega_x \vec{j} \tag{4.39}$$

For the ball to move in the direction as shown in Figure 4.6, $\omega_y$ should be positive, while $\omega_x$ should be negative. A positive rotation about an axis is a counter-clockwise rotation when looking from the arrowhead of the axis. We can also use the right-hand rule to determine the positive rotation around an axis. In an ideal situation, if the ball is perfectly symmetric, homogeneous, and spherical, and the surface is perfectly flat, Equation (4.39) will hold indefinitely. One will see the ball rolling forward but at the same time seemingly spinning as shown in Figure 4.6. Let's consider some special cases. If $\omega_x = \omega_y = 0$, while $\omega_z \neq 0$, the ball will spin vertically without rolling. If $\omega_x = \omega_z = 0$, while $\omega_y \neq 0$, the ball will roll along the $x$ axis. If If $\omega_y = \omega_z = 0$, while $\omega_x \neq 0$, the ball will move along the $y$ axis.

Apparently, a ball can roll in different directions more easily than a disk. If car tires are made of balls, instead of cylindrical tires, a car can move side ways, which would make road side parking really easy. However, the design and control of the rotation of a ball are quite difficult. For example, most electrical motors are cylindrical rotors, basically a thick disk. There are spherical motors, but their cost is

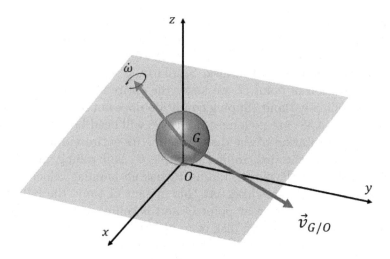

Figure 4.6:   The motion of a ball rolling and spinning.

quite high due to their complex designs. Ball tires are implemented in vacuum cleaners for easier maneuver and are a commercial success. In Figure 4.7, several examples of spherical motors were provided per the discussion above.

Finally, Equation (4.39) only predicts the speed of the ball at the current instant of time if the rolling without slipping condition applies. We do not know what would happen next. To predict the ball motion into the future, we will need the information related to the forces, in particular the contact forces of the ball and the ground.

### 4.2.3   *Case #3: General motion*

We consider the general motion of a rigid body in this section based on Equations (4.5) and (4.6). We need to know the motion of one point on the rigid body. Is there a preferred point on the rigid body? Apparently, the mass center of the rigid body is a good choice. In fact, in most cases, it is the best choice.

We will define the mass center in the context of dynamics. From statics, the mass center of a rigid body is defined as

$$M\vec{r}_{G/O} = \sum_i m_i \cdot \vec{r}_{i/O} \qquad (4.40)$$

| (a) | (b) | (c) |

Figure 4.7: (a) A design of a commercial spherical motor; (b) a moving robot based on a rolling spherical motor; and (c) a Dyson vacuum cleaner with a ball as the wheel.
*Source*: (a) https://i0.wp.com/metrology.news/wp-content/uploads/2019/08/tw o-Axis-Spherical-Motor.png?resize=800%2C445&ssl=1; (b) https://i.stack.img ur.com/sjYFw.jpg; and (c) https://3.bp.blogspot.com/_zqFoq3qej2c/ScLUpClO gTI/AAAAAAAAnyE/dkAysUc_2bA/s400/Picture+5.png.

where $M$ is the total mass of the rigid body, $m_i$ is the mass of a particle within the rigid body. For rigid bodies, we should use integration in Equation (4.40), but we use the summation sign for convenience. When $O$ is $G$, Equation (4.40) is reduced to

$$\sum_i m_i \cdot \vec{r}_{i/G} = 0 \qquad (4.41)$$

By deriving the time derivative of Equation (4.40), we have

$$M\vec{v}_{G/O} = \sum_i m_i \cdot \vec{v}_{i/O} = \sum_i m_i \cdot (\vec{v}_{G/O} + \vec{v}_{i/G})$$

$$= \left(\sum_i m_i\right) \vec{v}_{G/O} + \sum_i m_i \cdot \vec{v}_{i/G}$$

$$= M\vec{v}_{G/O} + \sum_i m_i \cdot (\vec{\omega} \times \vec{r}_{i/G})$$

$$= M\vec{v}_{G/O} + \vec{\omega} \times \left(\sum_i m_i \vec{r}_{i/G}\right) \tag{4.42}$$

Equation (4.42) leads to the result that $\sum_i m_i \cdot \vec{v}_{i/G} = 0$, which is the definition of the mass center of a rigid body in the context of dynamics. However, Equation (4.42) itself has a very important meaning in dynamics when the concept of linear momentum is introduced in later chapters.

To conduct the analysis for a rigid body in a general motion, it is preferred that we locate the mass center and try to measure the mass center motion and the rotation of the rigid body. From there, we know the motion of every point on the rigid body. However, if there is a point who's motion we can determine readily, then we will go by this one.

### 4.2.3.1  *Instantaneous center for a 2D rigid body*

For determining the angular velocity of a 2D rigid body, there is a graphical way if we know the velocities of two points. As shown in Figure 4.8. If we know there is a point, $C$, with zero velocity on the rigid body, then

$$\vec{v}_{i/O} = 0 + \vec{\omega} \times \vec{r}_{i/C} \tag{4.43}$$

$$\vec{v}_{j/O} = 0 + \vec{\omega} \times \vec{r}_{j/C} \tag{4.44}$$

Equations (4.43) and (4.43) indicate that $\vec{r}_{i/C}$ must be perpendicular to $\vec{v}_{i/O}$, while $\vec{r}_{j/C}$, perpendicular to $\vec{v}_{j/O}$. We can draw a line passing through the point perpendicular to its velocity. This zero-velocity point must reside on this perpendicular line. To satisfy both conditions, this zero-velocity point must be at the intersection of both perpendicular lines, as shown in Figure 4.8. This point may or may not be on the rigid body. We can imagine that the rigid body is now extended to include this intersection point. The original rigid body and the virtual enlarged rigid body will appear to rotate around this point at this instant in time. This point is denoted as the Instantaneous Center (I.C.). Note that point $A$ in Figure 4.5 is an I.C. Also, any stationary point on a rigid body is an I.C.

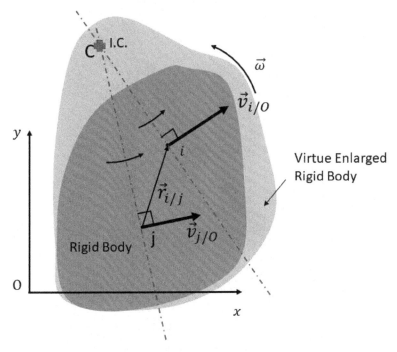

Figure 4.8: Identification of an instantaneous center for a 2D rigid body.

The I.C. offers a convenient way to determine the angular velocity of the rigid body. For example, if we know $v_{i/O}$ and the location of the I.C., the magnitude of the angular velocity becomes

$$\omega = \frac{v_{i/O}}{r_{i/C}} \tag{4.45}$$

## Example 4.1: The Identification of the Instantaneous Center

**Problem Statement:** As shown in Figure 4.9, a beam slides down against a wall. It is found that point $A$ is sliding down at a speed of 6 mm/s. What is the angular velocity of the beam and what is the velocity of point $B$? The unit of the positions is cm.

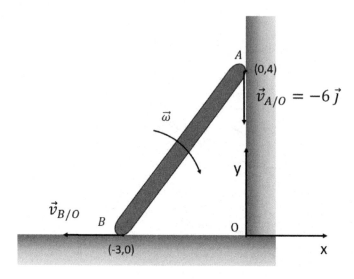

Figure 4.9: A beam slides down against a wall.

**Solution:**

**Force Analysis:** Skipped.

**Motion Analysis:** We define the reference frame as shown in Figure 4.9. Because point $A$ is going downward, point $B$ is to the left. We can draw the velocity vectors for both points even though we do not know the magnitude of $\vec{v}_{B/O}$.

**Governing Equations:** We will first identify the I.C. As shown in Figure 4.10, we draw two perpendicular lines which intersect at $(-3, 4)$, which is the location of the I.C. The angular velocity is then determined as

$$\omega = \frac{v_{A/O}}{r_{A/C}} = 6/30 = 0.2 \text{ (rad/s)} \qquad (4.46)$$

The vector expression of the angular velocity is $\vec{\omega} = -0.2\vec{k}$ because it rotates clockwise. The velocity of point $B$ is then determined as

$$\vec{v}_{B/O} = \vec{\omega} \times \vec{r}_{B/C} = (-0.2\vec{k}) \times (-40\vec{j}) = -8\vec{i} \text{ (mm/s)} \qquad (4.47)$$

If we do not use this method of I.C., we can still solve the problem by setting up the following equation:

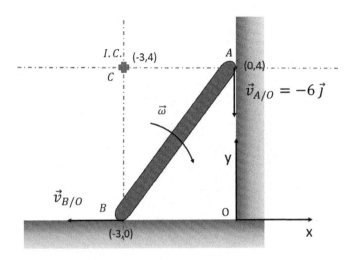

Figure 4.10: Identifying the I.C.

$$\vec{v}_{B/O} = v_{B/O}(-\vec{i}) = \vec{v}_{A/O} + \vec{\omega} \times \vec{r}_{B/A}$$

$$= -6\vec{j} + (\omega\vec{k}) \times (-30\vec{i} - 40\vec{j}) \qquad (4.48)$$

Equation (4.48) contains two scalar equations and two unknowns. We will reach the same result.

### 4.2.3.2 *Instantaneous axis of rotation for a 3D rigid body*

The concept and identification of the I.C. are more complex for 3D motion. In general, there could exist many instantaneous centers unless it is the rotation around one fixed point case, discussed earlier. We need to extend the concept of instantaneous center to the instantaneous axis of rotation, which is an axis with zero velocity and the entire rigid body rotates around this axis at this instant of time.

As shown in Figure 4.11, if we know the direction of $\vec{\omega}$ of a 3D rigid body, we can place a plane perpendicular to the direction of $\vec{\omega}$ across the 3D rigid body, such as plane $P$. On plane $P$, if we know the velocity of two points, $E$ and $F$, these two velocity vectors must reside on plane $P$ as well. We treat plane $P$ as a 2D rigid body and identify its I.C. as $(I.C.)_P$. Then, the vector of $\vec{\omega}$ must pass through $(I.C.)_P$ and we can draw an axis line, which is identified as the instantaneous axis of rotation of the 3D rigid body. As in the case of 2D rigid bodies, the axis of rotation may or may not pass through the 3D rigid body.

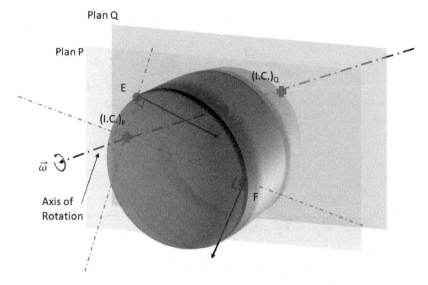

Figure 4.11:   Identifying the instantaneous axis of rotation of a 3D rigid body.

If we place a second perpendicular plane and repeat the same process, the instantaneous center $(I.C.)_Q$ on plane $Q$ must reside on the axis of rotation as well. The identification of the axis of rotation is more difficult in general than finding the I.C. If the rigid body is rotating around a stationary point, then the location of the axis of rotation can be defined easily as an arrow leaving the stationary point with a direction same as $\vec{\omega}$.

The method illustrated in Figure 4.11 is a natural extension of the 2D method to identify the axis of rotation. However, do we always have the I.C. in 2D cases or the axis of rotation in 3D cases? We will try to answer this question via the analysis below.

For a 3D rigid body, given the motion of a point $P$ on the rigid body (including its virtually enlarged rigid body) and the angular velocity of the rigid body, if we assume that an axis of rotation exists and let $(x_{IC}, y_{IC}, z_{IC})$ be a point on the axis of rotation, then the velocity of point $P$ can be expressed as

$$\vec{v}_{P/O} = \vec{\omega} \times \vec{r}_{P/IC}$$
$$= (\omega_x \vec{i} + \omega_y \vec{j} + \omega_z \vec{k}) \times [(x_P - x_{IC})\vec{i}$$
$$+ (y_P - y_{IC})\vec{j} + (z_P - z_{IC})\vec{k}] \qquad (4.49)$$

where the position of point $P$ is $(x_P, y_P, z_P)$.

Multiplying out Equation (4.49), we obtain the equations for all three directions,

$$v_{Px} = \omega_y z_P - \omega_y z_{IC} - \omega_z y_P + \omega_z y_{IC} \qquad (4.50)$$

$$v_{Py} = \omega_z x_P - \omega_z x_{IC} - \omega_x z_P + \omega_x z_{IC} \qquad (4.51)$$

$$v_{Pz} = \omega_x y_P - \omega_x y_{IC} - \omega_y x_P + \omega_y x_{IC} \qquad (4.52)$$

To verify if $(x_{IC}, y_{IC}, z_{IC})$ exists, we rearrange the above three equations into a matrix form,

$$\begin{bmatrix} v_{Px} - \omega_y z_P + \omega_z y_P \\ v_{Py} - \omega_z x_P + \omega_x z_P \\ v_{Pz} - \omega_x y_P + \omega_y x_P \end{bmatrix} = \begin{bmatrix} 0 & \omega_z & -\omega_y \\ -\omega_z & 0 & \omega_x \\ \omega_y & -\omega_x & 0 \end{bmatrix} \begin{bmatrix} x_{IC} \\ y_{IC} \\ z_{IC} \end{bmatrix} \qquad (4.53)$$

To ensure that $(x_{IC}, y_{IC}, z_{IC})$ exists, the determinant of the matrix must not be zero or the left side vector of Equation (4.53) needs to be zero.

$$\det \left( \begin{bmatrix} 0 & \omega_z & -\omega_y \\ \omega_z & 0 & \omega_x \\ \omega_y & -\omega_x & 0 \end{bmatrix} \right) = 2\omega_x \omega_y \omega_z \qquad (4.54)$$

Let's apply Equations (4.49)–(4.54) to a few special cases. First, let's revisit the rolling disk case of Figure 4.5.

### Example 4.2: The Identification of Axis of Rotation of a 2D Disk

**Problem Statement:** As shown in Figure 4.5, we have $\omega_x = \omega_y = 0$. Choose the mass center as the reference point, Equations (4.50)–(4.52) lead to

$$v_{Gx} = -\omega_z y_G + \omega_z y_{IC} \qquad (4.55)$$

$$v_{Gy} = \omega_z x_G - \omega_z x_{IC} \qquad (4.56)$$

$$v_{Gz} = 0 \qquad (4.57)$$

Because $v_{Gy} = 0$, we must have $x_G = x_{IC}$. Finally, because we know that $v_{Gx} = -\omega_z y_G$, then $y_{IC} = 0$. Therefore, $(x_{IC}, y_{IC}, z_{IC}) = (0, 0, 0)$, which is the contact point with the ground.

For the 2D cases, we can even obtain from Equations (4.55)–(4.57) that

$$x_{IC} = -(v_{Gy} - \omega_z x_G)/(\omega_z) \tag{4.58}$$

$$y_{IC} = (v_{Gx} + \omega_z y_G)/(\omega_z) \tag{4.59}$$

where $G$ can be any point on the rigid body. In other words, there is always an I.C. for a 2D rigid body motion. If $\omega_z = 0$, we have a translation motion. For example, if we replace the disc in Figure 4.5 with a block, then $y_{IC} \to \infty$ and $x_{IC}$ can be any value.

Let's consider the 3D case. In a general 3D rotation with $\omega_x \neq 0$, $\omega_y \neq 0$, and $\omega_z \neq 0$, we will be able to obtain a solution for $(x_{IC}, y_{IC}, z_{IC})$ by solving Equation (4.53). However, what happens if one of the angular velocity components is zero? We will consider a famous example for illustration.

## Example 4.3: The Identification of Axis of Rotation of a Rotating Cylinder

**Problem Statement:** As shown in Figure 4.12, a collar $A$ can slide up and down as well as rotate around a vertical shaft. A horizontal shaft is attached to the collar. A cylinder $B$ with a radius $R$ is fitted to the horizontal shaft and can rotate around the horizontal shaft but does not slide along it.

At the current instant shown, $R = 2$ cm, $a = 4$ cm, $\dot{\phi} = 2$ rad/s, $\dot{\psi} = 6$ rad/s, and $A$ is 5 cm over the base plate. Determine the location and direction of the axis of rotation of cylinder $B$.

**Solution:**

**Force Analysis:** Skipped.

**Motion Analysis:** We will first establish proper observation frames. Two observation frames, $O-X-Y-Z$ and $A-\hat{x}-\hat{y}-\hat{z}$, are defined as shown in Figure 4.12. At this instant of time, $O-X-Y-Z$ and $A-\hat{x}-\hat{y}-\hat{z}$ are aligned. However, while $O-X-Y-Z$ is an inertial reference frame, $A-\hat{x}-\hat{y}-\hat{z}$ is rotating and moving with collar $A$. The frame rotation of $A-\hat{x}-\hat{y}-\hat{z}$ is observed by $O-X-Y-Z$ as $\vec{\Omega}_{A/O} = \dot{\phi}\vec{K}$. With respect to $A-\hat{x}-\hat{y}-\hat{z}$, cylinder $B$ is observed to have an angular rotation, $\hat{\omega}_{B/A} = \dot{\psi}\hat{i}$. Using the additivity of angular velocity,

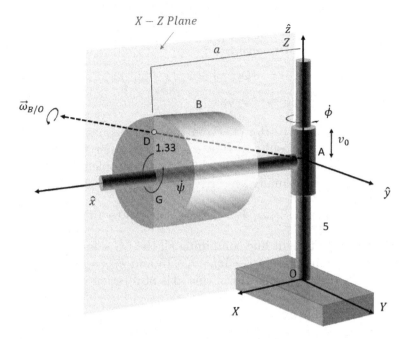

Figure 4.12: Identifying the axis of rotation of a rotating cylinder.

we determine the angular velocity of cylinder $B$ for this instant of time as

$$\vec{\omega}_{B/O} = \hat{\omega}_{B/A} + \vec{\Omega}_{A/O} = \dot{\psi}\hat{i} + \dot{\phi}\vec{K} = \dot{\psi}\vec{I} + \dot{\phi}\vec{K} = 6\vec{I} + 2\vec{K} \quad (4.60)$$

Equation (4.60) indicates that the angular velocity, $\vec{\omega}_{B/O}$, is aligned with a plane parallel to the $X-Z$ plane. Therefore, $\omega_y = 0$.

For any point $P$ on cylinder $B$, its velocity can be determined by Equations (4.50)–(4.52) as

$$v_{Px} = -\omega_z y_P + \omega_z y_{IC} \quad (4.61)$$

$$v_{Py} = \omega_z x_P - \omega_z x_{IC} - \omega_x z_P + \omega_x z_{IC} \quad (4.62)$$

$$v_{Pz} = \omega_x y_P - \omega_x y_{IC} \quad (4.63)$$

If we enlarge cylinder $B$ to include collar $A$ and choose $A$ as point $P$ for Equations (4.61)–(4.63), the above equation is further simplified to

$$0 = \omega_z y_{IC} \tag{4.64}$$

$$0 = -\omega_z x_{IC} - \omega_x z_A + \omega_x z_{IC} \tag{4.65}$$

$$v_{Az} = -\omega_x y_{IC} \tag{4.66}$$

where point $A$ is at $(0, 0, z_A)$. From the above equation, we obtain that $y_{IC} = 0$. This results in that $v_{Az} = 0$. In other words, point $A$ cannot move up and down as specified in the problem statement so that an I.C. exists. Finally, we arrive at

$$0 = -\omega_z x_{IC} - \omega_x z_A + \omega_x z_{IC} = 2x_{IC} + 30 - 6z_{IC} \tag{4.67}$$

which defines a straight line containing all the I.C.s, and it is the axis of rotation, defined by $x_{IC} - 3z_{IC} = -15$ and $y_{IC} = 0$, as shown in Figure 4.12. If $A$ does not move, this axis of rotation exists and $A$ is on the axis of rotation.

We can calculate and determine point $D$ where the axis of rotation leaves rigid body $B$. Point $D$ is 1.33 cm above the center $G$. Its coordinate with respect to $O-X-Y-Z$ is $(4, 0, 6.33)$.

As a verification, we can determine the velocity of $G$ using Equations (4.61)–(4.63). The result is $\vec{v}_{G/O} = 8\vec{J}$, where $\vec{r}_{G/O} = 4\vec{I} + 5\vec{K}$ and the I.C. can be any point on the axis of rotation.

From Equation (4.67), the orientation of the axis of rotation is determined by the ratio of $\frac{\omega_x}{\omega_z}$. In this particular case, there is a stationary point $A$ on the axis of rotation; therefore, it appears that the axis of rotation originates from point $A$. However, in general, this may not be the case. An axis of rotation could be like a swirling stick in space, not around a fixed point. In practical applications, we would prefer to design machines so that the axis of rotation is rotating around a fixed point.

### 4.2.4 *Space cone and body cone*

Let's consider the rolling without slipping disk of Figure 4.13(a). For this 2D case, we have an I.C. at any given instant of time. This I.C. can be extended in the $z$ direction and we have an axis of rotation. This axis of rotation moves along the surface, forming a straight line, indicated by $F'$, $E'$, $D'$, etc. On the other hand, those points, $F$, $E$, $D$, etc., on the disk form a circle. We can call the straight line of the

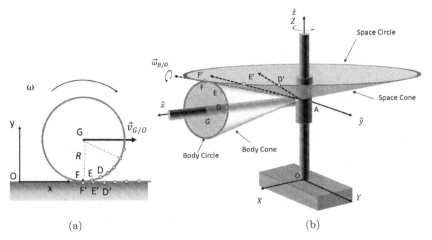

Figure 4.13:   (a) Identify the space line and the body circle; and (b) identify the space cone and the body cone.

surface a space line because it describes the trajectory of the axis of rotation in space. We call the circle as the body circle because it describes those points on the body which coincide with the axis of rotation once before. Stretching it to 3D along the $z$ direction, we have a space surface and a body cylinder. The space surface is actually a space cylinder with a diameter of $\infty$. In general, we can state that by tracing those mating points, we form a space cylinder and a body cylinder.

Now, let's examine the 3D case of Figure 4.12. At the end surface of the cylinder (Figure 4.13(b)), we can identify the mating pairs, $F/F'$, $E/E'$, $D/D'$, etc. Points $F'$, $E'$, $D'$, etc. form a horizontal space circle. Accordingly, points $F$, $E$, $D$, etc. form a body circle.

If we look at other points along the axis of rotation all the way to point $A$, a space cone and a body cone are formed. Typical statements from textbooks on Dynamics often provide statements such as

> *As the direction of the instantaneous axis of rotation changes in space, the locus of the axis generates a fixed space cone. If the change in the direction of this axis is viewed with respect to the rotating body, the locus of the axis generates a body cone.*

The statement above is quite confusing because it is not clear which reference frames are used in viewing the space cone or the body cone. Viewing through a rotating reference frame can have strange consequences as we discussed in Chapter 3. The space cone and body cone can be understood better by tracing the locus of the mating points coinciding the axis of rotation as discussed above. From this new way of discussion, we are able to relate the 2D and 3D cases, through the same discussion of instantaneous centers.

### 4.2.5 *Gear pairs designed by the space and body cones concept*

As stated above, the space and body cones are special cases when there is a fixed point on the axis of rotation. However, this special case provides a very good guideline for machine design with rotating bodies. As shown in Figure 4.14, a pair of mating bevel gears are designed based on the concept of the space cone and body cone.

Recall the rotating cylinder in Example (4.3), the axis of rotation has an angular velocity $\vec{\omega} = \hat{\omega}_x \hat{i} + \omega_z \vec{K}$. If we fix $\hat{i}$ to make it non-rotating $(\hat{i} \longrightarrow \vec{I})$ and give $\omega_x \vec{I}$ to one gear and give $\omega_z \vec{K}$ to the other gear, the axis of rotation becomes fixed, and is the mating axis of the bevel gear pair of Figure 4.14. The orientation of the mating

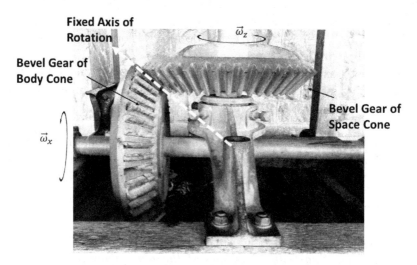

Figure 4.14:   Bevel gear design based on the concept of the space and body cones. *Source*: Modified image based on https://upload.wikimedia.org/wikipedia/commons/0/0c/Bevel_gears_on_grain_mill_at_Dordrecht%2C_Eastern_Cape.jpg.

Figure 4.15: A single-axis lathe provides an axis of rotation using a spindle. *Source*: Modified image based on http://1.bp.blogspot.com/-xfZcOG-NurU/ UQY6-Dh_5GI/AAAAAAAAAQg/EAFnGUN_r9E/s1600/Lathe-6.jpg.

axis depends on the ratio of $\omega_x$ and $\omega_z$, which in turn depends on the radius of the gears. In the case of Figure 4.14, both bevel gears have the same radius; therefore, the axis of rotation is at $45^o$ from the horizontal axis. The bevel gear shapes are the shapes of the space and body cones.

With this pair of bevel gears, we can transmit power from one axis to another at a different orientation.

### 4.2.6 *Applications of axis of rotation in machining*

For machine tools used for machining, such as turning, milling, or grinding, spindles are used for providing axes of rotations. In Figure 4.15, a single axis lathe is shown. A spindle is used to provide an axis of rotation for the turning operation. This axis of rotation is required to stay fixed in space in order to provide very high machining precision. The accuracy and the methods to measure the error of the axis of rotation are defined in international standards.[1] We will

---

[1]ANSI/ASME B89.3.4M – 1985. Axes of rotation: methods for specifying and testing; ISO 230-7:2006. Test code for machine tools — Part 7: geometric accuracy of axes of rotation.

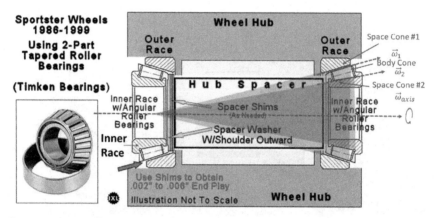

Figure 4.16: Different axes of rotation, body, and space cones of an automobile wheel.
*Source*: Modified image based on http://sportsterpedia.com/lib/exe/fetch. php/techtalk:evo:wheels:wheelbearing-timken.jpg.

not go into details of these standards but we should know that the subject of axis of rotation is of utmost importance in machining or manufacturing in general. Spindles are the components used to provide axes of rotations. The spindle shafts are supported by bearings. These spindle bearings are basically designed based on the space and body cone concept. For example, in Figure 4.16, an automobile wheel is shown. Tapered roller bearings are used to support a shaft to mate with the wheel. This shaft provides an axis of rotation $\vec{\omega}_{\text{axis}}$ for tire rotations.

The space cone and body cone concept are used inside the tapered roller bearing. The tapered roller is a body cone. This body cone pairs with the outer race as a space cone. The corresponding instantaneous axis of rotation is $\vec{\omega}_1$ as shown in Figure 4.16. The roller also pairs with the inner race, as another space cone, and a second instantaneous axis of rotation, $\vec{\omega}_2$, is defined with respect to the shaft. The axis of rotation of the shaft, $\vec{\omega}_{\text{axis}}$ is fixed in space for the tire, while the instantaneous axes of rotation, $\vec{\omega}_1$ and $\vec{\omega}_2$, revolve over their respective space cones. Both space cones originated from the same point at the tip. The body cone of the tapered roller revolves inside the space cone of the outer race, but outside of the space cone of the inner race.

## 4.3   Kinematics of 2D Rigid Body Assemblies

In this section, we extend the kinematic analysis from one rigid body to an assembly of rigid bodies. We already saw a few cases regarding how rigid bodies can be coupled together into an assembly. Basically, the coupling is either through contacts or constraints, as discussed in Sections 2.14 and 2.15, Chapter 2. Specifically, the contact will be rolling without slipping, as discussed in Section 4.2.2, or slipping contact. For constraints, the typical ones are pins or pined–pined rods. When a total constraint, such as welding or bolting, is used to join two rigid bodies, they become one rigid body. Some flexible constraints can also be used to join two rigid bodies such as springs or flexures.

### 4.3.1   *Degrees of freedom of 2D rigid body assemblies*

An unconstrained 2D rigid body has three degrees of freedom, as shown in Figure 4.17(a).

If we connect a second member (II) to member I via a pin, as shown in Figure 4.17(b), this assembly now has four degrees of freedom. Both members are free to rotate, but they are joined at point A. A is free to move in the $x–y$ plane. If we fix member II (as shown in Figure 4.17(c)), it becomes a fixed pin as shown in Figure 4.17(d). The assembly has only one rotation left. When we see a mechanism similar to that in Figure 4.17(d), we should remember that it is actually a two-member system as shown in Figure 4.17(c). Gruebler's equation is often used to calculate the degrees of freedom

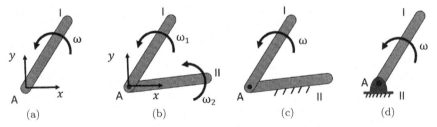

(a)          (b)          (c)          (d)

Figure 4.17:   The degrees of freedom of 2D assemblies: (a) a free 2D member; (b) two 2D members pinned together; (c) one fixed and one pinned members; and (d) a 2D member with a fixed pin.

of a rigid body assembly. Gruebler's equation is written as

$$M = 3\,n - 3j_3 - 2j_2 - 1\,j_1 \tag{4.68}$$

where $M$ is the degrees of freedom of the assembly, $n$ is the total number of members of the assembly, $j_3$ is the number of the joint that eliminates all three degrees of freedom, $j_2$ is the number of the joint that eliminates two degrees of freedom, and $j_1$ is the number of the joint that eliminates one degree of freedom.

Apparently, a fixed joint as shown in Figure 4.17(c) to fix member II is a $j_3$ joint. A fixed joint can also be achieved by bolting, welding, bonding, etc. A pin joint is an example of a $j_2$ type of joint, which eliminates two translational motions but allows for rotation. There are other types of $j_2$ joint, to be discussed later along with the $j_1$ joint. For the assembly shown in Figure 4.17(d), we only have one pin joint; therefore, the degree of freedom is calculated to be

$$M = 3\,(2) - 3(1) - 2(1) - 1\,(0) = 1 \tag{4.69}$$

Let us look at other examples of the $j_1$ and $j_2$ joints in 2D assemblies. In Figure 4.18, examples of the $j_2$ joint are shown. In Figure 4.18(a), rigid body A is confined in a straight groove and only allowed to move horizontally without rotation. Therefore, it has only one degree of freedom. The contact condition between rigid body A and the groove is a $j_2$ joint.

Applying Gruebler's equation, the degree of freedom of rigid body A is calculated to be

$$M = 3\,(2) - 3(1) - 2(1) - 1(0) = 1 \tag{4.70}$$

Note that the grove could be curved, as shown in Figure 4.18(b). When rigid body B is moving along the curved groove, it has both

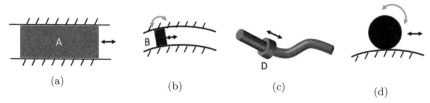

(a)          (b)          (c)          (d)

Figure 4.18:   Examples of the $j_2$ joint: (a) a block within a straight grove; (b) a block with a curved groove; (c) a collar over a curved shaft; and (d) a disk rolling without slipping over a curved surface.

horizontal and vertical movements, but the movement path is pre-
defined and rigid body B cannot deviate from this path. Thus, it
has only one degree of freedom in translation. Furthermore, even
though there is a rotation of rigid body B when it moves along this
curved groove, this rotation is not independent from the translation
movement, as indicated by its lightly shaded arrow sign. We know
exactly the orientation of rigid body B at any specific location along
the groove. Therefore, the joint of Figure 4.18(b) is the same as that
of Figure 4.18(a). Similarly, the $j_2$ joint could be a collar moving
along a rod, straight or curved, as shown in Figure 4.18(c). In Figure 4.18(d), a roller rolls without slipping over a surface (curved or
straight). Because there is no slipping, the linear motion of the roller
center and the rotation of the roller are related, not independent
(thus, the rotation is indicated as a lightly shaded arrow sign). As a
result, the roller in Figure 4.18(d) has only one degree of freedom.
A rolling-without-slipping joint is, therefore, a $j_2$ joint, which eliminates two degrees of freedom.

In Figure 4.19, examples of the $j_1$ joint are shown. In Figures 4.19(a) and 4.19(b), rigid body A is confined in a straight or
a curved groove and is only allowed to move along the groove and
rotate at the same time. The contact between rigid body A and the
groove is not rolling-without-slipping. Therefore, rigid body A has
two degrees of freedom because the rotational and the translational
movements are independent.

Note that we may not require the upper boundary of the groove. It
can be an open surface if the rigid body is heavy enough to maintain
contact with the surface, as shown in Figures 4.19(c) and 4.19(d).
The contact conditions in Figure 4.19 are called sliding contact or
contact with slip.

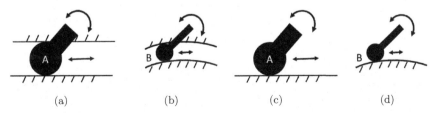

|       |       |       |       |
| :---: | :---: | :---: | :---: |
|  (a)  |  (b)  |  (c)  |  (d)  |

Figure 4.19:   Sliding contact as the $j_1$ joint: (a) a disk sliding inside a groove,
(b) a disk sliding inside a curved groove, (c) a disk sliding over a flat surface, and
(d) a disk sliding over a curved surface.

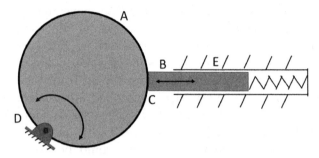

Figure 4.20: An example of the $j_1$ joint, similar to those used in internal combustion engines.

A rigid body assembly with a sliding contact is shown in Figure 4.20. A circular rigid body (A) is rotating around a pin (D), while a rod (B) is pressed against the circular body at point C by a spring. The rotation of A will cause rod B to move horizontally, and the spring ensures the contact between A and B. Rod B can only move horizontally because it is constrained by the slot E. A practical example of Figure 4.20 is the cam system commonly used in internal combustion engines to control the opening and closing of the intake and exhaust valves.

For the assembly of Figure 4.20, we have three rigid bodies (A, B, and D), two $j_2$ joints (D and E), and one $j_1$ joint (C). Applying Gruebler's equation, we have

$$M = 3(3) - 3(1) - 2(2) - 1(1) = 1 \tag{4.71}$$

Therefore, the assembly shown in Figure 4.20 has only one degree of freedom.

### 4.3.2 *Paradox of Gruebler's equation*

Although Gruebler's equation provides a convenient way to determine the degrees of freedom for rigid body assemblies, the calculation results of Gruebler's equation are not always correct. As shown in Figure 4.21, two examples are provided to demonstrate the flaw of Gruebler's equation. In Figure 4.21(a), the assembly could be a set of frictional wheels or a pair of engaged gears.

(a)                                                   (b)

Figure 4.21:   Rigid body assemblies which have more degrees of freedom than those predicted by Gruebler's equation: (a) a pair of friction wheels or gears; and (b) a platform supported by multiple parallel links.
*Source*: (a) https://www.divilabs.com/2013/06/gears-their-common-types-used-in.html.

Applying Gruebler's equation, the assembly of Figure 4.21(a) should have zero degree of freedom because

$$M = 3(3) - 3(1) - 2(3) - 1(0) = 0 \qquad (4.72)$$

where the three $j_2$ joints are the two pin joints and the rolling contact between the gears. Similarly, for the mechanism in Figure 4.21(b), Gruebler's equation predicts zero degrees of freedom, as

$$M = 3(5) - 3(1) - 2(6) - 1(0) = 0 \qquad (4.73)$$

However, both assemblies in Figure 4.21 have one degree of freedom. This paradox is caused by a fundamental flaw of Gruebler's equation, which is that Gruebler's equation does not consider the relationship between the joints. Some of the joints might be redundant, such as the links in Figure 4.21(b). If we remove one of the pined elements, there is no change in the motion of the assembly.

Another way to look at the paradox is that Gruebler's equation is a scalar equation, but an assembly is usually in 2D or 3D. As a result, similar joints in different orientations could have different effects in eliminating degrees of freedom. As shown in Figure 4.22, rigid body $D$ in assembly (a) has a partial rotational motion, while the rotation is completely eliminated in (b); however, Gruebler's equation will predict zero degrees of freedom for both,

$$M = 3(5) - 3(1) - 2(6) - 1(0) = 0 \qquad (4.74)$$

Note that there are two $j_2$ joints at point $O$ for pinning three rigid bodies, $A$, $B$, and $D$, together.

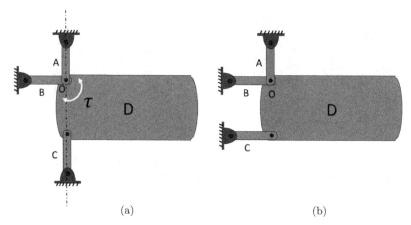

(a)                                    (b)

Figure 4.22: Examples to show the flaw of Gruebler's equation: (a) an incorrectly predicted case; and (b) a correctly predicted case.

However, the joints related to element $C$ in assemblies of Figures 4.22(a) and 4.22(b) have different constraining effects. As elements $A$, $B$, and $C$ are all pined–pined rods, they only provide constraint forces along their element directions. In assembly of (a), all three constraint forces will pass through point $O$. These elements cannot resist a moment, $\tau$, until they shift out of their current positions to make element $C$ not point to $O$ anymore. As a result, rigid body $D$ can have a small partial rotational motion. On the other hand, the elements in assembly of (b) can resist any forces or moments applied to rigid body $D$ to keep it from any movements. Because the arrangements of element $C$ are different, they have different constraining effects. However, Gruebler's equation, as a scalar equation, cannot account for these effects.

In conclusion, we need to use Gruebler's equation with caution. With attention to potential paradoxes, Gruebler's equation is still an effective tool for structural designs.

### 4.3.3 Velocity analysis of 2D linkage systems

In this section, we will discuss the famous four-bar linkage systems and their variations. Figure 4.23 illustrates a standard four-bar linkage system. Applying Gruebler's equation, this four-bar linkage system has only one degree of freedom. If the rotation of link $AB$ is

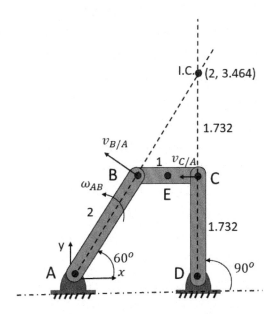

Figure 4.23: A standard four-bar linkage with link $AB$ as the driving element.

given, as the driving element, then the rotations of links $BC$ and $CD$ can be uniquely determined. Based on Figure 4.23, we define the angular velocities of all the links with respect to the same fixed reference frame $x-y$. We assume all rotations are counter-clock-wise when we formulate the kinematic equations. For the rigid body of link $AB$, we know point $A$ is stationary and we know its rotation, $\vec{\omega}_{AB} = 10\vec{k}$. As a result, we know the motion of every point on link $AB$. The velocity of point $B$ is found to be

$$\vec{v}_{B/A} = \vec{v}_A + \vec{\omega}_{AB} \times \vec{r}_{B/A}$$
$$= 0 + 10\vec{k} \times (2\cos 60° \ \vec{i} + 2\sin 60° \ \vec{j})$$
$$= -17.32\vec{i} + 10\vec{j} \tag{4.75}$$

We can determine the velocity of points $C$ with respect to either rigid body $BC$ or $CD$, which leads to

$$\vec{v}_{C/A} = \vec{v}_{B/A} + \vec{\omega}_{BC} \times \vec{r}_{C/B}$$
$$= \vec{v}_D + \vec{\omega}_{CD} \times \vec{r}_{C/D} = \vec{\omega}_{CD} \times \vec{r}_{C/D} \tag{4.76}$$

Equation (4.76) is a vector equation; therefore, it can be split into two scalar equations,

$$1.732 \, \omega_{CD} = 17.32$$

$$0 = 10 + \omega_{BC} \tag{4.77}$$

We obtain $\vec{\omega}_{BC} = -10\vec{k}$ and $\vec{\omega}_{CD} = 10\vec{k}$.

We can also invoke the I.C. method. The instantaneous center of link $BC$ can be found as shown in Figure 4.23. Both points $B$ and $C$ will appear to rotate around point $I.C.$; therefore,

$$\vec{v}_{B/A} = \vec{\omega}_{BC} \times \vec{r}_{B/IC}$$

$$\vec{v}_{C/A} = \vec{\omega}_{BC} \times \vec{r}_{C/IC} = \vec{\omega}_{CD} \times \vec{r}_{C/D} \tag{4.78}$$

We quickly determine $\omega_{BC} = v_{B/A}/r_{B/IC} = \sqrt{17.32^2 + 10^2}/2 = 10$ and it is a clock-wise rotation. The velocity of point $C$ can be determined as $\vec{v}_{C/A} = -17.32\vec{i}$.

For the four-bar linkage system of Figure 4.23, there is no rotation speed reduction from $\omega_{AB} = 10$ to $\omega_{CD} = 10$. However, this is only valid at the current orientation. Let's examine a general four-bar linkage system as shown in Figure 4.24.

Based on the instantaneous center, we can write down two equations to relate the rotations of the linkages similar to Equation (4.78).

$$-\omega_{BC} \, e = \omega_{AB} \, a$$

$$-\omega_{BC} \, f = \omega_{CD} \, c \tag{4.79}$$

Eliminating $\omega_{BC}$, we obtain $\omega_{CD} = (af/ce)\omega_{AB}$. At different orientations, the value of $f/e$ will be different; therefore, even when $\omega_{AB}$ is constant, $\omega_{CD}$ may not be a constant.

By choosing appropriate parameters, mainly the length of each link, the four-bar linkage system can be adapted for different applications. As shown in Figure 4.25(a), a four-bar linkage system can be designed as the mechanism for a windshield wiper. The link $AB$ can rotate a full circle, while link $CD$ only rotates a limited angle. In Figures 4.25(b) and 4.25(c), a double-A car suspension system is shown, which is a four-bar linkage system. The four joints are labeled as $M$, $E$, $G$, and $N$. Linkage $EG$ is connected to the tire/wheel assembly, while joints $M$ and $N$ are connected to the car body, considered as

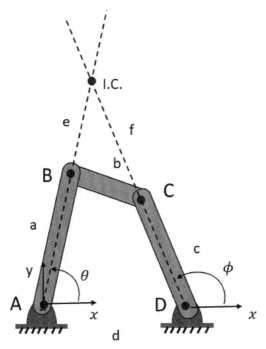

Figure 4.24: A more general four-bar linkage with link $AB$ as the driving element.

the foundation. The actual implementation of this four-bar linkage system in a 3D setup is shown in Figure 4.25(b). The links $ME$ and $NG$ look like the letter "A," which is why it is called a double-A suspension system.

If we take a standard four-bar linkage system and replace some pin joints, many different variations can be designed. In Figure 4.26(a), pin joint $D$ becomes a sliding joint, which is the $j_2$ joint shown in Figure 4.18. The resulting four-bar linkage is now a system similar to the piston and the crankshaft of an internal combustion engine. In Figure 4.26(b), joint $C$ is the connecting joint between the piston and the piston rod. Note that joint $C$ in Figure 4.26(a) can only move horizontally. The velocity equation regarding point $C$ is

$$\vec{V}_{C/O} = \vec{V}_{B/A} + \vec{\omega}_{BC} \times \vec{r}_{C/B}$$

$$= \vec{\omega}_{AB} \times \vec{r}_{B/A} + \vec{\omega}_{BC} \times \vec{r}_{C/B} = \dot{x}_C \vec{i} \qquad (4.80)$$

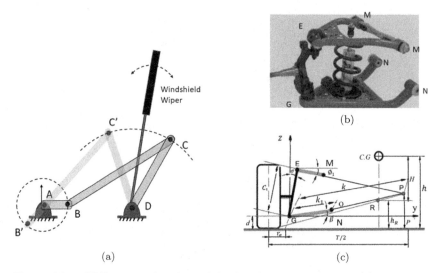

Figure 4.25: Different applications of the four-bar linkage system: (a) windshield wiper mechanism; (b) a double-A suspension system; and (c) the equivalent four-bar linkage system of the double-A suspension.

*Source*: (b) https://i.stack.imgur.com/RUjor.png; and (c) https://cdn.jvejourn als.com/articles/17862/xml/img4.jpg.

Equating the components of the above equations in both directions, we have

$$0 = a \cos \theta \, \omega_{AB} + b \cos \phi \, \omega_{BC} \qquad (4.81)$$

$$\dot{x}_C = b \omega_{BC} \sin \phi - a \omega_{AB} \sin \theta \qquad (4.82)$$

Also from the geometry, we have

$$x_C = a \cos \theta + b \cos \phi \qquad (4.83)$$

$$a \sin \theta = b \sin \phi \qquad (4.84)$$

There are four unknowns in Equations (4.81)–(4.84), namely $\phi$, $\dot{x}_c$, $x_c$, and $\omega_{BC}$, and we can solve for them to obtain

$$v_C = -a \sin \theta \, \omega_{AB} \left( 1 + \frac{a \cos \theta}{b \cos \phi} \right) \qquad (4.85)$$

$$\cos \phi = \sqrt{\frac{b^2 - a^2 \sin^2 \theta}{b^2}} \qquad (4.86)$$

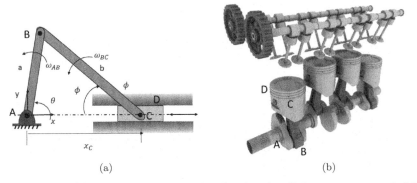

(a)                                                    (b)

Figure 4.26: (a) A slider system made of a four-bar linkage system; and (b) application of (a) as a piston and crankshaft system of an internal combustion engine.
*Source*: (b) https://upload.wikimedia.org/wikipedia/commons/6/69/Engine_movingparts.jpg.

Assigning a constant $\omega_{AB}$, we can plot the speed of $C$ based on the above equation, as shown in Figure 4.27. The sliding speed of the piston is not constant even though $\omega_{AB}$ is constant. It comes to a complete stop at $\theta = 0$ and $\theta = 180°$. The maximal sliding speed occurs at about $67°$ and $293°$, not $90°$ and $270°$.

For the application of the internal combustion engines, the sliding speed is the input while $\omega_{AB}$ is the output. A constant output of $\omega_{AB}$ is desirable. Therefore, we need to control piston speed, as shown in Figure 4.27. However, the sliding speed of the piston is determined by the explosion within the combustion chamber and the loading on the crankshaft. It is quite difficult to control the sliding speed as required by Figure 4.27 to render a constant crankshaft rotational speed. We can now understand why a motorcycle with one cylinder engine shakes a lot. To even out the rotational speed, multiple cylinders are used in car engines, and the higher the number of cylinders, the smoother the running of the engine. A V8 engine is smoother than a V6 or I6 (in-line six), which is better than a four cylinder engine.

If we convert point $C$ of a standard four-bar linkage system to the sliding joint, a new design is achieved, as shown in Figure 4.28(a). One famous application of this four-bar linkage variation is the widely used McPherson suspension system as shown in Figure 4.28(b). Both joints $A$ and $D$ are fixed on the car body. A tire is

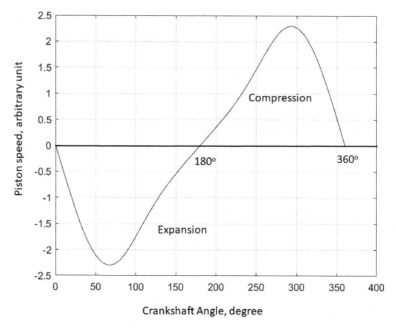

Figure 4.27: The sliding speed of a piston versus the crankshaft angle at a constant crankshaft rotational speed.

attached to Link $BC$. As the McPherson suspension system, joint $D$ has a limited rotation because it is typically a rubber mount. There are advantages and disadvantages between the double-A suspension of Figure 4.25(b) vs the McPherson suspension of Figure 4.28(b). Readers can research more to learn about them.

### 4.3.4 *Gear trains and planetary gear systems*

In this section, we focus on the rolling without slipping coupling. One critical application is for gear trains as shown in Figure 4.21(a). One of the most ingenious gear train systems is the planetary gear train system, as shown in Figure 4.29.

In a planetary gear train system, the shafts of the sun gear, ring gear, and the arm (also denoted as the carrier) are co-axial and centered at point $O$ as shown, while the planet gears have their shafts attached to the arm at point $P$. For the kinematic analysis of the planetary gear train system, a 2D diagram of Figure 4.30 is sufficient.

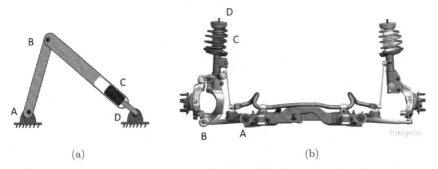

(a)                                                    (b)

Figure 4.28:  (a) A four-bar linkage system with joint $C$ a sliding joint; and (b) application of (a) as the McPherson suspension system.

*Source*:  (b)  https://espirituracer.com/archivos/2017/11/9999rpm-C%C3%B3 mo-se-dise%C3%B1a-una-suspensi%C3%B3n-opel-hiperstrut.jpg.

Figure 4.29:  3D structure of a planetary gear train system, a modified screen shot image from the following source.

*Source*: https://www.youtube.com/watch?v=ARd-Om2VyiE.

Without loss of generality, assuming that a planet gear and the arm are aligned vertically, we can define the velocities of points $A$ and $A'$ as well as $B$ and $B'$ with the kinematic equations we learned earlier. Point $A$ is on the sun gear, while point $A'$ is the corresponding mating point on the planet gear. Similarly, point $B$ is on the ring gear, while point $B'$ is on the planet gear. We also assume that all gears

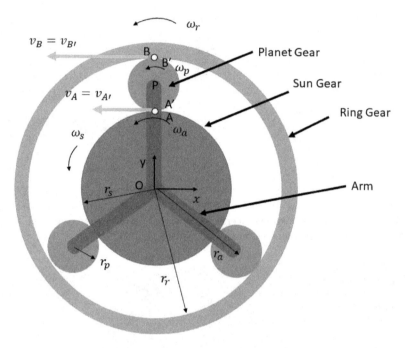

Figure 4.30:   A 2D representation of the planetary gear train system for analysis.

are in a counter-clockwise rotation and their corresponding rotational speeds are marked in Figure 4.30. From geometry, the radius of each gear is labeled in Figure 4.30 as well.

Due to gears being engaged via rolling-without-slipping coupling, we know $\vec{v}_{A/O} = \vec{v}_{A'/O}$ and $\vec{v}_{B/O} = \vec{v}_{B'/O}$. We then obtain the following two equations:

$$\vec{v}_{A/O} = \vec{\omega}_s \times \vec{r}_{A/O} = \vec{v}_{A'/O}$$
$$= \vec{v}_{P/O} + \vec{v}_{A'/P} = \vec{\omega}_a \times \vec{r}_{P/O} + \vec{\omega}_p \times \vec{r}_{A'/P} \qquad (4.87)$$
$$\vec{v}_{B/O} = \vec{\omega}_r \times \vec{r}_{B/O} = \vec{v}_{B'/O}$$
$$= \vec{v}_{P/O} + \vec{v}_{B'/P} = \vec{\omega}_a \times \vec{r}_{P/O} + \vec{\omega}_p \times \vec{r}_{B'/P} \qquad (4.88)$$

Working out Equations (4.87) and (4.88) and expressing them in scalars, we obtain

$$-\omega_s \cdot r_s = -\omega_a \cdot r_a + \omega_p \cdot r_p \qquad (4.89)$$
$$-\omega_r \cdot r_r = -\omega_a \cdot r_a - \omega_p \cdot r_p \qquad (4.90)$$

Equations (4.89) and (4.90) are the governing equations for the planetary gear train system. In addition, we have information related to the geometry,

$$r_a = r_s + r_p \tag{4.91}$$

$$r_r = r_a + r_p = r_s + 2r_p \tag{4.92}$$

Given the geometric information, there are four unknowns involved in Equations (4.89) and (4.90), namely $\omega_s$, $\omega_a$, $\omega_p$, and $\omega_r$. Because we only have two equations, we cannot solve for all four unknowns. We have to know two of them and determine the other two by solving Equations (4.89) and (4.90). Often the planetary gear train system is used as a power transmission device, and one of the four unknowns is assigned as the input (known) and one of them as the output (unknown, to be determined). It should be noted that the rotation of the planet gear $\omega_p$ typically cannot be assigned easily because it does not rotate around a fixed shaft. Therefore, we have to keep $\omega_p$ as an unknown. Finally, the last gear rotation, which is not $\omega_P$, not assigned as the input or the output, needs to be assigned a known value.

In mechanical design, one convenient way to assign a known value to a gear is to make the gear non-rotating by holding it with a clutch or a pin.

For illustration purposes, let's assume $r_p = 0.5r_s$; therefore, $r_a = 1.5r_s$ and $r_r = 2r_s$.

Let's consider a few arrangement examples of the planetary gear train system.

First, we assign the sun gear as the input and the arm as the output. We then attach a clutch to hold the ring gear from rotating so that $\omega_r = 0$.

Equations (4.89) and (4.90) can now be solved to determine the relationship between $\omega_a$ as a function of $\omega_s$.

By adding Equations (4.89) and (4.90) and letting $\omega_r = 0$, we obtain

$$-\omega_s \cdot r_s = -2\omega_a \cdot r_a \tag{4.93}$$

The rotation of the output arm gear is found to be

$$\omega_a = \frac{r_s}{2r_a}\omega_s = \frac{1}{3}\omega_s \tag{4.94}$$

The output rotational speed is one-third of the input rotational speed, which is a speed reduction. If we assume no power loss in the power transmission, then the output torque is amplified three times of the input torque. If the planetary gear train system is used in a vehicle transmission, this could be a first or a second gear.

If the sun gear is chosen as the input, the ring gear as the output, and the arm is locked so that $\omega_a = 0$, Equations (4.89) and (4.90) lead to

$$-\omega_s \cdot r_s - \omega_r \cdot r_r = 0 \qquad (4.95)$$

The rotation of the output ring gear is found to be

$$\omega_r = -\frac{r_s}{r_r}\omega_s = -\frac{1}{2}\omega_s \qquad (4.96)$$

The output speed is one half of the input rotational speed, but it is also reversed. The torque is amplified two times. This is similar to a reversed gear in a vehicle transmission. Note that a typical reverse gear of cars would have a higher gear ratio over 3.

If the arm is chosen as the input, the ring gear as the output, and $\omega_s = 0$, solving Equations (4.89) and (4.90), we obtain

$$-\omega_s \cdot r_s = -2\omega_a \cdot r_a \qquad (4.97)$$

which leads to

$$\omega_r = \frac{2r_a}{r_r}\omega_a = \frac{3}{2}\omega_a \qquad (4.98)$$

The output speed is 1.5 times the input rotational speed. The torque is reduced to two-third that of the input. This is similar to an overdrive in a vehicle transmission.

If two of the gears in a planetary gear system are locked to be at the same speed, then from Equations (4.89) and (4.90), all gears will be rotating at the same speed, making it one rotating rigid body.

The planetary gear train system is widely used in automatic transmission for automobiles. To allow for more design flexibility, several planetary gear trains are coupled. Figure 4.31 is an automatic transmission design based on the widely used Allison 1000[2].

---

[2]McGlothlin, M., 2010, "Allison 1000 Transmission — Inside The Allison 1000, A Look At The History, Strengths, And Weaknesses of GM's Iconic Transmission," Motortrend, March 1.

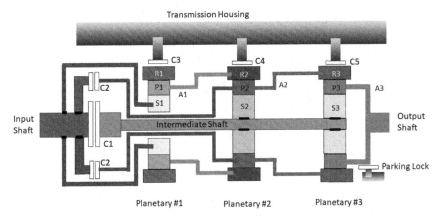

Figure 4.31: The planetary gear systems of an automatic transmission based on Allison 1000.

The transmission shown in Figure 4.31 has three planetary gear train systems. The input shaft from the engine is permanently connected to the sun gear of the first planetary system. When clutch $C1$ is engaged, the input shaft is connected to the sun gears of the second and third planetary systems. When clutch $C2$ is engaged, the input shaft is also connected to the arm of the second planetary system. In addition, the arm of the first planetary system is permanently connected to the ring gear of the second system. The arm of the second system is permanently connected to the ring gear of the third system.

Clutch $C3$, when engaged, keeps the ring gear of the first system from rotating. Similarly, clutch $C4$ controls the ring gear of the second system and clutch $C5$ controls the ring gear of the third system. There is a parking lock which can latch the output shaft when the transmission is in Park. By controlling different clutches, this transmission can produce five different gear ratios for the forward movement and one for reverse.

We have the governing equations for each planetary system as

$$-\omega_{s1} \cdot r_s = -\omega_{a1} \cdot r_a + \omega_{p1} \cdot r_p \tag{4.99}$$

$$-\omega_{r1} \cdot r_r = -\omega_{a1} \cdot r_a - \omega_{p1} \cdot r_p \tag{4.100}$$

$$-\omega_{s2} \cdot r_s = -\omega_{a2} \cdot r_a + \omega_{p2} \cdot r_p \tag{4.101}$$

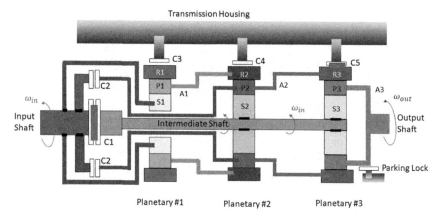

Figure 4.32:   The first gear setting of the transmission system of Figure (4.31).

$$-\omega_{r2} \cdot r_r = -\omega_{a2} \cdot r_a - \omega_{p2} \cdot r_p \qquad (4.102)$$

$$-\omega_{s3} \cdot r_s = -\omega_{a3} \cdot r_a + \omega_{p3} \cdot r_p \qquad (4.103)$$

$$-\omega_{r3} \cdot r_r = -\omega_{a3} \cdot r_a - \omega_{p3} \cdot r_p \qquad (4.104)$$

We will discuss how each gear setting works. We will use the same gear geometry defined by Equations (4.91) and (4.92).

### 4.3.4.1   *First gear of an automatic transmission*

When operating as the first gear, the clutches $C1$ and $C5$ are engaged, as shown in Figure 4.32. The input shaft rotation is fed to all sun gears.

Under this first gear setting, the only planetary system in action is the third system, for which $\omega_{s3} = \omega_{in}$ and $\omega_{r3} = 0$ with $\omega_{a3}$ as the output. Using Equations (4.103) and (4.104), we quickly obtain

$$\omega_{out} = \omega_{a3} = \frac{1}{3}\omega_{in} \qquad (4.105)$$

The actual first gear ratio of Allison 1000 is 3.10. The difference is due to the gear geometry being slightly different from those defined in Equations (4.91) and (4.92).

### 4.3.4.2 *Second gear of an automatic transmission*

At the second gear, clutches $C1$ and $C4$ are engaged, as shown in Figure 4.33. As a result, $\omega_{r2} = 0 = \omega_{a1}$, as well as $\omega_{a2} = \omega_{r3}$. Solving Equations (4.101) and (4.104), we obtain

$$\omega_{\text{out}} = \omega_{a3} = \frac{1}{1.8}\omega_{\text{in}} \qquad (4.106)$$

The actual second gear ratio of Allison 1000 is 1.81.

### 4.3.4.3 *Third gear of an automatic transmission*

For the third gear, clutches $C1$ and $C3$ are engaged, as shown in Figure 4.34. As a result, $\omega_{r1} = 0$, as well as $\omega_{a1} = \omega_{r2}$ and $\omega_{a2} = \omega_{r3}$. Solving Equations (4.99)–(4.104), we obtain

$$\omega_{\text{out}} = \omega_{a3} = \frac{19}{27}\omega_{\text{in}} = \frac{1}{1.42}\omega_{\text{in}} \qquad (4.107)$$

The actual second gear ratio of Allison 1000 is 1.41.

### 4.3.4.4 *Fourth gear of an automatic transmission*

For the fourth gear, clutches $C1$ and $C2$ are engaged, as shown in Figure 4.35. As a result, all planetary systems are running as a solid rigid body system to render the gear ratio as 1.0.

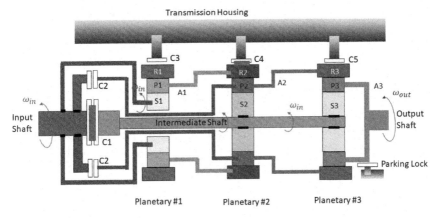

Figure 4.33: The second gear setting of the transmission system of Figure 4.31.

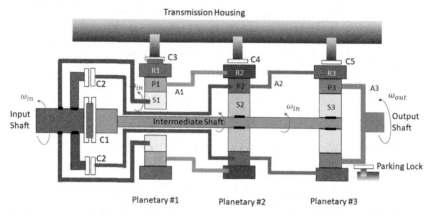

Figure 4.34: The third gear setting of the transmission system of Figure (4.31).

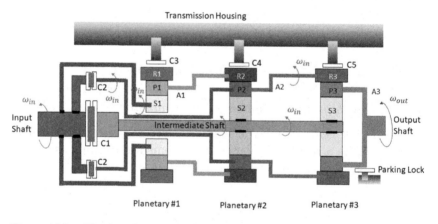

Figure 4.35: The fourth gear setting of the transmission system of Figure 4.31.

### 4.3.4.5   *Fifth gear of an automatic transmission*

In fifth gear, clutches $C2$ and $C3$ are engaged, as shown in Figure 4.36. Note that clutch $C1$ is disengaged. Therefore, $\omega_{a1} = \omega_{r2}$ and $\omega_{a2} = \omega_{r3}$. For the first planetary system, $\omega_{s1} = \omega_{\text{in}}$ and $\omega_{r1} = 0$. From Equations (4.99) and (4.100), we have

$$\omega_{a1} = \frac{1}{3}\omega_{\text{in}} \qquad (4.108)$$

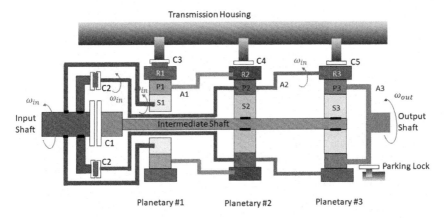

Figure 4.36: The fifth gear setting of the transmission system of Figure 4.31.

For the second planetary system, we have $\omega_{r2} = \omega_{a1} = \frac{1}{3}\omega_{\text{in}}$ and $\omega_{a2} = \omega_{\text{in}}$. Solving Equations (4.101)–(4.102), we obtain

$$\omega_{s2} = \frac{7}{3}\omega_{\text{in}} \qquad (4.109)$$

Note that for the second planetary system, both inputs are not zero.

For the third planetary system, $\omega_{r3} = \omega_{a2} = \omega_{\text{in}}$ and $\omega_{s3} = \omega_{s2} = \frac{7}{3}\omega_{\text{in}}$. Solving Equations (4.103)–(4.104), we finally obtain

$$\omega_{\text{out}} = \omega_{a3} = \frac{13}{9}\omega_{\text{in}} = \frac{1}{0.69}\omega_{\text{in}} \qquad (4.110)$$

The gear ratio at the fifth gear, which is the over-drive, for the Allison 1000 transmission is 0.71 .

### 4.3.4.6 *Reverse gear of an automatic transmission*

At reverse gear, clutch $C3$ and $C5$ are engaged, as shown in Figure 4.37. We have $\omega_{r1} = 0$ and $\omega_{r3} = 0$. Because the arm of the second system is connected to the ring gear of the third system, $\omega_{a2} = 0$. Solving Equations (4.99) and (4.100, then Equations (4.101) and (4.102, and finally Equations (4.103) and (4.104, we obtain

$$\omega_{\text{out}} = \omega_{a3} = \frac{-1}{4.5}\omega_{\text{in}} \qquad (4.111)$$

The actual reverse gear ratio for Allison 1000 is −4.49. Note that the gear reduction at reverse is the largest for all shifts.

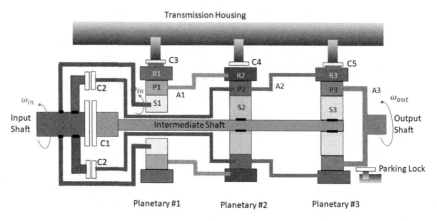

Figure 4.37: The reverse gear setting of the transmission system of Figure 4.31.

### 4.3.4.7 *The planetary system with two non-zero inputs*

From the above calculations, we notice that in some of the settings, both inputs to a planetary system are not zero. In fact, there is no requirement that the second input must be held stationary. The planetary gear system is like a mechanical calculator based on Equations (4.89) and (4.90). We can actually connect three devices to a planetary system. By assigning the rotations of two of them, the rotation of the third device can then be controlled based on a mathematical formula. A system, denoted as the power split device, with two non-zero inputs, was implemented as the transmission of Toyota Prius, a hybrid vehicle (Figure 4.38).

This power split device essentially is a standard planetary gear train system. The internal combustion engine (ICE) is connected to the arm, the first motor/generator (MG1) is connected to the sun gear, and the second motor/generator (MG1) is connected to the ring gear. Each motor/generator can be used either as an electric motor or a generator for charging the battery. The second motor/generator is connected to a shaft gear which is chained to a set of gear trains to the differential of the front wheels for propelling the vehicle. Equations (4.89) and (4.90) become

$$-\omega_{MG1} \cdot r_{MG1} = -\omega_{ICE} \cdot r_{ICE} + \omega_p \cdot r_p \qquad (4.112)$$

$$-\omega_{MG2} \cdot r_{MG2} = -\omega_{ICE} \cdot r_{ICE} - \omega_p \cdot r_p \qquad (4.113)$$

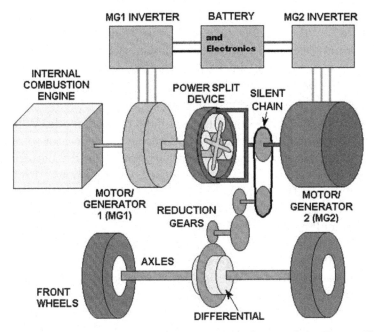

Figure 4.38: A planetary gear train system implemented in Toyota Prius, denoted as power split device, to connect internal combustion engine and two motor/generators together for different operation modes.

*Source*: https://www.google.com/url?sa=i&url=http.

Here is a summary of a few driving modes allowed by the planetary gear system.[3]

- **Light acceleration at low speeds:** At low speeds during light acceleration, the vehicle is powered by MG2 and ICE is shut off. Therefore, $\omega_{ICE} = 0$ and $\omega_{MG2}$ is the input, tied to the vehicle moving speed. From Equations (4.112) and (4.113), we have $\omega_{MG1} = -\frac{r_{MG2}}{r_{MG1}}\omega_{MG2}$, which is used to charge the battery. MG1 may not turn on and simply rotates at the rotation equivalent to the vehicle speed. As a result, all the power from the battery is used to drive the vehicle.

---

[3]https://prius.fandom.com/.

- **Normal driving:** During normal driving, the vehicle is mainly powered by ICE. Under this mode, $\omega_{ICE}$ will operate at a known rotational speed and $\omega_{MG2}$ is also known, tied to the vehicle speed. MG2 is not powered, simply rotates. The output will be $\omega_{MG1}$ to charge the battery. From Equations (4.112) and (4.113), we obtain $\omega_{MG1} = \frac{-r_{MG2}}{r_{MG1}}\omega_{MG2} + \frac{2r_{ICE}}{r_{MG1}}\omega_{ICE}$.

- **Regenerative braking:** This is a light braking mode. When the brake is applied by the driver, MG2 becomes a generator to convert the vehicle kinetic energy to electricity to charge the battery. ICE can continue to run at a known speed and the power from ICE charges the battery via MG1. Under this mode, $\omega_{MG1}$ is the output and is determined the same way as in the normal driving mode.

- **Full acceleration:** At full acceleration for climbing or passing, both ICE and MG2 drive the vehicle. It is similar to the normal driving mode with MG2 powered.

- **B-mode braking:** When the B-mode braking, MG1 is energized to drive ICE, while MG2 is known and tied to the vehicle speed. However, MG2 now functions as a generator to convert the kinetic energy of the vehicle to charge the battery. MG1 spins ICE to cause engine braking and drains the battery. If MG2 generates too much electric current than the battery allows, MG1 spins the ICE faster to consume the excessive electricity. Under this mode, both $\omega_{MG1}$ and $\omega_{MG2}$ are known inputs and $\omega_{ICE}$ is the output as $\omega_{ICE} = \frac{r_{MG1}}{2r_{ICE}}\omega_{MG1} + \frac{r_{MG2}}{2r_{ICE}}\omega_{MG2}$. Additional braking can be engaged via the regular rotors and brake pads.

### 4.3.5   *CAM systems*

In this section, we consider the sliding coupling for connecting two 2D rigid bodies. An important application of the sliding coupling is the CAM system of the internal combustion engine to control the intake and exhaust valves. In Figure 4.39, a CAM, which is a circular disk, rotates with respect to pin $O$ at its rim.

A CAM shaft is allowed to move only horizontally and it is pressed against the disk at points $A$ and $A'$ by a spring. A real CAM usually has a special shape for precise control of the movement of the CAM shaft.

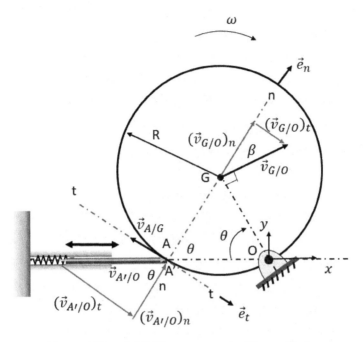

Figure 4.39:   A CAM arrangement with no offset.

If the CAM shaft is aligned with pin $O$, as shown in Figure 4.39, and the motion of the cam shaft can be easily solved via geometry.

$$x_{A'} = -2R\cos\theta \qquad (4.114)$$

Taking a derivative of Equation (4.114), we quickly obtain

$$v_{A'} = \dot{x}_{A'} = 2R\sin\theta\dot{\theta} = 2R\omega\sin\theta \qquad (4.115)$$

However, if the cam shaft is not aligned with pin $O$, such simple geometric information may not be available. The solution via the geometric relationship is an *ad hoc* approach. We would prefer to conduct the CAM motion analysis based on the rigid body kinematics we have presented in this chapter. The procedures might be more complex, but we will be able to solve for many different arrangements of CAM systems.

Based on Figure 4.39, we will first determine the velocity of point $A$ based on the rigid body of the disk. At the sliding contact of points $A$ and $A'$, we can define a tangential and a normal direction as shown in Figure 4.39.

to maintain the contact, the velocity component of $\vec{v}_A$ and $\vec{v}_{A'}$ in the normal direction should be the same,

$$(\vec{v}_{A'})_n = (\vec{v}_A)_n \tag{4.116}$$

However, there is no relationship between the tangential velocity components.

For this particular CAM system, we know that the direction of $\vec{v}_{A'}$ is horizontal. Therefore, knowing $(\vec{v}_{A'})_n$, we can determine $\vec{v}_{A'}$.

The velocity of point $A$ is

$$\vec{v}_{A/O} = \vec{v}_{G/O} + \vec{v}_{A/G} = \vec{\omega} \times \vec{r}_{G/O} + \vec{\omega} \times \vec{r}_{A/G} \tag{4.117}$$

Because $\vec{v}_{A/G}$ is only in the tangential direction, we have

$$(\vec{v}_{A/O})_n = (\vec{v}_{G/O})_n = R\,\omega\cos\beta\,\vec{e}_n = (\vec{v}_{A'/O})_n \tag{4.118}$$

where $\beta = 2\theta - \frac{\pi}{2}$ from the geometry.

The speed of point $A'$ becomes

$$v_{A'/O} = \frac{2R\,\omega\cos\beta}{\cos\theta} = 2R\,\omega\sin\theta \tag{4.119}$$

We obtain the same result as in Equation (4.115). However, we can obtain additional information based on this kinematic method. We can calculate the speed difference in the tangential velocities for points $A$ and $A'$. This difference is the sliding speed, which is critical to the wear between the CAM and the CAM shaft.

The tangential velocity of point $A$ is

$$(\vec{v}_{A/O})_t = (\vec{v}_{G/O})_t + \vec{v}_{A/G} = (R\omega\sin\beta - R\omega)\,\vec{e}_t \tag{4.120}$$

and the tangential velocity of point $A'$ is

$$(\vec{v}_{A'/O})_t = v_{A'/O}\sin\theta\,\vec{e}_t = 2R\,\omega\sin^2\theta\,\vec{e}_t \tag{4.121}$$

The sliding speed is found to be

$$v_{\text{sliding}} = |(\vec{v}_{A'/O})_t - (\vec{v}_{A/O})_t| = R\omega|(\cos(2\theta) + 1 + 2\sin^2\theta)| \tag{4.122}$$

The maximum sliding speed happens at $\theta = 0$ and $\frac{\pi}{2}$ as $2R\,\omega$.

We can also observe this sliding speed based on the relative motion analysis with a rotating reference frame. Note that if we establish a reference frame at point $A$ aligned with the normal and tangential direction at the instant shown in Figure (4.39), point $A'$ will appear to have a velocity along the tangential direction. However, note that this tangential-normal reference frame is rotating. Assume that the frame rotation is $\vec{\Omega}$ observed by an inertial reference frame, the motion of $A'$ can be established as

$$\vec{v}_{A'/O} = \vec{v}_{A/O} + \vec{\Omega} \times \hat{r}_{A'/A} + \hat{v}_{A'/A} \tag{4.123}$$

$$\vec{a}_{A'/O} = \vec{a}_{A/O} + \vec{\Omega} \times (\vec{\Omega} \times \hat{r}_{A'/A}) + \dot{\vec{\Omega}} \times \hat{r}_{A'/A} + 2\vec{\Omega} \times \hat{v}_{A'/A} + \hat{a}_{A'/A} \tag{4.124}$$

Note that $\hat{r}_{A'/A} = 0$, the above two equations are reduced to

$$\vec{v}_{A'/O} = \vec{v}_{A/O} + \hat{v}_{A'/A} = \vec{v}_{A/O} + (\vec{v}_{A'/A})_t \tag{4.125}$$

$$\vec{a}_{A'/O} = \vec{a}_{A/O} + 2\vec{\Omega} \times \hat{v}_{A'/A} + \hat{a}_{A'/A} \tag{4.126}$$

From Equation (4.126), we cannot define the acceleration relationship between $A'$ and $A$ based on the contact condition, only the velocity relationship.

Note that $\hat{v}_{A'/A} = (\vec{v}_{A'/A})_t$ The sliding speed can be easily determined with Equation (4.125) if we know $\vec{v}_{A/O}$ and $\vec{v}_{A'/O}$.

As for Equation (4.126), it is more complicated. In general, we do not know $\vec{\Omega}$, the frame rotating velocity and $\hat{a}_{A'/A}$. Note that even though the tangential–normal reference frame observes a tangential velocity, it does not mean that the relative motion to this frame is a straight-line motion. As a result, $\hat{a}_{A'/A}$ can have components along both the tangential and normal directions, while the Coriolis acceleration, $2\vec{\Omega} \times \hat{v}_{A'/A}$, is along the normal direction. As a result, Equation (4.126) is not very useful to determine the acceleration of $\vec{a}_{A'/O}$. However, if desired, we can simply take a time derivative of $\vec{v}_{A'/O}$ to obtain the acceleration.

From Equation (4.126), we cannot define the acceleration relationship between $A$ and $A'$ based on the contact condition, only the velocity relationship.

Now, let's consider the CAM of Figure 4.40. We can follow the same strategy to determine $\vec{v}_{A'/O}$ although there are a few extra

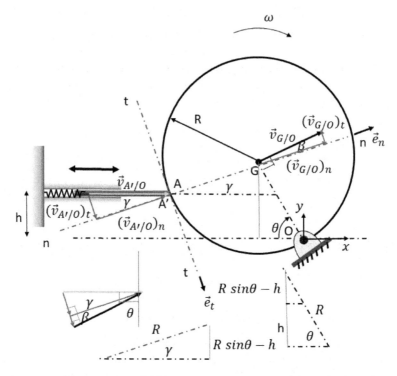

Figure 4.40: A CAM arrangement with an offset $h$.

angles involved. From Figure 4.40,

$$(\vec{v}_{A/O})_n = (\vec{v}_{G/O})_n = R\,\omega\cos\beta\,\vec{e}_n = (\vec{v}_{A'/O})_n \qquad (4.127)$$

and

$$v_{A'/O} = \frac{R\,\omega\cos\beta}{\cos\gamma} \qquad (4.128)$$

To obtain $\beta$ and $\gamma$, we need to refer to a few triangles shown in Figure 4.40. From them, we obtain $\gamma = \sin^{-1}\left(\frac{R\sin\theta - h}{R}\right)$ and $\beta = \frac{\pi}{2} - \gamma - \theta$.

To verify this result, if we let $h = 0$, we have $\gamma = \theta$ and $\beta = \frac{\pi}{2} - 2\theta$, as before. We need to change the sign of $\beta$ to keep it between $0°$ and $90°$.

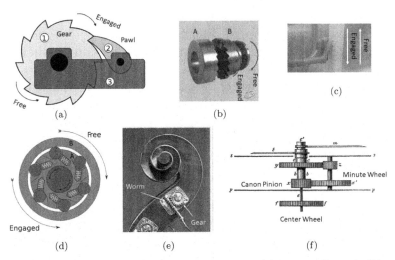

Figure 4.41: Different one-way coupling designs: (a) gear and pawl; (b) star ratchets; (c) cable tie; (d) sprag clutch; (e) worm and gear; and (f) canon pinion for clock work.

*Source*: (a) https://upload.wikimedia.org/wikipedia/commons/thumb/b/b7/Ratchet_Drawing.svg/1200px-Ratchet_Drawing.svg.png; (b) https://i.stack.imgur.com/3SMm2.jpg; (c) https://en.wikipedia.org/wiki/Ratchet_(device)#/media/File:Comealong.jpg; (d) https://upload.wikimedia.org/wikipedia/commons/thumb/c/c2/Freewheel_en.svg/1200px-Freewheel_en.svg.png; (e) https://upload.wikimedia.org/wikipedia/commons/c/c3/Worm_Gear.gif; and (f) https://en.wikipedia.org/wiki/Wheel_train.

### 4.3.6 *One way coupling*

There are many ingenious couplings developed over the years used in cars, watches, cameras, etc. Figure 4.41 lists a few interesting examples.

Figure 4.41(a) illustrates a gear and a pawl. The gear can only rotate in one direction. This mechanism is commonly used in ratchet wrenches of hand tools. Figure 4.41(b) shows a pair of star ratchets. If ratchet $A$ is fixed, then ratchet $B$ can only rotate in one way. This design is commonly used in bicycles and the winding system of mechanical watches. In Figure 4.41(c), it is a cable tie, allowing tightening in one direction. The sprag clutch of Figure 4.41(d) allows wheel $B$ to rotate freely in the clockwise direction with respect to wheel $A$, but is locked with $A$ when $B$ rotates counter clockwise. The tensioning devices used in musical instruments are a worm paired with a gear as shown in Figure 4.41(e). The worm can turn the gear, but not the other way around. Finally, Figure 4.41(f) illustrates the gear train of a clockwork. The minute hand is mounted on a canon pinion, which could be driven by the center wheel for time keeping, or be driven by a minute wheel for manual adjustment. The coupling between the center wheel and the canon pinion is via a frictional fit. When adjusting time of a clock or a watch, the canon pinion is forced to turn, overcoming the friction with the center wheel. One common problem with clocks and watches is a loose canon pinion. When the canon pinion is too loose, the clock or watch still runs properly, but the minute and hour hands would either not turn or turn too slowly. One way to improve this design is by using the planetary gear system to replace the frictional coupling so that the minute hand can be driven by two different rotations. However, such a design was probably not used in watches because of the complexity and cost.

## 4.4 Kinematics of 3D Rigid Body Assemblies

In the previous sections, we discussed mostly 2D rigid body assemblies. For some 2D assemblies, if we arrange the components in different orientations, they become 3D rigid body assemblies. We will discuss two important devices, differentials and universal joints, used in automobiles.

### 4.4.1 *Differentials*

A differential (Figure 4.42) is required in a vehicle to allow for proper cornering. The main components in a differential include: the driving gear, connected with the drive shaft of the vehicle; a crown gear, which is engaged with the driving gear; two half shafts, each connected to a wheel; two sun gears, each connected to a half shaft; and several planet gears between the two sun gears.

In fact, a differential is a 3D version of a planetary gear system. The conversion from a 2D planetary system to the 3D differential is shown in Figure 4.43. The arm is always defined as the input. This arm in a differential also contains a crown gear, engaged with the driving shaft. The planet gears face each other in a differential. The ring gear and the sun gear are both called sun gears and are of

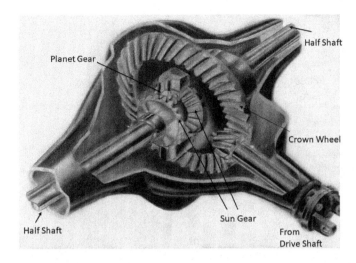

Figure 4.42:  The main components of an automobile differential.

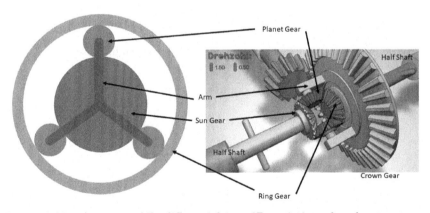

Figure 4.43:  An automobile differential is a 3D variation of a planetary gear train system.
*Source*: Modified image based on https://i.vimeocdn.com/video/463772792_640. jpg.

the same size. The sun gears in a differential are the outputs, each connected to a half shaft.

The kinematic analysis of a differential follows the similar analysis of a planetary gear system. As shown in Figure 4.44, the velocities of points $A$ and $A'$ must be the same, same with points $B$ and $B'$.

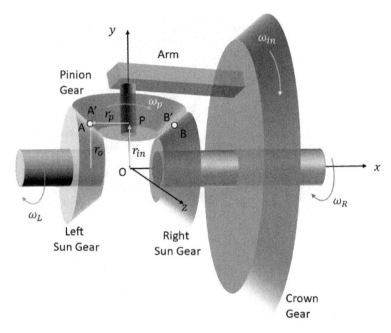

Figure 4.44:   The kinematic analysis of an open differential.

We obtain the following two equations:

$$\vec{v}_{A/O} = \vec{\omega}_L \times \vec{r}_{A/O} = \vec{v}_{A'/O} = \vec{v}_{P/O} + \vec{\omega}_p \times \vec{r}_{A'/P}$$
$$= \vec{\omega}_{\text{in}} \times \vec{r}_{P/O} + \vec{\omega}_p \times \vec{r}_{A'/P} \qquad (4.129)$$

$$\vec{v}_{B/O} = \vec{\omega}_R \times \vec{r}_{B/O} = \vec{v}_{B'/O} = \vec{v}_{P/O} + \vec{\omega}_p \times \vec{r}_{B'/P}$$
$$= \vec{\omega}_{\text{in}} \times \vec{r}_{P/O} + \vec{\omega}_p \times \vec{r}_{B'/P} \qquad (4.130)$$

Working out Equations (4.129) and (4.130) and express them in scalars, with all velocities in the $z$ direction, we obtain

$$\omega_L \cdot r_0 = \omega_{\text{in}} \cdot r_0 + \omega_p \cdot r_p \qquad (4.131)$$

$$\omega_R \cdot r_0 = \omega_{\text{in}} \cdot r_0 - \omega_p \cdot r_p \qquad (4.132)$$

where $r_{\text{in}} = r_0$.

Equations (4.131) and (4.132) contain three unknowns, $\omega_p$, $\omega_R$, and $\omega_L$. With only two equations, we cannot define all of them.

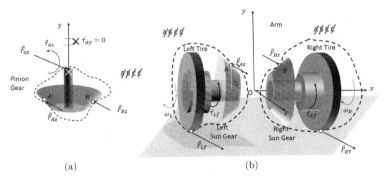

Figure 4.45: The force analysis of an open differential: (a) FBD for the pinion gear; and (b) FBD for the half shaft/tire.

Adding Equations (4.131) and (4.132), we have

$$\omega_R + \omega_L = 2\omega_{\text{in}} \qquad (4.133)$$

From the above equations, we know that if $\omega_p = 0$, then $\omega_R = \omega_L$. This is when the vehicle moves forward. If the vehicle makes a right turn, $\omega_L = \omega_{\text{in}} + \omega_p \frac{r_p}{r_0}$ and $\omega_R = \omega_{\text{in}} - \omega_p \frac{r_p}{r_0}$. The left tire will rotate faster while the right tire is reduced by the same amount. It is reversed when making a left turn.

The device shown in Figures 4.42–4.44 is called an OPEN DIFFERENTIAL. We need to conduct a force analysis to explain the reason for this terminology. As shown in Figure 4.45(a), an FBD is constructed using the ABCC method. We have two contacts at points $A$ and $B$ and one constraint on the arm.

The moment equation around the $y$ axis is

$$\sum \tau_{ay} = F_{Az} r_P - F_{Bz} r_P = I_P \alpha_P \qquad (4.134)$$

where $I_P$ and $\alpha_P$ are the moment of inertia and the angular acceleration of the planet gear, respectively.

Because the shaft of the planet gear can rotate freely inside the arm bore and the planet gear has a small moment of inertia, the total moment for the pinion gear in the $y$ direction can be assumed to be zero. As a result, for an open differential, we have

$$F_{Az} = F_{Bz} = F_z \qquad (4.135)$$

From Figure 4.45(b), Equation (4.135) dictates that the driving torque to each half shaft be the same.

$$\tau_R = \tau_L = F_z \cdot r_0 \qquad (4.136)$$

Neglecting the moment of inertia of the wheel/tire assembly, the total moment around the $x$ axis for each half shaft and tire assembly needs to be nearly zero. The tractive forces, $F_{RT}$ and $F_{LT}$, from ground to push the right and left tires are the same,

$$F_{RT} = F_{LT} = \frac{r_0}{r_t} F_z \qquad (4.137)$$

where $r_t$ is the tire radius.

Note that the tractive forces are related to the friction condition between the tire and the ground. When, for example, the left tire hits a puddle of oil, causing $F_{LT} \simeq 0$, $F_{Az} \simeq 0$, it results in $F_{Bz} \simeq 0$ and finally, $F_{RT} \simeq 0$. Because the right tire is in the rolling-without-slipping situation, it will maintain its speed (could be zero), while the left tire will spin fast, draining the power from the engine. The vehicle will not accelerate as a result.

In summary, an open differential allows the speeds of each half shaft to be different, but at the price of the torques being the same.

To remedy this situation, many limited-slip differentials were developed (4.46a). We will discuss the design of these limited-slip differentials in Chapter 9. Their underlined principle is to make $\sum \tau_{ay} \neq 0$ so that it allows for $F_{Az}$ to be different form $F_{Bz}$. To achieve this, the driver can activate a locking device so that the planet gear is not allowed to rotate, i.e., $\omega_p = 0$, which forces $\omega_R = \omega_L$. This is a locked differential, as opposed to the open differential. The driver should only lock the differential after the vehicle has come to a complete stop. The differential should be unlocked once the vehicle moves properly again.

Another approach is to make the planet gear not free rotating and not being locked up entirely. The most common approach for limited slip differentials is to transmit the driving torque to the left and the right half shaft through clutch plates, as shown in Figure 4.46(a). The speeds of the left and the right can still be different, but not by much. The drawback is the wear of the clutch plates over time.

One ingenious limited-slip differential design is the Torsen Differential Figure 4.46(b), which uses the worm/gear coupling to achieve

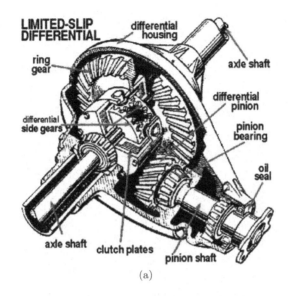

(a)

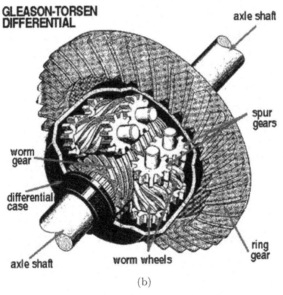

(b)

Figure 4.46: (a) A conventional limited-slip differential; and (b) Torsen differential.

*Source*: https://members.rennlist.com/951_racerx/PS84Gleason.html.

automatic mode switching between the open and locked differentials within one design.[4] The readers are encouraged to read the cited reference for this ingenious design. Note that the "worm wheels" in Figure 4.46(b) are actually the gear, while the "worm gear" is actually the worm in a worm/gear pair. If one side of the tire starts to slip, its half shaft (axle shaft) and the worm connected to this side of the tire will drive the connected spur gear, which in turn will drive the spur gear and the gear (worm wheel) of the other side. However, The gear (worm wheel) of the other side cannot drive its mating worm. The end result is that the tires are locked together without slipping. When making a turn, one side can increase while the other side will decrease its speed, the Torsen differential will function like an open differential. No control devices are needed and the Torsen differential will act like a locked differential or an open differential as needed. The Torsen differential can be considered as an art of mechanical design.

Finally, a limited-slip differential is also used to distribute power to the front and rear tires, achieving the all-wheel-drive capability.

### 4.4.2  *Universal joints and constant velocity joints*

When transmitting power from one device to another, often the shafts of these devices are not aligned to the same axis of rotation. Even if two shafts are designed to be co-axial, there could be misalignment due to lack of precision or due to distortion caused by forces and vibrations during operation. As a result, between shafts, a device, commonly known as universal joint, is used to connect the shafts, as shown in Figure 4.47.[5]

As shown in Figures 2.30 and 4.47, a drive shaft is placed in between the input and the output shafts via two universal joints. Even if both the input and output shafts rotate at the same constant speed, the drive shaft in between does not. The speed variation of

---

[4]Chocholek, S.E., 1988, "The development of a differential for the improvement of traction control," IMechE, C368/88, http://www.zhome.com/ZCMnL/tech/Torsen/Torsen.htm.

[5]Vesali, F., Rezvani, M.A., and Kashfi, M. Dynamics of universal joints, its failures and some propositions for practically improving its performance and life expectancy. *Journal of Mechanical Science and Technology* **26**(8) (2012) 2439–2449. DOI: 10.1007/s12206-012-0622-1.

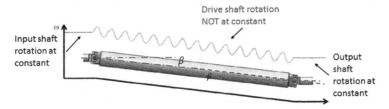

Figure 4.47: An input shaft and an output shaft are connected with a drive shaft fitted with a universal joint on each end.
*Source*: Vesali, F., Rezvani, M.A., and Kashfi, M. Dynamics of universal joints, its failures and some propositions for practically improving its performance and life expectancy. *Journal of Mechanical Science and Technology* **26**(8) (2012) 2439–2449. DOI 10.1007/s12206-012-0622-1.

the drive shaft causes vibrations and it gets worse with an increasing inclined angle, $\beta$.

We will conduct kinematic analysis for universal joints to explain why this could happen.

The design and force analysis of a universal joint are presented in Figures 2.29 and 2.30. The main component of an universal joint is the cross-member, also denoted as the journal cross, fitted with four bearings.

As shown in Figure 4.48, the input shaft axis ($z$) is perpendicular to arm $AB$. The rotation of arm $AB$ forms plane $AB$, which is also defined by the $x-y$ plane. Initially, arm $OA$ is aligned with the $x$ axis. A short time later, arm $AB$, rotating on plane $AB$, shifts by an angle $\theta_{\text{in}}$ from the $x$ axis.

Similarly, the output shaft axis, $\tilde{z}$, is perpendicular to the arm $CD$. The rotation of the arm $CD$ forms the plane $CD$, which is defined by the $\tilde{x}-\tilde{y}$ plane. Initially, arm $OC$ is aligned with the $\tilde{x}$ axis. A short time later, arm $CD$, rotating on plane $CD$, shifts by an angle $\theta_{\text{out}}$ from the $\tilde{x}$ axis.

The transformation between these two reference frames can be established as

$$
\begin{bmatrix} \vec{i} \\ \vec{j} \\ \vec{k} \\ 1 \end{bmatrix} = T_{tz}(-90^\circ) \cdot T_{tx}(-\beta) \begin{bmatrix} \tilde{i} \\ \tilde{j} \\ \tilde{k} \\ 1 \end{bmatrix}
\tag{4.138}
$$

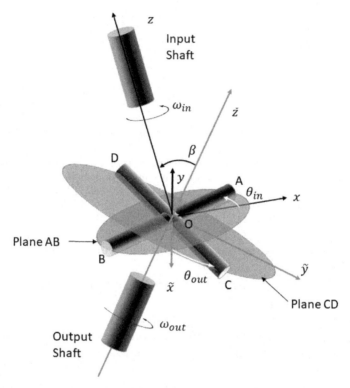

Figure 4.48: Kinematic analysis of a universal joint between the input axis and the output axis.

As discussed in Section 3.10.2, Equation (4.138) should be constructed by moving a little coordinate from $O-x-y-z$ to $O-\tilde{x}-\tilde{y}-\tilde{z}$. As indicated by Equation (4.138), we first rotate a little coordinate around the $z$ axis by $-90°$ to align $x$ of the little coordinate to $\tilde{x}$ ($T_{tz}(-90°)$). Next, the little coordinate rotates with respect to its current $x$ axis (i.e., $\tilde{x}$) by $-\beta$ ($T_{tx}(-\beta)$) so that the little coordinate is now aligned with $O - \tilde{x} - \tilde{y} - \tilde{z}$. After the second movement, we arrive at $[\tilde{i}\ \tilde{j}\ \tilde{k}\ 1]'$.

From Equation (4.138), the unit vectors of both reference frames are related by

$$\vec{i} = \cos\beta\tilde{j} + \sin\beta\tilde{k} \tag{4.139}$$

$$\vec{j} = -\tilde{i} \tag{4.140}$$

$$\vec{k} = -\sin\beta\tilde{j} + \cos\beta\tilde{k} \tag{4.141}$$

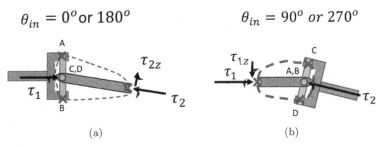

$$\theta_{in} = 0^o \, or \, 180^o \qquad\qquad \theta_{in} = 90^o \, or \, 270^o$$

(a)          (b)

Figure 4.49: Force analysis of a universal joint at different orientations: (a) 0 and 180°; and (b) 90 and 270°.

It is interesting to note that the normal direction of the cross-member, perpendicular to both arms $AB$ and $CD$, is not fixed in space. When $OA$ is aligned with the $x$ axis and $OC$ with the $\tilde{x}$ axis, the normal direction of the cross member is aligned with the input axis $z$. On the other hand, when $OA$ is aligned with the $y$ axis and $OC$ with the $\tilde{y}$ axis, the normal direction of the cross member is aligned with the output axis $\tilde{z}$. For all other orientations, the normal direction of the cross member will lie in between the input and the output axes.

Figure 4.49 illustrates the direction of the cross member, along with the resulting torques. As shown in Figure 4.49(a), when the cross member is aligned with the input shaft, the driving torque ($\tau_1$) is transmitted fully to the cross member, but because the drive shaft is not tilted at an angle, a bending moment, ($\tau_{2z}$), will be induced at the drive shaft in addition to the torque, $\tau_2$. Similarly, in Figure 4.49(b), when the cross member is aligned with the drive shaft, there will be no bending moment on the drive shaft, but the input shaft is now subjected to a bending moment, $\tau_{1z}$. The bending moment, $\tau_{2z}$ will cause the drive shaft to lash around if one end of the drive shaft breaks loose, which will be discussed later.

We would like to establish a relationship between $\omega_{\text{in}}$ and $\omega_{\text{out}}$. We will follow the same method used for rigid body motion analysis discussed in this chapter.

As defined in Figure 4.48, the reference frames $x-y-z$ and $\tilde{x}-\tilde{y}-\tilde{z}$ are both inertial references frames with the origin at point $O$, which is a stationary point.

There are three rigid bodies: the input shaft, the cross member, and the output shaft. The cross member is pinned to the input shaft at points $A$ and $B$, while pinned to the output shaft at points $C$ and $D$.

The input shaft rotates around a fixed axis, or its axis of rotation is fixed, defined as the $z$ axis. Similarly, the output shaft's fixed axis of rotation is the $\tilde{z}$ axis. The cross member is actually rotating around a fixed point, which is its center, $O$. We define the angular velocity of these three rigid bodies as $\omega_{\text{in}}$, $\omega_U$, and $\omega_{\text{out}}$, for the input shaft, the cross member, and the output shaft, respectively. Based on Figure 4.48, the angular velocity vectors are

$$\vec{\omega}_{\text{in}} = \omega_{\text{in}}\vec{k} \tag{4.142}$$

$$\vec{\omega}_U = \omega_{Ux}\vec{i} + \omega_{Uy}\vec{j} + \omega_{Uz}\vec{k} \tag{4.143}$$

$$\vec{\omega}_{\text{out}} = \omega_{\text{out}}\tilde{k} \tag{4.144}$$

Note that $\vec{\omega}_{\text{out}}$ is expressed in $\tilde{k}$.

We would like to establish a relationship between $\omega_{\text{out}}$ and $\omega_{\text{in}}$, which has been published and is widely available. It will be useful if we can also obtain an expression for $\vec{\omega}_U$, which is rarely addressed in the literature.

We will first establish the position vectors for points $A$ and $C$ using the moving coordinate technique discussed earlier.

$$\vec{r}_{A/O} = \begin{bmatrix} x_A \\ y_A \\ z_A \\ 1 \end{bmatrix} = T_{tz}(\theta_{\text{in}}) \cdot T_x(a) \begin{bmatrix} 0 \\ 0 \\ 0 \\ 1 \end{bmatrix} = \begin{bmatrix} a\cos\theta_{\text{in}} \\ a\sin\theta_{\text{in}} \\ 0 \\ 1 \end{bmatrix} \tag{4.145}$$

$$\vec{r}_{C/O} = \begin{bmatrix} x_C \\ y_C \\ z_C \\ 1 \end{bmatrix} = T_{tz}(-90^o) \cdot T_{tx}(-\beta) \cdot T_{tz}(\theta_{\text{out}}) \cdot T_x(a) \begin{bmatrix} 0 \\ 0 \\ 0 \\ 1 \end{bmatrix}$$

$$= \begin{bmatrix} a\sin\theta_{\text{out}}\cos\beta \\ -a\cos\theta_{\text{out}} \\ -a\sin\theta_{\text{out}}\sin\beta \\ 1 \end{bmatrix} \tag{4.146}$$

where $T_x(a)$ is a transformation matrix for moving the little reference coordinate along the current $x$ axis by a distance of $a$. The definitions of displacement transformation matrices are

$$T_x(a) = \begin{bmatrix} 1 & 0 & 0 & a \\ 0 & 1 & 0 & 0 \\ 0 & 0 & 1 & 0 \\ 0 & 0 & 0 & 1 \end{bmatrix} \tag{4.147}$$

$$T_y(a) = \begin{bmatrix} 1 & 0 & 0 & 0 \\ 0 & 1 & 0 & a \\ 0 & 0 & 1 & 0 \\ 0 & 0 & 0 & 1 \end{bmatrix} \tag{4.148}$$

$$T_z(a) = \begin{bmatrix} 1 & 0 & 0 & 0 \\ 0 & 1 & 0 & 0 \\ 0 & 0 & 1 & a \\ 0 & 0 & 0 & 1 \end{bmatrix} \tag{4.149}$$

Because $\vec{r}_{A/O}$ and $\vec{r}_{C/O}$ are always perpendicular, we obtain

$$\vec{r}_{A/O} \cdot \vec{r}_{C/O} = \cos\theta_{\text{in}} \sin\theta_{\text{out}} \cos\beta - \sin\theta_{\text{in}} \cos\theta_{\text{out}} = 0 \tag{4.150}$$

From Equation (4.150), we obtain a relationship between $\theta_{\text{in}}$ and $\theta_{\text{out}}$ as

$$\tan\theta_{\text{in}} = \cos\beta \tan\theta_{\text{out}} \tag{4.151}$$

By taking derivatives Equation (4.151), we can obtain the relationship between the input rotation and the output rotation as

$$\omega_{\text{out}} = \left( \frac{1}{1 - \sin^2\beta \cos^2\theta_{\text{in}}} \right) \cos\beta \, \omega_{\text{in}} \tag{4.152}$$

This is the well-known result via calculus. We would like to derive the relationship of Equation (4.152) using the same kinematic method we have discussed in this chapter and, along the way, we can also determine the angular velocity of the cross member, not available via the calculus method.

We first determine the velocity of point $A$ on the input shaft and $A'$ on the cross member. Points $A$ and $A'$ share the same position in space.

$$\vec{v}_{A/O} = \vec{\omega}_{\text{in}} \times \vec{r}_{A/O} = \vec{v}_{A'/O} = \vec{\omega}_U \times \vec{r}_{A'/O} \qquad (4.153)$$

Note that $\vec{r}_{A/O} = \vec{r}_{A'/O}$.
Introducing Equations (4.144) and (4.145) into (4.153), we obtain

$$-a\omega_{\text{in}} \sin \theta_{\text{in}} = -a\omega_{Uz} \sin \theta_{\text{in}} \qquad (4.154)$$

$$a\omega_{\text{in}} \cos \theta_{\text{in}} = a\omega_{Uz} \cos \theta_{\text{in}} \qquad (4.155)$$

$$0 = a\,\omega_{Ux} \sin \theta_{\text{in}} - a\,\omega_{Uy} \cos \theta_{\text{in}} \qquad (4.156)$$

From the above equations, we have

$$\omega_{Uz} = \omega_{\text{in}} \qquad (4.157)$$

$$\omega_{Uy} = \frac{\sin \theta_{\text{in}}}{\cos \theta_{\text{in}}}\omega_{Ux} \qquad (4.158)$$

We can now express $\vec{\omega}_U$ as

$$\vec{\omega}_U = \omega_{Ux}\vec{i} + \frac{\sin \theta_{\text{in}}}{\cos \theta_{\text{in}}}\omega_{Ux}\vec{j} + \omega_{\text{in}}\vec{k} \qquad (4.159)$$

We still need to determine $\omega_{Ux}$ in the above equation. Let's consider the velocity of point $C$.

$$\vec{v}_{C'/O} = \vec{\omega}_U \times \vec{r}_{C'/O} = \vec{v}_{C/O} = \vec{\omega}_{\text{out}} \times \vec{r}_{C/O} \qquad (4.160)$$

Introducing Equations (4.145), (4.151), and (4.159) into (4.160), we obtain

$$\vec{v}_{C'/O} = \left(-a \sin \beta \cos \beta \frac{\sin^2 \theta_{\text{out}}}{\cos \theta_{\text{out}}}\,\omega_{Ux} + a \cos \theta_{\text{out}}\,\omega_{\text{in}}\right)\vec{i}$$

$$+ \left(a \cos \beta \sin \theta_{\text{out}}\,\omega_{\text{in}} + a \sin \beta \sin \theta_{\text{out}}\,\omega_{Ux}\right)\vec{j}$$

$$+ \left(-a \cos \theta_{\text{out}}\,\omega_{Ux} - a \cos^2 \beta \frac{\sin^2 \theta_{\text{out}}}{\cos \theta_{\text{out}}}\,\omega_{Ux}\right)\vec{k}$$

$$= \vec{v}_{C/O} = -a \sin \theta_{\text{out}}\,\omega_{\text{out}}\tilde{i} + a \cos \theta_{\text{out}}\,\omega_{\text{out}}\tilde{j} \qquad (4.161)$$

We need to convert all the unit vectors in $x-y-z$ to those for $\tilde{x}-\tilde{y}-\tilde{z}$ for the above equation using Equation (4.141). The $\tilde{k}$ component of $\vec{v}_{C/O}$ is zero, which leads to

$$
\left( -a \sin\beta \cos\beta \frac{\sin^2\theta_{\text{out}}}{\cos\theta_{\text{out}}} \, \omega_{Ux} + a \cos\theta_{\text{out}} \, \omega_{\text{in}} \right) \sin\beta
$$

$$
+ \left( -a \cos\theta_{\text{out}} \, \omega_{Ux} - a \cos^2\beta \frac{\sin^2\theta_{\text{out}}}{\cos\theta_{\text{out}}} \, \omega_{Ux} \right) \cos\beta = 0 \quad (4.162)
$$

From Equation (4.162), the relationship between $\omega_{\text{in}}$ and $\omega_{Ux}$ is found to be

$$
\omega_{Ux} = \left( \frac{\sin\beta}{\tan^2\theta_{\text{in}} + \cos\beta} \right) \omega_{\text{in}} \quad (4.163)
$$

Finally, from the $\tilde{i}$ component of $\vec{v}_{C/O}$, we establish the relationship between $\omega_{\text{in}}$ and $\omega_{\text{out}}$ as

$$
\omega_{\text{out}} = \left( \cos\beta + \frac{\sin^2\beta}{\tan^2\theta_{\text{in}} + \cos\beta} \right) \omega_{\text{in}} \quad (4.164)
$$

Equations (4.164) and (4.152) look different, but it can be proven numerically that they are identical.

Figure 4.50 is the speed ratio of $\omega_{\text{out}}/\omega_{\text{in}}$ for different values of $\beta$, plotted using both Equations (4.164) and (4.152). The results are identical.

From Figure 4.50, it is apparent that higher inclined angles will contribute to higher speed variations of the output shaft. We can take one more derivative of Equations (4.164) or (4.152) to obtain the angular acceleration relationship, which will be an exercise for readers and it is well documented in the literature.

Based on the kinematic approach, we obtained Equation (4.163) and, therefore, the angular velocity of the cross member, which cannot be obtained by the calculus method. Introducing Equation (4.163) into (4.159), we have the explicit expression of $\vec{w}_U$. Because the center of the cross member, $O$, is stationary, the instantaneous axis of rotation will originate from point $O$ and has the same direction as $\vec{w}_U$. The unit directional vector of $\vec{w}_U$ is found as $\vec{w}_U/|\vec{w}_U|$. The instantaneous axis of rotation at different input angles is plotted in Figure 4.51. The angle of the instantaneous axis of rotation to the

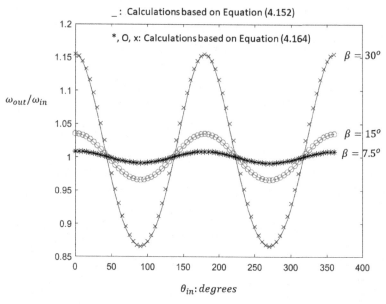

Figure 4.50: The speed variations of the output shaft at different inclined angles with a constant input shaft rotational speed.

$z$ is defined as $\phi_U$ and it varies at different input angles. The projection of the axis of rotation on the $x-y$ plane follows the input angle; therefore, it moves with arm $OA$. As shown in Figure 4.51, when $\theta_{\text{in}} = 0°$, $\phi_U = \beta = 30°$. The value of $\phi_U$ reduces to 0 when $\theta_{\text{in}} = 90°$. The pattern repeats.

We can also calculate the angular velocity components along arms $OA$ and $OC$. The rotation along the arm exhibits itself as a spinning motion of the arm inside the supporting bracket. To allow this spinning action, needle bearings are used. The spinning angular velocities can be found as the component of $\vec{\omega}_{\text{in}}$ along $\vec{r}_{OA}$ or $\vec{r}_{OC}$,

$$\omega_{sA} = \frac{\vec{r}_{OA}}{|\vec{r}_{OA}|} \cdot \vec{\omega}_{\text{in}}$$

$$\omega_{sC} = \frac{\vec{r}_{OC}}{|\vec{r}_{OC}|} \cdot \vec{\omega}_{\text{in}} \tag{4.165}$$

The values of $\omega_{sA}$ and $\omega_{sC}$, along with $\omega_{\text{out}}$ and $\omega_U$, are plotted in Figure 4.52.

As shown in Figure 4.52, the output shaft speed varies the same way as predicted in Figure 4.50 between $\omega_{\text{in}}/\cos\beta$ when $\theta_{\text{in}} = $

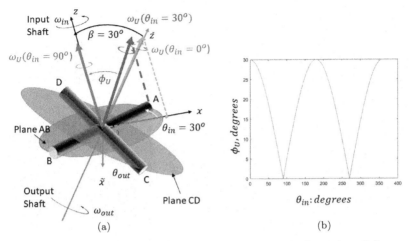

Figure 4.51: (a) The orientation of the instantaneous axis of rotation of the cross member and (b) the simulated orientation change with respect to the input shaft at different input rotating angles.

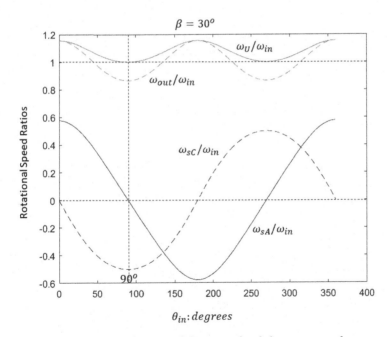

Figure 4.52: The bearing sliding speeds of the cross member.

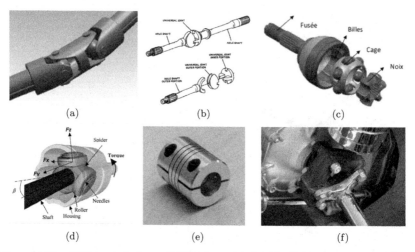

(a)  (b)  (c)

(d)  (e)  (f)

Figure 4.53:  Different designs of CV joints: (a) double Cardan CV joint; (b) the Tracta joint; (c) the Rzeppa joint; (d) the tripod joint; (e) the flexible bellow coupling; and (f) the flexible rubber disc coupling.

*Source*: (a) https://upload.wikimedia.org/wikipedia/commons/thumb/a/a2/Dou ble_Cardan_Joint_(animated).gif/240px-Double_Cardan_Joint_(animated).gif; (b) https://upload.wikimedia.org/wikipedia/commons/0/07/Tracta_Constant_Veloc ity_Joint.jpg; (c) https://journals.openedition.org/activites/docannexe/image/ 5007/img-2.png; (d) Tribology Transactions, 2005, (4): 505–514 DOI: https://www.buildyourcnc.com/images/IMG_7083-800.JPG 10.1080/05698190 500313528; (e) https://www.buildyourcnc.com/images/IMG_7083-800.JPG; and (f) http://upload.wikimedia.org/wikipedia/commons/a/ac/Matra_ms7_rag_ joint.jpg*.

$0°$or$180°$ and $\omega_{\text{in}} \cdot \cos \beta$ when $\theta_{\text{in}} = 90°$ or $270°$. The value of $\omega_U$ varies from $\omega_{\text{in}}/\cos \beta$ when $\theta_{\text{in}} = 0°$ or $180°$ and $\omega_{\text{in}}$ when $\theta_{\text{in}} = 90°$ or $270°$. The spinning rotation of arm $OA$, $\omega_{sA}$ is between $+\frac{1}{2}\omega_{\text{in}}/\cos \beta$ and $-\frac{1}{2}\omega_{\text{in}}/\cos \beta$, while $\omega_{sC}$ is between $+\frac{1}{2}\omega_{\text{in}}$ and $-\frac{1}{2}\omega_{\text{in}}$. It is interesting to note that $|\omega_{sC}|_{\max} = |\omega_{sA}|_{\max} \cdot \cos \beta$. The maximum spinning rate of arm $OA$ is higher than that of $OC$. This indicates that the bearing of arm $OA$ and $OB$ might fail earlier than those of $OC$ and $OD$.

The speed variation caused by the universal joint is highly undesirable because it leads to vibration, excessive forces, and energy losses. Over time, it could also lead to failures of the universal joint or the drive shaft. As shown in Figure 4.47, two universal joints with reversed inclined angles can be used so that the output shaft of the second universal joint will rotate at a constant speed identical to that of the first input shaft. The second universal joint cancels the

Figure 4.54:   A Thompson CV joint.
*Source*: https://grabcad.com/library/thompson-coupling-1-4pt.

speed variation due to the first universal joint. Note that the phases of the universal joints should be the same in order to achieve correct cancellation. Two universal joints and the drive shaft can be made into a short unit, denoted as Double Cardan joint, as shown in Figure 4.53(a). The double Cardan joint is one type of constant velocity (CV) joint used in machines. Figure 4.53 illustrates several different CV joints, such as (b) the Tracta joint, (c) the Rzeppa joint, (d) the tripod joint, (e) the flexible bellow coupling, and (f) the flexible rubber disc coupling.

The design of CV joints could be complex, but the idea is similar to the space cone and body cone concepts of Figure 4.13 and the bevel gears of Figure 4.14. Unlike the bevel gear with only one engaging point, a CV joint has many engaging points around the coupling. However, it is not possible to achieve rolling without slipping at all of these engaging points; as a result, friction is an issue for CV joints, in particular at high inclined angles. CV joints are typically heavily greased and protected with rubber boots. The CV joints are widely used in front-wheel drive vehicles and their rubber boots should be inspected regularly.

An advanced CV joint design, called the Thompson CV joint, likely can offer very high performance with reduced friction, in addition to enhanced coupling torque and axial forces (Figure 4.54).

Figure 4.55: (a)–(c): The knock or lash action of a conventional universal joint when one end of the output shaft is broken; (d) and (e): The action of the Thompson CV joint, appearing to be without lashing.
*Source*: https://www.youtube.com/watch?v=UEvaOg7glKk.

It appears to be a viable solution for high loading applications, such as for supporting the helicopter propellers.[6]

One safety concern of universal joints or CV joints is its potential to kick, lash, or knuckle when the shaft or one end of the coupling breaks. Under this circumstance, the drive shaft will swing and lash around, knocking at the nearby components and, some times, ruptures a fuel tank if the tank is in proximity. The Thompson CV joint appears to be safer in this aspect as demonstrated by the video (footnote 6). A few screen shots are shown in Figure 4.55. Figure 4.55(a) is a conventional universal joint in operation. When the hand releases the drive shaft, it starts to swing and lash around Figures 4.55(b) and 4.55(c), which is dangerous. On the other hand, the Thompson CV joint appears to be stable in its original orientation without lashing around.

---

[6]https://www.youtube.com/watch?v=UEvaOg7glKk.

## 4.5 Relative Motion Analysis for 2D Rigid Body Assemblies with Rotating Reference Frames

In Section 3.10, we discussed relative motion analysis using different reference frames, including rotating reference frames. In this section, we will apply the same analysis to rigid body assemblies. We will focus on only 2D assemblies, although the same analysis can be applied to 3D assemblies as well.

The two equations we will be using are presented in Section 3.10.4 with respect to Figure 3.33. In Figure 3.33, $O-x-y-z$ is typically an inertial reference frame, but it does not have to be, which will be explained in an example later. The reference frame $B-\hat{x}-\hat{y}-\hat{z}$ rotates with respect to $O-x-y-z$ with a frame rotation, $\Omega$ and $\dot{\Omega}$. Equations (3.128) and (3.142) relate the motion observation between these two reference frames. A successful relative motion analysis relies on defining these two reference frames properly.

One main guideline for defining the second reference frame is that the second reference frame should render the motions to be observed as simple motions, i.e., either straight line or perfect circular motions, as discussed in Section 3.10.5.

We will consider several examples and demonstrate how to apply the relative motion equations correctly.

### Example 4.4: Double Pendulum with Sliding Collar Coupling

**Problem Statement:** As shown in Figure 4.56, collar $B$ can slide within a rotating rod $OD$, which is pinned at point $O$. The sliding speed and acceleration of $B$ are measured by a linear encoder attached to rod $OD$ as $\hat{v}_B$ and $\hat{a}_B$, respectively. A second rod, $BC$, rotates around point $B$. The rotation of rod $OD$ is measured to be $\theta_1$, $\omega_1$, and $\alpha_1$ by a shaft encoder attached to $O$ with respect to an inertial reference frame $O-x-y$, while the rotation of rod $BC$, $\hat{\theta}_2$, $\hat{\omega}_2$, and $\hat{\alpha}_2$ is measured by a shaft encoder attached to $OD$ but sliding with collar $B$. Point $B'$ is a point on rod $OD$ overlapping with point $B$ at the instant shown in Figure 4.56. Determine the motions of point $B$ and $C$ with respect to $O-x-y$.

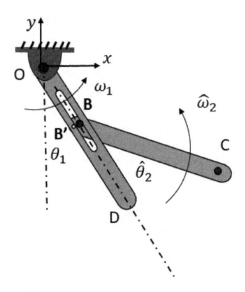

Figure 4.56:   Relative motion analysis for a 2D rigid body assembly.

## Solution:

**Force Analysis:** Skipped.

**Motion Analysis:** The motions of point $B$ and $C$ are too complicated for us to write down expressions for them directly. We will enlist a second observer to help observing them. We will focus on point $B$ first.

As shown in Figure 4.57(a), if we choose a second observer residing at point $B'$ with a rotating reference frame $\hat{x}-\hat{y}$ attached to rod $OD$, point $B$ will appear to move in a straight line motion. In other words, the motion of $B$ is now made SIMPLE! We have

$$\vec{v}_{B/O} = \vec{v}_{B'/O} + \vec{\Omega} \times \hat{r}_{B/B'} + \hat{v}_{B/B'}$$

$$= \vec{\omega}_1 \times \vec{r}_{B'/O} + \vec{\Omega} \times \hat{r}_{B/B'} + \hat{v}_B$$

$$= \vec{\omega}_1 \times \vec{r}_{B'/O} + \hat{v}_B \qquad (4.166)$$

where $\vec{\Omega} = \vec{\omega}_1$ because $\hat{x}-\hat{y}$ is attached to rod $OD$. Also, $\vec{r}_{B/B'} = 0$ because $B$ and $B'$ occupy the same point in space at this instant of time.

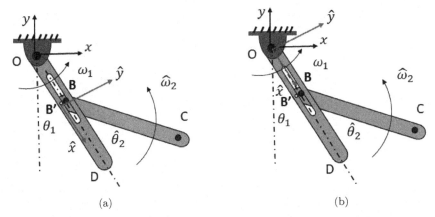

Figure 4.57: (a) A rotating frame at point $B'$ and (b) a rotating frame at point $O$.

From the relative acceleration equation,

$$\vec{a}_{B/O} = \dot{\vec{a}}_{B'/O} + \dot{\vec{\Omega}} \times \hat{r}_{B/B'} + \vec{\Omega} \times (\vec{\Omega} \times \hat{r}_{B/B'}) + 2\vec{\Omega} \times \hat{v}_{B/B'} + \hat{a}_{B/B'}$$

$$= \vec{\omega}_1 \times (\vec{\omega}_1 \times \vec{r}_{B'/O}) + \vec{\alpha}_1 \times \vec{r}_{B'/O} + 2\vec{\omega}_1 \times \hat{v}_B + \hat{a}_B \qquad (4.167)$$

Let's work out some numbers. Let $\theta_1 = 30°$, $\hat{\theta}_2 = 45°$, $OB' = 2$ cm, $BC = 5$ cm, $\omega_1 = 2$ rad/s, $\hat{\omega}_2 = 1$ rad/s, $\alpha_1 = 3$ rad/s$^2$, and $\hat{\alpha}_2 = 0$. We have $\vec{r}_{B'/O} = 2\sin 30°\vec{i} - 2\cos 30°\vec{j}$ (cm). In addition, let $\hat{v}_B = 10$ cm/s and $\hat{a}_B = -4$ cm/s$^2$. Expressing them in the vector form based on $\hat{x} - \hat{y}$, we have $\hat{v}_B = 10\,\hat{i}$ (cm/s) and $\hat{a}_B = -4\hat{i}$ (cm/s$^2$). The vector form for the rotations are $\vec{\omega}_1 = 2\vec{k}$ (rad/s), $\hat{\omega}_2 = 1\hat{k}$ (rad/s), $\vec{\alpha}_1 = 3\vec{k}$ (rad/s$^2$).

Introducing the above numbers into Equations (4.166) and (4.167), we have

$$\vec{v}_{B/O} = \vec{\omega}_1 \times \vec{r}_{B'/O} + \hat{v}_B$$

$$= 2\vec{k} \times (1\vec{i} - 1.732\vec{j}) + 10\hat{i}$$

$$= 3.364\vec{i} + 2\vec{j} + 10\hat{i}$$

$$= 8.464\vec{i} - 6.660\vec{j} \qquad (4.168)$$

where $\hat{i} = \sin 30°\vec{i} - \cos 30°\vec{j}$, $\hat{j} = \cos 30°\vec{i} + \sin 30°\vec{j}$, and $\hat{k} = \vec{k}$.

Equation (4.167) becomes

$$
\begin{aligned}
\vec{a}_{B/O} &= \vec{\omega}_1 \times (\vec{\omega}_1 \times (\vec{r}_{B'/O}) + \vec{\alpha}_1 \times \vec{r}_{B'/O} + 2\vec{\Omega} \times \hat{v}_B + \hat{a}_B \\
&= 2\vec{k} \times (2\vec{k} \times (1\vec{i} - 1.732\vec{j})) + 3\vec{k} \times (1\vec{i} - 1.732\vec{j}) + 2(2\vec{k}) \\
&\quad \times (10\hat{i}) - 4\hat{i} \\
&= 1.196\vec{i} + 9.924\vec{j} - 4\hat{i} + 40\hat{j} \\
&= 33.837\vec{i} + 33.392\vec{j}
\end{aligned}
\tag{4.169}
$$

We can obtain the motion of point $B$ based on a different second observer, as shown in Figure 4.57(b). The new expressions for the velocity and the acceleration of point $B$ become

$$
\begin{aligned}
\vec{v}_{B/O} &= \vec{v}_{O/O} + \vec{\Omega} \times \vec{r}_{B/O} + \hat{v}_{B/O} \\
&= 0 + \vec{\Omega} \times \hat{r}_{B/O} + \hat{v}_B \\
&= \vec{\omega}_1 \times \hat{r}_{B/O} + \hat{v}_B
\end{aligned}
\tag{4.170}
$$

$$
\begin{aligned}
\vec{a}_{B/O} &= \vec{a}_{O/O} + \dot{\vec{\Omega}} \times \hat{r}_{B/O} + \vec{\Omega} \times (\vec{\Omega} \times \hat{r}_{B/O}) + 2\vec{\Omega} \times \hat{v}_{B/O} + \hat{a}_{B/O} \\
&= 0 + \vec{\alpha}_1 \times \hat{r}_{B/O} + \vec{\omega}_1 \times (\vec{\omega}_1 \times (\hat{r}_{B/O})) + 2\vec{\omega}_1 \times \hat{v}_B + \hat{a}_B
\end{aligned}
\tag{4.171}
$$

We will obtain the same results. Note that $\hat{r}_{B/O} = 2\hat{i}$.

Let's consider the motion of point $C$. If we use the second observer of Figure 4.57(b) at point $O$, the motion of point $C$ is still too complicated, not a simple motion. Our choice of the second observer has to be the one similar to Figure 4.57(a). However, there is a subtle difference, we actually need to attach $\hat{x}-\hat{y}$ to point $B$, not point $B'$. By doing so, points $C$ and $B$ are on the same rigid body ($BC$) and point $C$ will appear as a perfect rotation with respect to point $B$ at $\hat{\omega}_2$.

$$
\begin{aligned}
\vec{v}_{C/O} &= \vec{v}_{B/O} + \vec{\Omega} \times \hat{r}_{C/B} + \hat{v}_{C/B} \\
&= \vec{v}_{B/O} + \vec{\omega}_1 \times \hat{r}_{C/B} + \hat{\omega}_2 \times \hat{r}_{C/B}
\end{aligned}
\tag{4.172}
$$

Using $\vec{v}_{B/O}$ obtained earlier, we can determine $\vec{v}_{C/O}$.

Similarly, the acceleration of point $C$ becomes

$$\vec{a}_{C/O} = \vec{a}_{B/O} + \dot{\vec{\Omega}} \times \hat{r}_{C/B} + \vec{\Omega} \times (\vec{\Omega} \times \hat{r}_{C/B}) + 2\vec{\Omega} \times \hat{v}_{C/B} + \hat{a}_{C/B}$$

$$= \vec{a}_{B/O} + \vec{\alpha}_1 \times \hat{r}_{C/B} + \vec{\omega}_1 \times (\vec{\omega}_1 \times \hat{r}_{C/B}) + 2\vec{\omega}_1 \times \hat{v}_{C/B}$$

$$+ \hat{\omega}_2 \times (\hat{\omega}_2 \times \hat{r}_{C/B}) + \hat{\alpha}_2 \times \hat{r}_{C/B} \qquad (4.173)$$

With $\vec{a}_{B/O}$ obtained earlier, we can determine $\vec{a}_{C/O}$. Note that $\hat{r}_{C/B} = |r_{BC}|(\cos\hat{\theta}_2\,\hat{i} + \sin\hat{\theta}_2\,\hat{j})$.

## Example 4.5: Triple Pendulum Motion Analysis

**Problem Statement:** As shown in Figure 4.58, a triple pendulum is formed by pinning three rods. The rotation of rod $OA$ is controlled by a motor attached to the base at $O$. With respect to $O-x-y-z$, the rotation of $OA$ is defined by $\theta_1$, $\omega_1$, and $\alpha_1$. The rotation of rod $AB$ is controlled by a motor attached to rod $OA$ at $A$. Therefore, the rotation of $AB$ is defined with respect to rod $OA$ as $\hat{\theta}_2$, $\hat{\omega}_2$, and $\hat{\alpha}_2$. Similarly, the rotation of rod $BC$ is controlled by a motor attached to rod $AB$ at $B$. Its rotation is defined with respect to rod $AB$ as $\hat{\theta}_3$, $\hat{\omega}_3$, and $\hat{\alpha}_3$.

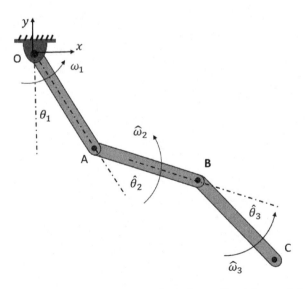

Figure 4.58:  Relative motion analysis for a 2D triple pendulum.

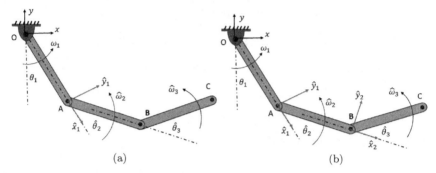

(a)                                        (b)

Figure 4.59:   Reference frames used for the analysis of a triple pendulum: (a) the reference frame pair of point $B$; and (b) the references pairs for point $C$.

Determine the motions of $B$ and $C$ with respect to $O-x-y-z$.

**Solution:**

**Force Analysis:** Skipped.

**Motion Analysis:** Let's conduct motion analysis for point $B$ first. With respect to $O-x-y-z$, point $B$ has a complicated motion for which we cannot write down an expression for its velocity and acceleration directly. If we enlist a second observer, $A-\hat{x}_1-\hat{y}_1$, attached to rod $OA$ as shown in Figure 4.59(a), to observe point $B$, the motion of $B$ becomes a perfect rotation. The rotational property of $A-\hat{x}_1-\hat{y}_1-\hat{z}_1$ is defined by $O-x-y-z$. Because $A-\hat{x}_1-\hat{y}_1-\hat{z}_1$ is attached to the rigid body $OA$, it has the same rotational property of $OA$. Therefore, $\vec{\Omega}_1 = \omega_1$ and $\dot{\Omega}_1 = \alpha_1$. Invoking the relative velocity equation, we have

$$\vec{v}_{B/O} = \vec{v}_{A/O} + \vec{\Omega}_1 \times \hat{r}_{B/A} + \hat{v}_{B/A}$$
$$= \vec{\omega}_1 \times \vec{r}_{A/O} + \vec{\omega}_1 \times \hat{r}_{B/A} + \hat{\omega}_2 \times \hat{r}_{B/A} \qquad (4.174)$$

where $\vec{r}_{A/O} = |OA|(\sin\theta_1\vec{i} - \cos\theta_1\vec{j})$ and $\hat{r}_{B/A} = |AB|(\cos\hat{\theta}_2\hat{i}_1 + \sin\hat{\theta}_2\hat{j}_1)$. These position vectors are defined so that counter clockwise rotations are considered as positive. In addition, the unit directional vectors of $A - \hat{x}_1 - \hat{y}_1 - \hat{z}_1$ are defined as

$$\hat{i}_1 = \sin\theta_1\vec{i} - \cos\theta_1\vec{j} \qquad (4.175)$$

$$\hat{j}_1 = \cos\theta_1\vec{i} + \sin\theta_1\vec{j} \qquad (4.176)$$

$$\hat{k}_1 = \vec{k} \qquad (4.177)$$

Invoking the relative acceleration equation, the acceleration of $B$ becomes

$$
\begin{aligned}
\vec{a}_{B/O} &= \vec{a}_{A/O} + \dot{\vec{\Omega}}_1 \times \hat{r}_{B/A} + \vec{\Omega}_1 \times (\vec{\Omega}_1 \times \hat{r}_{B/A}) + 2\vec{\Omega}_1 \times \hat{v}_{B/A} + \hat{a}_{B/A} \\
&= \vec{a}_{A/O} + \vec{\alpha}_1 \times \hat{r}_{B/A} + \vec{\omega}_1 \times (\vec{\omega}_1 \times \hat{r}_{B/A}) + 2\vec{\omega}_1 \times \hat{v}_{B/A} \\
&\quad + \hat{\omega}_2 \times (\hat{\omega}_2 \times (\hat{r}_{B/A})) + \hat{\alpha}_2 \times \hat{r}_{B/A} \\
&= \vec{\alpha}_1 \times \vec{r}_{A/O} + \vec{\omega}_1 \times (\vec{\omega}_1 \times \vec{r}_{A/O}) \\
&\quad + \vec{\alpha}_1 \times \hat{r}_{B/A} + \vec{\omega}_1 \times (\vec{\omega}_1 \times \hat{r}_{B/A}) + 2\vec{\omega}_1 \times \hat{v}_{B/A} \\
&\quad + \hat{\omega}_2 \times (\hat{\omega}_2 \times \hat{r}_{B/A}) + \hat{\alpha}_2 \times \hat{r}_{B/A}
\end{aligned} \tag{4.178}
$$

For point $C$, we have different ways to select the pair of reference frames for invoking relative motion equations. First, let's consider only rods $AB$ and $BC$. In this way, we pretend that $A-\hat{x}_1-\hat{y}_1-\hat{z}_1$ is an inertial reference frame even though it is not. The second reference frame, $B-\hat{x}_2-\hat{y}_2-\hat{z}_2$, is attached to $AB$ so that point $C$ now appears as a perfect circular motion. This is exactly the same as in the analysis of point $B$. The frame rotation of $B-\hat{x}_2-\hat{y}_2-\hat{z}_2$ with respect to $A-\hat{x}_1-\hat{y}_1-\hat{z}_1$ is $\hat{\Omega}_2 = \hat{\omega}_2$ and $\dot{\hat{\Omega}}_2 = \hat{\alpha}_2$. As a result, we have

$$
\begin{aligned}
\vec{v}^{*}_{C/A} &= \vec{v}^{*}_{B/A} + \hat{\Omega}_2 \times \hat{r}_{C/B} + \hat{v}_{C/B} \\
&= \hat{\omega}_2 \times \hat{r}_{B/A} + \hat{\omega}_2 \times \hat{r}_{C/B} + \hat{\omega}_3 \times \hat{r}_{C/B}
\end{aligned} \tag{4.179}
$$

where $\hat{r}_{C/B} = |BC|(\cos(\hat{\theta}_3)\hat{i}_2 + \sin(\hat{\theta}_3)\hat{j}_2)$.

We put an asterisk to $\vec{v}^{*}_{C/A}$ and $\vec{v}^{*}_{B/A}$ to remind ourselves that $A-\hat{x}_1-\hat{y}_1-\hat{z}_1$ is really not an inertial reference frame.

In addition, the unit directional vectors of $B-\hat{x}_2-\hat{y}_2-\hat{z}_2$ are defined as

$$
\hat{i}_2 = \cos\hat{\theta}_2 \hat{i}_1 + \sin\hat{\theta}_2 \hat{j}_1 \tag{4.180}
$$

$$
\hat{j}_2 = -\sin\hat{\theta}_2 \hat{i}_1 + \cos\hat{\theta}_2 \hat{j}_1 \tag{4.181}
$$

$$
\hat{k}_2 = \hat{k}_1 \tag{4.182}
$$

The acceleration of $C$ is

$$\vec{a}^*_{C/A} = \vec{a}^*_{B/A} + \dot{\hat{\Omega}}_2 \times \hat{r}_{C/B} + \hat{\Omega}_2 \times (\hat{\Omega}_2 \times \hat{r}_{C/B}) + 2\hat{\Omega}_2 \times \hat{v}_{C/B} + \hat{a}_{C/B}$$

$$= \hat{\omega}_1 \times (\hat{\omega}_1 \times \hat{r}_{B/A}) + \hat{\alpha}_1 \times \hat{r}_{B/A} +$$

$$+ \vec{\alpha}_2 \times \hat{r}_{C/B} + \hat{\omega}_2 \times (\hat{\omega}_2 \times \hat{r}_{C/B}) + 2\hat{\omega}_2 \times \hat{v}_{C/B}$$

$$+ \hat{\omega}_3 \times (\hat{\omega}_3 \times \hat{r}_{C/B}) + \hat{\alpha}_3 \times \hat{r}_{C/B} \qquad (4.183)$$

After working out Equations (4.179) and (4.183) to obtain $\vec{v}^*_{C/A}$ and $\vec{a}^*_{C/A}$, we realize that these two observations are with respect to $A-\hat{x}_1-\hat{y}_1-\hat{z}_1$, which is rotating. Therefore, $\hat{v}_{C/A} = \vec{v}^*_{C/A}$ and $\hat{a}_{C/A} = \vec{a}^*_{C/A}$.

We can pretend that we do not have $B-\hat{x}_2-\hat{y}_2-\hat{z}_2$. We are observing point $C$ using two reference frames $O-x-y-z$ and $A-\hat{x}_1-\hat{y}_1-\hat{z}_1$. The velocity and acceleration of $C$ with respect to $O-x-y-z$ are now derived as

$$\vec{v}_{C/O} = \vec{v}_{A/O} + \vec{\Omega}_1 \times \hat{r}_{C/A} + \hat{v}_{C/A}$$

$$= \vec{\omega}_1 \times \hat{r}_{A/O} + \vec{\omega}_1 \times \hat{r}_{C/A} + \hat{v}_{C/A} \qquad (4.184)$$

and

$$\vec{a}_{C/O} = \vec{a}_{A/O} + \dot{\vec{\Omega}}_1 \times \hat{r}_{C/A} + \vec{\Omega}_1 \times (\vec{\Omega}_1 \times \hat{r}_{C/A}) + 2\vec{\Omega}_1 \times \hat{v}_{C/A} + \hat{a}_{C/A}$$

$$= \vec{a}_{A/O} + \vec{\alpha}_1 \times \hat{r}_{B/A} + \vec{\omega}_1 \times (\vec{\omega}_1 \times \hat{r}_{C/A}) + 2\vec{\omega}_1 \times \hat{v}_{C/A} + \hat{a}_{C/A} \qquad (4.185)$$

We already know every term in Equations (4.184) and (4.185) except $\hat{r}_{C/A}$, which can be obtained simply as

$$\hat{r}_{C/A} = \hat{r}_{B/A} + \hat{r}_{C/B} \qquad (4.186)$$

We need to convert $\hat{r}_{C/B}$ to be represented by $\hat{i}_1$ and $\hat{j}_1$ for calculation.

There is a second way of obtaining the motion of point $C$ by using $O-x-y-z$ and $B-\hat{x}_2-\hat{y}_2-\hat{z}_2$. The frame rotation of $B-\hat{x}_2-\hat{y}_2-\hat{z}_2$ needs to be defined by $O-x-y-z$. We only know the frame rotation of $B-\hat{x}_2-\hat{y}_2-\hat{z}_2$ with respect to $A-\hat{x}_1-\hat{y}_1-\hat{z}_1$ and $A-\hat{x}_1-\hat{y}_1-\hat{z}_1$ to $O-x-y-z$. However, because of the additivity of angular rotation, we have

$$\vec{\Omega}_{2/O} = \hat{\Omega}_{2/1} + \hat{\Omega}_{1/O} = \hat{\omega}_2 + \vec{\omega}_1 \qquad (4.187)$$

The velocity and acceleration of point $C$ using the reference frame pair of $O-x-y-z$ and $B-\hat{x}_2-\hat{y}_2-\hat{z}_2$ become

$$\vec{v}_{C/O} = \vec{v}_{B/O} + \vec{\Omega}_{2/O} \times \hat{r}_{C/B} + \hat{v}_{C/B} \qquad (4.188)$$

and

$$\vec{a}_{C/O} = \vec{a}_{B/O} + \dot{\vec{\Omega}}_{2/O} \times \hat{r}_{C/B} + \vec{\Omega}_{2/O} \times (\vec{\Omega}_{2/O} \times \hat{r}_{C/B})$$
$$+ 2\vec{\Omega}_{2/O} \times \hat{v}_{C/B} + \hat{a}_{C/B} \qquad (4.189)$$

The approach using Equations (4.184) and (4.185), or Equations (4.188) and (4.189) will arrive at the same results, but they are based on different reference frame pairs. Either approach works. With this reference frame pair concept, we can conduct analysis for a quadruple pendulum or pendulums with many links.

### Example 4.6: Whirling of a Flexible Shaft

**Problem Statement:** As shown in Figure 4.60, a disk of mass $m$ is attached to a flexible shaft. The disk rotates at a constant angular speed, $\omega$. The geometric center of the disk is at point $S$, while the mass center of the disk is at point $G$ due to the presence of unbalanced mass. While the disk is rotating, a centrifugal force will cause the flexible shaft to bend, deviating from the shaft axis center, $O$. The disk center, $S$, will demonstrate a whirling motion with a path shown in Figure 4.60(b). In general, the whirling motion and the disk rotation are not synchronized. In some cases, the whirling might be synchronized with the disk, i.e., $\dot{\theta} = \omega$. Derive the velocity and acceleration of point $G$ for the unsynchronized whirling motion.

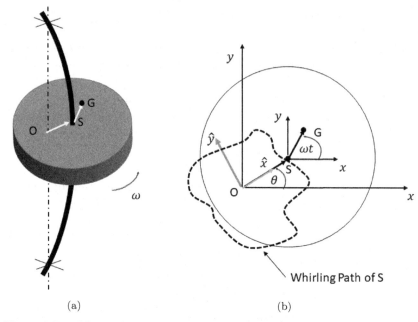

(a)  (b)

Figure 4.60: Motion analysis for a flexible shaft with whirling: (a) 3D sketch of the flexible shaft and disk; and (b) 2D sketch of the disk.

**Solution:**

**Force Analysis:** Skipped.

**Motion Analysis:** The motion of point $G$ is complicated. However, we know that $G$ rotates around $S$ because they belong to the same rigid body. The motion of $S$ is complicated as well, but we will pretend that we know it initially. The reference frame pair to be used are $O{-}x{-}y$ and $S{-}x{-}y$, both are non-rotating. The velocity and acceleration of $G$ become

$$\vec{v}_{G/O} = \vec{v}_{S/O} + \vec{v}_{G/S}$$
$$= \vec{v}_{S/O} + \vec{\omega} \times \vec{r}_{G/S} \tag{4.190}$$

and

$$\vec{a}_{G/O} = \vec{a}_{S/O} + \vec{a}_{G/S}$$
$$= \vec{a}_{S/O} + \dot{\vec{\omega}} \times \vec{r}_{G/S} + \vec{\omega} \times (\vec{\omega} \times \vec{r}_{G/S}) \tag{4.191}$$

We need to define $\vec{v}_{S/O}$ and $\vec{a}_{S/O}$ for the above two equations. For this purpose, lets use another reference frame pair, $O{-}x{-}y$ and

$O-\hat{x}-\hat{y}$. Note that $O-\hat{x}-\hat{y}$ is a rotating reference frame with respect to $O-x-y$. Because axis $\hat{x}$ is always aligned with $OS$, the frame rotation of $O-\hat{x}-\hat{y}$ is defined as $\Omega = \dot{\theta}$ and $\dot{\Omega} = \ddot{\theta}$.

With respect to $O-\hat{x}-\hat{y}$, point $S$ is in a simple straight line motion along the $\hat{x}$ axis; therefore,

$$\hat{v}_{S/O} = \dot{\hat{x}}_S \, \hat{i} \tag{4.192}$$

$$\hat{a}_{S/O} = \ddot{\hat{x}}_S \, \hat{i} \tag{4.193}$$

Invoking the relative motion equations based on the reference frame pair $O-x-y$ and $O-\hat{x}-\hat{y}$, the motion of $S$ can be defined as

$$\begin{aligned}
\vec{v}_{S/O} &= \vec{v}_{O/O} + \vec{\Omega} \times \hat{r}_{S/O} + \hat{v}_{S/O} \\
&= 0 + (\dot{\theta} \, \vec{k}) \times (\hat{x}_S \, \hat{i}) + \dot{\hat{x}} \, \hat{i} \\
&= (\dot{\theta} \cdot \hat{x}_S) \, \hat{j} + \dot{\hat{x}}_S \, \hat{i}
\end{aligned} \tag{4.194}$$

$$\begin{aligned}
\vec{a}_{S/O} &= \vec{a}_{O/O} + \dot{\vec{\Omega}} \times \hat{r}_{S/O} + \vec{\Omega} \times (\vec{\Omega} \times \hat{r}_{S/O}) + 2\vec{\Omega} \times \hat{v}_{S/O} + \hat{a}_{S/O} \\
&= 0 + \hat{x}_S\ddot{\theta}\hat{j} - \hat{x}_S\dot{\theta}^2\hat{i} + (2\dot{\theta} \cdot \dot{\hat{x}}_S) \, \hat{j} + \ddot{\hat{x}}_S \, \hat{i} \\
&= (\ddot{\hat{x}}_S - \hat{x}_S\dot{\theta}^2)\hat{i} + (\hat{x}_S\ddot{\theta} + 2\dot{\theta} \cdot \dot{\hat{x}}_S)\hat{j}
\end{aligned} \tag{4.195}$$

The above two equations are the same as the $r - \theta$ description discussed in Section 3.7.

Introducing Equations (4.194) and (4.195) into Equation (4.193), we completely define the motion of $G$ with respect to $O-x-y$. However, we still need to define $\hat{i}$ and $\hat{j}$, together with the information of $\theta$, $\dot{\theta}$, $\ddot{\theta}$, $\hat{x}$, and $\ddot{\hat{x}}$. It will require several sensors to measure these variables. Suitable sensors are discussed in Section 3.4.1.

In the case of synchronized whirling, the motions of $G$ and $S$ can be defined and the above equations are simplified.

## 4.6 Concluding Remarks

In this chapter, we conducted motion analysis for rigid body assemblies, denoted as rigid body kinematics. The rotation of a rigid body is an universal property of the rigid body. The most important concept is discussed in Section 4.2 and Equation (4.3). Basically, if we know the motion of one point and we know the rotation of the rigid body, we can define the motion of any point on the same rigid body. This concept is applicable to both 2D and 3D rigid bodies.

The motion of a 3D rigid body rotating around a fixed point is discussed in Section 4.2.1, which is important for gyro motion analysis.

The concept of instantaneous center for 2D rigid bodies is presented in Section 4.2.3.1, and its 3D version, denoted as instantaneous axis of rotation, is presented in Section 4.2.3.2. The instantaneous axis of rotation leads to the concept of space cone and body cone, which can be used for machine design (Sections 4.2.4 and 4.2.5).

Kinematics of 2D rigid body assemblies is discussed in Section 4.3. Several different couplings between two rigid bodies are illustrated. Several important 2D rigid body assemblies and their practical applications, such as the planetary gear train assembly, are presented in Section 4.3. The kinematics of 3D rigid body assembly is analyzed in Section 4.4. In particular, detailed analyses are conducted for differentials and universal joints. Many devices discussed in this chapter will be revisited with additional force analysis.

Finally, relative motion analysis with rotational reference frames for rigid body assemblies is presented in Section 4.5. Several examples are presented to illustrate the selection of the reference frame pair. A second reference frame is chosen so that the motion to be observed becomes a simple motion (straight line or perfectly circular motion).

With Chapters 3 and 4, we have acquired the skill to conduct motion analysis for 2D and 3D systems. In the next chapters, we will link forces and acceleration via *Newton's Second Law* and the treatment is denoted Kinetics.

Chapter 5

# Kinetics of Dynamic Systems

In this chapter, we will discuss *Newton's Second Law* for particles and rigid body systems. Instead of considering 2D systems first and then 3D systems, as normally done in most textbooks in Dynamics, we will start with general 3D systems and show how it could be simplified for 2D systems. In this way, we hopefully can prevent the situation that most students only learn dynamics up to 2D systems.

## 5.1 *Newton's Second Law*

We will start from Figure 3.1 and the corresponding Equation (3.1) for one particle to describe *Newton's Second Law*, as follows.

Based on an inertial reference frame, $O-x-y-z$, as shown in Figure 5.1, *Newton's Second Law* can be applied for object $B$ as

$$\sum \vec{F} = m\, \vec{a}_{B/O} \tag{5.1}$$

where $m$ is the mass of object $B$ and $\sum \vec{F}$ is the total force acting on B.

Equation (5.1) is general and particle $B$ could be a standalone particle or a particle within a group of particles. The information of $\sum \vec{F}$ typically is obtained independently via force analysis using the FBD method discussed in Chapter 2.

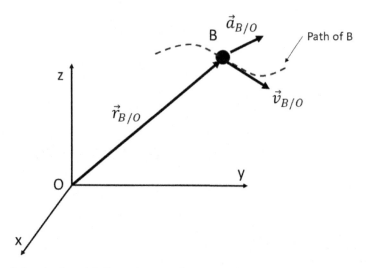

Figure 5.1: An inertial Cartesian coordinate used by observer $O$ to observe the motion of object $B$.

## 5.2 *Newton's Second Law* for a Particle

In this section, we consider a system which contains only one particle. However, the particle could be big or small as long as the motion of the particle can be defined uniquely by one position vector, one velocity vector, and one acceleration vector. For example, the entire Earth can be considered as one particle when discussing its orbit around the sun or a car can be considered as one particle in many engineering problems.

As discussed in Chapter 3, depending on which description is used, we can break down the vector equation of Equation (5.1) into those components in different directions.

For the $x-y-z$ description, Equation (5.1) becomes

$$\sum F_x = m \, \ddot{x}_{B/O} \tag{5.2}$$

$$\sum F_y = m \, \ddot{y}_{B/O} \tag{5.3}$$

$$\sum F_z = m \, \ddot{z}_{B/O} \tag{5.4}$$

Note that the subscription $B/O$ is used to emphasize that the acceleration must be with respect to an inertial reference frame, $O-x-y-z$.

For the $r-\theta-z$ description, Equation (5.1) becomes

$$\sum F_r = m\, a_r = m(\ddot{r} - r\dot{\theta}^2) \tag{5.5}$$

$$\sum F_\theta = m\, a_\theta = m(r\ddot{\theta} + 2\dot{r}\dot{\theta}) \tag{5.6}$$

$$\sum F_z = m\, a_z = m\ddot{z} \tag{5.7}$$

Note that technically, the $r-\theta-z$ reference is a rotating reference frame but the acceleration components in the above equation are with respect to an inertial coordinate which aligns with $r-\theta-z$ at the current instant of time, as discussed in Chapter 3.

For the $t-n-b$ description, Equation (5.1) becomes

$$\sum F_t = m\, a_t = m\dot{v} \tag{5.8}$$

$$\sum F_n = m\, a_n = m\frac{v^2}{\rho} \tag{5.9}$$

$$\sum F_b = 0 \tag{5.10}$$

Similarly, the $t-n-b$ reference frame is a rotating reference frame but the acceleration components in the above equation are with respect to an inertial coordinate which aligns with $t-n-b$ at the current instant of time.

For different problems, we simply choose a suitable expression of *Newton's Second Law* to solve the problem. Equations (5.2)–(5.4) are the most common and we should determine $\vec{a}_{B/O}$ based on the relative motion equation with the help of a rotating reference frame if the motion of B is not a simple motion.

We will present a few examples for illustration.

## Example 5.1: Projectile Motion Revisit

**Problem Statement:** As shown in Figure 5.2, a projectile flies through air. Consider the quadratic air drag and lift, determine the governing equations of the projectile.

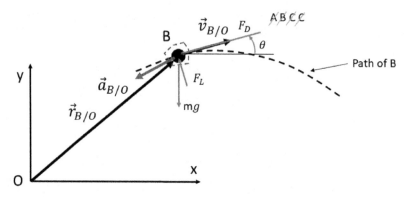

Figure 5.2: FBD and motion analysis of a projectile.

**Solution:**

**Force Analysis:** We should follow the ABCC procedure to construct an FBD for projectile $B$, as shown in Figure 5.2. There are no "Applied Forces"; there is one "Body Force" as the gravity, $mg$; as for "Contact Forces," the contact between the projectile and the air particles form a drag $F_D$ and a lift $F_L$, the drag opposes the velocity direction, while the lift is perpendicular to the velocity vector. Finally, there are no "Constraint Forces." A complete FBD is shown in Figure 5.2.

**Motion Analysis:** First, we establish an inertial reference, $O-x-y$, which is attached to the ground. We also decide that the $x-y$ description is more suitable in this case because $mg$ is always aligned with the $y$ axis. The $t-n$ description could be a viable option because $F_D$ is always aligned with the tangential direction, while $F_L$ is aligned with the normal direction. However, if we choose the $t-n$ description, we still need to work out angle $\theta$ with respect to the $x$ axis. Often, we would like to obtain the trajectory of the projectile, which is typically expressed in the $x-y$ coordinate. With the considerations above, it is decided to stay with the $x-y$ description.

**Governing Equations:** Based on the $x-y$ description, we will choose Equations (5.2)–(5.4) as the governing equation,

$$\sum F_x = -F_D \cos\theta - F_L \sin\theta = m\ddot{x}_{B/O} = m\ddot{x} \qquad (5.11)$$

$$\sum F_y = F_L \cos\theta - F_D \sin\theta - mg = m\,\ddot{y}_{B/O} = m\ddot{y} \qquad (5.12)$$

where $\theta = \tan^{-1}(\dot{y}/\dot{x})$, $v = \sqrt{\dot{x}^2 + \dot{y}^2}$, $F_D = \mu_D v^2$, and $F_L = \mu_L v^2$. The parameters $\mu_D$ and $\mu_L$ are the drag and lift coefficients, respectively, obtained by experiments.

**Solving for Unknowns:** Let's count the unknowns. The unknowns involved are $\ddot{x}$, $\ddot{y}$, $\dot{x}$, and $\dot{y}$. Therefore, we have four unknowns but only two equations. Let's examine the unknowns. We realize that we can relate the velocities and accelerations by

$$\ddot{x} = \frac{d}{dt}\dot{x} \tag{5.13}$$

$$\ddot{y} = \frac{d}{dt}\dot{y} \tag{5.14}$$

Therefore, we have four unknowns and four equations. We can solve for them. Because Equations (5.11)–(5.14) are nonlinear differential equations, analytical solutions are not easy to find in general. We will pursue numerical solutions using Matlab.

To use Matlab, we need to convert Equations (5.11)–(5.14) to a specific form, denoted as the state-space form, where the second-order differential equations are converted into a set of first-order differential equations. The following is a detailed procedure to accomplish this task.

First, we need to define the state variables as

$$x_1 = x \tag{5.15}$$

$$x_2 = y \tag{5.16}$$

$$x_3 = \dot{x} \tag{5.17}$$

$$x_4 = \dot{y} \tag{5.18}$$

Introducing Equations (5.15)–(5.18) to (5.11) and (5.12), we have

$$\dot{x}_3 = \ddot{x} = -\frac{\mu_D}{m}(x_3^2 + x_4^2)\cos\theta - \frac{\mu_L}{m}(x_3^2 + x_4^2)\sin\theta \tag{5.19}$$

$$\dot{x}_4 = \ddot{y} = \frac{\mu_L}{m}(x_3^2 + x_4^2)\cos\theta - \frac{\mu_D}{m}(x_3^2 + x_4^2)\sin\theta - g \tag{5.20}$$

where $\theta = \tan^{-1}(x_4/x_3)$.

Similarly, Equations (5.13) and (5.14) become

$$\dot{x}_1 = \dot{x} = x_3 \tag{5.21}$$

$$\dot{x}_2 = \dot{y} = x_4 \tag{5.22}$$

Putting Equations (5.19)–(5.22) together, we have the state-space form of the governing equation as

$$\dot{x}_1 = x_3 \tag{5.23}$$

$$\dot{x}_2 = x_4 \tag{5.24}$$

$$\dot{x}_3 = -\frac{\mu_D}{m}(x_3^2 + x_4^2)\cos\theta - \frac{\mu_L}{m}(x_3^2 + x_4^2)\sin\theta \tag{5.25}$$

$$\dot{x}_4 = \frac{\mu_L}{m}(x_3^2 + x_4^2)\cos\theta - \frac{\mu_D}{m}(x_3^2 + x_4^2)\sin\theta - g \tag{5.26}$$

The following Matlab script can be used to solve for Equations (5.23)–(5.26).

```
% Matlab program for the projectile motion
time=0:0.05:10;
time=time';
x0=[0;0;20*cos(pi/4);20*sin(pi/4)];
[t,x]=ode45('projsystem1',time,x0);
plot(x(:,1),x(:,2));
axis equal;
axis([0 45 0 12]);
xlabel('x: m');
ylabel('y: m');
title('Projectile Trajectories');
```

We can save the above command script as a command script file under the file name, for example, `projsim.m`. We need to construct a second Matlab file to define `projsystem1` in the above script. The file of `projsystem1.m` will be equivalent to Equations (5.23)–(5.26).

```
% projsystem1 is the state-space governing
equation
    function f=f(t,x)
    f=zeros(4,1)
```

```
theta = atan(x(4)/x(3));
mud=0.05;
mul=0.0;
m=10; % kg
g=9.81; % m/s2
v2=x(3)*x(3) + x(4)*x(4);
f(1)=x(3);
f(2)=x(4);
f(3)=-(mud/m)*v2*cos(theta) - (mul/m)*v2 *
sin(theta);
f(4)=(mul/m)*v2*cos(theta) -
(mud/m)*v2*sin(theta) - g;
```

We can use the above scripts to simulate the trajectories of a projectile at different wind drag and lift conditions with a launching speed of 20 m/s at 45°. As shown in Figure 5.3, curve $G$ is the trajectory without considering the wind effects, also known as the Galileo parabola. With the wind drag as $\mu_D = 0.05$ $(N \cdot s^2)/m^2$, the trajectory is curve $S$. Curves $G$ and $S$ matched the results of Figure 3.18. If we further assume $\mu_L = 0.05$ $(N \cdot s^2)/m^2$, the trajectory becomes curve $F$.

**Solution Verification:** We can verify the simulation with the equations in Section 3.5.1. The maximum height with the Galileo parabola is

$$h_{\max} = \frac{v_y^2(0)}{2g} = 10.194 \ (m) \tag{5.27}$$

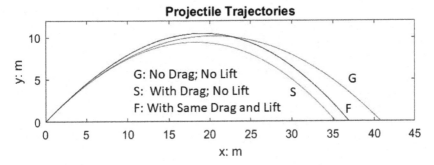

Figure 5.3: Projectile trajectory simulations with and without air resistance, similar to the results of Figure 3.18.

which matches curve $G$ of Figure 5.3. We should also check the initial condition and it is indeed aligned with 45°.

Finally, for a projectile with different shapes, we can assign different values of the drag and lift coefficients and use the same Matlab script to predict the trajectories.

## Example 5.2: Tightening Rope and Sliding Disk

**Problem Statement:** As shown in Figure 5.4, a heavy disk of mass $m = 50$ kg is pulled in by a rope, while rotating around a pillar. The rope is reeled in through the hole at $A$ over the pillar with a height $b$. The reeling-in rate of the rope is $\dot{a} = -0.5$ m/s. Initially, the radial distance of the disk is $r_0 = 3$ m and the angular speed is $\dot{\theta}_0 = 0.1$ rad/s. Determine the tension of the rope when $r = 2$ m, assuming that there is no friction between the disk and the surface.

**Solution:** This is a rather complicated problem. We will proceed step by step and think along the way to solve it.

**Force Analysis:** The FBD of the disk is shown in Figure 5.5. Following the ABCC method of Chapter 2, there are no "Applied forces," there is a "Body force" $mg$, there is a "Contact force," as the normal force $N$ and no frictional forces, and finally, there is a "Constraint

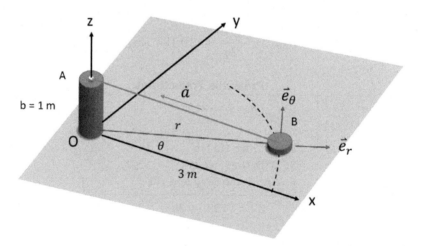

Figure 5.4:   A disk is reeled in by a rope through a pillar hole.

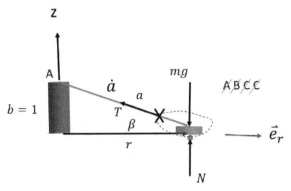

Figure 5.5: The free-body diagram analysis of the disk using the ABCC method.

force" as the rope tension. The normal contact force, $N$, is compressive; therefore, it points toward the boundary, while the tension of the rope, $T$, is tensile and it points away from the boundary. We construct the FBD as shown in Figure 5.5.

**Motion Analysis:** Based on Figure 5.4, this is a 2D general motion on the $x-y$ plane. We establish an inertial reference frame, $O-x-y-z$ and we also define the $r - \theta$ description accordingly. It is apparent that we should use the $r - \theta$ description to solve this problem. We can write down the velocity and acceleration equations accordingly as

$$\vec{v}_{B/O} = \vec{v} = \dot{r}\,\vec{e}_r + r\dot{\theta}\,\vec{e}_\theta \tag{5.28}$$

$$\vec{a}_{B/O} = \vec{a} = (\ddot{r} - r\dot{\theta}^2)\vec{e}_r + (r\ddot{\theta} + 2\dot{r}\dot{\theta})\vec{e}_\theta + 0\vec{k} \tag{5.29}$$

**Governing Equations:** We apply *Newton's Second Law* to link the accelerations of Equation (5.29) and the forces from Figure 5.5.

$$\sum F_r = -T\cos\beta = m(\ddot{r} - r\dot{\theta}^2) \tag{5.30}$$

$$\sum F_\theta = 0 = m(r\ddot{\theta} + 2\dot{r}\dot{\theta}) \tag{5.31}$$

$$\sum F_k = T\sin\beta - mg + N = 0 \tag{5.32}$$

**Solving for Unknowns:** Let's count the unknowns involved in Equations (5.30)–(5.32). When $r = 2\,\mathrm{m}$, we need to know $\dot{\theta}$, $\ddot{\theta}$, $\dot{r}$, $\ddot{r}$, $T$, and $N$. We know $\beta = \tan^{-1}(b/r)$.

There are six unknowns but we only have three equations. Because we know the rope reeling rate, $\dot{a}$, and from the geometry, we have $1 + r^2 = a^2$. Taking a derivative of this geometric relationship, we have

$$\dot{r} = \frac{a}{r}\dot{a} = \frac{\sqrt{r^2 + 1}}{r}\dot{a} \tag{5.33}$$

Taking another derivative of Equation (5.33), we have

$$\ddot{r} = -\frac{1}{r^3}\dot{a}^2 \tag{5.34}$$

So far, we have five equations and six unknowns. One last equation will be the relationship between $\dot{\theta}$ and $\ddot{\theta}$,

$$\ddot{\theta} = \frac{d\dot{\theta}}{dt} \tag{5.35}$$

Now, we have six equations to solve for six unknowns. We need to make a decision which equation will be used first. Equation (5.31) involves only $r$, $\theta$, and their derivatives, while Equation (5.32) involves the forces. We can do something with Equation (5.31) via some clever calculus manipulations. Note that

$$0 = (r\ddot{\theta} + 2\dot{r}\dot{\theta}) = \frac{1}{r}\left(\frac{d}{dt}(r^2\dot{\theta})\right) \tag{5.36}$$

From Equation (5.36), we have

$$r^2\dot{\theta} = C \tag{5.37}$$

where $C$ is a constant.

Plugging in the initial condition, we obtain $C = 0.9\,\mathrm{m}^2\,\mathrm{rad/s}$.

From Equation (5.37), we know that $\dot{\theta}$ is a function of $r$ and from Equation (5.35), we have

$$\ddot{\theta} = -\frac{2C}{r^3}\dot{r} \tag{5.38}$$

With Equations (5.33), (5.34), (5.37), and (5.38), we have all the information for the right-hand side of Equations (5.30)–(5.32). Let's work out numbers based on the given conditions,

$$\dot{\theta}|_{r=2} = \frac{C}{r^2} = 0.225 \ (\text{rad/s}) \tag{5.39}$$

$$\ddot{\theta}|_{r=2} = 0.126 \ (\text{rad/s}^2) \tag{5.40}$$

$$\dot{r}|_{r=2} = -0.559 \ (\text{m/s}) \tag{5.41}$$

$$\ddot{r}|_{r=2} = -0.0313 \ (\text{m/s}^2) \tag{5.42}$$

Plugging all these numbers into Equations (5.30) and (5.32), we obtain

$$T|_{r=2} = 7.41 \ (\text{N}) \tag{5.43}$$

$$N|_{r=2} = 487.19 \ (\text{N}) \tag{5.44}$$

**Solution Verification:** Let's verify the calculations using Equation (5.31),

$$2(0.12577) + 2(-0.559)(0.225) = 0.25155 - 0.25155 \simeq 0 \tag{5.45}$$

If we simplify this problem by letting $\dot{r} = -0.5\,\text{m/s}$ instead of $\dot{a} = -0.5\,\text{m/s}$, the problem becomes much easier. Now, $\ddot{r} = 0$, while $\dot{\theta}$ is the same. We obtain $T = 5.66\,\text{N}$, which is lower because $|\dot{r}| = 0.5\,\text{m/s}$ is less than $0.559\,\text{m/s}$ when $\dot{a} = 0.5\,\text{m/s}$.

From the practical control point of view, if we want to keep $\dot{r}$ as a constant, then, from Equation (5.33), the rope reeling-in rate has to be $\dot{a} = (r/\sqrt{r^2+1})\dot{r}$, which is not a constant. It is a harder reeling control task.

Finally, we can revise the script for Example 5.1 to determine all the variables as functions of time. This example can be modified for designing, e.g., an automatic fishing pole reeling-in control system. However, fishing will not be as fun with an automatic reeling system.

## Example 5.3: Race Track Bank Angle and Speed

**Problem Statement:** As shown in Figure 5.6(a), a circular race track is designed with a bank angle $\theta$. The radius of the track is $\rho$ while the static friction coefficient $\mu_s$ under the dry condition is 0.8. Determine the maximum speed allowed by the track so that the car will not slide toward the fence. Also determine the speed allowed if the car does not rely on friction to keep it from sliding toward the fence.

(a)

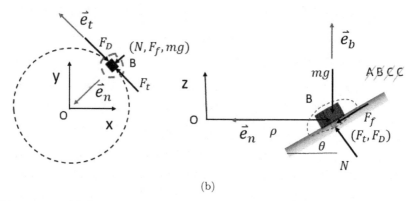

(b)

Figure 5.6: (a) A race track with a bank angle $\theta$; and (b) the motion and force analysis of a car on the race track.

*Source*: https://upload.wikimedia.org/wikipedia/commons/d/d4/Chicagoland_Speedway_ow.jpg.

## Solution:

**Force Analysis:** A useful free-body diagram is shown in Figure 5.6(b). We only have the body force and contact forces in this case. For the contact forces, there is a normal force, $N$, a frictional force, $F_f$ along the bank direction, a tractive force, $F_t$, pointing to the $\vec{e}_t$ direction, and the wind drag, $F_D$, in the opposite direction. Figure 5.6(b) is a 2D FBD representation of a 3D case with two 2D sketches.

**Motion Analysis:** Apparently, this is a 2D circular motion. We can apply either the $r - \theta$ or the $t-n$ description. The $t-n$ description is preferred because we do not care about the specific location of the car, only the speed and acceleration.

The motion equations are

$$\vec{v} = v\vec{e}_t \tag{5.46}$$

$$\vec{a} = \dot{v}\vec{e}_t + \frac{v^2}{\rho}\vec{e}_n + 0\vec{e}_b = 0\vec{e}_t + \frac{v^2}{\rho}\vec{e}_n + 0\vec{e}_b \tag{5.47}$$

Assuming that the car is running at its maximum speed, we have $\dot{v} = 0$.

**Governing Equations:** Applying *Newton's Second Law*, the governing equations are

$$F_t - F_D = 0 \tag{5.48}$$

$$N \sin\theta + F_f \cos\theta = m\frac{v^2}{\rho} \tag{5.49}$$

$$-mg + N\cos\theta - F_f \sin\theta = 0 \tag{5.50}$$

**Solving for Unknowns** In Equations (5.48)–(5.50), there are five unknowns, $F_t$, $F_D$, $F_f$, $N$, and $v$. Equation (5.48) is readily solvable for $F_t = F_D$, but we need to determine $F_D$ independently. We can assume the quadratic wind drag law to express $F_D$ as

$$F_D = \frac{1}{2}\rho_{\text{air}}C_D A v^2 \tag{5.51}$$

where $\rho_{\text{air}}$ is the air density, $C_D$ is the wind drag coefficient, and $A$ is the drag cross-section area of the vehicle. In a typical vehicle, the value of $C_D A$ is about $0.6\,\text{m}^2$ and $\rho_{\text{air}} = 1.225\,\text{kg/m}^3$.

Now, we have four equations for five unknowns. We need one more equation. Assuming that at the maximum speed, $v = v_{\text{max}}$, the car is on the verge of sliding, then, $N$, $F_D$, and $F_f$ are related as

$$\sqrt{F_t^2 + F_f^2} = \sqrt{F_D^2 + F_f^2} = \mu_s\, N = 0.8\ N \tag{5.52}$$

We now have five unknowns and five equations. We can solve for all unknowns and express them as functions of the bank angle.

Because the equations are nonlinear, it is quite messy to solve for them by hand. We will only seek numerical solutions. Before we do so, let's consider the case without wind drag. The solutions become simple and we obtain

$$N = \frac{mg}{\cos\theta - \mu_s \sin\theta} \tag{5.53}$$

$$F_f = \frac{\mu_s mg}{\cos\theta - \mu_s \sin\theta} \tag{5.54}$$

$$v_{\max} = \sqrt{\frac{\rho g(\sin\theta + \mu_s \cos\theta)}{\cos\theta - \mu_s \sin\theta}} \tag{5.55}$$

If the car does not rely on friction, the allowable speed $v^*_{\max}$ becomes

$$v^*_{\max} = \sqrt{\frac{\rho g \sin\theta}{\cos\theta}}. \tag{5.56}$$

The speed $v^*_{\max}$ is used for designing the race track. For example, if $v^*_{\max} = 150\,\text{km/hr}$ and $\rho = 250\,\text{m}$, then the bank angle is

$$\theta = \tan^{-1}\left(\frac{v^2_2}{\rho g}\right) = 35.3^{\circ}. \tag{5.57}$$

The bank angle of a typical NASCAR track is about $36^{\circ}$.

**Solution Verification:** Let's consider the famous race track in Indianapolis, USA, for Indy 500. The radius at the circular section is $250\,\text{m}$, the top speed is $400\,\text{km/hr}$. The bank angle is $9.2^{\circ}$. As a result, the maximum speed without relying on the friction is only about $72\,\text{km/hr}$, compared with the maximum speed at $187.4\,\text{km/hr}$ with $\mu_s = 0.8$. The race tires allow $\mu_s$ to reach 1.5, for which the maximum speed at the turn becomes $264.2\,\text{km/hr}$. To go at $400\,\text{km/hr}$, it has to be on the straight section of the track.

Let's consider the wind drag effect. The tractive force is needed to overcome wind drag, which then reduces available $F_f$ to keep the car from sliding. By considering the wind drag, we can use a symbolic equation solver of Matlab (see the following program script) to solve the governing equations. We obtain the maximum speed without sliding for $\mu_s = 0.8$ as $187.2\,\text{km/hr}$ vs $187.4\,\text{km/hr}$, without considering the wind drag. At $\mu_s = 1.5$, it is $263.8\,\text{km/hr}$ vs $264.2\,\text{km/hr}$ without wind drag. Finally, all the calculated forces are listed in the following table. It appears that it is alright to exclude the wind drag in the calculation. Note that the equation solver might return multiple answers and we need to choose the correct ones.

| $C_d A$ (m$^2$) | $\mu_s$ | $F_D$ (N) | $F_t$ (N) | $F_f$ (N) | $N$ (N) | $V_{max}$ (km/hr) |
|---|---|---|---|---|---|---|
| 0.6 | 0.8 | 993.5 | 993.5 | 13,659 | 17,119 | 187.2 |
| 0.0 | 0.8 | 0 | 0 | 13,701 | 17,126 | 187.4 |
| 0.6 | 1.5 | 1,973.2 | 1,973.2 | 29,449 | 19,676 | 263.8 |
| 0.0 | 1.5 | 0 | 0 | 29,536 | 19,691 | 264.2 |

The following Matlab script is used for calculation.

```
% Matlab program for maximum speed and bank
angle calculation
    % Ft is tractive force in N;
    % Fd is wind drag in N
    % N is the normal force in N
    % v is the speed in m/s
    % Ff is the lateral frictional force in N
    % theta is the bank angle in rad

    clear Ft Fd N v Ff
    syms Ft Fd N v Ff
    mu=1.5; % static friction coefficient
    m = 1500; % car mass in kg
    CdA = 0.6;% wind drag area coefficient in
m*m;
    rho=1.225; % air density in kg/(m*m*m)
    R=250; % race track radius in m;
    g=9.81; % gravitational acceleration in
m/(s*s)
    theta=9.2*pi/180;
    eq1 = Fd == 0.5*rho*CdA*v*v;
    eq2 = N*sin(theta) + Ff*cos(theta) ==
m*v*v/R;
    eq3 = -m*g + N*cos(theta) - Ff*sin(theta)==
0;
```

```
eq4 = Ft == Fd;
eq5 = Ft*Ft + Ff*Ff == (mu*N)*(mu*N);
eqns = [eq1,eq2,eq3,eq4,eq5];
S = solve(eqns);
maxspeed = double([S.v(4)])*3600/1000
normal = double([S.N(4)])
drag = double([S.Fd(4)])
tractive = double([S.Ft(4)])
lateral = double([S.Ff(4)])
```

We will consider another particle example related to *Newton's Second Law*. As shown in Figure 5.7(a), various orbits of satellites are shown. Most of these orbits are circular while a few are elliptical, such as Molnya, a Russia TV satellite. In Figure 5.7(b), the orbit inclinations of International Space Station (ISS) and Hubble Space Telescope (HST) are shown. In the following example, we will describe the governing equations for these orbits.

**Example 5.4: Satellite Trajectories**

**Problem Statement:** As shown in Figure 5.8, a satellite is in a general orbit, not necessarily circular. The $r - \theta$ description is used and the plane of the satellite orbit is defined as the $x - y$ plane. The $r - \theta$ description is chosen for good reasons. From the sensor point of view, despite the long distance, a radar can track a satellite very well by measuring the radial distance, $r$, via electromagnetic waves while $\theta$ is determined by the orientation of the radar. Because we need to track the position of the satellite, the $t - n$ description is not suitable for deriving the governing equations for the satellite.

Figure 5.7: (a) Various orbit heights above the Earth's surface; and (b) the inclinations of International Space Station (ISS) and Hubble Space Telescope (HST).

*Source:* (a) http://2.bp.blogspot.com/-GRQS3dANmNQ/TZlxG4_d-hI/AAAA AAAAAB4/dBaM8chIdjw/s1600/Satellite+orbits.jpg; and (b) https://upload. wikimedia.org/wikipedia/commons/thumb/a/ac/Comparison_ISS_HST_orbits_ globe_centered_in_Cape_Verde.svg/768px-Comparison_ISS_HST_orbits_globe_ centered_in_Cape_Verde.svg.png*-4pt.

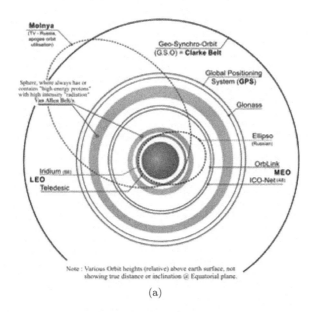

(a)

(b)

Figure 5.7:

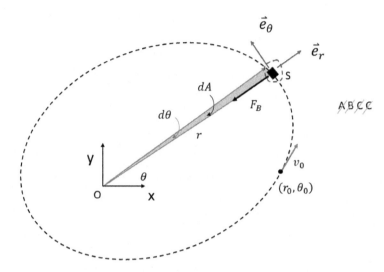

Figure 5.8:   Kinematic and kinetic analysis of a satellite.

## Solution:

**Force Analysis:** The FBD analysis shows that there is only one body force, $F_B$, pointing toward the center of the Earth if we ignore the gravitational forces from the sun, the moon, etc. This force condition is called a central force condition.

**Motion Analysis:** We establish an inertial reference frame, $O-x-y$, at the center of the Earth. Here, we know that the Earth is moving around the Sun; therefore, $O-x-y$ is really not an inertial reference frame. However, it takes a year for the Earth to go around the Sun. We should keep in mind that there will be inaccuracy in the analysis with this assumption.

**Governing Equations:** Applying *Newton's Second Law*, we have

$$-F_B = m(\ddot{r} - r\dot{\theta}^2) \tag{5.58}$$

$$0 = m(r\ddot{\theta} + 2\dot{r}\dot{\theta}) \tag{5.59}$$

Applying the same treatment of Equations (5.36) and (5.37) to Equation (5.59), we obtain

$$r^2\dot{\theta} = r^2\frac{d\theta}{dt} = h \tag{5.60}$$

where $h$ is a constant of integration.

From Figure 5.8, the instantaneous sweeping area, $dA$, is equal to $\frac{1}{2}r^2 d\theta$; therefore,

$$\frac{dA}{dt} = \frac{1}{2}r^2\frac{d\theta}{dt} = \frac{h}{2} \tag{5.61}$$

The term $dA/dt$ is denoted as the area speed, which is a constant for a satellite only subjected to a central force. This essentially is Kepler's Second Law of Planetary Motion.

Both a circular or an oval orbit will satisfy this constant area speed condition.

*Newton's gravitational law* defines the central force as

$$F_B = G\frac{M_e m}{r^2} \tag{5.62}$$

where $M_e$ is the mass of the Earth and $m$ is the mass of the satellite.

Introducing Equations (5.60) and (5.62) into (5.58), we obtain a governing equation as

$$\frac{d^2 r}{d\theta^2} + \frac{GM_e}{h^2}r^2 - r = 0 \tag{5.63}$$

When deriving the above equation, we also utilize the following identities,

$$\frac{dr}{dt} = \frac{dr}{d\theta}\frac{d\theta}{dt} = \frac{h}{r^2}\frac{dr}{d\theta} \tag{5.64}$$

$$\frac{d^2 r}{dt^2} = \frac{d}{dt}\left(\frac{h}{r^2}\frac{dr}{d\theta}\right) = \frac{d}{d\theta}\left(\frac{h}{r^2}\frac{dr}{d\theta}\right)\frac{d\theta}{dt}$$

$$= \left[\frac{d}{d\theta}\left(\frac{h}{r^2}\frac{dr}{d\theta}\right)\right]\frac{h}{r^2} \tag{5.65}$$

**Solving for Unknowns:** Equation (5.63) is a nonlinear differential equation and it is difficult to obtain a closed-form solution. We can pursue a numerical solutions using a similar program as in Example 5.2. However, by assigning $\xi = 1/r$ and via some clever calculus manipulation, we can convert Equation (5.63) into a linear differential equation.

$$\frac{d^2\xi}{d\theta^2} + \xi = \frac{GM_e}{h^2} \tag{5.66}$$

A homogeneous solution can be obtained for Equation (5.66) as

$$\xi = a_1 \cos \lambda\theta + a_2 \sin \lambda\theta \tag{5.67}$$

We find that $\lambda = \pm 1$. The total solution of Equation (5.66) becomes

$$\xi = a_1 \cos \lambda\theta + a_2 \sin \lambda\theta + \frac{GM_e}{h^2} \tag{5.68}$$

Taking a one time derivative of $\xi$, we have

$$\dot{\xi} = -\frac{1}{r^2}\dot{r} = -a_1\lambda \sin \lambda\theta \, \dot{\theta} + a_2\lambda \cos \lambda\theta \, \dot{\theta} \tag{5.69}$$

The values of $a_1$ and $a_2$ are determined by the initial condition when the satellite enters the orbit at $r_0$ with $v_0$. Note that we choose the $x - y$ frame so that at the entry point $\theta_0 = 0$. To enter the orbit at $r_0$ and $\theta_0$, we also need the entry velocity to be in the $\vec{e}_\theta$ direction only, i.e., $\dot{r} = 0$ at $r = r_0$. These initial conditions lead to $a_2 = 0$. The solution becomes

$$\xi = a_1 \cos \theta + \frac{GM_e}{h^2} \tag{5.70}$$

The velocity of the satellite becomes

$$\vec{v}_{S/O} = \dot{r}\,\vec{e}_r + r\dot{\theta}\,\vec{e}_\theta = ha_1 \sin \theta \,\vec{e}_r + \frac{h}{r}\,\vec{e}_\theta \tag{5.71}$$

Plugging the velocity condition at the entry point, we obtain

$$h = r_0 v_0 \tag{5.72}$$

Also, introducing $\xi = 1/r_0$ at the entry point to Equation (5.70), we obtain $a_1$ as

$$a_1 = \frac{1}{r_0} - \frac{GM_e}{h^2} = \frac{1}{r_0}\left(1 - \frac{GM_e}{r_0 v_0^2}\right) \tag{5.73}$$

**Solution Verification:** With Equations (5.70) and (5.73), we can determine the orbits for different entry speeds at the same entry position. We use an entry point similar to the orbit of ISS, at an altitude of 420 km and an entry speed of 17238 km/hr. The resulting orbit is a circular path, as shown in Figure 5.9. At 120% of the entry speed at the same entry point, an oval orbit (Orbit A) is generated as shown in Figure 5.9.

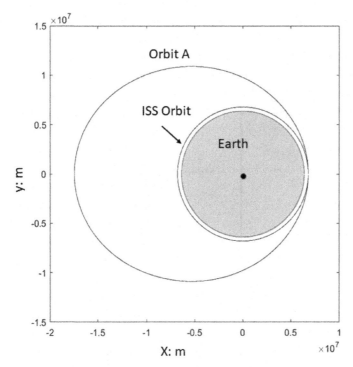

Figure 5.9: Simulated orbits for ISS and a higher entry speed.

## Example 5.5: A Simple Pulley Problem

**Problem Statement:** As shown in Figure 5.10(a), two objects are tied together with a cord around a pulley and held in place. At $t_0 = 0$, the objects are released from at rest. The mass of object $A$ is $10\,\mathrm{kg}$, while $20\,\mathrm{kg}$ for B. Determine the acceleration of the objects at the instant when they are released.

**Solution:**

**Force Analysis:** The FBD analysis for both objects is shown in Figure 5.10(b) using the ABCC method.

**Motion Analysis:** Both objects are in the straight line motion. The motions of the objects are dependent because one will go up while the other goes down at the same rate, defined as $\ddot{y}$. Because object $B$ has a larger mass, it will go down and object $A$ will be pulled up.

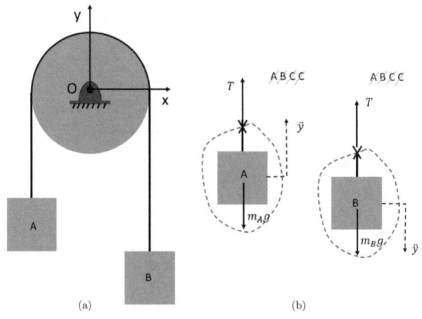

(a)                   (b)

Figure 5.10: (a) Acceleration of two blocks, tied around a pulley, released from the stationary state; and (b) the FBD analysis.

We mark the direction of the accelerations accordingly. Based on the inertial reference frame $O-x-y$, going up is considered as positive.

**Governing Equations:** We can write down two equations using *Newton's Second Law.*

$$T - m_A g = m_A \ddot{y} = 10\ddot{y} \tag{5.74}$$

$$T - m_B g = -m_B \ddot{y} = -20\ddot{y} \tag{5.75}$$

Note that in Equation (5.75), the right-hand side has a negative sign because the acceleration of $B$ is going downward.

**Solving for Unknowns:** There are only two unknowns, $T$ and $\ddot{y}$, involved in Equations (5.74) and (5.75). We can solve for them readily. We obtain $\ddot{y} = 3.27\,\text{m/s}^2$ and $T = 130.8\,\text{N}$.

**Solution Verification:** This is a very simple problem. The objective is to show that two FBDs are needed for establishing two governing equations. In addition, it is important to go with the sign convention, particularly for Equation (5.75). If we plug in the values of $T$ and $\ddot{y}$ back to Equation (5.75), we will see indeed that a negative sign is needed on the right-hand side of the equation.

## 5.3 *Newton's Second Law* for a Rigid Body

We will extend *Newton's Second Law* to a rigid body. From Section 2.3, Chapter 2, we learned how to move all the forces to a specific point of a rigid body and express the entire loading condition as one force and the resulting moment, as shown in Figure 2.10. We also learned how to represent the motion of a rigid body in Chapter 4. We simply need to link the force/moment to the motion using *Newton's Second Law*, as shown in Figure 5.11. The mass center of the rigid body is one preferred reference point to move all the forces to.

It turns out the extension of *Newton's Second Law* from a particle to a rigid body is not trivial. We need to introduce a few new concepts before we can link the motion of a rigid body to its loading condition.

### 5.3.1 *Linear momentum and angular momentum*

In this section, we introduce two important concepts, linear momentum and angular momentum, in dynamics. Referring to

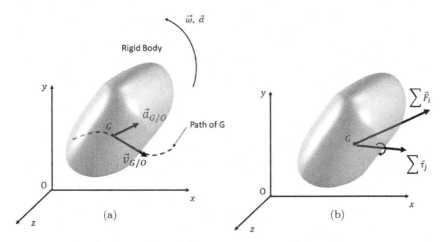

Figure 5.11: (a) The mass center and motion of a rigid body; and (b) the total force and moment on a rigid body with respect to the mass center.

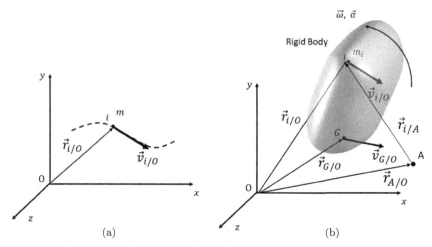

Figure 5.12: The definitions of linear momentum and angular momentum for (a) a particle; and (b) a rigid body.

Figure 5.12(a), the linear momentum of a particle with mass $m$ and velocity $\vec{v}_{i/O}$ with respect to an inertial reference frame is defined as

$$\vec{P}_{i/O} = m\vec{v}_{i/O} \tag{5.76}$$

Correspondingly, the angular momentum with respect to $O$ is

$$\vec{H}_{i/O} = \vec{r}_{i/O} \times m\vec{v}_{i/O} = \vec{r}_{i/O} \times \vec{P}_{i/O} \tag{5.77}$$

The significance of the linear momentum and the angular momentum will soon be clear. If we take a time derivative of Equation (5.76), we have

$$\dot{\vec{P}}_{i/O} = \dot{m}\vec{v}_{i/O} + m\dot{\vec{v}}_{i/O} = m\vec{a}_{i/O} = \sum \vec{F} \tag{5.78}$$

assuming that the mass of this particle is unchanged. It turns out that we can generalize *Newton's Second Law* as

$$\sum \vec{F} = \dot{\vec{P}}_{i/O} \tag{5.79}$$

We can allow the mass to change when we deal with multiple-particle systems.

Similarly, let's take a time derivative of Equation (5.77),

$$\dot{\vec{H}}_{i/O} = \dot{\vec{r}}_{i/O} \times m\vec{v}_{i/O} + \vec{r}_{i/O} \times m\dot{\vec{v}}_{i/O}$$

$$= \vec{r}_{i/O} \times m\dot{\vec{v}}_{i/O} = \vec{r}_{i/O} \times \sum \vec{F} \tag{5.80}$$

Note that $\dot{\vec{r}}_{i/O} \times m\vec{v}_{i/O} = 0$ because $\dot{\vec{r}}_{i/O} = \vec{v}_{i/O}$. We recognize that $\vec{r}_{i/O} \times \sum \vec{F}$ is the moment of $\sum \vec{F}$ with respect to point $O$. Therefore, we have

$$\sum \vec{\tau}_O = \vec{r}_{i/O} \times \sum \vec{F} = \dot{\vec{H}}_{i/O} \tag{5.81}$$

Equations (5.79) and (5.81) have a nice symmetry. If we apply Equation (5.81) to Example 5.4, we have

$$\sum \vec{\tau}_O = 0 = \dot{\vec{H}}_{S/O} \tag{5.82}$$

where

$$\vec{H}_{S/O} = \vec{r} \times m(\dot{r}\vec{e}_r + r\dot{\theta}\vec{e}_\theta) = \vec{r} \times (mr\dot{\theta}\vec{e}_\theta) \tag{5.83}$$

Equation (5.82) indicates that the angular momentum of a satellite with respect to the Earth is a constant, which is essentially the result of Equation (5.60).

### 5.3.2 *Linear momentum and angular momentum for a rigid body*

Let's extend the definition of the linear momentum and the angular momentum to a rigid body. We will use the summation operator rather than the integration operator in the derivation for the purpose of simpler illustration. As shown in Figure 5.12(b), an arbitrary point $i$ has a mass $m_i$, with velocity $\vec{v}_{i/O}$, and position vector $\vec{r}_{i/O}$. The total linear momentum of the rigid body is the sum of all points of the rigid body,

$$\vec{P} = \sum_i \vec{P}_{i/O} = \sum_i m_i \vec{v}_{i/O}$$

$$= \sum_i m_i (\vec{v}_{G/O} + \vec{\omega} \times \vec{r}_{i/G})$$

$$= \left( \sum_i m_i \right) \vec{v}_{G/O} + \vec{\omega} \times \left( \sum_i m_i \vec{r}_{i/G} \right)$$

$$= M \vec{v}_{G/O} \tag{5.84}$$

where $M = \sum_i m_i$ and $\sum_i m_i \vec{r}_{i/G} = 0$ by the definition of mass center, $G$.

Equation (5.84) indicates that the total linear momentum of a rigid body is equal to the total mass times the velocity of the mass center. It is as if the mass center is a particle with the total mass, carrying the total linear momentum of the entire rigid body.

Let's construct an FBD for one point of the rigid body, with $\vec{F}_i$ as the external force acting at point $i$ while $\vec{f}_i$ is the internal force due to the constraint forces of $i$ with surrounding particles. We apply *Newton's Second Law* to one particle and then add up all the equations,

$$\sum_i (\vec{F}_i + \vec{f}_i) = \sum_i \vec{F}_i = \sum_i m_i \vec{a}_{i/O}$$

$$= \sum_i m_i [\vec{a}_{G/O} + \vec{\alpha} \times \vec{r}_{i/G} + \vec{\omega} \times (\vec{\omega} \times \vec{r}_{i/G})]$$

$$= M \vec{a}_{G/O} \tag{5.85}$$

Because it is a rigid body, we can move all the external forces to the mass center and add them up. All the internal forces will be canceled because of *Newton's Third Law*.

Comparing Equations (5.84) and (5.85), we realized that Equation (5.79) can be applied to a rigid body with respect to the mass center. Therefore, we have

$$\sum \vec{F} = \dot{\vec{P}} = M \vec{a}_{G/O} \tag{5.86}$$

However, Equation (5.86) does not contain any information about the rigid body rotation. Therefore, it is not the complete dynamics description of the rigid body.

Let's define the total angular momentum of the rigid body. We will consider different reference points when defining the total angular momentum. As shown in Figure 5.12(b), we will define the total angular momentum with respect to point $O$ and $A$, respectively. Within the reference frame $O-x-y-z$, $O$ is a stationary point, while Point $A$ is an arbitrary point and we do not specify the motion property of Point $A$ for now.

For $m_i$, the angular momentum can be defined as

$$\vec{H}_{i/O} = \vec{r}_{i/O} \times m_i \vec{v}_{i/O} = \vec{r}_{i/O} \times \vec{P}_{i/O} \tag{5.87}$$

$$\vec{H}_{i/A} = \vec{r}_{i/A} \times m_i \vec{v}_{i/O} = \vec{r}_{i/A} \times \vec{P}_{i/O} \tag{5.88}$$

Note that only the position vectors are different in the above equations.

We add up the angular momentum of every point of the rigid body to determine the total angular momentum.

$$\vec{H}_O = \sum_i (\vec{r}_{i/O} \times m_i \vec{v}_{i/O}) \tag{5.89}$$

$$\vec{H}_A = \sum_i (\vec{r}_{i/A} \times m_i \vec{v}_{i/O}) = \sum_i [(\vec{r}_{i/O} - \vec{r}_{A/O}) \times m_i \vec{v}_{i/O}]$$

$$= \sum_i (\vec{r}_{i/O} \times m_i \vec{v}_{G/O}) + \sum_i (\vec{r}_{i/O} \times m_i \vec{v}_{i/G})$$

$$- \sum_i (\vec{r}_{A/O} \times m_i \vec{v}_{G/O}) - \sum_i (\vec{r}_{A/O} \times m_i \vec{v}_{i/G})$$

$$= M\vec{r}_{G/O} \times \vec{v}_{G/O} + \sum_i [(\vec{r}_{i/G} + \vec{r}_{G/O}) \times m_i \vec{v}_{i/G}]$$

$$- \vec{r}_{A/O} \times M\vec{v}_{G/O}$$

$$= M\vec{r}_{G/O} \times \vec{v}_{G/O} + \vec{H}_G - \vec{r}_{A/O} \times M\vec{v}_{G/O} \tag{5.90}$$

where $\vec{H}_G = \sum_i (\vec{r}_{i/G} \times m_i \vec{v}_{i/G})$, which is defined as the angular momentum with respect to the mass center.

In the above derivation, we recognize that $\sum_i (m_i \vec{v}_{i/G}) = 0$ and $\sum_i (m_i \vec{r}_{i/O}) = M\vec{r}_{G/O}$.

### 5.3.3 *Governing equations for a rigid body*

Let's take a derivative of $\vec{H}_A$, which leads to

$$\dot{\vec{H}}_A = M\vec{r}_{G/O} \times \dot{\vec{v}}_{G/O} + \sum_i (\vec{r}_{i/G} \times m_i \dot{\vec{v}}_{i/G})$$

$$- \dot{\vec{r}}_{A/O} \times M\vec{v}_{G/O} - \vec{r}_{A/O} \times M\dot{\vec{v}}_{G/O} \qquad (5.91)$$

The term $-\dot{\vec{r}}_{A/O} \times M\vec{v}_{G/O}$ can be eliminated if we require point $A$ to be a stationary point. This is highly important because without this requirement, the equation would be more complicated when we derive it further. Equation (5.91) now becomes

$$\dot{\vec{H}}_A = M\vec{r}_{G/O} \times \dot{\vec{v}}_{G/O} + \sum_i (\vec{r}_{i/G} \times m_i \dot{\vec{v}}_{i/G}) - \vec{r}_{A/O} \times M\dot{\vec{v}}_{G/O}$$

$$= M(\vec{r}_{G/O} - \vec{r}_{A/O}) \times \vec{a}_{G/O} + \sum_i [\vec{r}_{i/G} \times m_i(\vec{a}_{i/O} - \vec{a}_{G/O})]$$

$$= \vec{r}_{G/A} \times M\vec{a}_{G/O} + \sum_i (\vec{r}_{i/G} \times m_i \vec{a}_{i/O}) = \vec{r}_{G/A} \times M\vec{a}_{G/O} + \dot{\vec{H}}_G$$

$$= \vec{r}_{G/A} \times \sum_i \vec{F}_i + \sum_i [\vec{r}_{i/G} \times (\vec{F}_i + \vec{f}_i)]$$

$$= \vec{r}_{G/A} \times \sum_i \vec{F}_i + \sum_i (\vec{r}_{i/G} \times \vec{F}_i) = \vec{r}_{G/A} \times \sum_i \vec{F}_i + \sum \vec{\tau}_G$$

$$= \sum_i (\vec{r}_{i/A} \times \vec{F}_i) = \sum \tau_A \qquad (5.92)$$

where the internal force term is eliminated in the above equation because every internal force pair shares the same position vector. Moreover, the term $\sum_i (\vec{r}_{i/G} \times \vec{F}_i)$ is the sum of the moment of each external force with respect to the mass center, denoted as $\sum \vec{\tau}_G$.

By comparing the terms in Equation (5.92), we also realize that $\sum \vec{\tau}_G = \dot{\vec{H}}_G$.

Because $O$ is a stationary point, letter $A$ can be replaced with letter $O$ in Equation (5.92). There might be questions that the above derivations did not consider force couples, which are free vectors (see Section 2.3). In fact, we did. Because we will consider the force pair of a force couple as two individual external forces acting on two different points. Finally, we have

$$\sum \vec{\tau}_A = \dot{\vec{H}}_A = \sum_i (\vec{r}_{i/A} \times m_i \dot{\vec{v}}_{i/O}) \tag{5.93}$$

$$\sum \vec{\tau}_G = \dot{\vec{H}}_G = \sum_i (\vec{r}_{i/G} \times m_i \dot{\vec{v}}_{i/G}) \tag{5.94}$$

where point $A$ must be a stationary point.

To completely define the dynamics of a rigid body, we can pair Equation (5.86) with either Equation (5.93) or (5.94). We will express them together as

$$\sum \vec{F} = \dot{\vec{P}} = M \vec{a}_{G/O} \tag{5.95}$$

$$\sum \vec{\tau}_A = \dot{\vec{H}}_A \tag{5.96}$$

or

$$\sum \vec{F} = \dot{\vec{P}} = M \vec{a}_{G/O} \tag{5.97}$$

$$\sum \vec{\tau}_G = \dot{\vec{H}}_G \tag{5.98}$$

We can choose either Equations (5.95) and (5.96) or (5.97) and (5.98) as the generalized form of *Newton's Second Law* for a rigid body. In general, Equations (5.97) and (5.98) are easier to use. If Equations (5.95) and (5.96) are chosen, one must ensure point $A$ is a stationary point. However, for Equations (5.97) and (5.98), the mass center does not have to be a stationary point, which makes them less restrictive and easier to use.

### 5.3.4 *Moment of inertia*

Although Equations (5.96) and (5.98) are neat, the calculation of $\vec{H}_G$ or $\vec{H}_A$ often is difficult and even more so for $\dot{\vec{H}}_G$ or $\dot{\vec{H}}_A$.

Let's carry out the determination of $\vec{H}_A$ or $\vec{H}_G$ in a general form. We will use the integration operator instead of the summation this time.

From Equation (5.90), we know that we only need to define $\vec{H}_G$, which will then determine $\vec{H}_A$.

$$\vec{H}_G = \int [\vec{r}_{i/G} \times (dm \vec{v}_{i/G})] \tag{5.99}$$

where $\vec{r}_{i/G}$ and $\vec{v}_{i/G}$ are the position and velocity vectors of an arbitrary point on the rigid body with respect to the mass center, respectively. The reference frame attached to the mass center is the same as the inertial reference frame, $O-x-y-z$. The rigid body has an angular velocity $\vec{\omega}$. Therefore, the position and the velocity vectors are

$$\vec{r}_{i/G} = x_{i/G}\vec{i} + y_{i/G}\vec{j} + z_{i/G}\vec{k} = x\vec{i} + y\vec{j} + z\vec{k} \qquad (5.100)$$

$$\vec{v}_{i/G} = \vec{\omega} \times \vec{r}_{i/G} \qquad (5.101)$$

where $\vec{\omega} = \omega_x\vec{i} + \omega_y\vec{j} + \omega_z\vec{k}$. The variables, $x$, $y$, and $z$ are used for simpler notation for the position vector, but we should be aware that all definitions from here on are with respect to the mass center.

Introducing Equations (5.100) and (5.101) to (5.99) and carrying out the cross multiplication, we obtain

$$\vec{H}_G = \int [y^2\omega_x - xy\omega_y - xz\omega_z + z^2\omega_x]dm\ \vec{i}$$

$$+ \int [z^2\omega_y - yz\omega_z - xy\omega_x + x^2\omega_y]dm\ \vec{j}$$

$$+ \int [x^2\omega_z - xz\omega_x - yz\omega_y + y^2\omega_z]dm\ \vec{k} \qquad (5.102)$$

Rewriting Equation (5.102) in a matrix form gives

$$\vec{H}_G = \left( \int \begin{bmatrix} y^2 + z^2 & -xy & -xz \\ -xy & x^2 + z^2 & -yz \\ -xz & -yz & x^2 + y^2 \end{bmatrix} dm \right) \begin{bmatrix} \omega_x \\ \omega_y \\ \omega_z \end{bmatrix}$$

$$= \begin{bmatrix} I_{xx} & -I_{xy} & -I_{xz} \\ -I_{xy} & I_{yy} & -I_{yz} \\ -I_{xz} & -I_{yz} & I_{zz} \end{bmatrix} \begin{bmatrix} \omega_x \\ \omega_y \\ \omega_z \end{bmatrix}$$

$$= I_G\vec{\omega} \qquad (5.103)$$

where the moment of inertia terms are defined accordingly as

$$I_{xx} = \int (y^2 + z^2)\, dm \qquad (5.104)$$

$$I_{yy} = \int (x^2 + z^2)\, dm \qquad (5.105)$$

$$I_{zz} = \int (x^2 + y^2)\, dm \tag{5.106}$$

$$I_{xy} = \int xy\, dm \tag{5.107}$$

$$I_{yz} = \int yz\, dm \tag{5.108}$$

$$I_{xz} = \int xz\, dm \tag{5.109}$$

In practice, Equations (5.104)–(5.109) are of little use because these moment of inertia terms are functions of time and they keep changing when the rigid body rotates. However, if we choose a different reference frame attached to $G$, we might be able to make the moment of inertia matrix independent of time.

Let's use the rotating rigid body of Example 4.3 for illustration. Point $A$ is fixed in Figure 5.13.

Based on non-rotating reference frame, $X-Y-Z$ attached to $G$, the angular momentum of rigid body $B$ is

$$\vec{H}_G = I_G \vec{\omega}_B \tag{5.110}$$

where $\vec{\omega}_B$ is the angular velocity of $B$ with respect to a non-rotating reference frame.

We can calculate $I_G$ for the instant shown in Figure 5.13; however, when rigid body $B$ moves to a new position, $I_G$ will be different.

We can express $\vec{H}_G$ using an inertial reference frame, $G-\tilde{x}-\tilde{y}-\tilde{z}$, which overlaps $G-\hat{x}-\hat{y}-\hat{z}$ at the instant shown in Figure 5.13. Note that $G-\hat{x}-\hat{y}-\hat{z}$ is a non-inertial reference frame, moving with $B$ and observing a rigid body rotation as $\omega_{\hat{x}}\hat{i}$ only.

We obtain

$$\vec{H}_G = [\vec{i}\ \vec{j}\ \vec{k}] \begin{bmatrix} H_{Gx} \\ H_{Gy} \\ H_{Gz} \end{bmatrix} = [\vec{i}\ \vec{j}\ \vec{k}]\, I_G \begin{bmatrix} \omega_x \\ \omega_y \\ \omega_z \end{bmatrix}$$

$$= \tilde{H}_G = [\tilde{i}\ \tilde{j}\ \tilde{k}] \begin{bmatrix} \tilde{H}_{Gx} \\ \tilde{H}_{Gy} \\ \tilde{H}_{Gz} \end{bmatrix} = [\tilde{i}\ \tilde{j}\ \tilde{k}]\, \tilde{I}_G \begin{bmatrix} \tilde{\omega}_{\tilde{x}} \\ \tilde{\omega}_{\tilde{y}} \\ \tilde{\omega}_{\tilde{z}} \end{bmatrix} \tag{5.111}$$

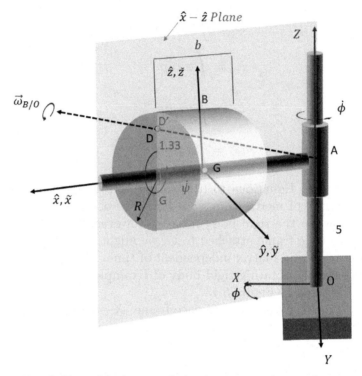

Figure 5.13: A 3D rigid body example for determining the angular momentum.

Note that, we introduce $[\vec{i}\ \vec{j}\ \vec{k}]$ and $[\tilde{i}\ \tilde{j}\ \tilde{k}]$ back in Equation (5.111) to avoid giving out a wrong indication that $[\tilde{H}_{Gx}\ \tilde{H}_{Gy}\ \tilde{H}_{Gz}]'$ is equal to $[H_{Gx}\ H_{Gy}\ H_{Gz}]'$. They are not equal in the scalar value but the same when the unit vectors are included.

The transformation between the two inertial reference frames, $G{-}X{-}Y{-}Z$ and $G{-}\tilde{x}{-}\tilde{y}{-}\tilde{z}$, is defined as

$$\begin{bmatrix} X \\ Y \\ Z \end{bmatrix} = T_t(\theta_x, \theta_y, \theta_z) \begin{bmatrix} \tilde{x} \\ \tilde{y} \\ \tilde{z} \end{bmatrix} \tag{5.112}$$

As discussed in Chapters 3 and 4, we also have

$$\begin{bmatrix} \vec{i} \\ \vec{j} \\ \vec{k} \end{bmatrix} = T_t(\theta_x, \theta_y, \theta_z) \begin{bmatrix} \tilde{i} \\ \tilde{i} \\ \tilde{k} \end{bmatrix} \tag{5.113}$$

We can also transpose Equation (5.113) to have

$$\begin{bmatrix} \vec{i} & \vec{j} & \vec{k} \end{bmatrix} = \begin{bmatrix} \tilde{i} & \tilde{k} & \tilde{k} \end{bmatrix} T_t'(\theta_x, \theta_y, \theta_z) \tag{5.114}$$

$$\begin{bmatrix} \vec{i} & \vec{j} & \vec{k} \end{bmatrix} (T_t')^{-1}(\theta_x, \theta_y, \theta_z) = \begin{bmatrix} \tilde{i} & \tilde{k} & \tilde{k} \end{bmatrix} \tag{5.115}$$

where $'$ is the matrix transpose operation. We will replace $T_t(\theta_x, \theta_y, \theta_z)$ as $T$ for simpler notation, as follows.

Note that the angular velocities observed by different inertial reference frames at the mass center are the same but expressed in different unit vectors and have different component values in those directions. Because they are still the same vector, we have

$$\vec{\omega} = \begin{bmatrix} \vec{i} & \vec{j} & \vec{k} \end{bmatrix} \begin{bmatrix} \omega_x \\ \omega_y \\ \omega_z \end{bmatrix}$$

$$= \begin{bmatrix} \tilde{i} & \tilde{j} & \tilde{k} \end{bmatrix} \begin{bmatrix} \tilde{\omega}_x \\ \tilde{\omega}_y \\ \tilde{\omega}_z \end{bmatrix}$$

$$= \begin{bmatrix} \vec{i} & \vec{j} & \vec{k} \end{bmatrix} (T')^{-1} \begin{bmatrix} \tilde{\omega}_x \\ \tilde{\omega}_y \\ \tilde{\omega}_z \end{bmatrix} \tag{5.116}$$

From Equation (5.116), we have

$$(T') \begin{bmatrix} \omega_x \\ \omega_y \\ \omega_z \end{bmatrix} = \begin{bmatrix} \tilde{\omega}_x \\ \tilde{\omega}_y \\ \tilde{\omega}_z \end{bmatrix} \tag{5.117}$$

Similarly, for the angular momentum, we have

$$\vec{H}_G = \begin{bmatrix} \vec{i} & \vec{j} & \vec{k} \end{bmatrix} \begin{bmatrix} H_{Gx} \\ H_{Gy} \\ H_{Gz} \end{bmatrix} = \begin{bmatrix} \vec{i} & \vec{j} & \vec{k} \end{bmatrix} I_G \begin{bmatrix} \omega_x \\ \omega_y \\ \omega_z \end{bmatrix}$$

$$= \tilde{H}_G = [\vec{i}\ \vec{j}\ \vec{k}] \begin{bmatrix} \tilde{H}_{Gx} \\ \tilde{H}_{Gy} \\ \tilde{H}_{Gz} \end{bmatrix} = [\vec{i}\ \vec{j}\ \vec{k}]\ (T')^{-1}\ \tilde{I}_G \begin{bmatrix} \tilde{\omega}_x \\ \tilde{\omega}_y \\ \tilde{\omega}_z \end{bmatrix}$$

$$= [\vec{i}\ \vec{j}\ \vec{k}]\ (T')^{-1}\ \tilde{I}_G T' \begin{bmatrix} \omega_x \\ \omega_y \\ \omega_z \end{bmatrix} \tag{5.118}$$

From Equation (5.118), we obtain

$$\begin{bmatrix} H_{Gx} \\ H_{Gy} \\ H_{Gz} \end{bmatrix} = I_G \begin{bmatrix} \omega_x \\ \omega_y \\ \omega_z \end{bmatrix}$$

$$= (T')^{-1}\ \tilde{I}_G \begin{bmatrix} \tilde{\omega}_{\tilde{x}} \\ \tilde{\omega}_{\tilde{y}} \\ \tilde{\omega}_{\tilde{z}} \end{bmatrix}$$

$$= (T')^{-1}\ \tilde{I}_G T' \begin{bmatrix} \omega_x \\ \omega_y \\ \omega_z \end{bmatrix} \tag{5.119}$$

From the Equation (5.119), we have

$$I_G = (T')^{-1}\tilde{I}_G\ T' \tag{5.120}$$

### 5.3.5 *Properties of $I_G$ and $T$*

Equation (5.103) indicates that $I_G$ is a symmetric matrix. Therefore, $I'_G = I_G$ and $\tilde{I}'_G = \tilde{I}_G$. Applying this property to Equation (5.120), we have

$$I_G = (T')^{-1}\ \tilde{I}_G\ T' = I'_G = T\ \tilde{I}_G\ ((T')^{-1})' \tag{5.121}$$

Equation (5.121) leads to the same identity that

$$T = (T')^{-1}$$
$$\implies T'T = I$$
$$\implies T' = T^{-1} \tag{5.122}$$

where $I$ is an identity matrix.

Equation (5.120) can now be rewritten in a simpler form,

$$I_G = T \; \tilde{I}_G \; T' = T \; \tilde{I}_G \; T^{-1} \tag{5.123}$$

It is also possible to choose a reference frame so that $\tilde{I}_G$ is a diagonal matrix, i.e., $I_{mn} = 0$ if $m \neq n$. Such a reference frame with respect to the rigid body is called the principal axes. It would be a preferred reference frame to define moment of inertia of a rigid body.

Figure 5.14 lists a few cases for some symmetric rigid bodies and the resulting moment of inertia matrices.

### 5.3.6 *Systematic method to determine angular momentum*

Representing $\vec{H}_G$ based on $\tilde{I}_G$ is easier because this inertia matrix is independent of time.

The following are the systematic procedures for establishing the angular momentum of a rigid body in a general motion. We will illustrate these procedures using Figure 5.13 as an example.

Step 1: Identify the rigid body and choose the principle axes as the reference frame $G - \hat{x} - \hat{y} - \hat{y}$.

Regarding Figure 5.13, $\hat{x}$ is the axial axis of the cylinder.

Step 2: Establish the relationship between the rigid body rotation and $G - \hat{x} - \hat{y} - \hat{y}$.

In general, with $G - \hat{x} - \hat{y} - \hat{y}$, we want to observe the rigid body to have only one simple rotation around one axis. This is a very useful strategy, similar to the relative motion analysis when we make a motion simple again, as discussed in Chapters 3 and 4.

Regarding Figure 5.13, we choose $G - \hat{x} - \hat{y} - \hat{y}$ so that the rigid body appears to have only one rotation, $\hat{\omega} = \dot{\psi}\hat{i}$. In other words, $G - \hat{x} - \hat{y} - \hat{y}$ is fixed to cylinder $B$, but not spinning with it.

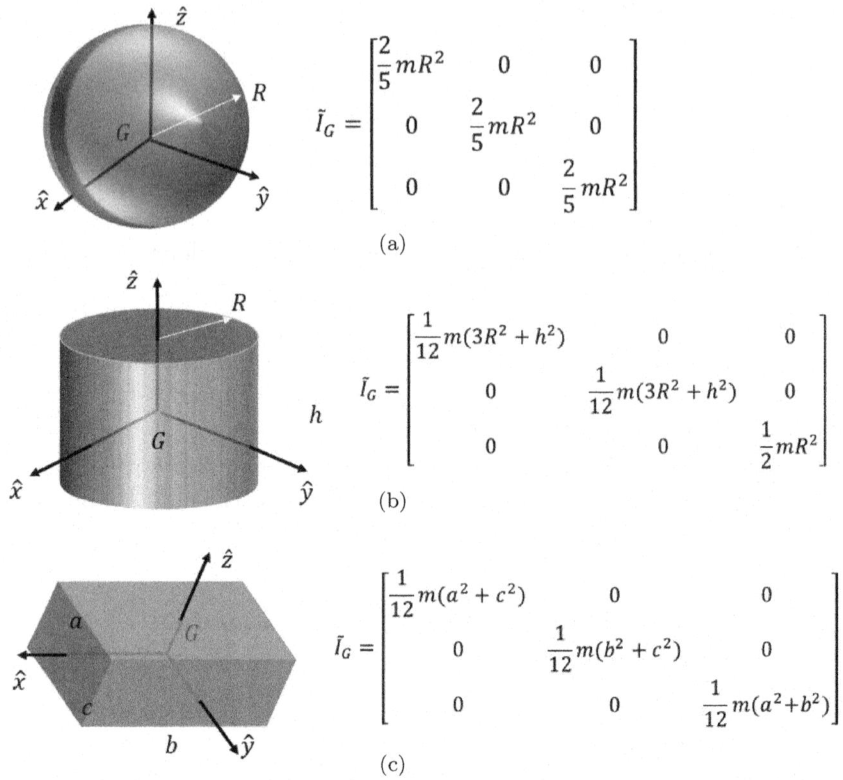

$$\tilde{I}_G = \begin{bmatrix} \frac{2}{5}mR^2 & 0 & 0 \\ 0 & \frac{2}{5}mR^2 & 0 \\ 0 & 0 & \frac{2}{5}mR^2 \end{bmatrix}$$

(a)

$$\tilde{I}_G = \begin{bmatrix} \frac{1}{12}m(3R^2 + h^2) & 0 & 0 \\ 0 & \frac{1}{12}m(3R^2 + h^2) & 0 \\ 0 & 0 & \frac{1}{2}mR^2 \end{bmatrix}$$

(b)

$$\tilde{I}_G = \begin{bmatrix} \frac{1}{12}m(a^2 + c^2) & 0 & 0 \\ 0 & \frac{1}{12}m(b^2 + c^2) & 0 \\ 0 & 0 & \frac{1}{12}m(a^2 + b^2) \end{bmatrix}$$

(c)

Figure 5.14: Moment of inertia for different geometries: (a) a sphere; (b) a cylinder; and (c) a cube.

Step 3: Establish the moment of inertia matrix with respect to $G-\hat{x}-\hat{y}-\hat{y}$.

For the case of Figure 5.13, based on Figure 5.14, we found

$$\hat{I}_G = \begin{bmatrix} \frac{1}{2}mR^2 & 0 & 0 \\ 0 & \frac{1}{12}m(3R^2 + b^2) & 0 \\ 0 & 0 & \frac{1}{12}m(3R^2 + b^2) \end{bmatrix} \qquad (5.124)$$

where $I_{xx} = \frac{1}{2}mR^2$ and $I_{yy} = I_{zz} = \frac{1}{12}m(3R^2 + b^2)$.

Step 4: Observe the frame rotation of $G-\hat{x}-\hat{y}-\hat{y}$.
We might need to establish a series of reference frames to do this task. Each additional reference frame should observe only one simple rotation around one axis. For Figure 5.13, the case is simple. With respect to an inertial reference frame, $G-X-Y-Z$, $G-\hat{x}-\hat{y}-\hat{y}$ has only one rotation as $\vec{\Omega} = \dot{\phi}\vec{K}$.

Step 5: Determine the rigid body angular velocity.
Using the additivity of angular velocity, we easily determine that

$$\vec{\omega}_{B/O} = \dot{\psi}\hat{i} + \dot{\phi}\vec{K} \tag{5.125}$$

Step 6: Establish an inertial reference frame $G-\tilde{x}-\tilde{y}-\tilde{y}$.
This inertial reference frame happens to overlap with $G-\hat{x}-\hat{y}-\hat{y}$ at this instant of time. We express $\vec{\omega}_{B/O}$ using the unit directional vectors defined by $G-\tilde{x}-\tilde{y}-\tilde{y}$.

$$\vec{\omega}_{B/O} = \tilde{\omega}_{B/O} = \dot{\psi}\hat{i} + \dot{\phi}\vec{K}$$
$$= \dot{\psi}\tilde{i} + \dot{\phi}\tilde{k} \tag{5.126}$$

Note that $\tilde{\omega}_{B/O}$ is the full description of the angular velocity, only expressed in $\tilde{i}$, $\tilde{j}$, and $\tilde{k}$ for this instant. However, we only need to recognize that $\tilde{i}$, $\tilde{j}$, and $\tilde{k}$ are not fixed in space and have a frame rotation, $\vec{\Omega} = \dot{\phi}\vec{K}$. This recognition is important when we try to take a time derivative of $\tilde{\omega}$.

Step 7: Establish the transformation between $G-X-Y-Z$ and $G-\tilde{x}-\tilde{y}-\tilde{z}$.
We should use the little moving coordinate concept discussed in Section 3.10.2 for establishing the transformation between $G-X-Y-Z$ and $G-\tilde{x}-\tilde{y}-\tilde{z}$. For the case of Figure 5.13,

$$\begin{bmatrix} X \\ Y \\ Z \end{bmatrix} = T_{tz}(\phi) \begin{bmatrix} \tilde{x} \\ \tilde{y} \\ \tilde{z} \end{bmatrix} = T \begin{bmatrix} \tilde{x} \\ \tilde{y} \\ \tilde{z} \end{bmatrix}$$
$$= \begin{bmatrix} \cos\phi & -\sin\phi & 0 \\ \sin\phi & \cos\phi & 0 \\ 0 & 0 & 1 \end{bmatrix} \begin{bmatrix} \tilde{x} \\ \tilde{y} \\ \tilde{z} \end{bmatrix} \tag{5.127}$$

With the transformation matrix, we can represent $\vec{\omega}_{B/O}$ with respect to $G-X-Y-Z$,

$$\vec{\Omega}_{B/O} = \dot{\psi}\cos\phi\vec{i} + \dot{\psi}\sin\phi\vec{j} + \dot{\phi}\vec{k} \qquad (5.128)$$

Step 8: Determine the angular momentum with respect to $G-\tilde{x}-\tilde{y}-\tilde{z}$.

First, we recognize that $\tilde{I}_G = \hat{I}_G$. The corresponding angular momentum of $B$ becomes

$$\vec{H}_G = [\vec{i} \quad \vec{j} \quad \vec{k}] \; I_G \begin{bmatrix} \dot{\psi}\cos\phi \\ \dot{\psi}\sin\phi \\ \dot{\phi} \end{bmatrix}$$

$$= \tilde{H}_G = [\tilde{i} \quad \tilde{j} \quad \tilde{k}] \; \tilde{I}_G \begin{bmatrix} \dot{\psi} \\ 0 \\ \dot{\phi} \end{bmatrix}$$

$$= [\tilde{i} \quad \tilde{j} \quad \tilde{k}] \begin{bmatrix} I_{xx}\dot{\psi} \\ 0 \\ I_{yy}\dot{\phi} \end{bmatrix}$$

$$= [\tilde{i} \quad \tilde{j} \quad \tilde{k}] \begin{bmatrix} \frac{1}{2}mR^2\dot{\psi} \\ 0 \\ \frac{1}{12}m(3R^2 + b^2)\dot{\phi} \end{bmatrix} \qquad (5.129)$$

Via Equation (5.123), we found that

$$I_G = T\hat{I}_G T^{-1} = T\hat{I}_G T'$$

$$= \begin{bmatrix} I_{xx}\cos^2\phi + I_{yy}\sin^2\phi & (I_{xx} - I_{yy})\cos\phi\sin\phi & 0 \\ (I_{xx} - I_{yy})\cos\phi\sin\phi & I_{xx}\sin^2\phi + I_{yy}\cos^2\phi & 0 \\ 0 & 0 & I_{yy} \end{bmatrix}$$

$$(5.130)$$

According to Equation (5.130), $I_G$ is a function of $\phi$; thus, a function of time. As a result, there is no advantage to express the angular momentum of $B$ with respect to $G-X-Y-Z$. The expression of $\tilde{H}_G$ is preferred.

### 5.3.7 *Derivative of angular momentum*

In order to use *Newton's Second Law* for rigid bodies, we need to know the derivative of the angular momentum. We will use the expression of $\tilde{H}_G$ because $\tilde{I}_G$ is not a function of time. Based on Equation (5.129), we have

$$\frac{d}{dt}(\tilde{H}_G) = \frac{d}{dt}\left( [\tilde{i} \quad \tilde{j} \quad \tilde{k}] \begin{bmatrix} \tilde{H}_{Gx} \\ \tilde{H}_{Gy} \\ \tilde{H}_{Gz} \end{bmatrix} \right)$$

$$= \frac{d}{dt}([\tilde{i} \quad \tilde{j} \quad \tilde{k}]) \begin{bmatrix} \tilde{H}_{Gx} \\ \tilde{H}_{Gy} \\ \tilde{H}_{Gz} \end{bmatrix} + [\tilde{i} \quad \tilde{j} \quad \tilde{k}]\frac{\partial}{\partial t}\left( \begin{bmatrix} \tilde{H}_{Gx} \\ \tilde{H}_{Gy} \\ \tilde{H}_{Gz} \end{bmatrix} \right)$$

$$= \vec{\Omega} \times ([\tilde{i} \quad \tilde{j} \quad \tilde{k}] \begin{bmatrix} \tilde{H}_{Gx} \\ \tilde{H}_{Gy} \\ \tilde{H}_{Gz} \end{bmatrix}) + [\tilde{i} \quad \tilde{j} \quad \tilde{k}]\frac{\partial}{\partial t}\left( \begin{bmatrix} \hat{H}_{Gx} \\ \hat{H}_{Gy} \\ \hat{H}_{Gz} \end{bmatrix} \right)$$

$$= \vec{\Omega} \times \tilde{H}_G + \frac{\partial}{\partial t}\tilde{H}_G \qquad (5.131)$$

We must point out again that $\tilde{H}_G$ is the full angular momentum description of the rigid body with respect to the inertial reference frame, $G-\tilde{x}-\tilde{y}-\tilde{z}$, which happens to align with $G-\hat{x}-\hat{y}-\hat{z}$ at this instant of time. The angular momentum observed by $G-\hat{x}-\hat{y}-\hat{z}$ is incomplete, and for Figure 5.13, it is only $[I_{xx}\dot{\psi} \ 0 \ 0]'$.

We can now revisit Equations (5.95–5.96) and (5.97–5.98) to define *Newton's Second Law* for a rigid body more explicitly. We will first consider 2D rigid bodies and then the 3D cases.

### 5.3.8 *Newton's Second Law for 2D rigid body cases*

For 2D rigid bodies, we assume that the body is moving on a $x-y$ plane. Therefore, the only rotation is around the $z$ axis. The angular velocity of the rigid body is

$$\vec{\omega} = \omega\vec{k} \qquad (5.132)$$

and the angular acceleration is

$$\vec{\alpha} = \dot{\omega}\vec{k} = \alpha\vec{k} \qquad (5.133)$$

Accordingly, the moment of inertial matrix is simplified to only $I_{zz}$. For a circular disk with radius $R$, $I_G = \frac{1}{2}mr^2$, simplified from Figure 5.14(b). For an $a \times b$ rectangular plate, $I_G = \frac{1}{12}m(a^2 + b^2)$, based on Figure 5.14(c). Similarly, for a slender rod with a length $a$, $I_G = \frac{1}{12}ma^2$.

The angular momentum is simplified to

$$\vec{H}_G = I_G\vec{\omega} \qquad (5.134)$$

As a result, Equation (5.97–5.98) becomes

$$\sum F_x = M\ddot{x}_{G/O} = Ma_{Gx} \qquad (5.135)$$

$$\sum F_y = M\ddot{y}_{G/O} = Ma_{Gy} \qquad (5.136)$$

$$\sum \tau_G = I_G\alpha \qquad (5.137)$$

Let's apply Equations (5.135)–(5.137) to a few examples.

## Example 5.6: A Circular Disk Rolling Down a Slope versus a Block

**Problem Statement:** As shown in Figure 5.15, a circular disk is rolling down a slope without slipping. Similarly, a block is sliding down the slope, without friction.
**Solution:**

**Force Analysis:** The FBD analysis for both objects is shown in Figure 5.16 using the ABCC method. Note that the block is assumed to be sliding down without friction. However, in general, the block is in contact with the slope over an area. Therefore, the normal force could be at a distance $x$ off the center.

**Motion Analysis:** The mass center of both objects are in a straight line motion. The circular disk has an angular velocity while the block does not. Because the disk is rolling without slipping, we know that $v_{A/O} = R\omega$ and $a_{A/O} = R\alpha$, as discussed in Section 4.2.2.

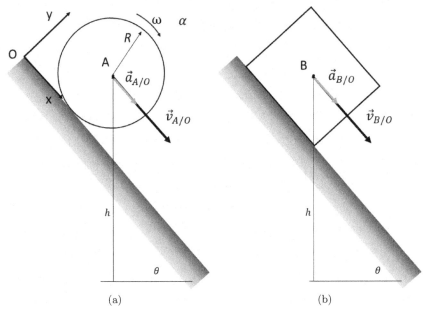

Figure 5.15: (a) A circular disk is rolling down a slope without slipping; and (b) a block of the same mass is sliding down the same slope without friction.

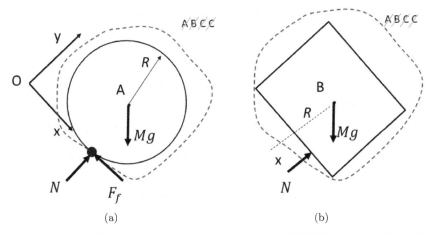

Figure 5.16: (a) FBD for the circular disk; and (b) FBD for the block for Example 5.6.

**Governing Equations:** Based on Equations (5.135)–(5.137), the governing equations for the disk are

$$\sum F_x = -F_f + Mg\sin\theta = Ma_{Ax} = MR\alpha \qquad (5.138)$$

$$\sum F_y = N - Mg\cos\theta = 0 \qquad (5.139)$$

$$\sum \tau_G = -RF_f = I_G(-\alpha) = \frac{1}{2}MR^2(-\alpha) \qquad (5.140)$$

For the block, the governing equations are

$$\sum F_x = Mg\sin\theta = Ma_{Bx} \qquad (5.141)$$

$$\sum F_y = N - Mg\cos\theta = 0 \qquad (5.142)$$

$$\sum \tau_G = xN = 0 \qquad (5.143)$$

**Solving for Unknowns:** Both sets of governing equations are solvable. We obtain $x = 0$; therefore, the normal force at the block is centered because there is no friction. The acceleration for the disk is

$$a_{Ax} = \frac{2}{3}g\sin\theta \qquad (5.144)$$

while the acceleration of the block is

$$a_{Bx} = g\sin\theta \qquad (5.145)$$

The block has a higher acceleration than the disk. Therefore, the block will reach the bottom of the slope faster. This seems strange. Should we use square tires rather than circular tires? Of course not. The block is faster because of zero friction. If we consider a sliding friction coefficient of $\mu_s$, the governing equations for the block become

$$\sum F_x = -\mu_s N + Mg\sin\theta = Ma_{Bx} \qquad (5.146)$$

$$\sum F_y = N - Mg\cos\theta = 0 \qquad (5.147)$$

$$\sum \tau_G = xN - \mu_s R = 0 \qquad (5.148)$$

Solving the above equations, we obtain

$$x = \frac{\mu_s R}{Mg \cos \theta} \qquad (5.149)$$

$$a_{Bx} = \left(1 - \mu_s \frac{\cos \theta}{\sin \theta}\right) g \sin \theta \qquad (5.150)$$

**Solution Verification:** Depending on the value of the friction coefficient, the block could slide down faster or slower than the disk. For example, if $\theta = 45°$, the disk and the block will have the same acceleration if $\mu_s = 1/3$. If the $\mu_s > 1/3$, the block will be slower.

### Example 5.7: Acceleration of a Motorcycle

**Problem Statement:** As shown in Figure 5.17, a motorcycle is accelerating. Determine the acceleration when the front wheel is lifted off the ground as shown and holding steady, ignoring the air resistance.

Figure 5.17: A motorcycle is accelerating with the front wheel lifted off the ground.
*Source*: https://static.pexels.com/photos/926/person-motor-racing-motorbike-motorcycle.jpg.

Figure 5.18: The corresponding force and motion analysis of the motorcycle with the front tire lifted.
*Source*: Modified image based on https://static.pexels.com/photos/926/person-motor-racing-motorbike-motorcycle.jpg.

**Solution:** First of all, riding a motorcycle with the front tire lifted is often called "wheelie" and is very dangerous. We will discuss its stability in Chapter 9.

**Force Analysis:** The FBD analysis for the motorcycle and the rider is shown in Figure 5.18 using the ABCC method. The mass center is located as shown.

**Motion Analysis:** The mass center of the motorcycle/rider is in a straight line motion horizontally. The angular velocity of the motorcycle/rider is zero.

**Governing Equations:** Based on Equation (5.135)–(5.137), the governing equations for the motorcycle/rider are

$$\sum F_x = F_f = Ma_{Gx} \tag{5.151}$$

$$\sum F_y = N - Mg = 0 \tag{5.152}$$

$$\sum \tau_G = -Nc\cos\theta + F_f(h\cos\theta + c\sin\theta) = 0 \tag{5.153}$$

**Solving for Unknowns:** There are three unknowns and we have three equations. We obtain the solutions as

$$N = Mg \tag{5.154}$$

$$F_f = \frac{Mgc\cos\theta}{h\cos\theta + c\sin\theta} \tag{5.155}$$

$$a_{Gx} = \frac{gc\cos\theta}{h\cos\theta + c\sin\theta} \tag{5.156}$$

**Solution Verification:** The solutions above appear quite straight forward, but could they be violating any physical laws? First, we should know that $F_f$ is limited by the friction coefficient $\mu_s$ between the tire and the ground. In other words, the maximum acceleration is limited to $\mu_s g$.

Let's consider some numbers. If we know that $\mu_s = 0.75$, $M = 250$ kg, $c = 0.4$ m and $h = 0.5$ m, the solutions become

$$N = 2452.5(\text{N}) \tag{5.157}$$

$$F_f = \frac{981.0\cos\theta}{0.5\cos\theta + 0.4\sin\theta}(\text{N}) \tag{5.158}$$

$$a_{Gx} = \frac{3.924\cos\theta}{0.5\cos\theta + 0.4\sin\theta}(\text{m/s}^2) \tag{5.159}$$

We can plot Equation (5.159) to verify the acceleration versus the lifting angle. From Figure 5.19, the acceleration is $0.8\,g$ when $\theta = 0°$ and it continues to drop as the angle increases. At $\theta = 4.764°$, the acceleration is $0.75\,g$.

This is what would have happened. When the rider pumps the gas to accelerate, if he keeps $\theta = 0°$, i.e., the front tire is just off the ground, the rear tire will skid. This could potentially cause the rider to lose control. The rider needs to lift the front tire up at $4.764°$ or about $5°$ to accelerate at the maximum possible acceleration without skidding.

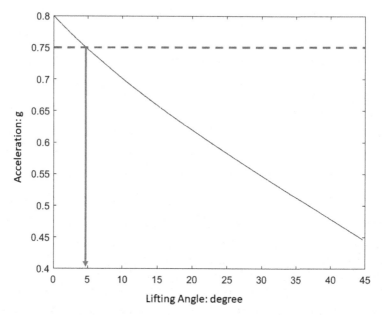

Figure 5.19: The acceleration of the motorcycle/rider with respect to the lifting angle.

The lifting angle can be higher, but the acceleration will be lower. However, the rider cannot continue to lift the front tire. Not only will the rider achieve less acceleration, but also the motorcycle could lose its stability to maintain a constant lifting angle. To understand this situation, we need to modify the governing equations to

$$\sum F_x = F_f = Ma_{Gx} \tag{5.160}$$

$$\sum F_y = N - Mg = Ma_{Gy} \tag{5.161}$$

$$\sum \tau_G = -Nc\cos\theta + F_f(h\cos\theta + c\sin\theta) = I_G\alpha \tag{5.162}$$

There are five unknowns. We can establish extra equations to relate the acceleration of point $A$, $\alpha$, $a_{Gx}$, and $a_{Gy}$. The problem is that when $\alpha$ is too high, the motorcycle will lift up without stabilizing to a specific angle, rendering the governing equations above invalid. It would happen fast and the motorcycle would flip over, a quite dangerous situation.

Finally, we also know that if $\mu_s$ is very low, then the motorcycle may not be lifted up. In addition, if $c \gg d$, it is hard to lift up because the front end is too heavy. This is typically the case in cars due to the weight of the car engine. It is rare to see a car moving with the front tires up in the air.

We might be able to assume a threshold acceleration condition related to the motorcycle's geometry. When, the acceleration exceeds the threshold value, the motorcycle would flip over. This problem is related to the roll-over problem of a vehicle and will be discussed in greater details in Chapter 9. The example below also address some issues related to if a box would flip over or not.

### Example 5.8: Acceleration of Stacked Boxes

**Problem Statement:** As shown in Figure 5.20, three boxes are stacked together. At $t = 0$, a force, $P$, is applied at box $C$. The static friction coefficient between the boxes and the ground is $\mu_s = 0.3$ and the sliding coefficient is $\mu_d = 0.2$. What is the maximum force of $P$

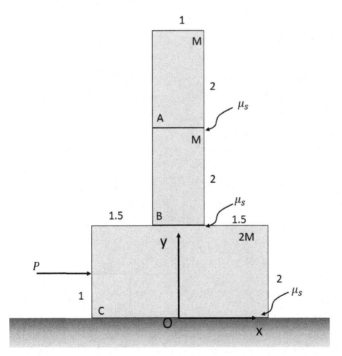

Figure 5.20:   Three boxes are stacked together and pushed by a force at box $C$.

so that box $C$ is sliding, while all three boxes are moving together without tipping over or sliding with respect to each? The moment of inertia with respect to the mass center each of the top two boxes is $10\,M\,\mathrm{kg} \cdot \mathrm{m}^2$.

**Solution:**

**Force Analysis:** To solve this problem, we need several FBDs. Assuming that the boxes are not tipping over or sliding with respect to each other, each box will have zero angular acceleration. Under this condition, we drew three FBDs as shown in Figure 5.21 using the ABCC method. We must observe the contact area between boxes and with the ground. When representing the distributed contact forces as a concentrated force, we must define the location of the concentrated forces.

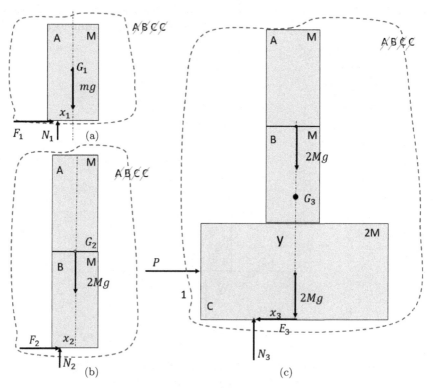

Figure 5.21: The corresponding FBDs: (a) for box $A$ only; (b) for boxes $A$ and $B$; and (c) for all three boxes.

**Motion Analysis:** The mass centers of all three boxes are in the straight line motion with zero angular acceleration. The combined mass center of all three boxes is at the height of 2.5 m, which can be easily determined.

**Governing Equations:** Based on Equation (5.134), the governing equations for the three FBDs are

$$\sum F_x = F_1 = Ma_{1x} \tag{5.163}$$

$$\sum F_y = N_1 - Mg = 0 \tag{5.164}$$

$$\sum \tau_G = F_1 \cdot 1 - N_1 \, x_1 = 0 \tag{5.165}$$

$$\sum F_x = F_2 = 2Ma_{2x} \tag{5.166}$$

$$\sum F_y = N_2 - 2Mg = 0 \tag{5.167}$$

$$\sum \tau_G = F_2 \cdot 2 - N_2 \, x_2 = 0 \tag{5.168}$$

$$\sum F_x = P - F_3 = 4Ma_{3x} \tag{5.169}$$

$$\sum F_y = N_3 - 4Mg = 0 \tag{5.170}$$

$$\sum \tau_G = P \cdot 1.5 - F_3 \cdot 2.5 - N_3 \, x_3 = 0 \tag{5.171}$$

**Solving for Unknowns**

First, based on the assumption, it is clear that

$$a_{1x} = a_{2x} = a_{3x} = a_x \tag{5.172}$$

We have nine equations but eleven unknowns. Because box $C$ is sliding, we can assume that

$$F_3 = \mu_d N_3 = 0.2 N_3 \tag{5.173}$$

Now we have ten equations but still eleven unknowns. We need to make another assumption for an additional equation. Because we want to know the maximum of force $P$, we can assume that $a_x$ is also at its maximum.

Examining Equations (5.163)–(5.165), the only force to accelerate box $A$ is $F_1$ while $F_1 \leq \mu_s N_1$.

Let's assume that $F_1 = \mu_s N_1$ for a maximum possible acceleration. We will review if this assumption is correct. With this assumption, we can now solve for all unknowns using Matlab's equation solver. We obtain

$$a_x = 0.3 \ g \tag{5.174}$$

$$x_1 = 0.3 \tag{5.175}$$

$$F_1 = 0.3 \ Mg \tag{5.176}$$

$$N_1 = Mg \tag{5.177}$$

$$x_2 = 0.6 \tag{5.178}$$

$$N_2 = 2 \ Mg \tag{5.179}$$

$$F_2 = 0.6 \ Mg \tag{5.180}$$

$$P = 2.0 \ Mg \tag{5.181}$$

$$x_3 = 0.25 \tag{5.182}$$

$$F_3 = 0.8 \ Mg \tag{5.183}$$

$$N_3 = 4 \ Mg \tag{5.184}$$

**Solution Verification:** We will first examine $x_1$, $x_2$, and $x_3$ to make sure that they are reasonable. The value of $x_3$ needs to be less than 2, which is correct. Both $x_1$ and $x_2$ need to be less than 0.5. However, $x_2 = 0.6 > 0.5$, which violates the assumption. As a result, we must discard the result above.

We now have to assume that $x_2 = 0.5$ and all other conditions stay the same. Reworking Equations (5.166)–(5.168), we have $F_2 = 0.5 \ Mg$ and $a_x = 0.25 \ g$. We then solve for all other unknowns as

$$a_x = 0.25 \ g \tag{5.185}$$

$$x_1 = 0.25 \tag{5.186}$$

$$F_1 = 0.25 \ Mg \tag{5.187}$$

$$N_1 = Mg \tag{5.188}$$

$$x_2 = 0.5 \tag{5.189}$$

$$N_2 = 2\ Mg \tag{5.190}$$

$$F_2 = 0.5\ Mg \tag{5.191}$$

$$P = 1.8\ Mg \tag{5.192}$$

$$x_3 = 0.175 \tag{5.193}$$

$$F_3 = 0.8\ Mg \tag{5.194}$$

$$N_3 = 4\ Mg \tag{5.195}$$

Both $F_1$ and $F_2$ are within the static friction limit, confirming the assumptions. The required force $P$ must be lower than 1.8 $Mg$.

What would happen if $P = 10\ Mg$? Quite likely, the three boxes will not stick together as assumed before. Or if the friction coefficient between boxes and surfaces is higher? The boxes may not stay upright and could tip over. We were lucky to assume one condition in the calculation above, but what if we were not lucky in our assumptions.

It would be better if we have a systematic way for conducting the analysis. As shown in Figure 5.22, for a box, there are four possible force and motion combinations. A box, under force $P$, could stay upright or flip. It could also remain stationary or sliding. There are four possible combinations. In Figure 5.22(a), the box stays stationary and not tilting. This is a static case. As $P$ increases, the frictional force, $F_f$, will increase up to the static friction limit, $\mu_s N$. The combination of $P$ and $F_f$ could enable the box to rotate clockwise and it has to be balanced by the moment due to $mg$ and $N$. The normal force $N$ needs to shift out of the center line to be a distance, $x_1$, off so that it can counter the moment by $P$ and $F_f$. However, $x_1$ is limited and cannot shift beyond the corner.

If force $P$ is large enough, the box is no longer stationary and it becomes a dynamic case. From the kinematics point of view, the box can only be in three conditions as shown in Figures 5.22(b)–5.22(d). In case (b), the box stays upright and is sliding over the surface. In case (c), the box tips over point $A$. Finally, the box could be tipping and sliding at the corner as shown in case (d). The box could hold in a specific tilting orientation, i.e., $\alpha = 0$, or sliding and flipping over, $\alpha > 0$.

For the problem of Figure 5.20, there would be 12 combinations for three boxes. It would be too time-consuming to go through every

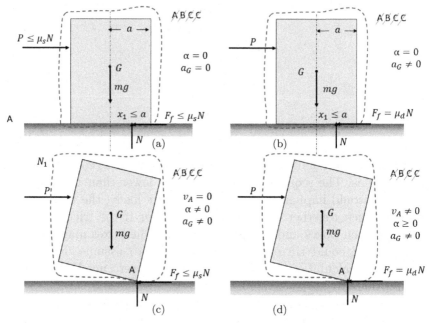

Figure 5.22: Four possible force/motion combinations: (a) a static case; (b) upright sliding case; (c) tilting around of fixed point at the corner; and (d) tilting and sliding.

possible case. We need to use the best judgment to assume a most likely situation.

Let us consider the case when $P = 10 \ Mg$ with all other contact conditions the same. First, we realize that box $C$ will have a very high acceleration which boxes $A$ and $B$ cannot achieve. Therefore, there will be sliding between boxes $B$ and $C$ to result in $F_2 = \mu_d N_2$. Because $F_2$ is lower than $P$, boxes $A$ and $B$ will have lower acceleration than in the previous case and less tendency to tilt. With these considerations, we can construct new FBDs of Figure 5.23 but change $F_2$ to the sliding frictional force.

The governing equations will become

$$\sum F_x = F_1 = Ma_{1x} \tag{5.196}$$

$$\sum F_y = N_1 - Mg = 0 \tag{5.197}$$

$$\sum \tau_G = F_1 \cdot 1 - N_1 \, x_1 = 0 \tag{5.198}$$

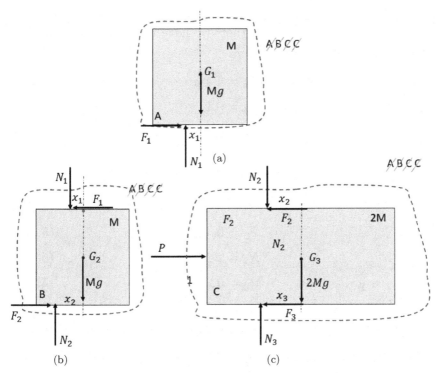

Figure 5.23: The corresponding FBDs for sliding boxes: (a) for box $A$ only; (b) for box $B$ only; and (c) for box $C$ only.

$$\sum F_x = F_2 - F_1 = Ma_{2x} \tag{5.199}$$

$$\sum F_y = N_2 - N_1 - Mg = 0 \tag{5.200}$$

$$\sum \tau_G = F_2 \cdot 1 - N_2 \, x_2 + N_1 \, x_1 + F_1 \cdot 1 = 0 \tag{5.201}$$

$$F_2 = 0.2N_2 \tag{5.202}$$

$$\sum F_x = 10 \, Mg - F_2 - F_3 = 2Ma_{3x} \tag{5.203}$$

$$\sum F_y = N_3 - N_2 - 2 \, Mg = 0 \tag{5.204}$$

$$\sum \tau_G = -F_3 \cdot 1 + N_2 \, x_2 + F_2 \cdot 1 - N_3 \, x_3 = 0 \tag{5.205}$$

$$F_3 = 0.2N_3 \tag{5.206}$$

We have 12 unknowns and 11 equations. Because we assume that boxes $A$ and $B$ will stay together, we have $a_{1x} = a_{2x}$. We can now solve for all the unknowns.

$$a_{3x} = 4.4\ g \tag{5.207}$$

$$x_3 = 0.1 \tag{5.208}$$

$$F_3 = 0.8\ Mg \tag{5.209}$$

$$N_3 = 4\ Mg \tag{5.210}$$

$$a_{2x} = 0.2\ g \tag{5.211}$$

$$x_2 = 0.4 \tag{5.212}$$

$$F_2 = 0.4\ Mg \tag{5.213}$$

$$N_2 = 2\ Mg \tag{5.214}$$

$$a_{1x} = 0.2\ g \tag{5.215}$$

$$x_1 = 0.2 \tag{5.216}$$

$$F_1 = 0.2\ Mg \tag{5.217}$$

$$N_1 = Mg \tag{5.218}$$

The above solutions are verified to be correct. Note that even the magnitude of $P$ is higher, the condition of the top two boxes will be the same because the external force is just the sliding friction on box $B$, which stays the same.

What if $\mu_s = 0.8$, and $\mu_d = 0.6$? We realize that $F_1$ and $F_2$ would be large and boxes $A$ and $B$ are unlikely to stay upright. We will assume both of them will be tilting at the corner. The corresponding FBDs are shown in Figure 5.24.

The corresponding governing equations are

$$\sum F_x = F_1 = Ma_{1x} \tag{5.219}$$

$$\sum F_y = N_1 - Mg = Ma_{1y} \tag{5.220}$$

$$\sum \tau_G = F_1 \cdot 1 - N_1\ 0.5 = I_{G1}\alpha_1 \tag{5.221}$$

$$\sum F_x = F_2 - F_1 = Ma_{2x} \tag{5.222}$$

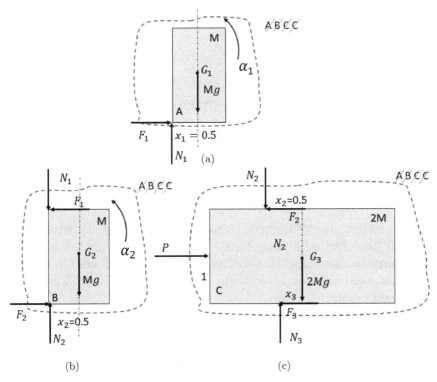

Figure 5.24: The corresponding FBDs with flipping boxes: (a) for box $A$ only; (b) for box $B$ only; and (c) for box $C$ only.

$$\sum F_y = N_2 - N_1 - Mg = Ma_{2y} \tag{5.223}$$

$$\sum \tau_G = F_2 \cdot 1 - N_2 \, (0.5) + N_1 \, (0.5)$$
$$+ \, F_1 \cdot 1 = I_{G2}\alpha_2 \tag{5.224}$$

$$\sum F_x = 3 \, Mg - F_2 - F_3 = 2Ma_{3x} \tag{5.225}$$

$$\sum F_y = N_3 - N_2 - 2 \, Mg = 0 \tag{5.226}$$

$$\sum \tau_G = -F_3 \cdot 1 + N_2 \, (0.5) + F_2 \cdot 1 - N_3 \, x_3 = 0 \tag{5.227}$$

$$F_3 = \mu_d N_3 = 0.6 N_3 \tag{5.228}$$

We have 14 unknowns but with only 10 equations. We need to construct more equations. Let's assume that box $B$ will slide and tilt over box $C$ while box $A$ will stick with box $B$ but tilting over. We then have one additional equation.

$$F_2 = \mu_d N_2 = 0.6 N_2 \qquad (5.229)$$

From motion analysis, we know that the mass center acceleration and the angular acceleration are related. We have the following additional equations

$$\vec{a}_{G1} = \vec{a}_A + \vec{\alpha}_1 \times \vec{r}_{G1/A} \qquad (5.230)$$

$$\vec{a}_{G2} = \vec{a}_B + \vec{\alpha}_2 \times \vec{r}_{G2/B} \qquad (5.231)$$

$$\vec{a}_A = \vec{a}_B + \vec{\alpha}_2 \times \vec{r}_{A/B} \qquad (5.232)$$

When we convert the above vector equations into scalar equations, we will have five additional equations but also introduce two extra unknowns, $a_A$ and $a_B$. All together, we now have 16 unknowns and 16 equations. We can solve them using a similar Matlab code as discussed before.

The solutions are

$$a_{3x} = 3.16 \; g \qquad (5.233)$$

$$x_3 = -0.044 \qquad (5.234)$$

$$F_3 = 2.441 \; Mg \qquad (5.235)$$

$$N_3 = 4.065 \; Mg \qquad (5.236)$$

$$\alpha_2 = 1.238 \; \text{rad/s}^2 \qquad (5.237)$$

$$F_2 = 1.239 \; Mg \qquad (5.238)$$

$$N_2 = 2.066 \; Mg \qquad (5.239)$$

$$\alpha_1 = 0.0517 \; \text{rad/s}^2 \qquad (5.240)$$

$$F_1 = 0.554 \; Mg \qquad (5.241)$$

$$N_1 = 1.003 \; Mg \qquad (5.242)$$

It seems, so far, that we are just lucky and guess right each time. In fact, it could be quite a trial-and-error process. Let's change the value of $P$ and see how the conditions will change.

Note that we have to change the governing equations and the additional equations each time we change the assumptions. For example, if we assume that both boxes on top are flipping but sticking to their lower corner, the additional equations will have to be revised as

$$\vec{a}_B = \vec{a}_C \tag{5.243}$$

$$\vec{a}_{G1} = \vec{a}_A + \vec{\alpha}_1 \times \vec{r}_{G1/A} \tag{5.244}$$

$$\vec{a}_{G2} = \vec{a}_B + \vec{\alpha}_2 \times \vec{r}_{G2/B} \tag{5.245}$$

$$\vec{a}_A = \vec{a}_B + \vec{\alpha}_2 \times \vec{r}_{A/B} \tag{5.246}$$

Note that no equation should be used to define $F_2$ and $F_1$ under such an assumption.

We can also define the conditions when the top two boxes are flipping as a rigid body while the lower corner of box $B$ sticks to box $C$. Finally, we also review the condition when both boxes $A$ and $B$ stay upright as before. It turns out if we increase $P$ from 3 mg to 10 mg, the top two boxes will transition from upright, flipping as one body, flipping separately but sticking at the corner, and finally to the condition when box $B$ will be flipping and sliding. All these conditions apply to $t = 0$. When $t > 0$, the flipping can cause different interactions between boxes and the governing equations will need to be re-derived.

The results are shown in Figure 5.25. Until $P < 3.4$ *Mg*, both top boxes will stay upright. When 3.4 *Mg* $< P < 5$ *Mg*, both boxes will flip as one rigid body and the lower corner of box $B$ sticks to box $C$. A larger $P$ force will cause the top boxes to have a higher angular acceleration. The friction coefficient, $\mu_2$, between box $B$ and $C$ is calculated and plotted in Figure 5.25. It is confirmed to be less than $\mu_s = 0.8$. When $P > 5$ *Mg*, the top two boxes are flipping separately, no longer as one rigid body. The transition is quite dramatic as $\alpha_2$ jumps to a value over 1 rad/s$^2$, while $\alpha_1$ starts near zero. Both angular acceleration would grow as $P$ becomes higher until when $P \approx 6.4$ *Mg*, $\mu_2$ would reach 0.8. Beyond that, box $B$ would start to slip. From here on, the conditions of the top two boxes would stay the same because $F_2$ stays constant.

We should be cautious about the results above. With different moment of inertia of the boxes, the results could be quite different.

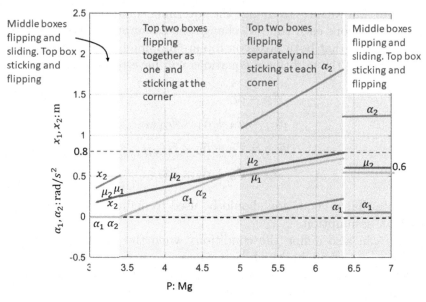

Figure 5.25:   The transition of top boxes when the driving force becomes higher.

We must check the results to ensure that they do not violate the assumptions. Finally, the transitions from one condition to another are not continuous in Figure 5.25. This is mainly due to the model we used for the frictional force, in which the transition from the static friction limit to the sliding friction is not continuous. In practice, it could be more continuous. However, the acceleration curves do not have to be continuous.

We want to point out one more observation of the results in Figure 5.25. When the top two boxes are flipping separately, $\alpha_2$ is much larger than $\alpha_1$. Some moment later, box $B$ will have a higher angular velocity than that of box $A$. This means that box $B$ will hit into box $A$. As a result, the governing equations derived above are only for $t = 0$. The values of $\alpha_1$ and $\alpha_2$ are related to their respective moment of inertia and the corner position to the mass center.

If the boxes in Figure 5.20 are homogeneous in density, the moment of inertia can be calculated as $I_G = 5M/12$, which is much lower than the values used in the above example. When $\mu_d = 0.6$ and $\mu_s = 0.8$, just a small force of $P$ will cause both boxes, $A$ and $B$, to flip and slide. Readers can verify this as an exercise.

### 5.3.9 Newton's Second Law for 2D rigid body with a fixed point

In this section, let's consider a special case in 2D rigid body dynamics. If there is a stationary point on the rigid body, we can simplify Equation (5.92) as

$$
\dot{\vec{H}}_A = \vec{r}_{G/A} \times \sum_i \vec{F} + \sum_i (\vec{r}_{i/G} \times \vec{F}_i)
$$

$$
= \vec{r}_{G/A} \times M\vec{a}_{G/O} + \sum_i ([\vec{r}_{i/G} \times m_i \vec{a}_{i/O}]
$$

$$
= \vec{r}_{G/A} \times M(\alpha \times \vec{r}_{G/A}) + \sum_i ([\vec{r}_{i/G} \times m_i(\vec{\alpha} \times \vec{r}_{i/G}]
$$

$$
= Mr_{G/A}^2 \vec{\alpha} + I_G \vec{\alpha} = (I_G + Mr_{G/A}^2)\vec{\alpha}
$$

$$
= I_A \vec{\alpha} \tag{5.247}
$$

From the above equation, we have

$$
I_A = I_G + Mr_{G/A}^2 \tag{5.248}
$$

Equation (5.248) is called the Parallel-Axis Theorem to relate $I_A$ and $I_G$. Note that $I_A$ has a real physical meaning only if $A$ is a fixed point on the rigid body. If $A$ is simply any fixed point in space, then we should use Equation (5.92) instead of (5.247).

Let's consider a few examples for the use of Equation (5.247).

### Example 5.9: Swinging Rigid Bodies

**Problem Statement:** As shown in Figure 5.26, a circular disk and a rod, with the same mass, are pinned on one end, respectively. At $t = 0$, the cord is cut and the disk or the rod will swing downward. Determine the angular acceleration of the disk and the rod. Also calculate the reaction forces at the pin.

**Solution:**

**Force Analysis:** The FBD analysis for both objects is shown in Figure 5.27 using the ABCC method after the cord is cut.

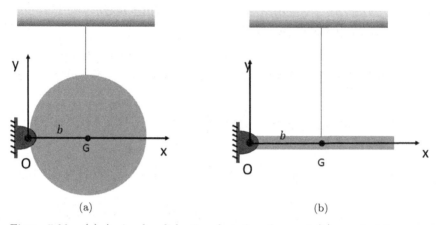

Figure 5.26: (a) A circular disk pinned at the edge; and (b) a rod of the same mass is pinned at one end. The distances from the mass center to the pin for both cases are the same.

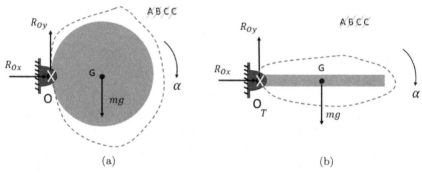

Figure 5.27: (a) FBD for the circular disk; and (b) FBD for the rod for Example 5.9.

**Motion Analysis:** The disk and the rod will swing down around pin $O$, as the rotation around a fixed point. At the time when the cord is cut, the velocity of the mass center and the angular velocity of the rigid body are zero.

**Governing Equations:** Based on Equation (5.92) or (5.247), the governing equations for the disk and the rod are the same,

$$\sum F_x = R_{Ox} = ma_{Ax} = mb\,\omega^2 = 0 \qquad (5.249)$$

$$\sum F_y = R_{Oy} - mg = mb(-\alpha) \tag{5.250}$$

$$\sum \tau_O = -b \cdot mg = I_O(-\alpha) = (I_G + mb^2)(-\alpha) \tag{5.251}$$

For the disk, the moment of inertia is $(I_O)_{\text{disk}} = mb^2 + 1/2\,mb^2 = 3/2\,mb^2$, while for the rod, $(I_O)_{\text{rod}} = mb^2 + 1/3\,mb^2 = 4/3\,mb^2$. We obtain

$$(\alpha)_{\text{disk}} = \frac{2g}{3b} \tag{5.252}$$

$$(\alpha)_{\text{rod}} = \frac{3g}{4b} \tag{5.253}$$

The rod will swing down faster. The reaction forces are found to be

$$(R_{Ox})_{\text{disk}} = 0; \quad (R_{Oy})_{\text{disk}} = \frac{1}{3}\,mg \tag{5.254}$$

$$(R_{Ox})_{\text{rod}} = 0; \quad (R_{Oy})_{\text{rod}} = \frac{1}{4}\,mg \tag{5.255}$$

The vertical force at the pin for the rod case is lower. Finally, if there is no pin and both rigid bodies are falling freely, the mass center acceleration will be $g$ and the angular acceleration zero.

### Example 5.10: Center of Percussion

**Problem Statement:** As shown in Figure 5.28(a), a rod swings down and hits a block at the vertical position. Determine the location of the block so that the horizontal reaction force upon impact is zero. The rod is not uniform in density. The mass center is located at $r_G$ from point $O$. The impact point is $d$ distance further down from the mass center. The moment of inertia with respect to the mass center is $I_G = mk_G^2$, where $k_G$ is called radius of gyration.

### Solution:

**Force Analysis:** The FBD analysis is shown in Figure 5.28(b) using the ABCC method. The impact force is considered as horizontal. The FBD of Figure 5.28(b) is for a general case.

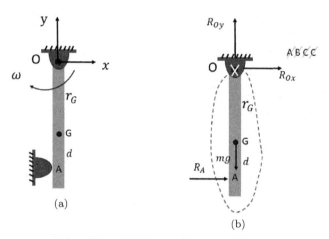

Figure 5.28: The location of the impact point affects the reaction force at the pin: (a) the impact when the rod hits the block at $A$; and (b) the corresponding FBD.

**Motion Analysis:** Upon impact, we assume that the rod has an angular velocity $\omega$, which is in the clockwise direction, and an angular acceleration $\alpha$, which is in the counterclockwise direction. Their values could be changing right before and during impact, but we are not interested in determining them. Again, this is the case of rotation around a fixed point at $O$.

**Governing Equations:** Based on Equation (5.92) or (5.247), the governing equations for the rod are

$$\sum F_x = R_{Ox} + R_A = ma_{Gx} = mr_G\alpha \qquad (5.256)$$

$$\sum F_y = R_{Oy} - mg = ma_{Gy} = mr_G\omega^2 \qquad (5.257)$$

$$\sum \tau_O = R_A(r_G + d) = I_O\alpha = (I_G + mr_G^2)\alpha \qquad (5.258)$$

There are four unknowns, $R_{Ox}$, $R_A$, $\alpha$, and $d$. Under the situation that $R_{Ox} = 0$, we will be able to express $R_A$ and $d$ as a function of $\alpha$. With $R_{Ox} = 0$, from Equation (5.256), we obtain

$$\alpha = \frac{R_A}{mr_G} \qquad (5.259)$$

Note that $R_A$ is changing during the impact; therefore, so is $\alpha$.

We got quite lucky that $d$ can be solved and it is not a function of $\alpha$.

Introducing the above result to Equation (5.258), we obtain

$$d = \frac{I_G}{mr_G} = \frac{mk_G^2}{mr_G} = \frac{k_G^2}{r_G} \qquad (5.260)$$

This special impact location renders the horizontal force at pin $O$ zero. The vertical force at pin $O$ depends on the value of $\omega$, which undergoes drastic changes from its maximum right before impact to nearly zero during impact.

From Equation (5.257), we know that

$$mg \leq R_{Oy} \leq mg + mr_G\omega^2 \qquad (5.261)$$

whose range is not a function of the impact point. As a result, the reaction force at $O$ is the smallest when the impact point is at $\frac{k_G^2}{r_G}$ below the mass center. This location is denoted Center of Percussion.

**Solution Verification:** If the rod is used to strike an object placed at the center of percussion, pin $O$ will last longer because the reaction force is the smallest. If the rod is a baseball bat and the pin represents the hands holding the bat, then if the ball hits the percussion center, the hands will experience the least amount of impact force. In other words, the hands sting the least. We often call this situation as hitting the sweet spot. However, the sweet spot and the center of percussion may not be exactly the same because the hands are not stationary. The idea is similar.

### 5.3.10 *Newton's Second Law for 3D rigid body cases*

When dealing with 3D rigid body dynamics, we need to use Equation (5.92) in its entirety. The situation could be complicated. For some special cases, for example, a rigid body rotating around a fixed point, the governing equations might be manageable.

We will consider the gyro dynamics in this section. In Section 4.2.1, we presented the motion analysis for a gyro with respect to Figure 4.4. This gyro is a rotating uniform disk fixed on a massless rod.

According to Figure 4.4, we choose an inertial reference frame $O{-}\tilde{x}{-}\tilde{y}{-}\tilde{z}$ which coincides with $O{-}\hat{x}_2{-}\hat{y}_2{-}\hat{z}_2$ at this moment. The non-inertial reference frame, $O-\hat{x}_2-\hat{y}_2-\hat{z}_2$, is attached to the gyro rod. Using $O-\hat{x}_2-\hat{y}_2-\hat{z}_2$, the gyro is observed to have a spinning motion, $\dot{\psi}$ only, which is not a complete observation. Based on the 3D motion equation of Section 4.2.1, the frame rotation of $O-\hat{x}_2-\hat{y}_2-\hat{z}_2$ is

$$\vec{\Omega} = \dot{\theta}\hat{i}_2 + \dot{\phi}\sin\theta_x\hat{j}_2 + \dot{\phi}\cos\theta_x\hat{k}_2 \tag{5.262}$$

The angular velocity of the gyro with respect to the inertial reference $O-\tilde{x}-\tilde{y}-\tilde{z}$ is

$$\begin{aligned}\vec{\omega}_B &= \dot{\theta}\hat{i}_2 + \dot{\phi}\sin\theta_x\hat{j}_2 + (\dot{\phi}\cos\theta_x + \dot{\psi})\hat{k}_2 \\ &= \dot{\theta}\tilde{i} + \dot{\phi}\sin\theta_x\tilde{j} + (\dot{\phi}\cos\theta_x + \dot{\psi})\tilde{k}\end{aligned} \tag{5.263}$$

Note that at this moment, $\hat{i}_2 = \tilde{i}$, $\hat{j}_2 = \tilde{j}$, and $\hat{k}_2 = \tilde{k}$.

As discussed in Section 5.3.6, the angular momentum of gyro $B$ with respect to the mass center becomes

$$\vec{H}_G = [\tilde{i}\ \tilde{j}\ \tilde{k}]\ \tilde{I}_G\vec{\omega}_B = [\tilde{i}\ \tilde{j}\ \tilde{k}]\tilde{I}_G \begin{bmatrix} \dot{\theta} \\ \dot{\phi}\sin\theta_x \\ \dot{\phi}\cos\theta_x + \dot{\psi} \end{bmatrix} \tag{5.264}$$

The moment of inertia matrix with respect to $O-\hat{x}_2-\hat{y}_2-\hat{z}_2$ is found to be

$$\tilde{I}_G = \begin{bmatrix} I_{xx} & 0 & 0 \\ 0 & I_{xx} & 0 \\ 0 & 0 & I_{zz} \end{bmatrix} \tag{5.265}$$

If the gyro is a thin and uniform disk, $I_{zz} = \frac{1}{2}mR^2$ and $I_{xx} = I_{yy} = \frac{1}{4}mR^2$, where $R$ is the radius of the disk. We do not have to assign specific values to $I_{xx}$ and $I_{zz}$ for the following derivation.

The angular momentum is simplified to

$$\vec{H}_G = [\tilde{i}\ \tilde{j}\ \tilde{k}] \begin{bmatrix} I_{xx}\dot{\theta} \\ I_{zz}\dot{\phi}\sin\theta_x \\ I_{zz}(\dot{\phi}\cos\theta_x + \dot{\psi}) \end{bmatrix} \tag{5.266}$$

From Section 5.3.7 and Equation (5.131), the total derivative of the angular momentum is found to be

$$\frac{d}{dt}(\tilde{H}_G) = \vec{\Omega} \times \tilde{H}_G + \frac{\partial}{\partial t}\tilde{H}_G$$

$$= [\hat{i} \; \hat{j} \; \hat{k}]\begin{bmatrix} I_{xx}(\ddot{\theta} - \dot{\phi}^2 \sin\theta_x \cos\theta_x) + I_{zz}\dot{\phi}\sin\theta_x(\dot{\phi}\cos\theta_x + \dot{\psi}) \\ I_{xx}(\ddot{\phi}\sin\theta_x + 2\dot{\phi}\dot{\theta}\cos\theta_x) - I_{zz}\dot{\theta}(\dot{\phi}\cos\theta_x + \dot{\psi}) \\ I_{zz}(\ddot{\psi} + \ddot{\phi}\cos\theta_x - \dot{\phi}\dot{\theta}\sin\theta_x) \end{bmatrix} \quad (5.267)$$

The gyro is assumed to be supported at point $O$ with a frictionless ball joint. However, this frictionless assumption can be eliminated, but we will have to consider more complicated frictional moments at point $O$.

Invoking Equation (5.98) and based on the FBD of Figure 5.29, we obtain the governing equations as

$$\sum \tau_{\tilde{x}} = (\tilde{r}_{O/G} \times \tilde{F}_y)_{\tilde{x}}$$

$$= I_{xx}(\ddot{\theta} - \dot{\phi}^2 \sin\theta_x \cos\theta_x)$$

$$+ I_{zz}\dot{\phi}\sin\theta_x(\dot{\phi}\cos\theta_x + \dot{\psi}) \quad (5.268)$$

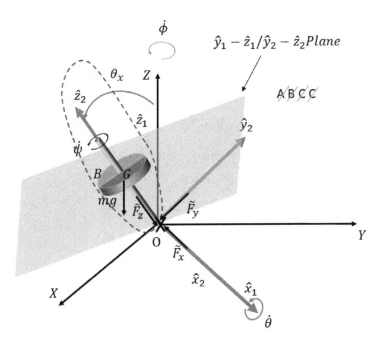

Figure 5.29: The FBD of a gyro of Figure 4.4, supported by a ball joint at point $O$.

$$\sum \tau_{\tilde{y}} = (\tilde{r}_{O/G} \times \tilde{F}_x)_{\tilde{y}}$$
$$= I_{xx}(\ddot{\phi}\sin\theta_x + 2\dot{\phi}\dot{\theta}\cos\theta_x)$$
$$-I_{zz}\dot{\theta}(\dot{\phi}\cos\theta_x + \dot{\psi}) \qquad (5.269)$$

$$\sum \tau_{\tilde{z}} = 0$$
$$= I_{zz}(\ddot{\psi} + \ddot{\phi}\cos\theta_x - \dot{\phi}\dot{\theta}\sin\theta_x)] \qquad (5.270)$$

In practice, a ball joint is never completely frictionless; therefore, a frictional torque component should be added to each equation above.

Let's consider the frictionless case only. Equations (5.268)–(5.270) are difficult to solve because of the unknown forces on the left-hand side.

To remove these unknown forces from consideration, we can derive the governing equations with respect to $O$. As a result, the only force to be considered is gravity.

Based on Equation (5.96), we have

$$\sum \tau_{\tilde{x}} = (\tilde{r}_{G/O} \times m\vec{g})_{\tilde{x}} \qquad (5.271)$$
$$= (I_{xx} + mb^2)(\ddot{\theta} - \dot{\phi}^2\sin\theta_x\cos\theta_x)$$
$$+ I_{zz}\dot{\phi}\sin\theta_x(\dot{\phi}\cos\theta_x + \dot{\psi})$$

$$\sum \tau_{\tilde{y}} = (\tilde{r}_{G/O} \times m\vec{g})_{\tilde{y}}$$
$$= (I_{xx} + mb^2)(\ddot{\phi}\sin\theta_x + 2\dot{\phi}\dot{\theta}\cos\theta_x)$$
$$- I_{zz}\dot{\theta}(\dot{\phi}\cos\theta_x + \dot{\psi}) \qquad (5.272)$$

$$\sum \tau_{\tilde{z}} = 0$$
$$= I_{zz}(\ddot{\psi} + \ddot{\phi}\cos\theta_x - \dot{\phi}\dot{\theta}\sin\theta_x) \qquad (5.273)$$

The moment of $mg$ with respect to $O$ can be determined as

$$(\tilde{r}_{G/O} \times m\vec{g}) = mg(-\vec{k}) \times (b\tilde{k})$$
$$= (-mg\sin\theta_x\tilde{j} - mg\cos\theta_x\tilde{k}) \times (b\tilde{k})$$
$$= mgb\sin\theta_x\tilde{i} \qquad (5.274)$$

The governing equations become

$$\sum \tau_{\tilde{x}} = mgb \sin \theta_x$$
$$= (I_{xx} + mb^2)(\ddot{\theta} - \dot{\phi}^2 \sin \theta_x \cos \theta_x) + I_{zz}\dot{\phi} \sin \theta_x (\dot{\phi} \cos \theta_x + \dot{\psi}) \tag{5.275}$$

$$\sum \tau_{\tilde{y}} = 0$$
$$= (I_{xx} + mb^2)(\ddot{\phi} \sin \theta_x + 2\dot{\phi}\dot{\theta} \cos \theta_x) - I_{zz}\dot{\theta}(\dot{\phi} \cos \theta_x + \dot{\psi}) \tag{5.276}$$

$$\sum \tau_{\tilde{z}} = 0$$
$$= I_{zz}(\ddot{\psi} + \ddot{\phi} \cos \theta_x - \dot{\phi}\dot{\theta} \sin \theta_x) \tag{5.277}$$

The solutions to the above equations for the general gyro motion are quite complicated . Readers can refer to many journal publications to find detailed derivations of solutions. At steady precession, both $\dot{\psi}$ and $\dot{\phi}$ are constant, and $\theta_x$ is also a constant. The governing equation can be further simplified.

Let's consider a special steady precession case with $\theta_x = 90°$. Equation (5.275) is reduced to

$$bmg = I_{zz}\dot{\phi}\dot{\psi} \tag{5.278}$$

The angular momentum of the gyro with respect to $O$ for this particular precession is

$$\vec{H}_O = [\tilde{i} \ \tilde{j} \ \tilde{k}] \begin{bmatrix} 0 \\ (I_{xx} + mb^2)\dot{\phi} \\ I_{zz}\dot{\psi} \end{bmatrix} \tag{5.279}$$

Because $\dot{\psi} >> \dot{\phi}$, the vector of $\tilde{H}_O$ will be mostly aligned with the $\tilde{z}_2$ axis. If the rotation of $\dot{\psi}$ points to the positive $\tilde{z}_2$ direction, $\tilde{H}_O$ will point to almost the same direction as shown in Figure 5.30(a). The moment due to gravity, $bmg$, points to the positive $\tilde{x}_2$ direction. The resulting precision will be counterclockwise with respect to $Z$, as shown in Figure 5.30(a). On the other hand, if $\dot{\psi}$ points to the negative $\tilde{z}_2$ direction, the same gravitation force will cause the precession to be clockwise, as shown in Figure 5.30(b). This result is also implied

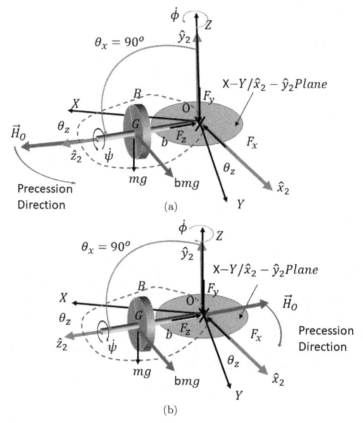

Figure 5.30: A gyro at steady precession with $\theta_x = 90°$ $O$: (a) the spinning of $\dot\psi$ is in the positive $\tilde z_2$ direction; and (b) the spinning of $\dot\psi$ is in the negative $\tilde z_2$ direction.

in Equation (5.278). If $\dot\psi$ is positive, $\dot\phi$ is positive (counterclockwise). However, if $\dot\psi$ is negative, $\dot\phi$ will be negative (clockwise). We should recognize this when watching a demo of the gyro precession because the precession direction is not always the same, depending on the direction of $\dot\psi$.

## 5.4 Concluding Remarks

We conclude this chapter by pointing out that *Newton's generalized Second Law* (Equations (5.95) and (5.96) or (5.97) and (5.98)) can be applied to both 2D and 3D cases. We must not forget that, for Equation (5.95–5.96), the reference point $A$ must be stationary when determining the angular momentum and its derivative. In general, we prefer Equation (5.97–5.98) because we are not limited to a stationary point. Only for the special case of rotating around a fixed point on the rigid body, the theorem of parallel axis (Equations (5.248) and (5.247)) can be used for some benefits.

Finally, we must learn to recognize different reference frames used for observing 3D rigid body dynamics. Just like in Chapter 4, we choose a rotating reference frame so that the rigid body appears to have only one simple rotation around one axis. We must distinguish which one is an inertial reference frame and which one is not.

## Chapter 6

# Linear and Angular Momentum Equations for Dynamic Systems

In Chapter 5, we introduced the concepts of linear momentum and angular momentum. Based on their derivatives, we derived the general form of *Newton's Second Law* for both 2D and 3D systems. By applying *Newton's Second Law*, we are able to determine the accelerations of a dynamic system at the instant of time under consideration. To extrapolate the result to future time, we would either have to solve a differential equation or to integrate the acceleration variables over time. In this chapter, we will integrate the generalized form of *Newton's Second Law* over time, which leads to a new form of *Newton's Second Law*. We denote them as the linear momentum and angular momentum equations.

## 6.1 General Linear and Angular Momentum Equations

As presented in Chapter 5, the general form of *Newton's Second Law* is

$$\sum \vec{F} = \dot{\vec{P}} = M\vec{a}_{G/O} \tag{6.1}$$

$$\sum \vec{\tau}_A = \dot{\vec{H}}_A \tag{6.2}$$

where point $A$ is a fixed point within an inertial reference frame. If the system is a rigid body, we also have

$$\sum \vec{F} = \dot{\vec{P}} = M\vec{a}_{G/O} \tag{6.3}$$

$$\sum \vec{\tau}_G = \dot{\vec{H}}_G \tag{6.4}$$

where $G$ is the mass center of the rigid body.

We will start with the rigid body instead of a particle. The integration form of Equations (6.1)–(6.2) and (6.3)–(6.4) are

$$\int_{t1}^{t2} \sum \vec{F} dt = \int_{t1}^{t2} \dot{\vec{P}} dt = \vec{P}(t_2) - \vec{P}(t_1) = \vec{P}_2 - \vec{P}_1 \tag{6.5}$$

$$\int_{t1}^{t2} \sum \vec{\tau}_A dt = \int_{t1}^{t2} \dot{\vec{H}}_A dt = \vec{H}_A(t_2) - \vec{H}_A(t_1)$$

$$= \vec{H}_{A2} - \vec{H}_{A1} \tag{6.6}$$

and

$$\int_{t1}^{t2} \sum \vec{F} dt = \int_{t1}^{t2} \dot{\vec{P}} dt = \vec{P}(t_2) - \vec{P}(t_1) = \vec{P}_2 - \vec{P}_1 \tag{6.7}$$

$$\int_{t1}^{t2} \sum \vec{\tau}_G dt = \int_{t1}^{t2} \dot{\vec{H}}_G dt = \vec{H}_G(t_2) - \vec{H}_G(t_1)$$

$$= \vec{H}_{G2} - \vec{H}_{G1} \tag{6.8}$$

For the above equations to be valid, we must express all the vector variables in the above equations using an inertial reference frame for the integration. In other words, it is important that the unit vectors of the reference frame are fixed during $t_1 \leq t \leq t_2$.

In Chapter 3 (Sections 3.6), we presented how we can take a derivative of a unit vector whose direction is changing. In Chapter 5 (Section 5.3.7), we also learned how to take a derivative of a vector expressed in an inertial reference frame, $O-\tilde{x}-\tilde{y}-\tilde{z}$, which is only valid for the instant of time. We are able to take derivatives of a vector when the vector is expressed in non-fixed unit directional vectors for the current instant of time.

However, integration of a non-fixed unit directional vector over a period of time is quite different. Either, the unit directional vectors have to be fixed during the time period of consideration or the time period must be very short so that the unit vectors are essentially pointing in the same directions.

Equation (6.5) or (6.7) is denoted as the *Linear Momentum Equation*. We typically do not like to express an equation with negative signs; therefore, we rewrite it as

$$\vec{P}_1 + \int_{t1}^{t2} \sum \vec{F} dt = \vec{P}_2 \tag{6.9}$$

Equation (6.9) is the standard form of the linear momentum equation.

Similarly, Equations (6.6) and (6.8) are rewritten, respectively, as

$$(\vec{H}_A)_1 + \int_{t1}^{t2} \sum \vec{\tau}_A \, dt = (\vec{H}_A)_2 \tag{6.10}$$

$$(\vec{H}_G)_1 + \int_{t1}^{t2} \sum \vec{\tau}_G \, dt = (\vec{H}_G)_2 \tag{6.11}$$

Equations (6.10) and (6.11) are the *Angular Momentum Equation* with respect to a fixed point $A$ or the mass center $G$, respectively.

### 6.1.1 *Linear momentum equation*

Equation (6.9) is the integration form of *Newton's Second Law* to account for the mass center motion.

The linear momentum equation allows for the consideration of two instants of time if we know $\int_{t1}^{t2} \sum \vec{F} dt$. However, unless we know $\sum \vec{F}$ as a function of time, we typically cannot carry out the integration.

In most textbooks, the term $\int_{t1}^{t2} \vec{F} dt$ is denoted as *Impulse* and the term $\int_{t1}^{t2} \sum \vec{F} dt = \sum \int_{t1}^{t2} \vec{F} dt$ is called the sum of impulses. The linear momentum equation is then denoted as the *Principle of Linear Impulse and Momentum*. In this book, we reserve the term *Impulse* to define a special kind of force, not the time integration of any force. We simply refer $\int_{t1}^{t2} \vec{F} dt$ as the time integration of a force, NOT calling it as "*Impulse*". The special kind of *Impulse* force will be presented in the next section.

### 6.1.2 What is an impulse?

If we are lucky that $\int_{t1}^{t2} \sum \vec{F} dt = 0$, Equation (6.9) is reduced to

$$\vec{P_1} = \vec{P_2} \tag{6.12}$$

Equation (6.12) is called *Conservation Equation of Linear Momentum*. Under this circumstance, the linear momentum does not change for $t_1 \leq t \leq t_2$. We will state that conservation of linear momentum holds for $t_1 \leq t \leq t_2$. However, we must prove that $\int_{t1}^{t2} \sum \vec{F} dt = 0$ before we can cite the conservation equation of linear momentum. In other words, we must remember that only under some special situations is the conservation equation of linear momentum valid and we must prove that.

One condition to result in $\int_{t1}^{t2} \sum \vec{F} dt = 0$ is that $\sum \vec{F} = 0$ during $t_1 \leq t \leq t_2$. In other words, all external forces acting on the system sum to zero.

For a particle with unchanged mass, this means that the velocity of the particle is constant. For a system of particles, conservation of linear momentum only indicates that the total linear momentum is constant, or the velocity of the mass center is constant. The velocity of an individual particle within the system can change, but collectively, the total linear momentum is constant, or conserved.

Essentially, a constant linear momentum implies *Newton's First Law* of motion, which states

> *An object at rest remains at rest, and an object in motion remains in motion at constant speed and in a straight line unless acted on by an unbalanced force.*

What if the time interval between $t_1$ and $t_2$ is very short? Is $\int_{t1}^{t2} \sum \vec{F} dt \approx 0$ if $t_2 - t_1 \approx 0$?

If $t_2 - t_1 = dt \approx 0$, and the force magnitude during this short time interval can be essentially considered as constant, then the integration of the force becomes

$$\int_{t1}^{t1+dt} \sum \vec{F} dt = \left( \sum \vec{F} \right) \cdot (dt) \tag{6.13}$$

Let's just consider a 1D case. The question becomes: "Is $(\sum F) \cdot (dt) \approx 0$ because $dt \approx 0$?" In fact, we can ask this question in several ways,

$$Is \quad 5 \cdot 0 = 0?$$

$$Is \quad 5000 \cdot 0 = 0?$$

$$Is \quad 50000000000 \cdot 0 = 0?$$

$$Is \quad \infty \cdot 0 = 0?$$

We all know the answers to the first three questions are "Yes!", but most people are not certain about the fourth question. When infinity times zero, does the product become zero? We need to be more rigorous to answer this question.

First, how do we mathematically define $\infty$? We cannot assign a specific number to $\infty$. We just know that it is very, very, very big, but we cannot put a number to it. The mathematical definition of $\infty$ is quite complicated and we will not go there. Let's just recognize that not all infinities are created equal. Similarly, not all zeros are created equal.

For a real number, $x$, if it keeps moving to the right on a real number line, eventually, we will say $x$ will approach $\infty$, denoted as $x \to \infty$. With this "simple" infinity, we will define a few more different infinities,

$$\lim_{x \to \infty} y_1(x) = \lim_{x \to \infty} 5x = \infty \qquad (6.14)$$

$$\lim_{x \to \infty} y_2(x) = \lim_{x \to \infty} x^2 = \infty \qquad (6.15)$$

$$\lim_{x \to \infty} y_3(x) = \lim_{x \to \infty} e^x = \infty \qquad (6.16)$$

Similarly, we can define different zeros,

$$\lim_{x \to \infty} z_1(x) = \lim_{x \to \infty} \frac{1}{x} = 0 \qquad (6.17)$$

$$\lim_{x \to \infty} z_2(x) = \lim_{x \to \infty} \frac{2}{x^3} = 0 \qquad (6.18)$$

$$\lim_{x \to \infty} z_3(x) = \lim_{x \to \infty} \frac{1}{e^{2x}} = 0 \qquad (6.19)$$

When one of the infinities multiplies one of the zeros defined above, what would happen? For example,

$$\lim_{x \to \infty} y_1 \cdot z_1 = \lim_{x \to \infty} 5x \cdot \frac{1}{x} = 5 \tag{6.20}$$

$$\lim_{x \to \infty} y_1 \cdot z_2 = \lim_{x \to \infty} 5x \cdot \frac{2}{x^3} = \lim_{x \to \infty} \frac{10}{x^2} = 0 \tag{6.21}$$

$$\lim_{x \to \infty} y_3 \cdot z_1 = \lim_{x \to \infty} e^x \frac{1}{x} = \infty \tag{6.22}$$

Therefore, when an infinity times a zero, the answer could be a finite number, zero, or infinity, depending on the natures of the infinity and the zero.

From the above discussion, we realize that $\int_{t1}^{t1+dt} \sum \vec{F} dt$ is not necessarily zero.

Let's consider a type of force which lasts for a fixed period of time at a constant magnitude as shown in Figure 6.1.

The forces shown in Figure 6.1 are called force pulses, or pulses. In Figure 6.1(a), the pulse lasts for $c$ time with a magnitude of $1/c$. We can define the function of the force pulse as

$$F(t) = \frac{1}{c}, \quad 0 \le t \le c$$

$$F(t) = 0, \quad t > c \tag{6.23}$$

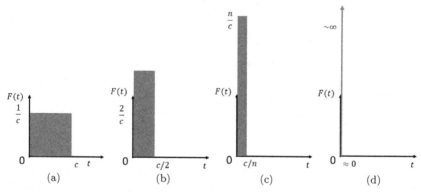

Figure 6.1:   Profiles of different pulses: (a) a square force pulse; (b) a force pulse with half the duration but twice the magnitude; (c) a force pulse with one n-th of the time duration and n times the magnitude; and (d) an impulse.

If we integrate the above $F(t)$ over time, we obtain

$$\int_0^\infty F(t)dt = \frac{1}{c}c = 1 \qquad (6.24)$$

In Figure 6.1(b), we half the pulse time duration but double the force magnitude, the integration of $F(t)$ over time is the same

$$\int_0^\infty F(t)dt = \frac{1}{2c}2c = 1 \qquad (6.25)$$

In Figure 6.1(c), we reduce the pulse duration to its n-th $(c/n)$, while increase the magnitude to n times. The integration of the pulse is still the same.

Now, let $n$ go to $\infty$, we have

$$F(t) = \lim_{n\to\infty} \frac{n}{c}, \quad 0 \le t \le \lim_{n\to\infty} c/n$$
$$F(t) = 0, \quad t > \lim_{n\to\infty} c/n \qquad (6.26)$$

With $n \to \infty$, the force pulse looks very strange, as shown in Figure 6.1(d). Its magnitude is now approaching $\infty$, but it lasts for almost no time, nearly zero. However, the integration over time for such a strange force pulse is still the same.

$$\int_0^\infty F(t)dt = \lim_{n\to\infty} \left(\frac{n}{c}\frac{c}{n}\right) = 1 \qquad (6.27)$$

The force pulse of Figure 6.1(d) is quite impossible in the real world. Such a seemingly unreal force pulse is denoted as an *Impulse*. We can consider it as an *im*possible *pulse*, i.e., an *impulse*. In addition, because the time integration of this impulse is 1, we will call it an *unit* impulse, or as commonly known as a delta function, expressed as $\delta(t)$. The function $\delta(t)$ is zero in time other than $t = 0$.

We can shift a delta function to a different time instant,

$$\delta(t-a) \equiv \infty, \quad t = a$$
$$\delta(t-a) = 0, \quad t \ne a \qquad (6.28)$$

In the above equation, we use the sign $\equiv$ for $\infty$ because its value is not rigorously defined. We can also have a different integration value for an impulse by multiplying a constant,

$$\int_0^\infty h \cdot \delta(t-a)dt = h \int_{a^-}^{a^+} \delta(t-a)dt = h \qquad (6.29)$$

$a^-$ means the time instant immediately before $t = a$ and $a^+$ is the time instant immediately after $t = a$.

In nature, it is quite impossible to have a force pulse truly as an impulse. However, some of them could be quite "impulse-like", having a huge (but not infinity) force magnitude and lasting for very short periods of time (but not zero). In a high energy accelerator for particle physics research, such as the one at CERN (the European Organization for Nuclear Research), two sub-atomic particles are accelerated to travel nearly at the speed of light and they then collide. The collision lasts in an extremely short amount of time at an enormous amount of force magnitude.

For the more routine mechanical world, when two cars hit into each other at high speeds, we also have something similar to an impulse but the time duration is a lot longer than the sub-atomic particles collision cases.

### 6.1.3  *Conservation and pseudo-conservation of linear momentum*

Let's consider the situations where the time integration of forces could be zero or nearly zero. First, if the time interval is long, i.e., $t_2 - t_1 \gg 0$, then $\int_{t_1}^{t_2} \sum \vec{F}(t)dt = 0$ only if $\sum \vec{F}(t) = 0$. In such a case, we have a conservation of linear momentum,

$$\vec{P_1} + \int_{t_1}^{t_2} \sum \vec{F}(t)dt = \vec{P_1} = \vec{P_2}, \quad \sum \vec{F}(t) = 0, \quad t_1 < t < t_2$$

$$(6.30)$$

If $t_2 - t_1 \approx 0$ and $\sum \vec{F}(t)$ is NOT HUGE, in other words, NOT impulse-like, then

$$\vec{P_1} + \int_{t_1}^{t_1+c} \sum \vec{F}(t)dt \approx \vec{P_1} = \vec{P_2}, \quad c \approx 0 \qquad (6.31)$$

Equation (6.31) is denoted as a *Pseudo-Conservation of Linear Momentum*. It is important to note that Equation (6.31) is only valid at $t = t_1$ and shortly after.

If $t_2 - t_1 \approx 0$ but $\sum \vec{F}(t)$ is HUGE, in other words, very impulse-like, then

$$\vec{P}_1 + \int_{t_1}^{t_1+c} \sum \vec{F}(t)dt = \vec{P}_2 \neq \vec{P}_1, \quad c \approx 0 \qquad (6.32)$$

There will be a sudden change in the linear momentum from $t = t_1$ to $t = t_1 + c$ despite $c \approx 0$. This can only happen if $\sum \vec{F}(t)$ is huge, impulse-like. We will discuss applications of Equations (6.30)–(6.32) later.

### 6.1.4   *Conservation and pseudo-conservation of angular momentum*

Similar to the discussion in the last section, we also have cases for conservation of angular momentum, pseudo-conservation of angular momentum, and sudden change of angular momentum. First, if the time interval is long, i.e., $t_2 - t_1 \gg 0$, then $\int_{t_1}^{t_2} \sum \vec{\tau}_A(t)dt = 0$ only if $\sum \vec{\tau}_A(t) = 0$. In such case, we have a conservation of angular momentum with respect to point $A$,

$$\vec{H}_{A1} + \int_{t_1}^{t_2} \sum \vec{\tau}_A(t)dt = \vec{H}_{A1} = \vec{H}_{A2}, \sum \vec{\tau}_A(t) = 0, \quad t_1 < t < t_2$$

$$(6.33)$$

If $t_2 - t_1 \approx 0$ and $\sum \vec{\tau}_A(t)$ is NOT HUGE, in other words, NOT impulse-like, then

$$\vec{H}_{A1} + \int_{t_1}^{t_1+c} \sum \vec{\tau}_A(t)dt \approx \vec{H}_{A1} = \vec{H}_{A2}, \quad c \approx 0 \qquad (6.34)$$

Equation (6.34) is denoted as the *Pseudo-Conservation of Angular Momentum with respect to point $A$*. It is important to note that Equation (6.34) is only valid at $t = t_1$ and shortly after, and is with respect to point $A$.

If $t_2 - t_1 \approx 0$ but $\sum \vec{\tau}_A(t)$ is HUGE (impulse-like), then

$$\vec{H}_{A1} + \int_{t_1}^{t_1+c} \sum \vec{\tau}_A(t)dt = \vec{H}_{A2} \neq \vec{H}_{A1}, \quad c \approx 0 \quad (6.35)$$

There will be a sudden change in the angular momentum from $t = t_1$ to $t = t_1 + c$ despite $c \approx 0$. This can only happen if $\sum \vec{\tau}_A(t)$ is huge, impulse-like. We will discuss applications of Equations (6.33)–(6.35) later.

## 6.2 Applications Using Conservation Law of Total Linear Momentum

We consider several cases for which we can apply some form of momentum conservation equations. Again, it is very important to note that conservation of momentum, linear or angular or both, only happens in special cases and we must verify the force conditions carefully to ensure that the conservation law indeed is valid.

### 6.2.1 *One-dimensional impact of two particles*

We first consider an idealized 1D impact case with two particles traveling along the same straight line. Long before impact, Particle $A$ travels at $(v_{A/O})_1$ and Particle $B$ at $(v_{B/O})_1$, both to the right, as shown in Figure 6.2(a). Particle $A$ has a mass $m_A$ and $B$ with $m_B$. Because $v_{A/O} > v_{B/O}$, particle $A$ will catch up and hit particle $B$ at $t = a$. Right before the impact, $t = a^-$, the speed of $A$ is $(v_{A/O})_1$ and $B$ is $(v_{B/O})_1$, as shown in Figure 6.2(b). The subscript, "1", is used to indicate the condition for the time instant immediately before the impact.

The impact then occurs during $a^- < t < a^+$ (Figure 6.2(c)). The impact could bring in drastic changes and we often are not capable of observing what is happening during the impact.

We then observe what happens immediately after the impact at $t = a^+$, as shown in Figure 6.2(d). The speed of $A$ is observed to be $(v_{A/O})_2$ and $B$ as $(v_{B/O})_2$. The subscript, "2", is used to indicate the condition for the time instant immediately after the impact.

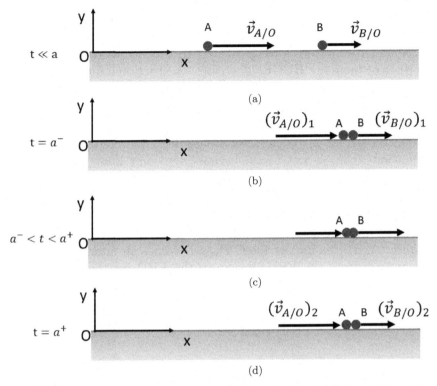

Figure 6.2: Two particle, one-dimensional impact case: (a) long before the impact; (b) immediately before the impact; (c) during impact; and (d) immediately after the impact.

The corresponding FBDs for $a^- \leq t \leq a^+$ are shown in Figure 6.3. If we consider an FBD for each particle throughout the impact duration, the linear momentum equation for particle $A$ along the $x$ direction can be determined as

$$P_{A1x} + \int_{a^-}^{a^+} \sum_x F(t)\,dt = m_A(v_{A/O})_1 + \int_{a^-}^{a^+} (-F_A - F_{imp})\,dt$$

$$= m_A \cdot (v_{A/O})_1 - \int_{a^-}^{a^+} F_A\,dt - \int_{a^-}^{a^+} F_{imp}\,dt$$

$$= m_A(v_{A/O})_1 - 0 - \int_{a^-}^{a^+} F_{imp}\,dt$$

$$= P_{A2x} = m_A \cdot (v_{A/O})_2 \qquad (6.36)$$

Since the impact time is very short, $\int_{a^-}^{a^+} F_A dt = 0$ because the friction force, $F_A$, is not huge. The friction force is related to the normal force and the normal force is related to the weight of the particle. Neither of them are HUGE. Therefore, the frictional force is limited, NOT huge. However, the impact force, as discussed before, could be impulse-like. As a result, we do not know the value of $\int_{a^-}^{a^+} F_{imp} dt$ even though the impact time duration is very short.

If $(v_{A/O})_1$ is given and we would like to determine $(v_{A/O})_2$, Equation (6.36) is not useful because we don't know the value of $\int_{a^-}^{a^+} F_{imp} dt$.

We can derive the linear momentum equation for particle $B$ as well, but we will have the same difficulty.

If we draw an FBD for both particles as a system, as shown in Figure 6.3(d), the impact force, $F_{imp}$, becomes an internal force and disappears from the FBD. Based on the FBD for the two-particle system, the linear momentum equation becomes

$$P_{A1x} + P_{B1x} + \int_{a^-}^{a^+} \sum_x F(t) dt$$

$$= m_A \cdot (v_{A/O})_1 + m_B \cdot (v_{B/O})_1 + \int_{a^-}^{a^+} (-F_A - F_B) dt$$

$$= m_A (v_{A/O})_1 + m_B (v_{B/O})_1$$

$$= P_{A2x} + P_{B2x} = m_A \cdot (v_{A/O})_2 + m_B \cdot (v_{B/O})_2 \qquad (6.37)$$

We now have the pseudo-conservation of linear momentum for the two-particle system during $a^- \le t \le a^+$, despite that we still don't know $(v_{A/O})_2$ and $(v_{B/O})_2$, individually. Recalling that the mass center carries the entire linear momentum of a system, we have

$$m_A \cdot (v_{A/O})_1 + m_B \cdot (v_{B/O})_1$$

$$= m_A \cdot (v_{A/O})_2 + m_B \cdot (v_{B/O})_2 = (m_A + m_B) v_{G/O} \qquad (6.38)$$

Equation (6.38) indicates that the mass center velocity stays constant throughout the impact. Because the mass center is moving at a constant velocity throughout the impact, we can establish an inertial reference frame based on the mass center and have a new perspective to observe this two-particle impact problem. The velocities of

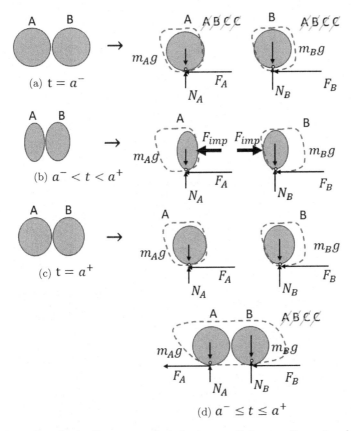

Figure 6.3:   Free-body diagram analysis for two particle; one-dimensional impact case: (a) FBDs for each particle immediately before the impact; (b) FBDs for each particle during impact; (c) FBDs for each particle immediately after the impact; and (d) FBD for both particles as a system throughout the impact.

the particles with respect to the mass center, $G$, become

$$(v_{A/O})_i = v_{G/O} + (v_{A/G})_i \tag{6.39}$$

$$(v_{B/O})_i = v_{G/O} + (v_{B/G})_i \tag{6.40}$$

where $i = 1$ or $2$ for the velocities immediately before and after the impact, respectively.

It is rather easy to show that $(v_{A/O})_1 > v_{G/O} > (v_{B/O})_1$. As a result, the relative velocities observed by the mass center become those shown in Figure 6.4.

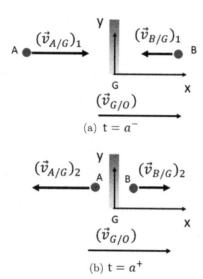

Figure 6.4:   Mass center perspective for the two particle, one-dimensional impact case: (a) velocities observed by the mass center immediately before the impact; and (b) velocities observed by the mass center immediately after the impact.

With the mass center perspective, both particles will hit into the mass center and then bounce out of the mass center. Because we use the same $x-y$ reference frame, pointing to the right is positive and to the left negative, particle $A$ has a positive linear momentum immediately before impact and its linear momentum becomes negative immediately after the impact. It is the opposite for particle $B$.

The total linear momentum with respect to the mass center is now zero and the total linear momentum equation becomes

$$m_A \cdot (v_{A/G})_1 - m_B \cdot (v_{B/G})_1$$
$$= -m_A \cdot (v_{A/G})_2 + m_B \cdot (v_{B/G})_2 = 0 \qquad (6.41)$$

From the above equation we also have

$$m_A \cdot (v_{A/G})_1 = m_B \cdot (v_{B/G})_1 \qquad (6.42)$$

$$m_A \cdot (v_{A/G})_2 = m_B \cdot (v_{B/G})_2 \qquad (6.43)$$

$$m_A((v_{A/G})_1 + (v_{A/G})_2) = m_B((v_{B/G})_1 + (v_{B/G})_2) \qquad (6.44)$$

Let's define the following speed ratios for the before and after impact speeds, $e_A = (v_{A/G})_2/(v_{A/G})_1$ and $e_B = (v_{B/G})_2/(v_{B/G})_1$ and rewrite Equation (6.44) as

$$m_A(v_{A/G})_1 \left[ 1 + \frac{(v_{A/G})_2}{(v_{A/G})_1} \right] = m_A(v_{A/G})_1[1 + e_A]$$

$$= m_B(v_{B/G})_1 \left[ 1 + \frac{(v_{B/G})_2}{(v_{B/G})_1} \right] = m_B(v_{B/G})_1[1 + e_B] \quad (6.45)$$

Introducing Equation (6.42) to (6.45), we obtain

$$e_A = \frac{(v_{A/G})_2}{(v_{A/G})_1} = e_B = \frac{(v_{B/G})_2}{(v_{B/G})_1} = e \quad (6.46)$$

The speed ratios based on the mass center reference frame are the same. This speed ratio is denoted as *Restitution Coefficient*. This result is very important because we realize for the two-particle impact as shown in Figure 6.4, both sides of the mass center are the same in terms of the restitution behavior. If we know the value of $e$, then we can determine $(v_{A/G})_2$ and $(v_{B/G})_2$. After that, we can use Equations (6.39) and (6.40) to calculate $(v_{A/O})_2$ and $(v_{B/O})_2$.

We now have a completely new perspective to interpret the two-particle, 1D impact problem. We only need to consider one side of the impact. The mass center reference frame is like a solid wall. We will call it the G-wall in this book. The impact problem is like a ball hitting a solid wall and bouncing off. There are two balls hitting the wall but we only need to consider one side. We can re-draw Figure 6.4 as 6.5.

### 6.2.2  *Properties of impact restitution coefficient*

The restitution coefficient is a characteristic of an impact. It is related to the combined material properties of the two particles. Unfortunately, just like the friction coefficient, the restitution coefficient is not quite predictable by theories and is mostly obtained by experiments. If we throw a very bouncy elastic ball to a solid wall, the ball bounces back almost at the same speed as the approaching speed toward the wall. In this case, the restitution coefficient is nearly one, i.e., $e \approx 1$. On the other hand, if we throw a sandbag to a solid

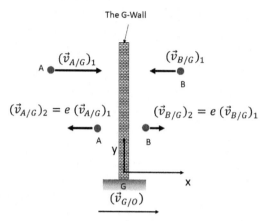

The G-Wall

$(\vec{v}_{A/G})_1$    $(\vec{v}_{B/G})_1$

$(\vec{v}_{A/G})_2 = e\,(\vec{v}_{A/G})_1$    $(\vec{v}_{B/G})_2 = e\,(\vec{v}_{B/G})_1$

$(\vec{v}_{G/O})$

Figure 6.5:   Analogy of the two-particle, 1D impact case as a ball bouncing off a solid wall.

wall, it does not bounce back at all, i.e., $e \approx 0$. A tennis ball has a higher restitution coefficient, about $0.6-0.7$, than that of a baseball, at about $0.1-0.2$. Remember that the restitution coefficient is related to the properties of both particles. If an elastic ball hits a sandbag, it does not bounce back much, indicating a very low restitution coefficient. The same low restitution coefficient should be used to determine the speeds of the elastic ball and the sandbag after impact.

Let's consider a numerical example for illustration.

## Example 6.1: Two-particle, 1D Impact Example

**Problem Statement:** Refer to Figures 6.2–6.4 for this example. Based on $o-x-y$, particle $A$, $10\,\mathrm{kg}$ in mass, is traveling at $10\,\mathrm{m/s}$ when it hits particle $B$, $5\,\mathrm{kg}$, at $5\,\mathrm{m/s}$. Determine the velocities of $A$ and $B$ immediately after impact. The restitution coefficient is 0.6.

**Solution:**

**Force Analysis:** With an FBD to include both particles, the impulse-like impact force is eliminated from the FBD.

**Motion Analysis:** We will establish an inertial reference frame based on the mass center, knowing that conservation of linear momentum can be applied throughout the impact.

**Governing Equations:** We will use the conservation equation of total linear momentum to solve this problem.

**Solving for Unknowns:** We will use the following steps to work out all the required information.

**Step #1: Determine the mass center velocity:** Based on Equation (6.38), the mass center velocity is found to be

$$(v_{G/O}) = [m_A \cdot (v_{A/O})_1 + m_B \cdot (v_{B/O})_1]/(m_A + m_B)$$

$$= (10 \cdot 10 + 5 \cdot 5)/(10 + 5) = 8.333 \, (\text{m/s}) \qquad (6.47)$$

**Step #2: Translate the impact condition to the mass center perspective:** The velocities before impacts for both particles are calculated based on Equations (6.39) and (6.40).

$$(v_{A/O})_1 = 10 = v_{G/O} + (v_{A/G})_1 = 8.333 + (v_{A/G})_1 \quad (6.48)$$

$$(v_{B/O})_1 = 5 = v_{G/O} + (v_{B/G})_1 = 8.333 + (v_{B/G})_1 \quad (6.49)$$

Solving the above equations, we obtain $(v_{A/G})_1 = 1.667 \, \text{m/s}$ and $(v_{B/G})_1 = -3.333 \, \text{m/s}$.

Let's confirm that based on the reference frame on the mass center, the total linear momentum is zero.

$$m_A \cdot (v_{A/G})_1 + m_B \cdot (v_{B/G})_1 = 10 \cdot 1.667 + 5 \cdot (-3.333) \approx 0$$
$$(6.50)$$

**Step #3: Draw the G-wall diagram similar to Figure 6.4:** The G-wall diagram for this problem is shown in Figure 6.6.

**Step #4: Calculate the bounce back speeds:** With $e = 0.6$, the bounce back speed for mass $A$ is found to be $(v_{A/G})_2 = 0.6 \cdot 1.667 = 1.000 \, \text{m/s}$ and that of $B$ is $(v_{B/G})_2 = 0.6 \cdot 3.333 = 2.000 \, \text{m/s}$. Note that $(v_{A/G})_2$ points to the left, while $(v_{B/G})_2$ points to the right.

**Step #5: Translate back to the initial reference frame:** We will use Equations (6.39) and (6.40) again,

$$(v_{A/O})_2 = v_{G/O} + (v_{A/G})_2 = 8.333 - 1.000 = 7.333 \, (\text{m/s}) \quad (6.51)$$

$$(v_{B/O})_2 = v_{G/O} + (v_{B/G})_2 = 8.333 + 2.000 = 10.333 \, (\text{m/s}) \quad (6.52)$$

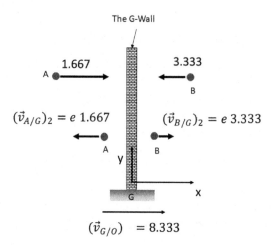

Figure 6.6: The two-particle, 1D impact example expressed from the mass center perspective.

Let's confirm if the total linear momentum is conserved with respect to the initial inertial reference frame.

$$m_A \cdot (v_{A/O})_2 + m_B \cdot (v_{B/O})_2 = 10 \cdot 7.533 + 5 \cdot 10.333 = 125.0$$

$$= m_A(v_{A/O})_1 + m_B(v_{B/O})_1 = 10 \cdot 10 + 5 \cdot 5 = 125 \qquad (6.53)$$

### 6.2.3   *Energy consideration of two-particle 1D impact*

We will discuss the energy concept in greater detail in the next chapter. Here, let's introduce the basic kinetic energy concept. When a particle has a speed, it is considered to possess a kinetic energy as $T = \frac{1}{2}mv^2$. With different inertial reference frames, the observed velocities could be different, as are the kinetic energies.

Let's examine the kinetic energy of the Example (6.1).

Before and after impact, the kinetic energies for $A$ and $B$ are

$$(T_{A/O})_1 = \frac{1}{2}m_A(v_{A/O})_1^2 = 500 \,(\text{kg m}^2/\text{s}^2) \qquad (6.54)$$

$$(T_{A/O})_2 = \frac{1}{2}m_A(v_{A/O})_2^2 = 268.889 \,(\text{kg m}^2/\text{s}^2) \qquad (6.55)$$

$$(T_{B/O})_1 = \frac{1}{2}m_B(v_{B/O})_1^2 = 62.500 \,(\text{kg m}^2/\text{s}^2) \qquad (6.56)$$

$$(T_{B/O})_2 = \frac{1}{2}m_A(v_{A/O})_2^2 = 266.944 \,(\text{kg m}^2/\text{s}^2) \qquad (6.57)$$

$$(T_{total})_1 = 562.500 \, (\text{kg m}^2/\text{s}^2) \tag{6.58}$$

$$(T_{total})_2 = 535.833 \, (\text{kg m}^2/\text{s}^2) \tag{6.59}$$

From the numbers above, we observed that particle $A$ lost energy due to impact, while particle $B$ actually gained energy. However, there is a loss of $26.667 \, \text{kg m}^2/\text{s}^2$ in total energy due to impact.

Let's redo the calculations, but from the mass center perspective.

$$(T_{A/G})_1 = \frac{1}{2} m_A (v_{A/G})_1^2 = 13.894 \, (\text{kg m}^2/\text{s}^2) \tag{6.60}$$

$$(T_{A/G})_2 = \frac{1}{2} m_A (v_{A/G})_2^2 = 5.000 \, (\text{kg m}^2/\text{s}^2) \tag{6.61}$$

$$(T_{B/O})_1 = \frac{1}{2} m_B (v_{B/O})_1^2 = 27.772 \, (\text{kg m}^2/\text{s}^2) \tag{6.62}$$

$$(T_{B/O})_2 = \frac{1}{2} m_A (v_{A/O})_2^2 = 10.000 \, (\text{kg m}^2/\text{s}^2) \tag{6.63}$$

$$(T_{total})_1 = 41.666 \, (\text{kg m}^2/\text{s}^2) \tag{6.64}$$

$$(T_{total})_2 = 15.000 \, (\text{kg m}^2/\text{s}^2) \tag{6.65}$$

Based on the inertial reference frame attached to the mass center, we observe that both particles lost energy due to impact. The total energy loss is $26.666 \, \text{kg m}^2/\text{s}^2$, which is essentially the same as the loss calculated based on the stationery inertial reference frame.

If we take the ratio of the remaining energy over the initial energy, we notice

$$\frac{(T_{A/G})_2}{(T_{A/G})_1} = 0.36 = (0.6)^2 = e^2 \tag{6.66}$$

$$\frac{(T_{B/G})_2}{(T_{B/G})_1} = 0.36 = (0.6)^2 = e^2 \tag{6.67}$$

$$\frac{(T_{total})_2}{(T_{total})_1} = 0.36 = (0.6)^2 = e^2 \tag{6.68}$$

The remaining energy is $e^2$ of the original energy. Actually, this result is not surprising and is foretold by Figures 6.5 and 6.6. The energy loss is $1 - e^2$ of the original energy. It should be noted that the definition of the restitution coefficient is based on the mass center perspective.

When $e = 1$, the energy loss is zero, and this type of impact is denoted as *Elastic Impact.* when $e = 0$, there is a total energy loss and this type of impact is denoted as *Plastic Impact.* These definitions must be based on the mass center perspective.

### 6.2.4 *Impact force consideration of two particles, 1D impact*

Let's consider the linear momentum equation for each particle and introduce the restitution coefficient based on Figure 6.5.

$$m_A(v_{A/G})_1 - \int_{a^-}^{a^+} (F_{imp})dt = m_A[-(v_{A/G})_2] \qquad (6.69)$$

Introducing the restitution coefficient and considering the average impact force, $(F_{imp})_{ave}$, and impact time, $\Delta t$, we have

$$m_A(v_{A/G})_1 + m_A[e \cdot (v_{A/G})_1] = \int_{a^-}^{a^+} (F_{imp})dt = (F_{imp})_{ave}\Delta t$$
$$(6.70)$$

The average impact force becomes

$$(F_{imp})_{ave} = (1 + e)m_A \frac{(v_{A/G})_1}{\Delta t} \qquad (6.71)$$

From the above equation, if the impact time remains the same, the average impact force is higher with a higher restitution coefficient. The impact force of an elastic impact will be twice that of an plastic impact if the impact time is the same. If a lower impact force is desired, a plastic impact is preferred.

We can also calculate the average acceleration from Equation (6.71) for both particles,

$$(a_{A/G})_{ave} = \frac{(F_{imp})_{ave}}{m_A} = (1 + e)\frac{(v_{A/G})_1}{\Delta t} \qquad (6.72)$$

$$(a_{B/G})_{ave} = \frac{(F_{imp})_{ave}}{m_B} = (1 + e)\frac{(v_{B/G})_1}{\Delta t}$$

$$= (1 + e)\frac{m_A}{m_B}\frac{(v_{A/G})_1}{\Delta t} \qquad (6.73)$$

From the above two equations, we observe that the average acceleration is higher for a higher restitution coefficient. More interestingly, the average acceleration of particle $B$ is $\frac{m_A}{m_B}$ that of particle $A$. If $m_A > m_B$, then $(a_{B/G})_{ave} > (a_{A/G})_{ave}$. In other words, the particle with a lower mass will exhibit a higher acceleration due to impact.

### 6.2.5 *Practical examples of two-particle 1D impact*

Let's consider some practical impact examples in real life.

### Example 6.2: Vehicle Collision

First, let's consider a head-to-head collision of two cars. In car designs, the top design consideration is to protect the drivers if collision occurs. Therefore, the impact force should be kept minimum. With this consideration, plastic impact is preferred and modern cars are designed with crumple zones to achieve plastic impact. When we see a car badly smashed due to an accident, we should appreciate the design excellence for achieving the plastic impact.

Many accidents on highways were due to tailgating, in which a car follows too close to the car in front. Let's consider an SUV of 2000 kg, at 140 km/hr, hits a sedan of 1000 kg at 100 km/hr from behind. Use the calculations above with $e = 0$. The mass center velocity is found to be 126.667 km/hr. Because $e = 0$, both cars would be stuck with the mass center and move at the same speed.

However, the sedan will experience an acceleration twice that of the SUV because it has one half of the mass. Since the driver of the sedan moves with the sedan, he or she will experience a similar acceleration. As a result, the impact force experienced by the driver is

$$
\begin{aligned}
F_{imp/driver} &= m_{driver} \cdot (a_{B/G})_{ave} \\
&= m_{driver} \cdot \frac{m_A}{m_B}(a_{A/G})_{ave} \\
&= m_{driver} \cdot 2 \cdot (a_{A/G})_{ave}
\end{aligned}
\tag{6.74}
$$

If both drivers are at similar body mass, then the sedan driver will experience twice as much impact force than the SUV driver. We conclude that a heavier car in general is safer than a lighter car when a collision happens.

To improve the safety design, car companies will measure the impact forces to the driver using dummies instrumented with sensors. They will slam a car at a speed into a solid wall for the test. It is interesting to ask which of the following collision conditions are dynamically equivalent.

- *Case #A:* A car at 50 km/hr hits the same car parked.
- *Case #B:* A car at 50 km/hr hits the same car coming from the opposite direction at the same speed.
- *Case #C:* A car at 50 km/hr hits a solid wall.
- *Case #D:* A car at 25 km/hr hits a solid wall.

If we translate the impact condition to the mass center perspective, it becomes clear that the solid wall essentially is the G-wall. Therefore, *Case #A* is equivalent to *Case #D*, while *Case #B* is equivalent to *Case #C*. The mass center perspective, again, is superior for the impact analysis.

**Example 6.3: Newton's Cradle**

A popular toy, called Newton's cradle, is shown in Figure 6.7. When the ball on the right swings down to hit the stationary balls, the hitting ball will come to stop instantly, while the leftest ball will

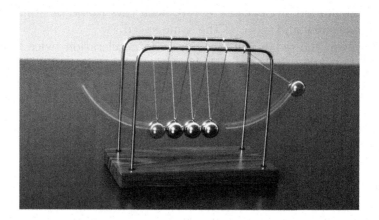

Figure 6.7:   Newton's cradle when elastic steel balls hit each other.
*Source:*   https://i0.wp.com/sefiks.com/wp-content/uploads/2017/02/newtons-cradle.png?resize=1024%2C576&ssl=1.

shoot out at the same time. The ball on the left then swings back to hit the stationary balls, the process repeats itself.

In this case, the restitution coefficient $e \approx 1$. We can explain Newton's cradle with only two balls of the same mass. When one hits the other at stationary, the mass center speed is one half. Therefore, from the mass center perspective, both balls are hitting the mass center wall at the same speed (see Figure 6.5). Because $e = 1$, they will leave the wall at the same speed. Converting the velocities back to the original inertial reference frame, the speed of the first will be zero after impact, while the ball at stationary initially will have the same velocity as the first ball before impact.

## Example 6.4: Tennis Hitting: Racket Head Speed, Racket Weight, and String Tension

We will not discuss the fine art of tennis hitting. We often hear that with a lower string tension, a player can hit with more power but less control. On the other hand, with a higher tension, it would be less power but better control. It sounds a bit counterintuitive with this statement. Let's review it with the knowledge we have gained from the two-particle 1D impact problem. Let's consider the serve, which could be one of the hardest shots to learn in tennis, but is the simplest, dynamically speaking. A racket hits a tennis ball which is essentially stationary. The racket is much heavier than the ball. In Figure 6.8, a sequence of images from a high speed photography video of a 142 *mph* serve posted on youtube are shown.

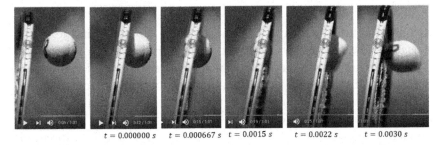

| $t = 0.000000\,s$ | $t = 0.000667\,s$ | $t = 0.0015\,s$ | $t = 0.0022\,s$ | $t = 0.0030\,s$ |

Figure 6.8:   Images of high speed photography of a tennis racket hitting a tennis ball.

*Source*: https://www.youtube.com/watch?v=VHV1YbeznCo.

As shown in Figure 6.8, the entire impact took about 0.003 s or 3 ms and the tennis ball gained a speed of 142 mph or 63.48 m/s. The weight of a tennis ball is typically 2 *oz* or about 57 g. A strung tennis racket is about 10.6 *oz* or 300 g. The weight ratio of racket to ball is easily over 5.

From the images, it is clear that modern tennis rackets are very stiff and they do not deform much at all. The tennis string deforms and the tennis ball deforms, quite substantially. In general, the tennis string is more elastic. When the string deforms, there is not much energy loss. Just like a spring, the deformed string stores energy and then releases it. However, it is different for a tennis ball. If we drop a tennis ball to the ground, it never rises to the same height. When a tennis ball deforms due to impact, it loses energy. The more it deforms, the more energy it loses. Dynamically speaking, the restitution coefficient is lower if the tennis ball deforms more. We will discuss the energy aspect of the impact problem in greater detail in Section 7.4, Chapter 7.

We can now explain the power aspect of the statement above. With a lower string tension, the string will deform more while the tennis ball deforms less by sinking into the deformed string. As a result, the restitution coefficient is higher for the same racket swing for a lower energy loss; therefore, it feels more powerful. When a string breaks during a point, even with a slight swing by the player, the ball could sail far out.

The control aspect of the statement is more complex. From the point of view of linear momentum, the racket tries to control the speed and direction of the tennis ball. Due to conservation of linear momentum, the gained linear momentum of the ball comes from the linear momentum of the racket before impact. A higher racket head speed provides a higher linear momentum. However, if the string tension is low, the ball will sail out, being very hard to control. With a higher string tension, a player can achieve a high racket head speed without hitting the ball out, thus, allowing better control.

Finally, we should know that the speed of the tennis ball after impact depends in part on the racket linear momentum right before impact. The racket head speed is the key, not how hard the force was exerted by the hand at the moment of impact. That force practically does not do much because the impact time is quite short and the hand force is limited. The same consideration applies to the baseball hitting, golf ball hitting, etc. The key is to accelerate early and to swing through so that a high speed of the bat or the golf club can be achieved before impact. Intentionally applying a higher force upon impact does not help. Finally, throwing a punch works the same way. The famous one-inch punch by the late martial art legend, Bruce Lee, likely was not real because with only one inch of travel, the fist of Bruce Lee likely cannot reach a high speed even with his extraordinary ability.

A lighter racket allows a player to swing much faster to generate a more powerful shot. Modern rackets are made with light and strong materials, such as graphite, to achieve this goal. In earlier years, tennis was played with wooden rackets, which were heavy and deform easily upon impact. Moreover, wooden rackets did not allow for very high tennis string tension. As a result, the shots were simply less powerful and players could not swing as fast as modern tennis players.

### Example 6.5: Cannon Firing and the Recoil Force

In the examples above, we presented so far on 1D impact problems with two particles hitting into each other. In fact, the impact problem could be reversed with the same conservation law of linear momentum. There are many such examples, such as firing a cannon and the resulting recoil force. As shown in Figure 6.9(a), an old cannon is on wheels. We might think that the wheels are only used to transport the cannon; however, the wheels are also important to absorb the recoil force when the cannon is fired. A similar situation is found in the firing of a fire arm, as shown in Figure 6.9(b).

At $t = 0$, as shown in Figure 6.9(b), the shot is fired. Before the firing, the bullet and the breech were stationary. Immediately after the firing, the bullet develops a large forward velocity $(v_{bullet})_{0+}$, while the breech develops a lower backward velocity $(v_{breech})_{0+}$. With the FBD including both the bullet and the breech, we have the pseudo-conservation of linear momentum for the short time period between immediately before $(t = 0^-)$ and after $(t = 0^+)$ the firing,

$$0 = m_{bullet}(v_{bullet})_{t=o+} - m_{breech}(v_{breech})_{t=o+} \qquad (6.75)$$

where $m_{bullet}$ and $m_{breech}$ are the masses for the bullet and the breech, respectively. In general, $m_{breech} \gg m_{bullet}$.

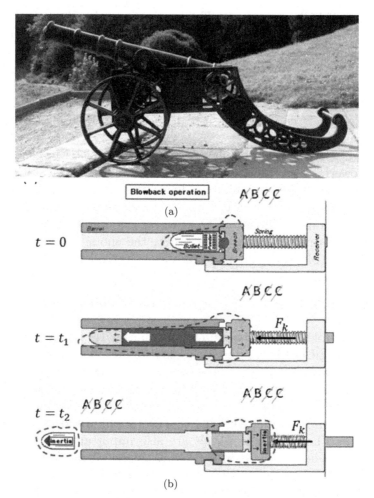

Figure 6.9: (a) A cannon on wheels; and (b) recoil force absorption mechanism. *Source*: (a) https://s0.geograph.org.uk/geophotos/02/04/96/2049638_a5b94d60. jpg; and (b) https://upload.wikimedia.org/wikipedia/commons/1/1f/SBBK-SB FWD_piston_cylinder.PNG.

At $t = t_1$, the bullet continues to move forward at the same speed without being impeded, while the breech moves backward but is resisted by the spring force $F_k$. The spring force grows bigger when the spring is compressed further. Due to this spring force, the conservation of linear momentum is no longer valid for the system for both the bullet and the breech after $t_1$. Instead, we have

$$m_{bullet}(v_{bullet})_{t=o^+} - m_{breech}(v_{breech})_{t=o^+} + \int_{0^+}^{t_1} F_k \, dt$$

$$= m_{bullet}(v_{bullet})_{t=t_1} - m_{breech}(v_{breech})_{t=t_1} \qquad (6.76)$$

The term $\int_{0^+}^{t_1} F_k \, dt$ is not zero because $t_1$ is NOT short. We can draw separate FBDs for the bullet and for the breech, respectively, as shown in the bottom figure of Figure 6.9(b). There are no forces acting on the bullet; therefore, its velocity stays as $v_{bullet})_{t=o^+}$, but the velocity of the breech will change and eventually come to a complete stop at $t = t_2$. For the breech, we have the linear momentum equation as

$$-m_{breech}(v_{breech})_{t=o^+} + \int_{0^+}^{t_2} F_k dt = m_{breech}(v_{breech})_{t=t_2} = 0$$

$$(6.77)$$

This spring force is the recoil force that the shooter will experience. From Equation (6.77) and Hooke's Law, we know that $F_k$ will grow and reach a maximum at $t = t_2$, but it is not impulse-like.

What if we hold the breech and the receiver rigidly without a spring? This is similar to bolting the cannon rigidly to the ground. In this case, because the breech stays stationary, it is like the mass center wall. The firing or the explosion force, which is impulse-like, will act on the breech/receiver rigid body. As a result, a huge, impulse-like, recoil force will occur. After a few shots, the structure to hold the breech/receiver (or the cannon) could be damaged. To reduce the recoil force, many recoil force damping mechanisms were developed. In addition to springs, hydraulic cushion is also used to further reduce the recoil force.

This reversed impact process also applies to the propulsion of rockets or spacecrafts in space.

### 6.2.6 *Two-dimensional impact problems*

Let's extend the analysis of the 1D impact problem to 2D. As shown in Figure 6.10(a), two particles hit into each other. At the contact point, we can define the tangential direction and the normal direction as before. In Figure 6.10(b), an FBD is constructed for each particle. It is important to know that the normal force, $N$, is the impact

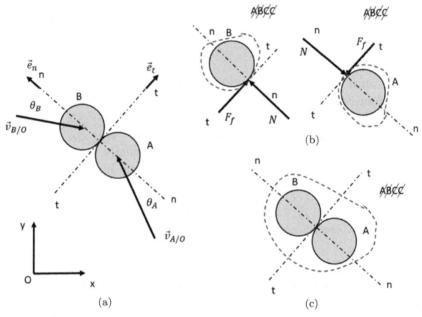

Figure 6.10:   Two-particle, 2D impact problem: (a) 2D impact with the normal and tangential directions; (b) FBDs for each particle; and (c) FBD for both particles as a system.

force, which would be impulse-like with a huge magnitude. On the other hand, the tangential force, $F_f$, is more related to friction and is NOT huge with a finite size. Finally, in Figure 6.10(c), an FBD is constructed for both particles as a system. There are no external forces acting on the two-particle system.

Based on the FBD analysis of Figure 6.10(c), the conservation of linear momentum can be applied to the system with both particles.

$$m_A \cdot (\vec{v}_{A/O})_1 + m_B \cdot (\vec{v}_{B/O})_1$$
$$= m_A \cdot (\vec{v}_{A/O})_2 + m_B \cdot (\vec{v}_{B/O})_2 \qquad (6.78)$$

For each particle, we have

$$m_A(\vec{v}_{A/O})_1 + \int_{a^-}^{a^+} (\vec{N} + \vec{F}_f)dt = m_A \cdot (\vec{v}_{A/O})_2 \qquad (6.79)$$

If we split the above equation into its respective components along the tangential and the normal directions, we have

$$m_A(v_{A/O})_1 \cos\theta_A - \int_{a^-}^{a^+} N\,dt = m_A \cdot (v_{A/O})_{2n} \qquad (6.80)$$

$$m_A(v_{A/O})_1 \sin\theta_A - \int_{a^-}^{a^+} F_f\,dt = m_A \cdot (v_{A/O})_{2t} \qquad (6.81)$$

$$-m_B(v_{B/O})_1 \cos\theta_B + \int_{a^-}^{a^+} N\,dt = m_B \cdot (v_{A/O})_{2n} \qquad (6.82)$$

$$m_B(v_{B/O})_1 \sin\theta_B + \int_{a^-}^{a^+} F_f\,dt = m_B \cdot (v_{B/O})_{2t} \qquad (6.83)$$

Because $F_f$ is not huge, NOT impulse-like, Equations (6.81) and (6.83) can be further simplified to

$$m_A(v_{A/O})_1 \sin\theta_A = m_A \cdot (v_{A/O})_{2t} \qquad (6.84)$$

$$m_B(v_{B/O})_1 \sin\theta_B = m_B \cdot (v_{B/O})_{2t} \qquad (6.85)$$

In other words, the velocities in the tangential direction will remain the same.

From this observation, we can break down a 2D impact problem into two directions. For the tangential direction, there is no impact and no changes to the velocity component for each particle. Along the normal direction, it resembles the 1D impact problem.

We then combine the velocity component of each direction after impact to determine the total velocity. We will use billiard shooting for illustration.

## Example 6.6: Billiard Shooting

As shown in Figure 6.11, a player tries to knock the #1 ball into the corner pocket of a pool table. Describe the strategy of the player to achieve this goal.

We redraw the billiard shooting as a 2D impact problem in Figure 6.12. Before shooting, the targeted ball $B$ is stationary, i.e.,

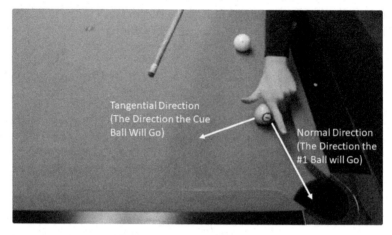

Figure 6.11: How to shoot a ball into the pocket of a pool game.
*Source*: https://www.basicbilliards.com/stun.php.

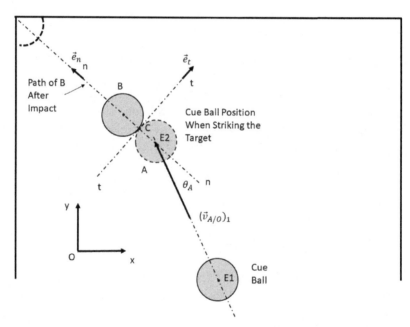

Figure 6.12: The strategy of shooting pool based on 2D impact dynamics.

$(\vec{v}_{B/O})_1 = 0$. The cue ball is some distance away. Ball $B$ has to follow the path which links its current position to the pocket, as shown in Figure 6.12. Draw this path across the $B$ ball, we find the impact point $C$. The ball $B$ path line, therefore, is the normal direction line for the 2D impact. The tangential line intersects the normal line at impact point $C$. Along the normal line, a radius distance from point $C$, we can identify where the center of the cue ball should be when it hits $B$. This new center position is marked as $E2$. The current center position of the cue ball is $E1$.

The player needs to aim the cue ball along the line connecting $E1$ and $E2$. The velocity in the normal direction before impact for the cue ball is $(\vec{v}_{A/O})_{1n} = v_0 \cos\theta_A \vec{e}_n$, where $\vec{e}_n$ is the unit direction vector along the normal line. After impact, ball $B$ should have a velocity $(\vec{v}_{B/O})_2 = (v_{B/O})_2 \vec{e}_n$, along the normal direction.

We can now apply the impact analysis based on conservation of linear momentum to complete the calculation. Note that all billiard balls have the same mass and the impact is considered as elastic impact.

Along the normal direction, it is a 1D impact problem with $e = 1$. Because ball $B$ is stationary before impact, along the normal direction, it is the same as Newton's cradle. The linear momentum of the cue ball along the normal direction will transfer to ball $B$ completely. Therefore, we have

$$(v_{B/O})_{2n} = (v_{A/O})_{1n} = v_0 \cos\theta_A \qquad (6.86)$$

$$(v_{A/O})_{2n} = 0 \qquad (6.87)$$

Along the tangential direction, there are no changes; therefore,

$$(v_{B/O})_{2t} = 0 \qquad (6.88)$$

$$(v_{A/O})_{2t} = v_0 \sin\theta_A \qquad (6.89)$$

Combining the velocity components after impact, we obtain the total velocities of both balls after impact,

$$(\vec{v}_{A/O})_2 = v_0 \sin\theta_A \vec{e}_t \qquad (6.90)$$

$$(\vec{v}_{B/O})_2 = v_0 \cos\theta_A \vec{e}_n \qquad (6.91)$$

where $\vec{e}_t$ is the unit directional vector along the tangential line. Note that, after impact, both balls will move perpendicular to each other, with the cue ball along the tangential direction and the target along the normal direction. This is why in Figure 6.11 the thumb and the index finger, forming a right angle, are used to indicate the directions of the balls, respectively.

In the above analysis, we do not consider the spin of the ball. To consider the spin, we have to consider the balls as rigid bodies, not particles. To consider the spin, we will need to consider the angular momentum, to be discussed in the next section.

## 6.3 Applications Using Conservation Law of Angular Momentum

As discussed earlier, Equations (6.33)–(6.34) represent the conservation or pseudo-conservation of angular momentum with point $A$ being a stationary point.

The case for $\sum \vec{\tau}_G = 0$ will require all external forces acting on the rigid body to be on the mass center, such as the satellite case in Chapter 5. However, such cases are not common. On the other hand, often we can find a stationary point $A$ to which $\sum \vec{\tau}_A = 0$.

Because Equations (6.33) and (6.34) are vector equations, sometimes, we might have the zero moment only in one or two directions. As a result, the angular momentum can be conserved along those directions. We then have partial conservation of angular momentum.

### 6.3.1 *Classical examples of conservation of angular momentum*

Let's consider a few examples for illustration.

### Example 6.7: Center of Percussion Revisit

Let's revisit the Center of Percussion discussed in Example 5.10. We will change the problem slightly to have ball $B$, traveling at speed $(v_B)_1$, hit the rod $A$ at a spot located $d$ distance from the mass center $G$. We would like to determine a specific hitting location so that the reaction force, $R_{ox}$, is zero. We will solve this problem by using the linear and angular momentum equations.

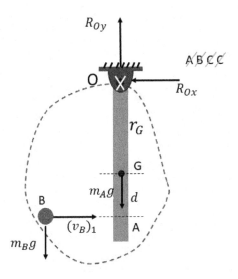

Figure 6.13:   Center of Percussion Revisit.

**Force Analysis:** We construct an FBD as shown in Figure 6.13 to include the ball and the rod.

**Motion Analysis:** For the time duration immediately before and after the ball hits the rod, the ball still travels horizontally while the rod will rotate around the pin after impact.

**Governing Equations:** Instead of using *Newton's Second Law* as in Chapter 5, we will use the linear and angular momentum equations. We will construct the angular momentum equation with respect to the stationary point $O$ and for the mass center $G$.

$$m_B \cdot (v_B)_1 - \int_{0^-}^{0^+} R_{ox}\, dt = m_B(v_B)_2 + m_A \omega_A r_G \qquad (6.92)$$

Essentially, we use a stationary point overlapping center $G$, not assuming that the mass center for the combined ball and rod is at $G$.

$$m_B \cdot (v_B)_1 \cdot (r_G + d) = m_B \cdot (v_B)_2 \cdot (r_G + d) + I_A \omega_A \qquad (6.93)$$

$$m_B \cdot (v_B)_1 \cdot d + \int_{0^-}^{0^+} R_{ox} \cdot r_G\, dt = I_G \cdot \omega_A + m_B(v_B)_2 \cdot d \qquad (6.94)$$

where $(v_B)_2$ is the ball speed immediately after impact. We still assume that it is moving to the right. The rod will gain a counterclockwise angular velocity $\omega_A$, immediately after impact.

Note that there is a conservation of angular momentum with respect to point $O$, but not for $G$ in general.

**Solving for Unknowns:** To define Center of Percussion, we let $R_{ox} = 0$ and solve for $d$. We can use either Equations (6.92) and (6.93) or (6.92) and (6.94).

With $R_{ox} = 0$, solving Equation (6.92), we have

$$m_B \cdot (v_B)_2 = m_B (v_B)_1 - m_A \omega_A r_G \tag{6.95}$$

Introducing Equation (6.95) into (6.93), we obtain

$$(R_G + d) m_A \omega_A r_G = I_A \omega_A$$
$$\implies (r_G + d) m_A r_G = I_A = I_G + m_A r_G^2$$
$$\implies d = \frac{I_G}{m_A r_G} \tag{6.96}$$

We obtain the same result of Example 5.10. We can solve the same problem with Equations (6.92) and (6.94).

**Solution Verification:** Unfortunately, we cannot solve for $(v_B)_2$ and $\omega_A$ unless we know the restitution coefficient of the impact. In practice, we can measure $\omega_A$ quite easily. The velocity of the ball after impact can then be determined using Equation (6.95).

### Example 6.8: A Disk Hitting a Bump

As shown in Figure 6.14(a), a disk rolls without slipping toward the left and hits the bump $A$ with a height $h$ at $t = t_2$. It is observed that the disk does not bounce back but instead pivots around point $A$. The disk angular velocity before impact is $\omega_1$. Determine the disk angular velocity, $\omega_2$, immediately after impact.

**Force Analysis:** We construct an FBD as shown in Figure 6.14(b). The impact force at $A$ is unknown and could be impulse-like during $t_2^- \leq t \leq t_2^+$.

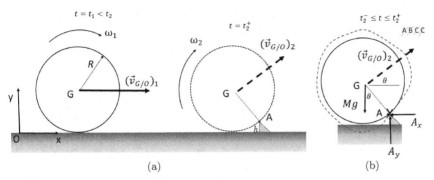

(a)                                                                    (b)

Figure 6.14: (a) A disk is rolling to the left and hits a bump; and (b) the disk then pivots around the bump.

**Motion Analysis:** As shown in Figure 6.14(a), the disk is rolling without slipping before hitting the bump and is in a rotation around a fixed point immediately after the impact.

**Governing Equations:** The governing equations we have learned so far include *Newton's Second Law*, the linear momentum equation, and the angular momentum equation. We do not know readily which one to use and we should go through all of them to ponder on which one might be useful for solving the problem.

First, let's consider *Newton's Second Law* for the governing equations. If we solve for them, then we will obtain the angular acceleration and the mass center acceleration. However, in order to determine the velocity, the accelerations need to be functions of time. It appears that might not be feasible. Therefore, we will NOT consider *Newton's Second Law* for the governing equations.

As for the linear momentum equation for this 2D problem, it will contain two equations; however, because the impact forces, $A_x$ and $A_y$, are unknown and impulse-like, we will have at least three unknowns, i.e., $A_x$, $A_y$, and $(v_{G/O})_2$. We may come back to it later.

Finally, let's consider the angular momentum equation. The advantage of the angular momentum equation is that we can choose a fixed reference point to eliminate some unknown forces. Apparently, if we choose $A$ as the reference point, the impact forces, $A_x$ and $A_y$, will not appear in the governing equations. With less unknowns, the chance of solving the governing equation is better. However, it should be noted that for this 2D problem, the angular momentum equation is a scalar equation; therefore, it can only contain one unknown to be solvable.

We decide to establish the angular momentum equation with respect to $A$ for $t_2^- \leq t \leq t_2^+$,

$$(H_A)_1 + \int_{t_2^-}^{t_2^+} \sum \tau_A \, dt = (H_A)_2$$

$$\Longrightarrow -(R-h)M(v_{G/O})_1 - I_G \omega_1 + \int_{t_2^-}^{t_2^+} MgR \sin \theta \, dt$$

$$= -I_A \omega_2 \qquad (6.97)$$

where $\theta = \cos^{-1}(R-h)/R$.

Because the impact duration is very short and $Mg$ is NOT HUGE, we can safely assume that $\int_{t_2^-}^{t_2^+} MgR \sin \theta \, dt \approx 0$. Equation (6.97) is reduced to

$$-(R-h)M(v_{G/O})_1 - I_G \omega_1 = -I_A \omega_2 \qquad (6.98)$$

where $\omega_1 = (v_{G/O})_1/R$. We can solve Equation (6.98) to obtain $\omega_2$.

We can write down the linear momentum equations as

$$M(v_{G/O})_1 - \int_{t_2^-}^{t_2^+} A_x dt = M(v_{G/O})_2 \cos \theta \qquad (6.99)$$

$$\int_{t_2^-}^{t_2^+} A_y dt = M(v_{G/O})_2 \sin \theta \qquad (6.100)$$

where $(v_{G/O})_2 = R\omega_2$.

With $\omega_2$, we can determine $\int_{t_2^-}^{t_2^+} A_x dt$ and $\int_{t_2^-}^{t_2^+} A_y \, dt$.

**Solving for Unknowns:** Let's solve this problem with a numerical example. Let $M = 10$ kg, $I_G = 0.156$ kg m$^2$, $R = 0.2$ m, $h = 0.03$ m, and $(v_{G/O})_1 = 10$ m/s.

With $I_A = I_G + mR^2 = 0.556$ kg m$^2$, we can solve Equation (6.98) to obtain $\omega_2$ as

$$\omega_2 = 44.604 \text{ rad/s} \qquad (6.101)$$

The mass center speed after impact becomes

$$(v_{G/O})_2 = R\omega_2 = 8.921 = 0.8921(v_{G/O})_1 \qquad (6.102)$$

Note that $(\vec{v}_{G/O})_2$ is no longer horizontal. There are sudden changes in the speed and the direction of the mass center velocity due to the impact forces. From Equations (6.99) and (6.100), we obtain

$$\int_{t_2^-}^{t_2^+} A_x \, dt = 24.17 \, (\text{Ns}) \tag{6.103}$$

$$\int_{t_2^-}^{t_2^+} A_y \, dt = 46.99 \, (\text{Ns}) \tag{6.104}$$

**Solution Verification:** Let's consider the effect of $h$ on the impact. What if $h$ is higher, for example, $h = 0.1 \, \text{m}$? Repeating the calculations, we obtain

$$\omega_2 = 32.014 \, \text{rad/s} \tag{6.105}$$

$$\int_{t_2^-}^{t_2^+} A_x \, dt = 67.99 \, (\text{Ns}) \tag{6.106}$$

$$\int_{t_2^-}^{t_2^+} A_y \, dt = 55.44 \, (\text{Ns}) \tag{6.107}$$

From the results above, it is obvious that if the bump is higher the impact forces will be higher, resulting in less mass center and lower angular velocities. In other words, the impact will be more severe. These results match our driving experience, if our car hits a bump. a higher bump results in a larger impact for the same speed.

On the other hand, if we consider the situation of a car hitting a person. The height of car compared with the mass center of a person is critical. In this case, we consider the car as the bump and the person hits the car with a speed.

A higher car height will cause a more severe injury of a pedestrian if all other conditions are the same. Based on this observation, the pedestrian safety laws in Europe regulate the hood/bumper height to reduce injuries to pedestrians and cyclists when frontal collisions with a vehicle happens.[1]

---

[1]Crandall, J.R., Bhalla, K.S., and Madeley, N.J. Designing road vehicles for pedestrian protection, *BMJ* **324**(7346) (2002) 1145–1148. Doi: 10.1136/bmj.324.7346.1145.

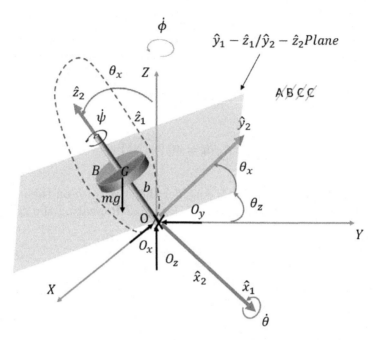

Figure 6.15:   The steady state precession of a gyro.

## Example 6.9: Steady-state Gyro Precession Revisit

As shown in Figure 6.15, a gyro is undergoing a steady-state precession, i.e., $\ddot{\phi} = \ddot{\theta} = \ddot{\psi} = 0$ and $\dot{\theta} = 0$.

**Force Analysis:** We construct an FBD as shown in Figure 6.15. The constraint forces at $O$ are defined with respect to the inertial reference frame, $X-Y-Z$.

**Motion Analysis:** The general motion analysis of a precessing gyro was conducted in Chapter 4. From Equations (4.23) and (4.32), the angular velocity of the gyro can be expressed in terms of $\hat{x}_2-\hat{y}_2-\hat{z}_2$ or $X-Y-Z$ as

$$\vec{\omega}_B = \dot{\theta}\hat{i}_2 + \dot{\phi}\sin\theta_x\hat{j}_2 + (\dot{\phi}\cos\theta_x + \dot{\psi})\hat{k}_2 \qquad (6.108)$$

$$= (\dot{\psi}\sin\theta_x\sin\theta_z + \dot{\theta}\cos\theta_z)\vec{I}$$

$$+(-\dot{\psi}\sin\theta_x\cos\theta_z + \dot{\theta}\sin\theta_z)\vec{J}$$

$$+(\dot{\psi}\cos\theta_x + \dot{\phi})\vec{K} \qquad (6.109)$$

Accordingly, the angular momentum of the gyro was derived in Chapter 5. From Section 5.3.4, we have

$$
\begin{bmatrix} H_{GX} \\ H_{GY} \\ H_{GZ} \end{bmatrix} = I_G \begin{bmatrix} \omega_X \\ \omega_Y \\ \omega_Z \end{bmatrix}
$$

$$
= T \begin{bmatrix} \hat{H}_{Gx} \\ \hat{H}_{Gy} \\ \hat{H}_{Gz} \end{bmatrix} = T\hat{I}_G \begin{bmatrix} \hat{\omega}_x \\ \hat{\omega}_y \\ \hat{\omega}_z \end{bmatrix} \qquad (6.110)
$$

where $T$ is the transformation matrix between $O-X-Y-Z$ and $O-\hat{x}_2-\hat{y}_2-\hat{z}_2$ as defined in Section 5.3.4, in Chapter 5. Here, we recognize $O-\hat{x}_2-\hat{y}_2-\hat{z}_2$ has a frame rotation, but the angular velocities expressed in the two equations above are complete, with respect to inertial reference frames, only expressed in different directional unit vectors.

**Governing Equations:** Because the gyro is undergoing steady-state precessing, we will consider both *Newton's Second Law* and Angular Momentum Equation for analysis. We also need to determine which reference frame to use. For *Newton's Second Law*, an inertial reference frame aligned with $\hat{x}_2-\hat{y}_2-\hat{z}_2$ is more convenient. The corresponding equations are derived in Section 5.3.10, Chapter 5, with respect to point $O$.

For steady-state precessing, the governing equations are reduced to

$$
\sum \tau_{\tilde{x}} = mgb \sin \theta_x
$$

$$
= (I_{xx} + mb^2)(-\dot{\phi}^2 \sin \theta_x \cos \theta_x) + I_{zz}\dot{\phi} \sin \theta_x (\dot{\phi} \cos \theta_x + \dot{\psi})
$$
$$
(6.111)
$$

For the angular momentum equation, we will choose $X-Y-Z$ as the reference frame because we would like to know if there is a conservation of angular momentum with respect to a fixed direction or directions. From Equation (6.11), the angular momentum equation is

$$(\vec{H}_O)_1 + \int_{t1}^{t2} \sum \vec{\tau}_O \, dt$$

$$= (\vec{H}_O)_1 + \int_{t1}^{t2} \vec{r}_{G/O} \times (-mg\vec{K}) dt$$

$$= (\vec{H}_O)_1 + \int_{t1}^{t2} [(mg \sin \theta_x \cos \theta_z)\vec{I} + (mg \sin \theta_x \sin \theta_z)\vec{J} + (0)\vec{K}] dt$$

$$= (\vec{H}_O)_2 \tag{6.112}$$

where $\vec{r}_{G/O} = (\sin \theta_x \sin \theta_z)\vec{I} - (\sin \theta_x \cos \theta_z)\vec{J} + \cos \theta_x \vec{K}$.

From Equation (6.112), because $\sum \tau_{OZ} = 0$, we obtain that

$$(\vec{H}_{OZ})_1 = (\vec{H}_{OZ})_2 \tag{6.113}$$

for all time during the steady-state precessing. In other words, we have a conservation of angular momentum, but only along the $Z$ axis. The angular momentum components along $X$ and $Y$ are not constant.

From Equation (6.110), we can derive $H_{OZ}$.

$$\begin{bmatrix} H_{OX} \\ H_{OY} \\ H_{OZ} \end{bmatrix} = I_O \begin{bmatrix} \omega_X \\ \omega_Y \\ \omega_Z \end{bmatrix}$$

$$= T \begin{bmatrix} \hat{H}_{Ox} \\ \hat{H}_{Oy} \\ \hat{H}_{Oz} \end{bmatrix} = T \hat{I}_O \begin{bmatrix} \hat{\omega}_x \\ \hat{\omega}_y \\ \hat{\omega}_z \end{bmatrix}$$

$$= T \begin{bmatrix} (I_{xx} + mb^2) & 0 & 0 \\ 0 & (I_{xx} + mb^2) & 0 \\ 0 & 0 & I_{zz} \end{bmatrix} \begin{bmatrix} \hat{\omega}_x \\ \hat{\omega}_y \\ \hat{\omega}_z \end{bmatrix} \tag{6.114}$$

Working out the above equation, we obtain

$$(H_{OZ}) = (I_{xx} + mb^2)\dot{\phi} \sin^2 \theta_x + I_{zz} \cos \theta_x (\dot{\phi} \cos \theta_x + \dot{\psi}) \tag{6.115}$$

From the above expression, $H_{OZ}$ is indeed a constant because $\dot{\phi}$, $\dot{\psi}$, and $\theta_x$ are all constants.

For specific $\dot{\psi}$ and $\theta_x$, can we obtain $\dot{\phi}$ from Equation (6.115)? The answer is no because we do not know $H_{Oz}$ at different precessing angles $\theta_x$. To determine the precessing angular speed, $\dot{\phi}$, we can use Equation (6.111).

**Solving for Unknowns:** Let's work out the problem with some numbers. Let $m = 0.5$ kg, $b = 0.05$ m, $\dot{\psi} = 100$ rad/s, $\theta_x = 60°$, $I_{xx} = 0.001$ kg m$^2$, and $I_{zz} = 0.0045$ kg m$^2$.
Equation (6.111) becomes

$$\sum \tau_{\hat{x}} = 0.5 \cdot 9.81(0.05) \sin 60°$$

$$= (0.001 + 0.5(0.05)^2)(-\dot{\phi}^2 \sin 60° \cos 60°)$$
$$+ (0.00045)\dot{\phi} \sin 60°(\dot{\phi} \cos 60° + 100) \qquad (6.116)$$

The above equation becomes a quadratic equation as

$$\dot{\phi}^2 - 50.00\dot{\phi} + 272.53 = 0 \qquad (6.117)$$

The solutions to Equation (6.117) are $\dot{\phi} = 6.23$ rad/s or 43.77 rad/s. The low precession is generally observed. Based on $\dot{\phi} = 6.23$ rad/s, we obtain $H_{OZ} = 0.0337$ kg $\cdot$ m$^2$/s. During the steady state precession at this particular precession angle and speed, this angular momentum in $Z$ is a constant. However, at a different $\theta_x$, both $\dot{\phi}$ and $H_{OZ}$ will be different. Let's consider three different angles, $\theta_x = 90°$, $60°$, and $0°$. At $\theta_x = 0°$, $H_{OZ} = I_{zz}\dot{\psi} = 0.045$ kg $\cdot$ m$^2$/s. Using the same number conditions, we found $\dot{\phi} = 5.45$ rad/s for $\theta_x = 90°$. The corresponding $H_{OZ} = 0.0123$ kg $\cdot$ m$^2$/s. The value of $H_{OZ}$ becomes lower as $\theta_x$ becomes higher (i.e., the gyro is more horizontal). To change from one precession angle to the next, we have to either add or reduce $H_{OZ}$. For example, to go from $\theta_x = 90°$ to $\theta_x = 60°$, we need to add angular momentum along $Z$ as

$$(H_{OZ})_{\theta_x=90°} + \int \tau_z \, dt = (H_{OZ})_{\theta_x=60°} \qquad (6.118)$$

The needed $\int \tau_z \, dt$ can be added by, for example, pushing the gyro along its precession direction, attempting to speed up $\dot{\phi}$. If we indeed add the correct amount of angular momentum, the gyro will "suddenly" rise up from $\theta_x = 90°$ to $60°$, a large angle change (33%);

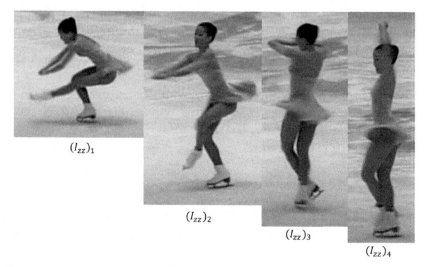

$(I_{zz})_1$

$(I_{zz})_2$

$(I_{zz})_3$

$(I_{zz})_4$

Figure 6.16:   The performance images of Michelle Kwan in 1998 Winter Olympics in Nagano, Japan.

*Source*: https://www.youtube.com/watch?v=UdYPFKO0X1k.

however, the precessing speed would only go from from 5.45 rad/s to 6.23 rad/s, a 14.3% increase.

**Solution verification:** Let's consider the case when $\theta_x = 0°$. There is no precessing, only a vertically spinning gyro. We have

$$(H_{OZ})_{\theta_x=0°} = I_{zz}\dot{\psi} \qquad (6.119)$$

If the gyro stays vertically but we reduce $I_{zz}$ to one half of its original value, in order to keep $(H_{OZ})_{\theta_x=0°}$ constant due to conservation of angular momentum along the $Z$ axis, the value of $\dot{\psi}$ needs to double. This is how a figure skater increases his or her spinning speed during performance, as shown in Figure 6.16. When a figure skater goes into a spin, the whole body is supported by one leg; therefore, no moment could be added by pushing with another leg. The spinning speed is changed by changing the moment of inertia via tugging in the arms and legs, as shown in Figure 6.16, in which $(I_{zz})_1 > (I_{zz})_2 > (I_{zz})_3 > (I_{zz})_4$. As a result, $(\dot{\psi})_1 < (\dot{\psi})_2 < (\dot{\psi})_3 < (\dot{\psi})_4$.

It is very important that the skater maintains his or her mass center vertically so that precession is not induced to make the body wobble. This takes a lot of practice and training to accomplish.

(a)                                          (b)

Figure 6.17: Merry-go-around prank: (a) all kids lean outward; and (b) all kids, except one, suddenly steps inward to substantially increase the spinning speed of the merry-go-around.

*Source*: https://www.youtube.com/watch?v=Oq2WHeZmzRM.

For a free rotating table, we can use bearings to ensure $\tau_z = 0$ and maintaining a upright position, as shown in Figure 6.17.

The merry-go-around initially rotates slowly when all kids lean outward to increase $I_{zz}$. They then all step in, except one, to substantially reduce $I_{zz}$. As a result, the merry-go-around suddenly spins a lot faster. The one not stepping inward will experience a sensation of a much higher centrifugal force. This is a common prank played by kids in the play ground.

When asked why this could happen, the answer is often "Conservation of angular momentum!" However, from the discussion above, we should know that conservation of angular momentum only holds along the $z$ direction.

### Example 6.11: A Person on a Free-Spinning Chair Holding a Spinning Wheel

Another common demonstration for "conservation of angular momentum" is shown in Figure 6.18, which are screen shot images from a youtube video produced by Flipping Physics. A person sitting on a free rotating chair (only around the vertical axis) is holding a bicycle wheel. Initially, the chair is not rotating while the wheel is rotating counterclockwise, i.e., the wheel has an angular momentum point to positive $z$ axis (Figure 6.18(a)). The total angular momentum around the $z$ axis is $I_w\dot{\psi}$, where $I_w$ is the moment of

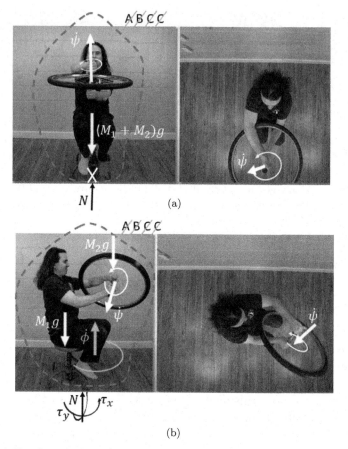

Figure 6.18: Images from a youtube video produced by Flipping Physics: (a) initial condition with a spinning wheel, pointing upward, while the person is motionless; and (b) the wheel was turned vertically and the person rotates. The wheel spinning axis is now horizontal.

*Source*: https://www.youtube.com/watch?v=vM6G-NgN0PY.

inertia of the wheel and $\dot{\psi}$ is the wheel spinning speed. The FBD for the initial condition shows that the weights of the person and the wheel and the normal force from the floor reach equilibrium.

Scenario #2: When the person flips the wheel vertically, with the spinning axis pointing horizontally, the wheel no longer has a spinning angular momentum around the $z$ axis (Figure 6.18(b)). From the corresponding FBD, there will be resulting moments, $\tau_x$ and $\tau_y$, from the floor on the chair, resulting from the angular momentum changes when the wheel axis points to a different direction.

Because the total angular momentum of the person and the wheel around the $z$ axis are conserved, the chair will turn counter-clockwise at a lower speed because the combined person/chair/wheel has a higher moment of inertia with respect to $z$.

Scenario #3: If the wheel is now turned upside down (not shown in Figure 6.18, with its spinning angular momentum pointing downward (clockwise), the chair will spin faster in the counter-clockwise direction. We can explain scenario #2 and #3 with the angular momentum equation below,

$$\begin{aligned}
H_{Pz} + H_{Wz} &= (H_{Pz})_1 + (H_{Wz})_1 = 0 + I_W \cdot \dot{\psi} \\
&= (H_{Pz})_2 + (H_{Wz})_2 = I_P \cdot \dot{\phi}_2 + a \cdot m_W(a \cdot \dot{\phi}_2) \\
&= (H_{Pz})_3 + (H_{Wz})_3 = I_P \cdot \dot{\phi}_3 + a \cdot m_W(a \cdot \dot{\phi}_3) - I_W\dot{\psi}
\end{aligned}$$
$$(6.120)$$

where $H_{Pz}$ and $H_{Wz}$ are the angular momentum of the person/chair and the wheel, respectively, $\dot{\phi}_2$ and $\dot{\phi}_3$ are the spinning speeds of the chair for the second and third scenarios, and finally, $a$ is the distance of the wheel mass center to the $z$ axis.

When the chair turns, the wheel will have an orbiting angular momentum with respect to the $z$ axis.

From Equation (6.120), we found that

$$\dot{\phi}_2 = \frac{I_W\dot{\psi}}{I_P + m_W a^2} \tag{6.121}$$

$$\dot{\phi}_3 = \frac{2I_W\dot{\psi}}{I_P + m_W a^2} \tag{6.122}$$

## 6.4   Gyroscope for Navigation

We will discuss the gyroscope used for navigation for airplanes, ships, satellite, etc. (Figure 6.19).

### Example 6.11: Gyroscope Instrument

As shown in Figure 6.19(a), a gyroscope contains the spinning rotor, supported by gimbal (2), which is then supported by gimbal (1). Finally, a frame (element (0)) is used to support gimbal (1).

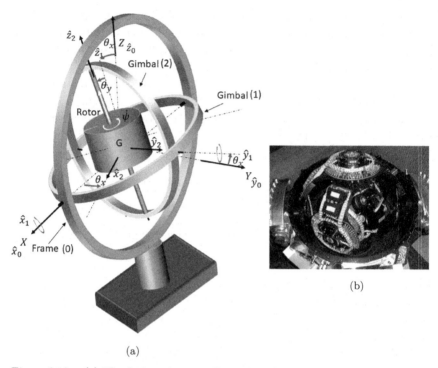

(a)

(b)

Figure 6.19: (a) The design of a general gyroscope with two gimbals and a frame; and (b) a gyroscope used for navigation.

*Source*: (b) https://upload.wikimedia.org/wikipedia/commons/4/44/ST-124_uncovered_(IMGP3445).JPG.

An actual gyroscope for navigation is a highly complicated instrument (Figure 6.19(b)) and we will not attempt to discuss its intricated design. Instead, we want to know how conservation of angular momentum is applied to a gyroscope.

It is a gross simplification to state that with the gimbals, the rotor of a gyroscope is torque-free; thus, its angular momentum is conserved.

**Motion Analysis:** As shown in Figure 6.19(a), we establish four reference frames to observe the rotation of the rotor. Every frame, except the inertial reference frame $G - X - Y - Z$, is attached to one of the gyroscope elements. The mass center of the rotor is $G$. Reference frame $G - \hat{x}_0 - \hat{y}_0 - \hat{z}_0$ is attached to the frame.

At this instant, $G-\hat{x}_0-\hat{y}_0-\hat{z}_0$ and $G-X-Y-Z$ are aligned but $G-\hat{x}_0-\hat{y}_0-\hat{z}_0$ could be pointing to different directions in the next moment while the axis directions of $G-X-Y-Z$ are fixed in space.

Reference frame $G-\hat{x}_1-\hat{y}_1-\hat{z}_1$ is attached to gimbal (1), aligned with the principle axes of gimbal (1). Gimbal (1) is attached to the frame with two frictionless pins. To transform from $G-\hat{x}_0-\hat{y}_0-\hat{z}_0$ to $G-\hat{x}_1-\hat{y}_1-\hat{z}_1$, we rotate around $\hat{x}_0$ by $\theta_x$. This rotation is the allowable rotation of gimbal (1) with respect to the frame. Reference frame $G-\hat{x}_2-\hat{y}_2-\hat{z}_2$ is attached to gimbal (2). To transform from $G-\hat{x}_1-\hat{y}_1-\hat{z}_1$ to $G-\hat{x}_2-\hat{y}_2-\hat{z}_2$, we rotate around $\hat{y}_1$ by $\theta_y$. Similarly, this is the allowable rotation of gimbal (2) with respect to gimbal (1). The transformations between these reference frames are

$$
\begin{bmatrix} X \\ Y \\ Z \end{bmatrix} = \begin{bmatrix} \hat{x}_0 \\ \hat{y}_0 \\ \hat{z}_0 \end{bmatrix} = T_{tx}(\theta_x) \begin{bmatrix} \hat{x}_1 \\ \hat{y}_1 \\ \hat{z}_1 \end{bmatrix}
\tag{6.123}
$$

$$
\begin{bmatrix} \hat{x}_1 \\ \hat{y}_1 \\ \hat{z}_1 \end{bmatrix} = T_{ty}(\theta_y) \begin{bmatrix} \hat{x}_2 \\ \hat{y}_2 \\ \hat{z}_2 \end{bmatrix}
\tag{6.124}
$$

Accordingly, the unit direction vectors are related as

$$
\vec{I} = \hat{i}_0; \quad \vec{J} = \hat{j}_0; \quad \vec{K} = \hat{k}_0
\tag{6.125}
$$

$$
\vec{I} = \hat{i}_1; \quad \vec{J} = \cos\theta_x \hat{j}_1 - \sin\theta_x \hat{k}_1;
$$

$$
\vec{K} = \sin\theta_x \hat{j}_1 + \cos\theta_x \hat{k}_1
\tag{6.126}
$$

$$
\hat{i}_1 = \cos\theta_y \hat{i}_2 + \sin\theta_y \hat{k}_2; \quad \hat{j}_1 = \hat{j}_2;
$$

$$
\hat{k}_1 = -\sin\theta_y \hat{i}_2 + \cos\theta_y \hat{k}_2
\tag{6.127}
$$

From the above relations, we further derive

$$
\vec{I} = \cos\theta_y \hat{i}_2 + \sin\theta_y \hat{k}_2
\tag{6.128}
$$

$$
\vec{J} = \sin\theta_x \sin\theta_y \hat{i}_2 + \cos\theta_x \hat{j}_2 - \sin\theta_x \cos\theta_y \hat{k}_2
\tag{6.129}
$$

$$
\vec{K} = -\cos\theta_x \sin\theta_y \hat{i}_2 + \sin\theta_x \hat{j}_2 + \cos\theta_x \cos\theta_y \hat{k}_2
\tag{6.130}
$$

Based on these reference frames, the rotor rotation can be observed as follows:

(i) With respect to $G-\hat{x}_2-\hat{y}_2-\hat{z}_2$, the rotor has a simple rotation $\dot{\psi}$ around $\hat{z}_2$, expressed as

$$\hat{\omega}_{rotor/gimbal2} = \dot{\psi}\hat{k}_2 \tag{6.131}$$

(ii) The rotation of gimbal (2) to gimbal (1) is

$$\hat{\omega}_{gimbal2/gimbal1} = \omega_{2y}\hat{j}_1 = \omega_{2y}\hat{j}_2 \tag{6.132}$$

(iii) The rotation of gimbal (1) with respect to the frame is

$$\hat{\omega}_{gimbal1/frame} = \omega_{1x}\hat{i}_1 = \omega_{1x}\vec{I} \tag{6.133}$$

(iv) The rotation of the frame to the inertial reference frame is

$$\hat{\omega}_{frame} = \vec{\omega}_f = \omega_{fx}\vec{I} + \omega_{fy}\vec{J} + \omega_{fz}\vec{K} \tag{6.134}$$

Combined, with the additivity of rotation, the rotor rotation with respect to $G-X-Y-Z$ is

$$\vec{\omega}_{rotor} = \dot{\psi}\hat{k}_2 + \omega_{2y}\hat{j}_1 + \omega_{1x}\hat{i}_0 + \vec{\omega}_f \tag{6.135}$$

Expressing $\vec{\omega}_{rotor}$ in $\hat{i}_2$, $\hat{j}_2$, and $\hat{k}_2$, we have

$$
\begin{aligned}
\vec{\omega}_{rotor} = {} & (\omega_{1x}\cos\theta_y + \omega_{fx}\cos\theta_y + \omega_{fy}\sin\theta_x\sin\theta_y \\
& - \omega_{fz}\cos\theta_x\sin\theta_y)\hat{i}_2 + (\omega_{2y} + \omega_{fy}\cos\theta_x + \omega_{fz}\sin\theta_x)\hat{j}_2 \\
& + (\dot{\psi} + \omega_{1x}\sin\theta_y + \omega_{fx}\sin\theta_y - \omega_{fy}\sin\theta_x\cos\theta_y \\
& + \omega_{fz}\cos\theta_x\cos\theta_y)\hat{k}_2 = \hat{\omega}_{rx}\hat{i}_2 + \hat{\omega}_{ry}\hat{j}_2 + \hat{\omega}_{rz}\hat{k}_2
\end{aligned} \tag{6.136}
$$

where $\hat{\omega}_{rx}$, $\hat{\omega}_{ry}$, and $\hat{\omega}_{rz}$ are defined accordingly per the equation above. We should note that each component of the rotor rotation is a function of the rotations of the gimbals and the frame, as well as the orientations of the gimbals and the frame.

The angular momentum of the rotor becomes

$$\vec{H}_{rotor} = \begin{bmatrix} \hat{i}_2 & \hat{j}_2 & \hat{k}_2 \end{bmatrix} \bar{I}_{Gr} \, \vec{\omega}_{rotor}$$

$$= \begin{bmatrix} \hat{i}_2 & \hat{j}_2 & \hat{k}_2 \end{bmatrix} \begin{bmatrix} I_{xx} & 0 & \\ 0 & I_{xx} & 0 \\ 0 & 0 & I_{zz} \end{bmatrix} \begin{bmatrix} \hat{\omega}_{rx} \\ \hat{\omega}_{ry} \\ \hat{\omega}_{rz} \end{bmatrix}$$

$$= \begin{bmatrix} \hat{i}_2 & \hat{j}_2 & \hat{k}_2 \end{bmatrix} \begin{bmatrix} (\hat{H}_{rotor})_x \\ (\hat{H}_{rotor})_y \\ (\hat{H}_{rotor})_z \end{bmatrix} \tag{6.137}$$

where $I_{xx}$ and $I_{zz}$ are moment of inertia based on $G - \hat{x}_2 - \hat{y}_2 - \hat{z}_2$, which is the principal axes of the rotor. Note that in general $\vec{H}_{rotor}$ and $\vec{\omega}_{rotor}$ are not pointing to the same direction.

Similarly, the rotation of gimbal (2) to the inertial reference frame becomes

$$\vec{\omega}_{gimbal2} = \omega_{2y}\hat{j}_2 + \omega_{1x}\hat{i}_1 + \vec{\omega}_{frame}$$

$$= \hat{\omega}_{2x}\hat{i}_2 + \hat{\omega}_{2y}\hat{j}_2 + \hat{\omega}_{2z}\hat{k}_2 \tag{6.138}$$

where $\hat{\omega}_{2x}$, $\hat{\omega}_{2y}$, and $\hat{\omega}_{2z}$ are defined as per the equation above. We should note that each component of the rotor rotation is a function of the rotations of the gimbals and the frame, as well as the gimbals and frame orientations.

The difference between $\vec{\omega}_{rotor}$ and $\vec{\omega}_{gimbal2}$ is $\dot{\psi}\hat{k}_2$.

The angular momentum of gimbal (2) with respect to $G - X - Y - Z$, but expressed in $\hat{i}_2 - \hat{j}_2 - \hat{k}_2$ is

$$\vec{H}_{gimbal2} = \begin{bmatrix} \hat{i}_2 & \hat{j}_2 & \hat{k}_2 \end{bmatrix} \bar{I}_{Gg2} \, \vec{\omega}_{gimbal2}$$

$$= \begin{bmatrix} \hat{i}_2 & \hat{j}_2 & \hat{k}_2 \end{bmatrix} \begin{bmatrix} I_{g2x} & 0 & \\ 0 & I_{g2z} & 0 \\ 0 & 0 & I_{g2z} \end{bmatrix} \begin{bmatrix} \hat{\omega}_{2x} \\ \hat{\omega}_{2y} \\ \hat{\omega}_{2z} \end{bmatrix}$$

$$= \begin{bmatrix} \hat{i}_2 & \hat{j}_2 & \hat{k}_2 \end{bmatrix} \begin{bmatrix} (\hat{H}_{gimbal2})_x \\ (\hat{H}_{gimbal2})_y \\ (\hat{H}_{gimbal2})_z \end{bmatrix} \tag{6.139}$$

where $I_{g2x}$ and $I_{g2z}$ are moment of inertia of gimbal (2) with respect to $\hat{x}_2-\hat{y}_2-\hat{z}_2$. Because gimbal (2) is a thin ring, we obtain $I_{g2x} = m_2 R_2^2$, where $R_2$ is the radius of gimbal (2) and $I_{g2z} = \frac{1}{2} m_2 R_2^2$.

For gimbal (1), its angular velocity, expressed in $\hat{i}_1-\hat{j}_1-\hat{k}_1$, is

$$\vec{\omega}_{gimbal1} = \omega_{1x}\hat{i}_1 + \vec{\omega}_{frame}$$

$$= (\omega_{1x} + \omega_{fx})\hat{i}_1$$

$$+(\omega_{fy}\cos\theta_x + \omega_{fz}\sin\theta_x)\hat{j}_1$$

$$+(-\omega_{fy}\sin\theta_x + \omega_{fz}\cos\theta_x)\hat{k}_1$$

$$= \omega_{1x}^*\hat{i}_1 + \omega_{1y}^*\hat{j}_1 + \omega_{1z}^*\hat{k}_1 \qquad (6.140)$$

In the above expression, a superscript $*$ is used to distinguish the components expressed in $\hat{i}_1-\hat{j}_1-\hat{k}_1$.

The angular momentum is

$$\vec{H}_{gimbal1} = \begin{bmatrix} \hat{i}_1 & \hat{j}_1 & \hat{k}_1 \end{bmatrix} \hat{I}_{Gg1} \, \vec{\omega}_{gimbal1}$$

$$= \begin{bmatrix} \hat{i}_1 & \hat{j}_1 & \hat{k}_1 \end{bmatrix} \begin{bmatrix} I_{g1x} & 0 & 0 \\ 0 & I_{g1x} & 0 \\ 0 & 0 & I_{g1z} \end{bmatrix} \begin{bmatrix} \omega_{1x}^* \\ \omega_{1y}^* \\ \omega_{1z}^* \end{bmatrix}$$

$$= \begin{bmatrix} \hat{i}_1 & \hat{j}_1 & \hat{k}_1 \end{bmatrix} \begin{bmatrix} (H_{gimbal1})_x^* \\ (H_{gimbal1})_y^* \\ (H_{gimbal1})_z^* \end{bmatrix} \qquad (6.141)$$

where $I_{g1x}$ and $I_{g1z}$ are moment of inertia of gimbal (1) with respect to $\hat{x}_1-\hat{y}_1-\hat{z}_1$. Because gimbal (1) is a thin ring as well, we obtain $I_{g1x} = \frac{1}{2} m_1 R_1^2$, where $R_1$ is the radius of gimbal (1) and $I_{g1z} = m_1 R_1^2$.

It will be soon clear why $\hat{x}_1-\hat{y}_1-\hat{z}_1$ is used to express the angular momentum of gimbal (1).

In general, $I_{g2x}$, $I_{g2z}$, $T_{g1x}$, and $I_{g1z}$ are much smaller than $I_{xx}$ and $I_{zz}$.

**Force Analysis:** As shown in Figure 6.20, three FBDs are constructed. The rotor is supported by gimbal (2) with two frictionless pins, $A$ and $B$, aligned with $\hat{z}_2$. Of course, it is nearly impossible to have a true frictionless pin, but let's accept this assumption for now. Gimbal (2) is attached to gimbal (1) with two frictionless pins, $C$

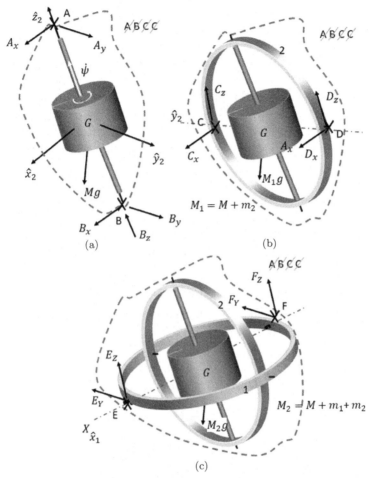

Figure 6.20:   Force analysis for the gyroscope: (a) FBD for the rotor; (b) FBD for the rotor and gimbal (2); and (c) FBD for the rotor, gimbals (2) and (1).

and $D$, aligned in $\hat{y}_2$. Gimbal (1) is attached to the frame via two frictionless pins, $E$ and $F$, aligned with $\hat{i}_0$ or $X$ at this moment.

In Figure 6.20(a), based on the FBD for the rotor only, the external forces involved are the gravitational force at the mass center and the forces at the frictionless pins, $A$ and $B$.

**Governing Equations:** From Figure 6.20(a), for the FBD for the rotor alone, we know that the moment with respect to axis $\hat{z}_2$ is zero because all the pin forces are passing through $\hat{z}_2$.

Similarly, for Figure 6.20(b), the FBD for the rotor and gimbal (2) assembly shows that the moment with respect to axis $\hat{y}_2$ is zero because all the pin forces are passing through $\hat{y}_2$.

Finally, for Figure 6.20(c), the FBD for the rotor, gimbal (2) and gimbal (1) assembly shows that the moment with respect to axis $X$ ($\hat{x}_1$) is zero because all the pin forces are passing through $X$.

The important result for the above force analysis is that the rotor is subjected to moments in the $\hat{x}_2$ and $\hat{y}_2$ directions; therefore, we cannot simply state that the angular moment of the rotor is conserved.

Figure 6.21 presents the complementary FBDs with respect to Figure 6.20 to show how external moments could be applied to the gyroscope and the resulting rotations of the frame and the gimbals. We would like to see if the rotor shaft $\hat{z}_2$ could be pointing to a fixed direction despite being subjected to external moments at the frame of the gyroscope.

**Governing Equations:** Based on the FBD of Figure 6.20(a), the angular momentum and moment equation are

$$\sum (\vec{\tau}_{rotor})_G = \begin{bmatrix} \hat{i}_2 & \hat{j}_2 & \hat{k}_2 \end{bmatrix} \begin{bmatrix} 2R_2(A_y - B_y) \\ 2R_2(A_x - B_x) \\ 0 \end{bmatrix}$$

$$= \dot{\vec{H}}_{rotor} = \vec{\Omega}_2 \times \vec{H}_{rotor} + \frac{\partial}{\partial t} \vec{H}_{rotor}$$

$$= \begin{bmatrix} \hat{i}_2 & \hat{j}_2 & \hat{k}_2 \end{bmatrix} \begin{bmatrix} (\dot{\vec{H}}_{rotor})_x \\ (\dot{\vec{H}}_{rotor})_y \\ (\dot{\vec{H}}_{rotor})_z \end{bmatrix} \tag{6.142}$$

where $\vec{\Omega}_2$ is the frame rotation of $\hat{x}_2$-$\hat{y}_2$-$\hat{z}_2$; thus, $\vec{\Omega}_2 = \vec{\omega}_{gimbal2}$

The expressions for $(\dot{\vec{H}}_{rotor})_x$, $(\dot{\vec{H}}_{rotor})_y$, and $(\dot{\vec{H}}_{rotor})_z$ are quite complicated as functions of orientations, angular velocities, and angular accelerations of the gimbals and the frame. We will not express them explicitly here. The derivation will be tedious but not difficult.

From Equation (6.142), we have

$$0 = (\dot{\vec{H}}_{rotor})_z \tag{6.143}$$

If the supporting pins are truly frictionless, the angular momentum in the $\hat{k}_2$ direction is conserved, but not the other two directions because in general $A_x \neq B_x$ and $A_y \neq B_y$. The reaction forces on $A$ and $B$ are related to the external moments applied to the gyroscope.

Equation (6.143) does not guarantee the conservation of the angular momentum of the rotor. The rotor can still change directions despite its component with respect to $\hat{z}_2$ being constant.

Let's apply the same analysis to the assembly of the rotor and gimbal (2), as well as the assembly of the rotor, gimbal (2), and gimbal (1), based on the FBDs of Figures 6.20(b) and 6.20(c), respectively. We obtain

$$\sum(\vec{\tau}_{rotor-gimbal2})_G = \begin{bmatrix} \hat{i}_2 & \hat{j}_2 & \hat{k}_2 \end{bmatrix} \begin{bmatrix} 2R_2(D_z - C_z) \\ 0 \\ 2R_2(C_x - D_x) \end{bmatrix}$$

$$= \dot{\vec{H}}_{rotor} + \dot{\vec{H}}_{gimbal2} = \vec{\Omega}_2 \times \vec{H}_{rotor} + \frac{\partial}{\partial t}\vec{H}_{rotor}$$

$$+ \vec{\Omega}_2 \times \vec{H}_{gimbal2} + \frac{\partial}{\partial t}\vec{H}_{gimbal2}$$

$$= \begin{bmatrix} \hat{i}_2 & \hat{j}_2 & \hat{k}_2 \end{bmatrix} \begin{bmatrix} (\dot{\vec{H}}_{rotor})_x + (\dot{\vec{H}}_{gimbal2})_x \\ (\dot{\vec{H}}_{rotor})_y + (\dot{\vec{H}}_{gimbal2})_y \\ (\dot{\vec{H}}_{rotor})_z + (\dot{\vec{H}}_{gimbal2})_z \end{bmatrix} \quad (6.144)$$

From Equation (6.144), we have, with respect to $\hat{y}_2$,

$$0 = (\dot{\vec{H}}_{rotor})_y + (\dot{\vec{H}}_{gimbal2})_y \quad (6.145)$$

Based on the FBD of Figure 6.20(c), we have

$$\sum(\vec{\tau}_{rotor-gimbal2-gimbal1})_G = \begin{bmatrix} \hat{i}_1 & \hat{j}_1 & \hat{k}_1 \end{bmatrix} \begin{bmatrix} 0 \\ 2R_2(E_z - F_z) \\ 2R_2(E_y - F_y) \end{bmatrix}$$

$$= \dot{\vec{H}}_{rotor} + \dot{\vec{H}}_{gimbal2} + \dot{\vec{H}}_{gimbal1}$$

$$= \vec{\Omega}_1 \times [T_{ty}^{-1}(\theta_y)\vec{H}_{rotor}] + \frac{\partial}{\partial t}[T_{ty}^{-1}(\theta_y)\vec{H}_{rotor}]$$

$$+\vec{\Omega}_1 \times [T_{ty}^{-1}(\theta_y)\vec{H}_{gimbal2}] + \frac{\partial}{\partial t}[T_{ty}^{-1}(\theta_y)\vec{H}_{gimbal2}]$$

$$+\vec{\Omega}_1 \times \vec{H}_{gimbal1} + \frac{\partial}{\partial t}\vec{H}_{gimbal1}$$

$$= \begin{bmatrix} \hat{i}_1 & \hat{j}_1 & \hat{k}_1 \end{bmatrix} \begin{bmatrix} (\dot{\vec{H}}_{rotor})_x^* + (\dot{\vec{H}}_{gimbal2})_x^* + (\dot{\vec{H}}_{gimbal1})_x^* \\ (\dot{\vec{H}}_{rotor})_y^* + (\dot{\vec{H}}_{gimbal2})_y^* + (\dot{\vec{H}}_{gimbal1})_y^* \\ (\dot{\vec{H}}_{rotor})_z^* + (\dot{\vec{H}}_{gimbal2})_z^* + (\dot{\vec{H}}_{gimbal1})_z^* \end{bmatrix} \quad (6.146)$$

From Equation (6.146), we have, with respect to $\hat{x}_1$ or $X$,

$$0 = (\dot{\vec{H}}_{rotor})_x^* + (\dot{\vec{H}}_{gimbal2})_x^* + (\dot{\vec{H}}_{gimbal1})_x^* \quad (6.147)$$

Equations (6.143), (6.145), and (6.147) define the conservation of angular momentum for the gyroscope, not the rotor alone. These equations are heavily related to the gimbal rotations as shown in Figure 6.21. However, for a gyroscope to function as desired, we need to achieve conservation of angular momentum for the rotor, not the entire gyroscope.

We will discuss if the rotor can point to a fixed direction within a certain accuracy.

If the initial condition of the gyroscope is that the rotations of the gimbals and the frame are zero, and the gimbals and the frame are perpendicular to each other, thus, $\theta_x = 0$ and $\theta_y = 0$, the initial angular velocity of the rotor based on Equation (6.136) is

$$\vec{\omega}_{rotor} = \dot{\psi}\hat{k}_2 \quad (6.148)$$

If we can avoid subjecting the gyroscope to high moments so that $(\dot{\vec{H}}_{gimbal1})_x^*$, $(\dot{\vec{H}}_{gimbal2})_x^*$, and $(\dot{\vec{H}}_{gimbal2})_y$ are small, then, Equations (6.143), (6.145), and (6.147) become

$$(\dot{\vec{H}}_{rotor})_x^* \approx 0 \quad (6.149)$$

$$(\dot{\vec{H}}_{rotor})_y \approx 0 \quad (6.150)$$

$$(\dot{\vec{H}}_{rotor})_z = 0 \quad (6.151)$$

As long as $\hat{x}_1$ is not on the plane of $\hat{y}_2 - \hat{z}_2$, $(\dot{\vec{H}}_{rotor})_x^* \approx 0$ can still achieve $(\dot{\vec{H}}_{rotor})_x \approx 0$, along $\hat{x}_2$.

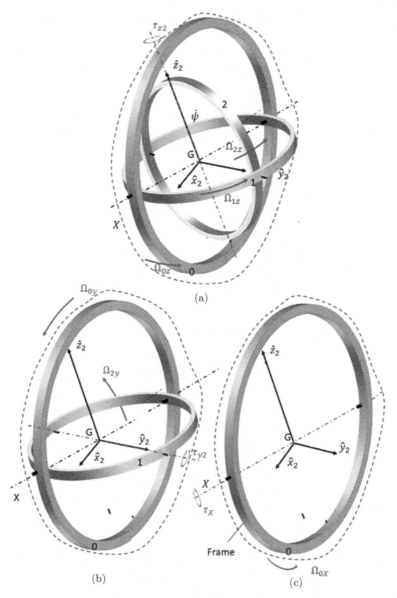

Figure 6.21:   Rotation analysis of the gyroscope: (a) The rotation of gimbal (2), gimbal (1), and the frame when a moment $\tau_{z2}$ is applied around the $\hat{z}_2$ axis; (b) The rotations of gimbal (1) and the frame when a moment, $\tau_{y2}$, is applied around the $\hat{y}_2$ axis; and (c) the rotation of the frame when a moment, $\tau_X$, is applied around the $X$ axis.

As a result, the angular momentum of the rotor can be conserved, within certain accuracy. We then have a constant angular momentum of the rotor with a limited time period as

$$\vec{H}_{rotor} \approx I_{zz}\dot{\psi}\hat{k}_2, \quad 0 < t < t_1 \qquad (6.152)$$

Equation (6.152) indicates that the direction of $\hat{k}_2$ is fixed in space; therefore, it can be used for navigation guidance. For example, if we align the gyro at North Pole pointing to the vertical direction and then move the gyroscope to the Equator, the gyro will now be pointing to the horizontal direction. This is one way to prove that the Earth is round, not flat.

However, after some time, the rotor will drift due to external moments and the friction of the supporting pins of gimbals. The direction of $\hat{k}_2$ will change. Calibration is then needed.

The analysis above points out that the gyroscope is a delicate machine and it cannot handle very large moments, in particular when the condition of gimbal lock happens.

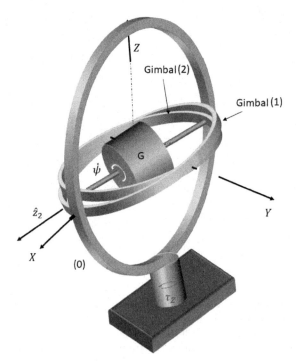

Figure 6.22: Gimbal lock situation of a gyroscope, in which gimbal (2) and gimbal (1) are almost aligned.

Figure 6.22 shows a gimbal lock situation in which gimble (2) turns with its $\hat{z}_2$ axis almost aligned with $X$. Under such conditions, the gyroscope cannot defend a moment applied in the $Z$ direction. For the condition shown in Figure 6.22, the rotor will rise up toward $Z$ direction when a large $\tau_Z$ is applied. Therefore, the gyroscope can no longer provide a fixed direction for guidance.

Another consequence of the gimbal lock is that gimbals (1) and (2) and the rotor could spin together as one rigid body, which could damage the gyroscope. When this happens, it will take a long time for the gyroscope to stop spinning to allow for reset if it is not damaged.

Let's review how a gyroscope would maintain a fixed direction when subjected to mild external moments. Several screen shots of a youtube video demonstrating the operation of a home-made gyroscope are presented in Figures 6.23 and 6.24. A moment, $\tau_z$, is

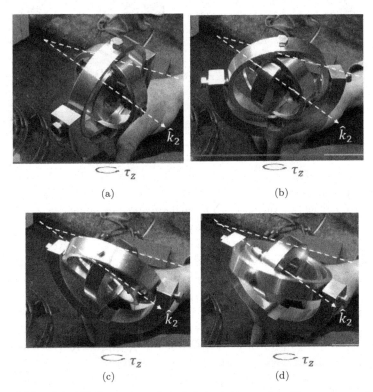

(a)             (b)

(c)             (d)

Figure 6.23: (a)–(d): Sequences of gimbals' rotations when the gyroscope is subjected to a moment at the frame. Note that the rotor maintains its orientation. *Source*: https://www.youtube.com/watch?v=oj7v3MXJL3M.

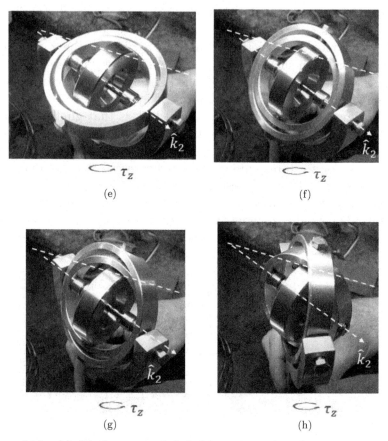

Figure 6.24: (e)–(h): Sequences of gimbals' rotations when the gyroscope is subjected to a moment at the frame. Note that the rotor maintains its orientation. *Source*: https://www.youtube.com/watch?v=oj7v3MXJL3M.

applied at the frame, which would result in moments, $\tau_{z2}$, $\tau_{y2}$, and $\tau_X$, defined in Figure 6.21, to rotate the gimbals and the frame. Notice that a background object is used as a reference and the $\hat{z}_2$ axis of the rotor of the first screen shot is drawn to define the $\hat{k}_2$ direction. The same $\hat{k}_2$ direction line is then copied over to other screen shots to check if the rotor maintains its orientation. Apparently, the rotor does maintain its orientation with this basic check. It might have drifts very slightly, but we cannot measure it with this simple check. In the first screen shot (Figure 6.23(a)), the rotor, gimbal (1), and gimbal (2) are roughly perpendicular to each other. For the next two screen shots, the frame rotates, so do the gimbals (Figures 6.23(b) and 6.23(c)). In Figures 6.23(d)–6.24(f), gimbals (1) and (2) are almost aligned, which is the condition of gimbal lock. During the gimbal lock, gimbals (1) and (2) appear to rotate together, at a faster rotation. Then, in Figure 6.24(g), the gimbal lock is resolved and gimbals (1) and (2) become perpendicular again. From Figure 6.24(h), it would repeat to the condition of Figure 6.23(a) again. Figures 6.23 and 6.24 also imply how a gyroscope can be used to determine the motion of an object based on the motion of the gimbals and the frame.

If we attach the frame of the gyroscope to an object, and when the object undergoes motion, the gimbals will rotate accordingly. By observing the motions of the gimbals with respect to the frame, we can define the motion of the object. We will use a commercial aviation gyroscope for illustration.

### 6.4.1 *Aviation applications*

To use a gyroscope for aviation guidance, we need to be able to determine the motion of the airplane. There are three rotations related to an airplane, pitch, roll, and yaw, as shown in Figure 6.25. At least two gyroscopes are needed to obtain the data of all three rotations. Attitude Indicator is used to detect the pitch and the roll, while Heading Indicator is used to determine the yaw. The question arises why a gyroscope cannot detect all the rotations?

To answer the aforementioned question, let's consider an Attitude Indicator for illustration. Referring to Figures 6.19–6.21, the frame of a gyroscope is attached to an airplane. With respect to Figure 6.25, the pitch is the rotation around the $X$ axis, the roll is the rotation around the $\hat{y}_2$ axis, while the yaw is the rotation around the $\hat{z}_2$ if the principal axes of the airplane are aligned with the above gyroscope axes, respectively. We need to find a gimbal which does not move during these rotations. From Figures 6.20(a) and 6.21(a), we know that only the rotor does not move if the airplane changes directions with a yaw movement. From Figures 6.20(b) and 6.21(b), we know that the rotor and gimbal (2) do not move during a pitch rotation for the airplane to climb or descend. From Figures 6.20(c) and 6.21(c), we know the rotor, gimbal (1), and gimbal (2) do not move when the airplane rolls to achieve a bank angle in order to make a turn. From the above analysis, we realize that with this setup, the gyroscope

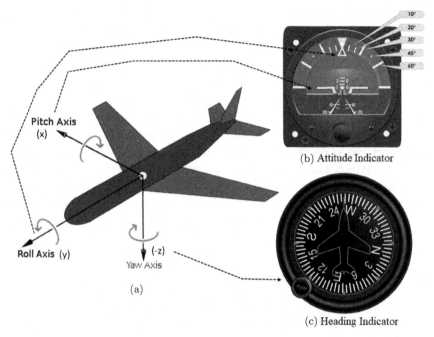

Figure 6.25: (a): Airplane rotations — pitch, roll, and yaw; (b) Attitude Indicator for detecting the pitch and the roll; and (c) Heading indicator for detecting the yaw.

*Source*: (a) https://upload.wikimedia.org/wikipedia/commons/thumb/c/c1/ Yaw_Axis_Corrected.svg/500px-Yaw_Axis_Corrected.svg.png; (b) https://upload .wikimedia.org/wikipedia/commons/1/1d/Attitude_Indicator.png; and (c) https:// upload.wikimedia.org/wikipedia/commons/thumb/a/ae/Heading_indicator.svg/ 1024px-Heading_indicator.svg.png.

cannot detect the yaw rotation around the $\hat{z}_2$ axis. For the pitch and the roll, gimbal (2) and the rotor are the only ones which do not move during these two rotations. An attitude indicator, with an indicator attached to the pins of gimbal (2), can then be used to detect the pitch and the roll. We cannot attach anything to the rotor because it is spinning fast.

The internal configuration of an attitude indicator is shown in Figure 6.26. In an attitude indicator, gimbal (1) is supported by the frame with one journal bearing. This is different from the gyroscope in Figure 6.19 in which gimbal (1) is supported by the frame with two pins. Both designs achieve the goal to allow free rotation around the $X$ axis. This is similar to a simply supported beam vs a cantilever beam.

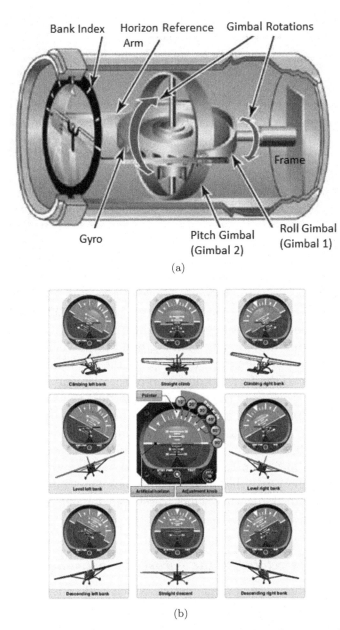

(a)

(b)

Figure 6.26: (a): Internal configuration of an altitude indicator; and (b) the display of an altitude during different pitch and roll maneuvers.

*Sources*: (a) http://i.stack.imgur.com/LQdKf.jpg; and (b) http://www.aero-mech anic.com/wp-content/uploads/2010/09/7-24.jpg.

As shown in Figure 6.26, a scope display with the roll angle marking and a airplane level line is fixed on the attitude indicator housing, which is fixed to the airplane body. For a pilot, this plate does not move. An Earth horizon display with ground, the sky, and a vertical triangle marking is attached to gimbal (2). As shown in Figure 6.26(b), if an airplane is undergoes a straight climb or straight descent, the scope display together with the frame and gimbal (1) (roll gimbal) will move together, rotating with respect to gimbal (2), rendering the airplane level marking shiting relative to the Earth horizon display. For an inertial observer, the Earth horizon display does not move but the scope display shifts. For the pilot, it is the opposite. When climbing, the airplane level line is in the sky region of the Earth horizon display. In descending, it is in the ground region. This is intuitive to pilots.

When an airplane is undergoing a level roll, such as level left bank or level right bank, the scope display, attached to the gyroscope frame of the attitude indicator will rotate while the rotor, gimbal (1) and gimbal (2) do not. For the pilot, the Earth horizon display will rotate in a way similar to how the ground will turn, seen by the pilot if the airplane is flying low enough for the ground to be visible.

To design a heading indicator, we place the rotor horizontally and perpendicular to the forward direction of the airplane. Gimbal (1) is allowed to rotate around the frame in the vertical axis of the airplane. The frame is fixed to the airplane body as before. The frame rotation relative to the assembly of gimbal (1), gimbal (2) and the rotor will indicate the heading direction of the airplane. A gear is attached to gimbal (1) and then to a compass card, overlapping an airplane outline. For the pilot, the airplane outline is fixed but the compass card will rotate, indicating which direction, for example, north or northwest, the airplane is heading.

### 6.4.2   *Issues related to the gyroscope instrument*

In practice, there are several factors which will cause a gyroscope to lose its ability to maintain a fixed direction in space. From the analysis above, we know that a gyroscope is sensitive to high moment loading. There is a general saying that a gyroscope instrument should be handled like an egg, in a very gentle way.

The issues related to the gyroscope instrument include:

- *Gyro drifts*: Based on Equations (6.143)–(6.147), the angular momentum of the rotor could change (drift) due to external moment and the friction at the gimbal pins. The drift rate will be small if the external loading and the friction are small, but over time, the drift becomes substantial.

  Another issue related to the drift is the gyro precession. If a moment is applied at the gimbal, not at the frame, a precession will occur, just like the case of Example (6.9). Precession can be used for aviation guidance as well but it could also have unintended consequences, such as leading to the gimbal lock.

- *Gimbal lock*: Gimbal lock is a serious issue for the gyroscope. For the altitude indicator of Figure 6.26, If the airplane is undergoing an extreme descent, gimbal (1) will rotate along with the instrument housing to become more in alignment with gimbal (2). The Earth horizon display is designed to stop moving after 20 degrees, but gimbal (1) can continue to rotate. If gimbal (2) and gimbal (1) are aligned, as in Figures 6.22 and 6.24(d)–6.24(f), the spinning axis of the rotor is now aligned to the roll axis ($X$ or $\hat{x}_1$) of gimbal (1). The high speed spinning motion of the rotor will transfer energy to gimbal (1) via friction, the Earth horizon display then spins continuously, commonly known as gyro tumble. This is a rather dangerous situation because the attitude indicator loses its ability for guidance. In the fateful Apollo 13 mission, gimbal lock actually happened. Modern gyroscopes are designed to give warnings or to prevent gimbal lock.

Gimbal lock can also happen when the rotor is powering down from the operating speed of approximately 20,000 rpm after it has landed. When the rotor spinning speed is lower, its ability to remain fixed in space is weakened and it could tilt. When the rotor, gimbal (1), and gimbal (2) are not in a mutually perpendicular configuration, the ability of the gyroscope to handle external loading is weakened. It becomes more susceptible to precession. In the gimbal lock situation, the gyroscope loses its ability to defend the moment loading in the yaw rotation for an Attitude Indicator.

If the airplane makes a sharp turn on the ground when the rotor has tilted, it could render the rotor to tilt further (denoted as topple or spill), hitting gimbal stops to cause damages. It could also lead to the gimbal lock and the tumble. The spinning caused by the tumble can take a long time to resolve.

## 6.5 Torque Free Motion

In the construction of a gyroscope, the objective is to render the rotor torque-free. From our discussion, we know that practically, it is not possible due to the gimbal design and frictions. However, in nature, there are true torque-free motions, such as the rotation of an artificial satellite or a meteoroid in space when the only forces are gravitational and they are acting only at the mass center. If we neglect the drifts of the gyroscope, its rotor has only one rotation and points to a fixed direction. However, this is not true for a general torque-free motion.

When the external forces only act on the mass center of a rigid body, the moment with respect to the mass center is zero. As a result,

$$\sum \vec{\tau}_G = \dot{\hat{H}}_G = \vec{\Omega} \times \hat{H}_G + \frac{\partial}{\partial t}\hat{H}_G = 0 \tag{6.153}$$

where $\hat{H}_G$ is the total angular momentum with respect to an inertial reference frame, but expressed in the principal axes of the rigid body.

### 6.5.1 *The torque-free rotation of an axisymmetric body*

As shown in Figure 6.27, an axisymmetric body $B$, such as planets and artificial satellites, has the principle axes as $\hat{x}_2$, $\hat{y}_2$, and $\hat{z}_2$. Because $\vec{H}_G$ is constant, it points to a fixed direction and we chose this direction as the $Z$ axis for an inertial reference frame. As shown in Figure 6.27, the axes $Z$, $\hat{y}_2$, and $\hat{z}_2$ are on the same plane.

The angular momentum can be expressed in $X-Y-Z$ and then convert to $\hat{x}_2-\hat{y}_2-\hat{z}_2$ as

$$\begin{aligned}
\vec{H}_G &= H_G \vec{K} \\
&= H_G(\sin\theta_x \hat{j}_2 + \cos\theta_x \hat{k}_2) \\
&= H_G \sin\theta_x \hat{j}_2 + H_G \cos\theta_x \hat{k}_2
\end{aligned} \tag{6.154}$$

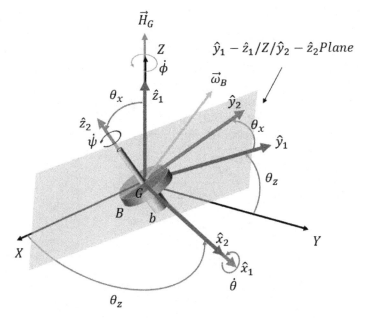

Figure 6.27:   The torque-free motion of an axisymmetric object.

From Equation (5.102), we have

$$\vec{H}_G = \begin{bmatrix} \hat{i}_2 & \hat{j}_2 & \hat{k}_2 \end{bmatrix} \bar{I}_G \hat{\omega}_B$$
$$= I_{yy}\hat{\omega}_x \hat{i}_2 + I_{yy}\hat{\omega}_y \hat{j}_2 + I_{zz}\hat{\omega}_z \hat{k}_2 \qquad (6.155)$$

where $\hat{\omega}_B = \vec{\omega}_B = \hat{\omega}_x \hat{i}_2 + \hat{\omega}_y \hat{j}_2 + \hat{\omega}_z \hat{k}_2$ is the angular velocity of the rigid body with respect to an inertial reference frame but expressed with respect to the principal axes of the rigid body. In the text below, we might use $\vec{\omega}_B$ and $\hat{\omega}_B$ interchangeably, but they are the same complete description of the angular velocity of the rigid body.

With Equating Equations (6.154) and (6.155), we found that

$$\hat{\omega}_x = 0 \qquad (6.156)$$

$$\hat{\omega}_y = \frac{H_G \sin \theta_x}{I_{yy}} \qquad (6.157)$$

$$\hat{\omega}_z = \frac{H_G \cos \theta_x}{I_{zz}} \qquad (6.158)$$

Equations (6.156)–(6.158) indicate that $\hat{\omega}_B$ lies on the $\hat{y}_2$–$\hat{z}_2$ plane, along with $\vec{H}_G$. However, $\vec{H}_G$ and $\hat{\omega}_B$ are not aligned. When we observe this axisymmetric object with an inertial reference frame, we observe $\vec{\omega}_B$ but not $\vec{H}_G$. As a result, even though $\vec{H}_G$ is a constant, $\hat{\omega}_y$ and $\hat{\omega}_z$ can change, rendering the rigid body not to have a fixed axis of rotation, despite its angular momentum being fixed.

Equating Equations (6.156)–(6.158) with (4.23), we can determine the spin ($\dot{\psi}$), nutation ($\dot{\theta}$), and precession ($\dot{\phi}$) of the torque-free motion as

$$\dot{\psi} = \frac{I_{yy} - I_{zz}}{I_{yy} \, I_{zz}} H_G \cos \theta_x \qquad (6.159)$$

$$\dot{\theta} = 0 \qquad (6.160)$$

$$\dot{\phi} = \frac{H_G}{I_{yy}} \qquad (6.161)$$

From the above equations, we know that $\theta_x$ is a constant, and $\dot{\phi}$ and $\dot{\psi}$ are related to each other. Because the spinning rotation is typically much higher, we typically perceive the effect of the slower $\dot{\phi}$. We can express $\dot{\phi}$ as a function of $\dot{\psi}$ as

$$\dot{\phi} = \frac{I_{zz}}{(I_{yy} - I_{zz}) \cos \theta_x} \dot{\psi} \qquad (6.162)$$

While $\vec{H}_G$ is fixed in space, $\vec{\omega}_B$ is generally not. Because the components in Equations (6.156)–(6.158) are all constant, $\vec{\omega}_B$ is fixed with respect to $\hat{x}_2$–$\hat{y}_2$–$\hat{z}_2$, which is rotating with respect to an inertial observer. As discussed in Chapter 4, $\vec{\omega}_B$ provides the instantaneous axis of rotation for the 3D body; therefore, the space cone and body cone concept can be used. The trajectory of $\vec{\omega}_B$ defines the space cone. The body cone is then defined accordingly by tracing a point on B overlapping the axis of $\vec{\omega}_B$ to spin around the $\hat{z}_2$ axis.

Note that the relative orientations of $\vec{\omega}_B$ and $\vec{H}_G$ to the $\hat{z}_2$ axis depend on the magnitudes of $\dot{\psi}$, $\dot{\phi}$, $I_{yy}$, and $I_{zz}$. For cases when $I_{yy} \gg I_{zz}$ and $\dot{\psi} \gg \dot{\phi}$, $\vec{\omega}_B$ will be closer to the $\hat{z}_2$ axis than $\vec{H}_G$, unlike a general case in Figure 6.27.

It is not easy to visualize this torque-free motion based on the equations above. Our human brain typically can only observe one rotation at a time. When two rotations are mixed, we see its combined effect and are unable to distinguish each rotation separately.

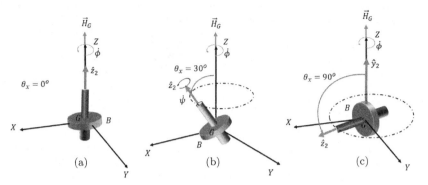

Figure 6.28: The torque-free motion of an axisymmetric object at different precession angles: (a) $0°$; (b) $30°$; and (c) $90°$.

However, torque-free motion for an axisymmetric body could be quite different in its rotations around different principal axes. We will examine the three cases, when $\theta_x = 0°$, $\theta_x = 30°$, and $\theta_x = 90°$, as shown in Figure 6.28.

From Equation (4.13) and $\dot{\theta} = 0$, we observe the frame rotation, $\vec{\Omega}_{2/O}$ of $\hat{x}_2 - \hat{y}_2 - \hat{z}_2$, more distinctively, which is

$$\vec{\Omega}_{2/O} = \dot{\phi}\sin\theta_x \hat{j}_2 + \dot{\phi}\cos\theta_x \hat{k}_2 = \dot{\phi}\vec{K} \tag{6.163}$$

Note that when $\theta_x = 0$, $\dot{\psi}$ and $\dot{\phi}$ cannot be distinguished, the torque-free object is simply spinning along the $Z$, fixed in space. This is a perfect gyroscope. When $\theta_x = 30°$ and with $\dot{\psi}$ known, $\dot{\phi}$ is determined and the torque-free motion is similar to a tilted gyro in precession induced by gravity. For a gyro, the precession speed is typically much slower than the spinning speed. On the other hand, for a torque-free motion, the torque-free precession is typically higher. We will examine the rotations of the spiral pass of an American football for illustration.

## Example 6.12: The Spiral Pass of a Football

The tight spiral pass of an American football by the strong arm of a quarterback is something of beauty to watch. The dynamics of the spiral pass has been studied extensively with wind tunnel testings

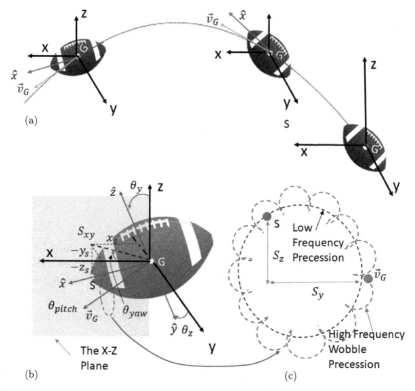

Figure 6.29: The motion of an tight pass of American football: (a) the flight trajectory; (b) the motion axes; and (c) the front view of different precession patterns.

and several theoretical models have been proposed.[2,3] Instrumented smart footballs can now be used to record the football trajectory and spinning motion to help players improve their throws.

As shown in Figure 6.29(a), a long pass is launched. The mass center of the football will follow a parabolic trajectory similar to the ones of Figures (3.18) and (3.19) in Section 3.5 of Chapter 3. This trajectory is on the $X - Z$ plane of an inertial reference frame.

[2]Soodak, H. A geometric theory of rapidly spinning tops, tippe tops, and footballs, *American Journal of Physics* **70**(2002) 815. Doi: 10.1119/1.1479741.

[3]Price, R.H., Moss, W.C., and Gay, T.J. Paradox of the tight spiral pass in American football: A simple resolution, *American Journal of Physics* **88**(9) (2020). Doi: 10.1119/10.0001388.

Interestingly, the spinning principal axis of the football ($\hat{x}$) also more or less follows the tangential line of the trajectory, which is also the velocity direction of the mass center. From this fact, it is clear that the football spiral pass is not an idealized torque-free motion and air plays an important role.

The alignment of the football spinning axis is mainly due to the shape of the football. Through a negative feedback mechanism, it aligns itself more or less but not exactly with the mass center velocity vector ($\vec{v}_G$) to reduce the wind drag in a tight spiral pass.

Figure 6.29(b) illustrates the alignment of the football with $\vec{v}_G$ during the flight. The spin principal axis, $\hat{x}$, of the football is not aligned exactly with $\vec{v}_G$. The tip of the football, point $S$, could precess in two frequencies as shown in Figure 6.29(c). When viewing from the $X$ axis, the trajectory of $S$ is not symmetric to the $X - Z$ plane, as shown in Figure 6.29(c). The misalignment, $S_y$, along the $Y$ direction generates a yaw angle, $\theta_{yaw}$, while the one along the $Z$ direction, $S_z$, is related to the pitch, $\theta_{pitch}$.

In a tight pass, the tip of the football undergoes a low frequency precession, as shown in Figure 6.29(c). If the throw is not tight, an additional high frequency wobble could be observed, on top of the low frequency precession. The precession of $S$ changes the pitch and yaw constantly, which in turn changes the wind drag and the lift on the football. The drag and lift produce a moment, $\tau_G$, with respect to the mass center. The low frequency precession is the result of the moment due to the drag and lift, just like the gyro precession due to the gravity. The moment induced precession, $\omega_p$, is proportional to the ratio of the moment to the angular momentum, $\tau_G/H_G$.[2] On the other hand, the high frequency wobble for a less-than-tight spiral pass is related to the torque-free motion and its angular velocity, $\dot{\phi}$, can be calculated by Equation (6.162). In a typical tight spiral pass, the spinning rate, $\dot{\psi}$, could reach over $60\,\text{rad/s}$ or over $550\,\text{rpm}$, while the moment induced precession rate, about $20\,\text{rpm}$.

Let's do a calculation for a less-than-tight spiral pass. If $\dot{\psi} = 60$ rad/s, $\theta_x = 15°$, $I_{yy} = 3I_{zz}$, from Equation (6.162), we obtain $\dot{\phi} = 31\,\text{rad/s}$.

From the above example, it is clear that the high frequency precession rate is much higher than that of the moment-induced low frequency precession.

### 6.5.2 *The torque-free rotation of a symmetric but not axisymmetric body*

In the last section, we discussed the rotations of axisymmetric 3D bodies, such as an American football. Because it is axisymmetric, we can always align $\vec{H}_G$, $\vec{\omega}_B$, $\hat{y}_2$, and $\hat{z}_2$ on the same plane, as shown in Figure 6.27. Due to this alignment, the rotation around $\hat{x}_2$ is always zero. However, for a symmetric but not axisymmetric body, such as a tennis racket, this is not the case. When we set a non-axisymmetric body into rotation, the rotations could occur around all three principal axes.

As shown in Figure 6.30, a tennis racket has three principal axes with different magnitudes of moment of inertia with respect to the mass center. We define principal axis #1 as the one with the largest moment of inertia and axis #3 for the one with the smallest moment of inertia. Therefore, $I_1 > I_2 > I_3$. Axis #2 is often called the intermediate axis. It can be demonstrated that if we rotate the tennis racket around axis #1 only, the rotation is stable, similar to the one in Figure 6.28(c). Similarly, if we rotate it around axis #3 only, it would be stable as well, similar to Figure 6.28(a). However, if one tries to rotate a tennis racket around axis #2 only, it turns out that it is impossible because a rotation around axes #1 and #3 will also

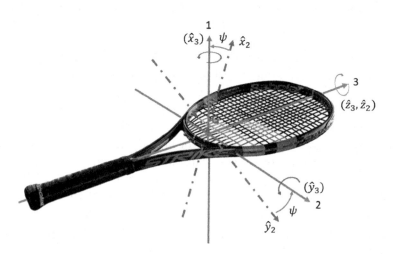

Figure 6.30: The three principal axes of a tennis racket.
*Source*: http://pngimg.com/uploads/tennis/tennis_PNG10391.png.

be induced. This phenomenon is often denoted as the tennis racket theorem or the instability of intermediate axis rotation.

The tennis racket theorem can be found in many videos and animations on the web.[4] It is also referred to as the Dzhanibekov Effect. A theoretical treatment of the Dzhanibekov Effect or the tennis racket theorem is presented in many journal articles.[5] Historical background of the Dzhanibekov effect is recounted as follows[6]:

*In 1985, cosmonaut Vladimir Dzhanibekov had been tasked with rescuing the Salyut 7 space station. He unpacked supplies sent from Earth, which were locked down with a wingnut. When the wingnut spun off the bolt, he noticed how the wingnut maintained its orientation for a short time, and then it flipped 180 degrees.*

In fact, what Dzhanibekov observed was not new, but it was more dramatic in the zero gravity condition, because the tennis racket theorem had been known for years.

To model the tennis racket theorem, let's transform from an inertial reference $G - X - Y - Z$ to $\hat{x}_2 - \hat{y}_2 - \hat{z}_2$, similar to Figure 6.27, to describe the rotation of a tennis racket. However, with $\hat{z}_2$ aligned with principal #3 axis of the racket, $\hat{x}_2$ and $\hat{y}_2$ may not align with principal axes #1 and #2. We need to rotate an additional $\psi$ around $\hat{z}_2$ to $\hat{x}_3 - \hat{y}_3 - \hat{z}_3$, which aligns with the principal axes of the racket, as shown in Figure 6.30. The transformation between $\hat{x}_2 - \hat{y}_2 - \hat{z}_2$ and $\hat{x}_3 - \hat{y}_3 - \hat{z}_3$ is

$$\begin{bmatrix} \hat{i}_2 \\ \hat{j}_2 \\ \hat{k}_2 \end{bmatrix} = T_{tz}(\psi) \begin{bmatrix} \hat{i}_3 \\ \hat{j}_3 \\ \hat{k}_3 \end{bmatrix} \tag{6.164}$$

[4]https://blog.wolfram.com/2020/12/15/simulating-zero-gravity-to-demonstrate-the-dzhanibekov-effect-and-other-surprising-physics-models/.

[5]Murakami, H., Rios, O., and Impelluso, T.J. A theoretical and numerical study of the Dzhanibekov and tennis racket phenomena, *Journal Applied Mechanics* **83**(11) (2016) 111006. Doi:10.1115/1.4034318.

[6]Trivailo, P.M. and Kojima, H. "Utilization of the "Dzhanibekov Effect" for the possible future space missions," 26th International Symposium on Space Flight Dynamics ISSFDAt: Matsuyama, Japan, 2017.

We can then represent the angular velocity of the tennis racket in $\hat{i}_3$, $\hat{j}_3$, and $\hat{k}_3$ accordingly. The angular velocity of the tennis racket with respect to an inertial reference frame aligned with $\hat{x}_3-\hat{y}_3-\hat{z}_3$ becomes

$$\begin{aligned}
\vec{\omega}_B &= \dot{\theta}\hat{i}_2 + \dot{\phi}\sin\theta_x\hat{j}_2 + (\dot{\phi}\cos\theta_x + \dot{\psi})\hat{k}_2 \\
&= (\dot{\theta}\cos\psi + \dot{\phi}\sin\theta_x\sin\psi)\hat{i}_3 \\
&\quad + (\dot{\phi}\sin\theta_x\cos\psi - \dot{\theta}\sin\psi)\hat{j}_3 \\
&\quad + (\dot{\phi}\cos\theta_x + \dot{\psi})\hat{k}_3 \\
&= \omega_1\hat{i}_3 + \omega_2\hat{j}_3 + \omega_3\hat{k}_3 \\
&= \begin{bmatrix} \hat{i}_3 & \hat{j}_3 & \hat{k}_3 \end{bmatrix} \begin{bmatrix} \omega_1 \\ \omega_2 \\ \omega_3 \end{bmatrix} = \begin{bmatrix} \hat{i}_3 & \hat{j}_3 & \hat{k}_3 \end{bmatrix} \omega_B
\end{aligned} \tag{6.165}$$

Note that we skip the $\hat{\phantom{x}}$ sign in the above equation for the angular velocity components to simply the expressions.

In Equation (6.165), $\theta_x$ is often expressed as $\theta$, one of the Euler angles. The three Euler angles are $\phi$ for precession, $\psi$ for spin, and $\theta$ for nutation.

Note that $\hat{x}_3-\hat{y}_3-\hat{z}_3$ is attached to the tennis racket; therefore, its frame rotation is the same as the angular velocity of the tennis racket.

$$\vec{\Omega} = \vec{\omega}_B = \omega_1\hat{i}_3 + \omega_2\hat{j}_3 + \omega_3\hat{k}_3 \tag{6.166}$$

Because there is no torque applied to the tennis racket, we have conservation of angular momentum with respect to its mass center,

$$\dot{\vec{H}}_G = 0 = \vec{\Omega} \times \vec{H}_G + \frac{\partial}{\partial t}H_G \tag{6.167}$$

where $\vec{H}_G = \begin{bmatrix} \hat{i}_3 & \hat{j}_3 & \hat{k}_3 \end{bmatrix} I_G\omega_B$ and $I_G = \begin{bmatrix} I_1 & 0 & 0 \\ 0 & I_2 & 0 \\ 0 & 0 & I_3 \end{bmatrix}$.

Introducing Equations (6.165) and (6.166) into (6.167), we obtain

$$\dot{\omega}_1 = \frac{1}{I_1}(I_2 - I_3)\omega_2\omega_3 \tag{6.168}$$

$$\dot{\omega}_2 = \frac{1}{I_2}(I_3 - I_1)\omega_1\omega_3 \tag{6.169}$$

$$\dot{\omega}_3 = \frac{1}{I_3}(I_1 - I_2)\omega_1\omega_2 \tag{6.170}$$

Equations (6.168)–(6.170) are nonlinear differential equations. We cannot readily check for its stability. Let's verify the stability of the rotation around different principal axes. Taking one more derivative for Equations (6.168–6.170), we have

$$\ddot{\omega}_1 = \frac{1}{I_1}(I_2 - I_3)(\dot{\omega}_2\omega_3 + \omega_2\dot{\omega}_3) \tag{6.171}$$

$$\ddot{\omega}_2 = \frac{1}{I_2}(I_3 - I_1)(\dot{\omega}_1\omega_3 + \omega_1\dot{\omega}_3) \tag{6.172}$$

$$\ddot{\omega}_3 = \frac{1}{I_3}(I_1 - I_2)(\dot{\omega}_1\omega_2 + \omega_1\dot{\omega}_2) \tag{6.173}$$

Introducing Equations (6.168)–(6.170) to Equations (6.171)–(6.173), we have

$$\ddot{\omega}_1 = \frac{(I_2 - I_3)(I_3 - I_1)}{I_1 I_2}\omega_3^2\omega_1 + \frac{(I_2 - I_3)(I_1 - I_2)}{I_1 I_3}\omega_2^2\omega_1 \tag{6.174}$$

$$\ddot{\omega}_2 = \frac{(I_3 - I_1)(I_2 - I_3)}{I_1 I_2}\omega_3^2\omega_2 + \frac{(I_3 - I_1)(I_1 - I_2)}{I_2 I_3}\omega_1^2\omega_2 \tag{6.175}$$

$$\ddot{\omega}_3 = \frac{(I_1 - I_2)(I_2 - I_3)}{I_1 I_3}\omega_2^2\omega_3 + \frac{(I_1 - I_2)(I_3 - I_1)}{I_2 I_3}\omega_1^2\omega_3 \tag{6.176}$$

If we linearize Equations (6.174)–(6.176) around $\omega_1 = \omega_0$, $\omega_2 = 0$, and $\omega_3 = 0$, it becomes

$$\delta\ddot{\omega}_1 = 0 \tag{6.177}$$

$$\delta\ddot{\omega}_2 = \frac{(I_3 - I_1)(I_1 - I_2)}{I_2 I_3}\omega_0^2\,\delta\omega_2 = c_2\,\delta\omega_2 \tag{6.178}$$

$$\delta\ddot{\omega}_3 = \frac{(I_1 - I_2)(I_3 - I_1)}{I_2 I_3}\omega_0^2\,\delta\omega_3 = c_3\,\delta\omega_3 \tag{6.179}$$

Note that both $c_2$ and $c_3$ are negative; therefore, $\omega_2$ and $\omega_3$ will fluctuate like harmonic functions with their magnitudes bounded.

In other words, we can conclude that the rotation around axis #1 is stable. We reach the same conclusion for the rotation around axis #3.

However, if we linearize Equations (6.174)–(6.176) around $\omega_1 = 0$, $\omega_2 = \omega_0$, and $\omega_3 = 0$, it becomes

$$\delta\ddot{\omega}_1 = \frac{(I_2 - I_3)(I_1 - I_2)}{I_1\,I_3}\omega_0^2\,\delta\omega_1 = d_1\delta\omega_1 \qquad (6.180)$$

$$\delta\ddot{\omega}_2 = 0 \qquad (6.181)$$

$$\delta\ddot{\omega}_3 = \frac{(I_1 - I_2)(I_2 - I_3)}{I_1\,I_3}\omega_0^2\,\delta\omega_3 = d_3\delta\omega_3 \qquad (6.182)$$

Note that both $d_1$ and $d_3$ are positive; therefore, $\omega_1$ and $\omega_3$ will grow exponentially. In other words, the rotation around axis #2, the intermediate axis, is unstable, as stated before.

If we are interested in simulating the motion of the tennis racket, we can rewrite Equation (6.165) as

$$\begin{bmatrix} \sin\theta\sin\psi & \cos\psi & 0 \\ \sin\theta\cos\psi & -\sin\psi & 0 \\ \cos\theta & 0 & 1 \end{bmatrix}\begin{bmatrix} \dot{\phi} \\ \dot{\theta} \\ \dot{\psi} \end{bmatrix} = \begin{bmatrix} \omega_1 \\ \omega_2 \\ \omega_3 \end{bmatrix} \qquad (6.183)$$

Equation (6.183) can be further rearranged as

$$\begin{bmatrix} \dot{\phi} \\ \dot{\theta} \\ \dot{\psi} \end{bmatrix} = \begin{bmatrix} \sin\theta\sin\psi & \cos\psi & 0 \\ \sin\theta\cos\psi & -sin\psi & 0 \\ \cos\theta & 0 & 1 \end{bmatrix}^{-1}\begin{bmatrix} \omega_1 \\ \omega_2 \\ \omega_3 \end{bmatrix}$$

$$= \begin{bmatrix} \csc\theta\sin\psi & \csc\theta\cos\psi & 0 \\ \cos\psi & -sin\psi & 0 \\ -\cot\theta\sin\psi & -\cot\theta\cos\psi & 1 \end{bmatrix}\begin{bmatrix} \omega_1 \\ \omega_2 \\ \omega_3 \end{bmatrix} \qquad (6.184)$$

Equations (6.174)–(6.176) and (6.184) can be combined into one state-space representation as the model for the torque-free tennis racket rotation. Numerical simulation can then be carried out using Matlab. We will not go into the programming here but refer to abundant animations available on the web.[7]

---

[7]https://blog.wolfram.com/2020/12/15/simulating-zero-gravity-to-demonstrate-the-dzhanibekov-effect-and-other-surprising-physics-models/.

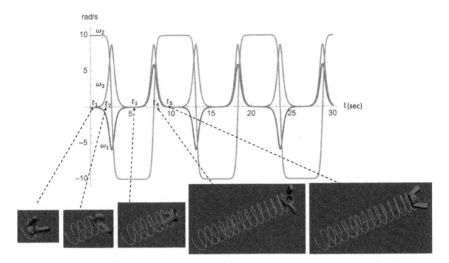

Figure 6.31: The simulation and animation for the Dzhanibekov effect.
*Source*: https://blog.wolfram.com/2020/12/15/simulating-zero-gravity-to-demon
strate-the-dzhanibekov-effect-and-other-surprising-physics-models/.

We can replace the tennis racket with a wing nut to demonstrate the Dzhanibekov effect. Many of the experiments have been conducted on International Space Station to repeat the phenomenon observed by Dzhanibekov (see footnote 4).

We recite the simulation results from the Wolfram website (see footnote 6) in Figure 6.31. At $t_1$, the wing nut mainly shows the rotation of $\omega_2$ around the intermediate axis. Due to small perturbations of $\omega_1$ and $\omega_3$, and the instability of the intermediate rotation, both $\omega_1$ and $\omega_3$ will grow in their magnitudes, while $\omega_2$ will reduce, as shown in the curves at $t_2$. The total angular momentum will be the same, without changes. The wing nut appears to flip and at $t_3$, it completes the flip and now rotates with the wings pointing backwards. Even though $\omega_2$ is still the same with respect to an inertial observer, it is now a negative value because $\hat{j}_3$ has reversed its direction. At $t_4$, the wing nut starts to flip again and at $t_5$, it completes the flip and rotates the same way as at $t_1$. If we hold a tennis racket at its handle and toss it up to rotate around axis #2, and catch the handle again when it comes down, we will notice that the racket has flipped around axis #3 by 180°. This is equivalent to the sequence of $t_1$ to $t_3$ in Figure 6.31.

## 6.6    Eccentric Impact of Rigid Bodies

In this section, we revisit the impact problem but generalize it to eccentric impact of two rigid bodies. We will use 2D rigid bodies for illustration, but the analysis can be easily extended to 3D rigid bodies. As shown in Figure 6.32(a), two 2D rigid bodies hit each other at one point. The impact points are labeled as $A$ and $B$, respectively. The mass center of rigid body #1 has a mass center velocity $(\vec{v}_{G1/O})_1$

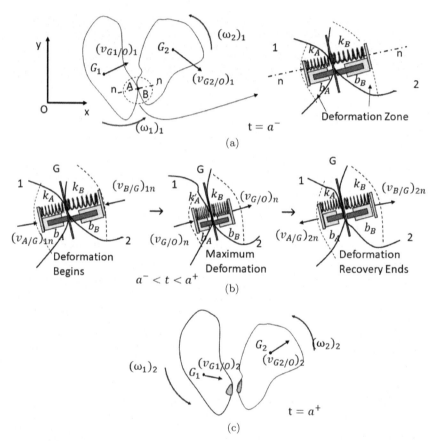

Figure 6.32:   (a) Two 2D rigid bodies hit each other at $C$; (b) the deformation zone changes during impact; and (c) the motions of the rigid bodies immediately after the impact.

right before impact, while #2 has $(\vec{v}_{G2/O})_1$. The rotation of #1 right before impact is $(\vec{\omega}_1)_1$ and for #2, it is $(\vec{\omega}_2)_1$. Even though we denote these two bodies as rigid bodies, there are local deformations at the impact point. The deformation zone is modeled as a combination of an ideal spring and a damper. As discussed before, we identify the normal direction of the impact, marked as line $n - n$. As shown in Figure 6.32(b), at the maximal deformation, the springs and dampers are compressed the most. At this moment, the deformation zones of $A$ and $B$ are moving at the same speed along the normal direction. We also assume that both $A$ and $B$ reach the maximal deformation at the same time. From this moment on, the deformation zones go into the recovery phrase. Depending on the material properties, the deformation might be fully recovered (elastic impact, with springs only without dampers), partially recovered, and zero recovery (plastic impact, with dampers only without springs).

Unlike the 1D, two-particle impact discussed in Section 6.2.1, we cannot define an inertial observer attached to the mass center of the two-particle system. If we only focus on the local deformation zone, we will adopt the same concept and define a virtual common mass center, $G$, as shown in Figure 6.32(b). The velocity of $G$ is assumed to be constant, $v_G$, during the short impact period. With this assumption, we can define the coefficient of restitution, $e$, as before,

$$e = \frac{|(v_{A/G})_{2n}|}{|(v_{A/G})_{1n}|} = \frac{|(v_{B/G})_{2n}|}{|(v_{B/G})_{1n}|} \tag{6.185}$$

where $(v_{A/G})_{in}$ and $(v_{B/G})_{in}$, with $i = 1$ or $2$ being the velocities along the impact normal direction before or after the impact. These velocities are the velocities of the point of the rigid body just outside of the deformation zone. Because we do not know the velocity of virtual mass center $G$, the expression of Equation (6.185) is not very useful. We will convert it to be the one with respect to an inertial reference frame $O - x - y$. From relative motion analysis, we have

$$\vec{v}_{A/G} = \vec{v}_{A/O} - \vec{v}_{G/O} \tag{6.186}$$

$$\vec{v}_{B/G} = \vec{v}_{B/O} - \vec{v}_{G/O} \tag{6.187}$$

Introducing Equations (6.186) and (6.187) into (6.185) with some manipulations, we obtain

$$
\begin{aligned}
e &= \frac{|(v_{A/O})_2 - v_{G/O}|_n}{|(v_{A/O})_1 - v_{G/O}|_n} = \frac{|(v_{B/O})_2 - v_{G/O}|_n}{|(v_{B/O})_1 - v_{G/O}|_n} \\
&= \frac{|(v_{A/O})_2 - v_{G/O} - [(v_{B/O})_2 - v_{G/O}]|_n}{|(v_{A/O})_1 - v_{G/O} - [(v_{B/O})_1 - v_{G/O}]|_n} \\
&= \frac{|(v_{A/O})_2 - (v_{B/O})_2|_n}{|(v_{A/O})_1 - (v_{B/O})_1|_n}
\end{aligned}
\tag{6.188}
$$

Note that the subscript "n" in the above equation indicates that all velocities are referring to those in the normal direction.

The velocity of points $A$ and $B$ can be obtained via 2D rigid body kinematics as

$$
(v_{A/O})_n = (\vec{v}_{G1/O} + \vec{\omega}_1 \times \vec{r}_{A/G1}) \cdot \vec{e}_n
\tag{6.189}
$$

$$
(v_{B/O})_n = (\vec{v}_{G2/O} + \vec{\omega}_2 \times \vec{r}_{B/G2}) \cdot \vec{e}_n
\tag{6.190}
$$

where $\vec{e}_n$ is the unit directional vector along the normal impact line.

Similarly, we can define the unit directional vector of the tangential line, $\vec{e}_t$, and define the velocity along the tangential line,

$$
(v_{A/O})_t = [(\vec{v}_{G1/O} + \vec{\omega}_1 \times \vec{r}_{A/G1})] \cdot \vec{e}_t
\tag{6.191}
$$

$$
(v_{B/O})_t = [(\vec{v}_{G2/O} + \vec{\omega}_2 \times \vec{r}_{B/G2})] \cdot \vec{e}_t
\tag{6.192}
$$

Let's consider the impact between a bowling ball and a pin to illustrate how the above equations can be used to determine the after-impact velocities.

## Example 6.13: Impact between a Bowling Ball and a Pin

As shown in Figure 6.33(a), a bowling ball, 7.3 kg, is moving along a straight line to hit a bowling pin. Right before impact, the bowling ball has a mass center velocity of 8 m/s along the $x$ axis, and an angular velocity, $\omega_1 = 30$ rad/s, around the $y$ axis. The ball hits the pin with the normal impact line 60° from the $x$ axis. The bowling pin has a mass of 1.6 kg and its mass center is 25 mm above the

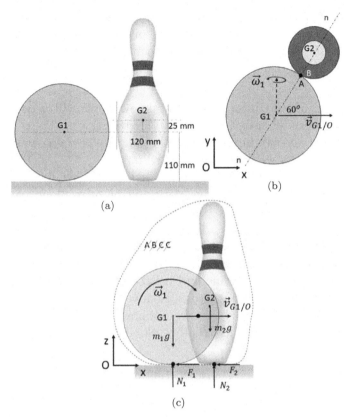

(a)

(b)

(c)

Figure 6.33: (a) A bowing ball hits a pin on the side; (b) the top view; and (c) the side view and FBD.

*Source*: https://i.stack.imgur.com/TR8Sd.jpg.

widest diameter barrel. The radius of the bowling ball is 110 mm. The moment of inertia of the pin along the $x$ or $y$ axis is $I_{2xx} = 0.016 \, \text{kg} \cdot \text{m}^2$.

**Problem Statement:** Determine the mass center velocities and angular velocities of the bowling pin and the bowling ball, immediately after the impact.

**Force Analysis:** The FBD for the bowling ball and the pin as a system is shown in Figure 6.33(c). The impact force is not shown in the FBD, but it is assumed that the impact force is only along the impact normal direction.

**Motion Analysis:** The pin is initially at rest. Immediately after impact, both the bowling ball and the pin are in general 3D motion. After impact, we need to determine the mass center velocities and the rotations of the ball and the pin. There are potentially 12 unknowns in total.

**Governing Equations:** With 12 unknowns, we will need 12 independent governing equations to solve for them.

Our strategy will be to reduce the number of unknowns via the laws in dynamics we have discussed so far.

First, we recognize that, with the ball in contact with the track at all times, its mass center velocity will be on a plane parallel to the $x-y$ plane without a $z$ component. This reduces the unknowns to 11.

Similarly, immediately after the impact, the mass center velocity of the pin will also be on the $x-y$ plane because the impact force is along the impact normal line and it is on the $x-y$ plane. All other forces are not HUGE enough to cause a velocity change immediately after the impact. This reduces the unknowns to 10.

We can define the mass center velocities immediately after the impact as

$$(\vec{v}_{G1/O})_2 = a\vec{i} + b\vec{j} \tag{6.193}$$

$$(\vec{v}_{G2/O})_2 = c\vec{i} + d\vec{j} \tag{6.194}$$

where $(\vec{v}_{G1/O})_1 = 8\vec{i}$ and $(\vec{v}_{G2/O})_2 = 0$ and a, b, c, and d are four unknowns.

The remaining six unknowns are the angular velocity components of the ball and the pin.

We recognize that $(\vec{\omega}_1)_1 = (\vec{\omega}_1)_2$ because the impact force passes through the mass center of the bowling ball. As a result, there are only three unknowns related to the angular velocity of the pin after impact.

Based on the angular momentum equation for the pin with respect to $G2$, we have

$$(\vec{H}_2)_1 + \int_{0^-}^{0^+} \vec{r}_{B/G2} \times \vec{F}_{imp} dt = (\vec{H}_2)_2$$

$$= 0 + \int_{0^-}^{0^+} \vec{r}_{B/G2} \times \vec{F}_{imp} dt$$

$$= \int_{0^-}^{0^+} (0.025 \sin 60° \vec{i} - 0.025 \cos 60° \vec{j}) F_{imp} dt$$

$$= \begin{bmatrix} \vec{i} & \vec{j} & \vec{k} \end{bmatrix} \begin{bmatrix} I_{2xx} & 0 & 0 \\ 0 & I_{2xx} & 0 \\ 0 & 0 & I_{2zz} \end{bmatrix} \begin{bmatrix} (\omega_{2x})_2 \\ (\omega_{2y})_2 \\ (\omega_{2z})_2 \end{bmatrix}$$

$$= I_{2xx}((\omega_{2x})_2 \vec{i} + (\omega_{2y})_2 \vec{j}) + I_{2zz}(\omega_{2z})_2 \vec{k} \qquad (6.195)$$

where $\vec{r}_{B/G2} = -0.06 \cos 60° \vec{i} - 0.06 \sin 60° \vec{j} - 0.025 \vec{k}$ and $\vec{F}_{imp} = F_{imp}(\cos 60° \vec{i} + \sin 60° \vec{j})$. Note that the principal axes of the pin align with the $x-y-z$ coordinate immediately before, during, and immediately after the impact.

From the above equation, we conclude that $(\vec{\omega}_2)_2$ will be along the direction of $\sin 60° \vec{i} - \cos 60° \vec{j}$.

Therefore,

$$(\vec{\omega}_2)_2 = (\omega_2)_2(\sin 60° \vec{i} - \cos 60° \vec{j}) \qquad (6.196)$$

where $\omega_{2x} = (\omega_2)_2 \sin 60°, \omega_{2y} = -(\omega_2)_2 \cos 60°$, and $\omega_{2z} = 0$.

There is only one unknown $(\omega_2)_2$ involved with the angular velocity of the ball and the pin after impact.

The total unknowns are reduced to five. We will need five independent governing equations. We will consider the linear momentum equation and the angular momentum equations for the combined system of the ball and the pin.

From the FBD of Figure 6.33(c), we recognize that conservation of total linear momentum and total angular momentum could be applied for the system with both the bowling ball and the pin as a system for the duration between immediately before and after the impact. Because the impact force is on the $x - y$ plane, the normal forces ($N_1$ and $N_2$) on the ball and the pin will not be huge, neither are the horizontal forces ($F_1$ and $F_2$).

For the conservation of total linear momentum, the governing equation is

$$m_1(\vec{v}_{G1/O})_1 + m_2(\vec{v}_{G2/O})_2 = m_1(\vec{v}_{G1/O})_2 + m_2(\vec{v}_{G2/O})_2 \quad (6.197)$$

From the above equation and introducing the related velocities, we have

$$\text{along } x: \ 58.4 = 7.3a + 1.6c \tag{6.198}$$

$$\text{along } y: \ 0 = 7.3b + 1.6d \tag{6.199}$$

Based on conservation of total angular momentum with respect to a stationary point coinciding with point $G2$, we have

$$\vec{r}_{G1/G2} \times m_1(\vec{v}_{G1/O})_1 + (\vec{H}_1)_1$$

$$= \vec{r}_{G1/G2} \times m_1(\vec{v}_{G1/O})_2 + (\vec{H}_1)_2 + (\vec{H}_2)_2 \tag{6.200}$$

where $\vec{r}_{G1/G2} = -0.17\cos 60°\vec{i} - 0.17\sin 60° - 0.025\vec{k}$, $(\vec{H}_1)_1 = (\vec{H}_1)_2 = I_1(\vec{\omega}_1)_1$, and $(\vec{H}_2)_2 = I_{2xx}(\vec{\omega}_2)_2$.

Rearranging the above equation, we obtain

$$\text{around } x: \ 0.6205b + 1.0745(8 - a) = 0 \tag{6.201}$$

$$\text{around } y: \ -0.1825b - 0.866I_{2xx}(\omega_2)_2 = 0 \tag{6.202}$$

$$\text{around } z: \ 0.1825(8 - a) - 0.5I_{2xx}(\omega_2)_2 = 0 \tag{6.203}$$

Because $(\vec{\omega}_2)_2$ is on the $x - y$, the above three equations are not fully independent. It can be shown that Equations (6.202) and (6.203) can be combined and become Equation (6.201). As a result, we will not use Equation (6.201) as one of the governing equations. Without it, the number of the governing equations is 4. We need an additional equation. This will come from the impact condition.

From Equation (6.188), assuming elastic impact, we have

$$e = 1 = \frac{|(v_{A/O})_2 - (v_{B/O})_2|_n}{|(v_{A/O})_1 - (v_{B/O})_1|_n} \tag{6.204}$$

We need to determine the velocities in the above equation based on the impact normal direction. We first find out the velocities of $A$ and $B$ immediately before and after impact.

$$(\vec{v}_{A/O})_2 = (\vec{v}_{G1/O})_2 + \vec{\omega}_1 \times \vec{r}_{A/G1}$$

$$= a\vec{i} + b\vec{j} - 1.65\vec{k} \tag{6.205}$$

$$(\vec{v}_{B/O})_2 = (\vec{v}_{G2/O})_2 + (\vec{\omega}_2)_2 \times \vec{r}_{B/G2}$$

$$= [c + 0.025 \cos 60°\,(\omega_2)_2]\vec{i}$$

$$+ [d + 0.025 \sin 60°\,(\omega_2)_2]\vec{j} - 0.060(\omega_2)_2\vec{k} \quad (6.206)$$

From the above equation, we can determine the velocity components along the impact normal direction, whose unit directional vector is $\vec{e}_n = \cos 60°\vec{i} + \sin 60°\vec{j}$. Based on Equations (6.189) and (6.190), we have

$$(v_{A/O})_{1n} = 8 \cos 60° \qquad (6.207)$$

$$(v_{A/O})_{2n} = a \cos 60° + b \sin 60° \qquad (6.208)$$

$$(v_{B/O})_{2n} = 0.025(\omega_2)_2 + c \cos 60° + d \sin 60° \qquad (6.209)$$

Introducing the above equations to Equation (6.204), we obtain

$$0.025(\omega_2)_2 + 0.5c + 0.866d - 0.5a - 0.866b = 4 \qquad (6.210)$$

In summary, we have Equations (6.198)–(6.199), (6.201)–(6.203), and (6.210) as governing equations, a total of six equations. The unknowns involved are $a$, $b$, $c$, $d$, and $(\omega_2)_2$, five unknowns. One of the equations will not be independent, but it can be used for verification. We set aside Equation (6.201) and solve for the remaining five equations to obtain

$$(\omega_2)_2 = 15.61 \text{ rad/s} \qquad (6.211)$$

$$a = 7.315 \text{ m/s} \qquad (6.212)$$

$$b = -1.185 \text{ m/s} \qquad (6.213)$$

$$c = 3.122 \text{ m/s} \qquad (6.214)$$

$$d = 5.405 \text{ m/s} \qquad (6.215)$$

Introducing the values of $a$ and $b$ to Equation (6.201), it is confirmed that it is satisfied.

With above results, we finally have

$$(\vec{v}_{G1/O})_1 = 8\vec{i}\,(\text{m/s}) \qquad (6.216)$$

$$(\vec{v}_{G1/O})_2 = 7.315\vec{i} - 1.185\vec{j}\,(\text{m/s}) \qquad (6.217)$$

$$(\vec{v}_{G2/O})_2 = 3.122\vec{i} + 5.405\vec{j} \text{ (m/s)} \qquad (6.218)$$

$$(\vec{\omega}_1)_1 = (\vec{\omega}_1)_2 = 30\vec{j} \text{ (rad/s)} \qquad (6.219)$$

$$(\vec{\omega}_2)_2 = 13.518\vec{i} - 7.805\vec{j}$$

$$= 15.609(\sin 60°\vec{i} - \cos 60°\vec{j}) \text{ (rad/s)} \qquad (6.220)$$

From the above results, we conclude that the pin will be bounced off along positive $x$ and $y$ directions, while the ball will reduce slightly the forward speed and gain a small velocity in the negative $y$ direction. In the mean time, the angular velocity of the ball does not change, while the pin will have an angular velocity about one half that of the ball, moving along a direction perpendicular to the impact normal direction. All these results match the actual experience of bowling.

**Verification:** We can also examine the case when the bowling ball hits the pin straight on. All the equations above still apply with the angle changed to 0°. With the same bowling ball motion before hitting the pin, the pin will be bouncing off at 12.48 m/s with a rotation at 31.21 rad/s, while the ball will slow down to 5.26 m/s. If we do not use the eccentric impact model, but consider the ball and the pin as two particles, the results will show that the pin will be bouncing off at 13.12 m/s without rotation and the ball will slow down to 5.12 m/s, without any rotations. The errors of the mass center velocities with the 1D impact assumption are surprisingly low for this particular case. However, the 1D impact model will be way off for the original eccentric impact bowling case.

## 6.7   Concluding Remarks

In this chapter, we converted *Newton's Second Law* into an integration form with respect to time. From there, the momentum concept was derived. Momentum can be either linear or angular. The determinations of these two momenta for a particle, a system of particles, and for rigid bodies were presented. For linear momentum, we only have to consider the motion of the mass center; however, for the angular momentum, it is more difficult and we must to choose a reference point, which can only either be the mass center or a stationary point within an inertial reference frame. For symmetric 3D rigid bodies, we usually choose an inertial reference which coincides with the principal axes of the rigid body for easier expression.

The new governing equations for solving dynamic problems are the linear momentum and angular momentum equations. If we can confirm either or both momenta are conserved, the governing equations become quite simple and powerful to solve many practical problems. However, we must VERIFY via proper force analysis to confirm that the momentum is conserved.

We presented detailed discussions on the impact problems, from 1D, two-particle impact to 3D eccentric impact between two rigid bodies. The concept of coefficient of restitution (COR) was introduced. In fact, COR is related to the concept of energy, which is the topic of the next Chapter. We will revisit the impact problem with the discussion of energy once it is introduced in the next chapter.

# Chapter 7

# Kinetic Energy and Work
# for Dynamic Systems

In this chapter, we will derive a new form of *Newton's Second Law*, from which two new concepts, kinetic energy and work, will be introduced. We will start with a general 3D rigid body and then simplify it to a particle in our discussion because we are already familiar with complicated rigid body motions from Chapters 5 and 6.

## 7.1 The Energy Form of *Newton's Second Law*

As shown in Figure 7.1, a rigid body undergoes a general 3D motion under the influences of some forces and moments. The expression of *Newton's Second Law* for this rigid body are presented in Chapter 5. We will use the moment equation with respect to the mass center $G$ for the following derivations. The axes, $\tilde{x}$–$\tilde{y}$–$\tilde{z}$, are the principal axes attached to the mass center and rotate with the rigid body. In other words, this rotating frame is like a skeleton representing the entire rigid body. Therefore, the frame rotation is the same as the rigid body rotation. We have

$$\sum \vec{F}_i = M\vec{a}_{G/O} \tag{7.1}$$

$$\left(\sum \vec{\tau}_j\right)_G = \dot{\vec{H}}_G = \vec{\Omega} \times \tilde{I}_G\tilde{\omega} + \tilde{I}_G\dot{\tilde{\omega}} \tag{7.2}$$

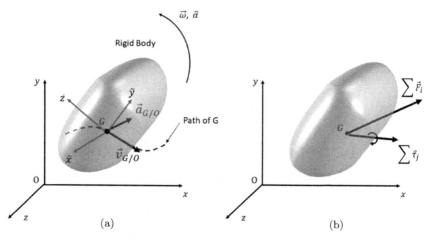

Figure 7.1: A rigid body undergoing 3D general motion under the influence of forces and moment: (a) the rigid body motion is characterized by its mass center velocity and acceleration along with the rotation; and (b) all external forces are moved to the mass center and the resulting total moment is a free vector.

where $\tilde{\omega} = \vec{\Omega} = [\tilde{\omega}_x \ \tilde{\omega}_y \ \tilde{\omega}_y]'$ and $\tilde{I}_G\tilde{\omega}$ is the angular momentum of the rigid body with respect to the mass center and an inertial reference coinciding with $\tilde{x}$–$\tilde{y}$–$\tilde{z}$ at this moment, but expressed in $\tilde{i}$, $\tilde{j}$, and $\tilde{k}$.

For Equation (7.1), we will dot multiply both sides with the vector, $d\vec{r}_{G/O}$, and then take an integration, which yields

$$\int \sum \vec{F}_i \cdot d\vec{r}_{G/O} = \int M\vec{a}_{G/O} \cdot d\vec{r}_{G/O} \qquad (7.3)$$

Equation (7.3) becomes a scalar equation. The integration of the left-hand side of Equation (7.3) generally cannot be carried out unless we know those forces as functions of $\vec{r}_{G/O}$. We will leave it as it is. For the right-hand side, we will do some calculus manipulations. Equation (7.3) becomes

$$\int \left( \sum \vec{F}_i \right) \cdot d\vec{r}_{G/O} = \int M\vec{a}_{G/O} \cdot d\vec{r}_{G/O}$$

$$= \int M\frac{d\vec{v}_{G/O}}{dt} \cdot d\vec{r}_{G/O}$$

$$= \int M\frac{d\vec{r}_{G/O}}{dt} \cdot d\vec{v}_{G/O}$$

$$= \int M\vec{v}_{G/O} \cdot d\vec{v}_{G/O}$$

$$= \int M\vec{v}_{G/O} \cdot ([d\vec{v}_{G/O}]_t + [d\vec{v}_{G/O}]_n)$$

$$= \int Mv_{G/O}\, dv_{G/O}$$

$$= \frac{1}{2}M[(v_{G/O})_2^2 - (v_{G/O})_1^2] \tag{7.4}$$

Note that $[d\vec{v}_{G/O}]_t$ is the tangential component of $d\vec{v}_{G/O}$, while $[d\vec{v}_{G/O}]_n$ is the normal component of $d\vec{v}_{G/O}$. In fact, because $[d\vec{v}_{G/O}]_t$ and $\vec{v}_{G/O}$ are in the same direction and $[d\vec{v}_{G/O}]_n$ is perpendicular to $\vec{v}_{G/O}$, we are able to simplify Equation (7.4) to its final form.

We will apply a similar treatment to Equation (7.2),

$$\int \left(\sum \vec{\tau}_j\right)_G \cdot d\vec{\theta} = \int \dot{\vec{H}}_G \cdot d\vec{\theta}$$

$$= \int \frac{d\vec{H}_G}{dt} \cdot d\vec{\theta} = \int \frac{d\vec{\theta}}{dt} \cdot d\vec{H}_G = \int \tilde{\omega} \cdot d(\tilde{I}_G\tilde{\omega})$$

$$= \int \tilde{\omega} \cdot d(\tilde{I}_G\tilde{\omega}) = \int \tilde{\omega} \cdot [\tilde{\Omega} \times \tilde{I}_G\tilde{\omega} + \tilde{I}_G\dot{\tilde{\omega}}]dt$$

$$= \int \tilde{\omega} \cdot [\tilde{\omega} \times \tilde{I}_G\tilde{\omega}]dt + \int \tilde{\omega} \cdot \tilde{I}_G\dot{\tilde{\omega}}dt$$

$$= \int \tilde{\omega} \cdot \tilde{I}_G\, d\tilde{\omega}$$

$$= \frac{1}{2}[\tilde{I}_x(\tilde{\omega}_x)_2^2 + \tilde{I}_y(\tilde{\omega}_y)_2^2 + \tilde{I}_z(\tilde{\omega}_z)_2^2]$$

$$- \frac{1}{2}[\tilde{I}_x(\tilde{\omega}_x)_1^2 + \tilde{I}_y(\tilde{\omega}_y)_1^2 + \tilde{I}_z(\tilde{\omega}_z)_1^2]$$

$$= \frac{1}{2}\tilde{\omega}_2'\tilde{I}_G\,\tilde{\omega}_2 - \frac{1}{2}\tilde{\omega}_1'\tilde{I}_G\,\tilde{\omega}_1 \tag{7.5}$$

In the derivation above, we have $\tilde{\omega} \cdot [\tilde{\omega} \times \tilde{I}_G\tilde{\omega}] = 0$ because $\tilde{\omega}$ is perpendicular to $\tilde{\omega} \times \tilde{I}_G\tilde{\omega}$. The term $\frac{1}{2}\tilde{\omega}'\tilde{I}_G\,\tilde{\omega}$ is in the matrix form for $\frac{1}{2}[\tilde{I}_x(\tilde{\omega}_x)^2 + \tilde{I}_y(\tilde{\omega}_y)^2 + \tilde{I}_z(\tilde{\omega}_z)^2]$. The symbol $'$ is the transpose operator. Note that $\tilde{I}_G$ is a diagonal matrix.

### 7.1.1  *Kinetic energy*

We will define the kinetic energy $T$ for a rigid body as

$$T = \frac{1}{2} M v_{G/O}^2 + \frac{1}{2} \tilde{\omega}' \tilde{I}_G \, \tilde{\omega} \tag{7.6}$$

The kinetic energy contains a translational component $\frac{1}{2} M v_{G/O}^2$ and a rotational component $\frac{1}{2} \tilde{\omega}' \tilde{I}_G \, \tilde{\omega}$. $T_1$ is the kinetic energy at $t = t_1$ and $T_2$ is at $t = t_2$.

Note that, we must define the kinetic energy with respect to an inertial reference, just like the case for *Newton's Second Law*. However, for different inertial reference frames, the kinetic energy are different if the origins are moving at different constant velocities. We must stay with the same inertial reference frame when we use the energy equation to solve for problems.

For different inertial reference frames with different orientations but with the same velocity of their origins, the expression of the kinetic energy can be obtained by transformation,

$$\begin{aligned}
T &= \frac{1}{2} M v_{G/O}^2 + \frac{1}{2} \tilde{\omega}' \tilde{I}_G \, \tilde{\omega} \\
&= \frac{1}{2} M v_{G/O}^2 + \frac{1}{2} (\hat{T}^{-1} \vec{\omega})' \, (\hat{T}^{-1} I_G \hat{T})(\hat{T}^{-1} \, \vec{\omega}) \\
&= \frac{1}{2} M v_{G/O}^2 + \frac{1}{2} \vec{\omega}' I_G \, \vec{\omega}
\end{aligned} \tag{7.7}$$

where $\hat{T}$ is the transformation matrix as defined in Chapter 5.

The above equation indicates that the rotational component of the kinetic energy can be based on different reference frames attached to the mass center. Using the principal axes of the symmetric rigid body leads to the simplest expression of the kinetic energy.

### 7.1.2  *Work*

We define the left-hand side term in Equations (7.4) and (7.5) as *Work*. The work done by the total force at the mass center is

$$W_F = \int \left( \sum \vec{F}_i \right) \cdot d\vec{r}_{G/O} \tag{7.8}$$

The work done by the total moment is

$$W_\tau = \int \sum \vec{\tau}_j \cdot d\vec{\theta} \tag{7.9}$$

Note that the total moment which includes the moments resulting from moving forces to the mass center.

Because the total force is on the mass center, the work is the total force multiplied by the infinitesimal displacement of the mass center.

We can extend this concept to define the work done by an individual force $\vec{F}_i$ more generally as

$$W_{F_i} = \int \vec{F}_i \cdot d\vec{r}_i \tag{7.10}$$

where $d\vec{r}_i$ is the infinitesimal displacement of the point of application of $\vec{F}_i$. We can extend the same concept to the moment by defining the work done by a moment $\vec{\tau}_j$ as

$$W_{\tau_j} = \int \vec{\tau}_j \cdot d\vec{\theta}_j \tag{7.11}$$

where $d\vec{\theta}_j$ is the infinitesimal angular displacement of the point of application of $\vec{\tau}_j$.

With the work definitions of Equations (7.10) and (7.11), we can show that

$$\int \sum \vec{F}_i \cdot d\vec{r}_{G/O} + \int \sum \vec{\tau}_j \cdot d\vec{\theta} = \int \sum (\vec{F}_i \cdot d\vec{r}_i) + \int \sum (\vec{\tau}_j \cdot d\vec{\theta}_j) \tag{7.12}$$

where $d\vec{\theta}_j = d\vec{\theta}$ for a rigid body.

In fact, the definition of work by Equations (7.10) and (7.11) does not require the rigid body assumption.

### 7.1.3 *Energy and work equations*

We can now add Equations (7.4) and (7.5) to arrive at the following expression:

$$T_1 + \int \sum (\vec{F}_i \cdot d\vec{r}_i) + \int \sum (\vec{\tau}_j \cdot d\vec{\theta}_j)$$
$$= T_1 + \int \sum (\vec{F}_k \cdot d\vec{r}_k) = T_2 \tag{7.13}$$

Note that we can break out each moment as a force couple and treat each force of the force couple separately. In this way, we only have to express the work done by forces, without mentioning moments.

Based on Equation (7.13), we perceive that the effects of the forces and moments are to change the kinetic energy of the system.

### 7.1.4 *Kinetic energy of a 2D rigid body*

Equation (7.6) can be readily simplified to a 2D rigid body case. If the rigid body is moving on the $x$–$y$ plane and can only rotate around the $z$ axis, the total kinetic energy is simplified to

$$T = \frac{1}{2}Mv_{G/O}^2 + \frac{1}{2}I_z\omega_z^2 \tag{7.14}$$

where $z$ is always a principal axis for a 2D rigid body.

### 7.1.5 *Kinetic energy of a particle*

The kinetic energy for a particle is further simplified because there is no rotation associated with a particle. The kinetic energy becomes

$$T = \frac{1}{2} Mv^2 \tag{7.15}$$

## 7.2 Work Done by Different Forces

The work and energy equation of Equation (7.13) is rather simple compared with *Newton's Second Law*, the linear momentum equation, and the angular momentum equation we learned in previous chapters. The main reason is that the work and energy equation (will be simply denoted as the energy equation from now on) is a scalar equation, not a vector equation. However, this also means that we can only have one unknown involved with the energy equation if we want to solve it. If we have more than one unknown, then the energy equation needs to be coupled with other governing equations.

Examining Equation (7.13), it is clear that the work term is the hardest to determine in general. To carry out the integration to calculate the work, the forces need to be functions of positions.

In a special case, if the work is zero, then $T_1 = T_2$, i.e., the kinetic energy is conserved. This can happen if $\sum \vec{F} = 0$, $d\vec{r} = 0$, $\sum \vec{F}$ is perpendicular to $d\vec{r}$, or the integration happens to be zero. Let's examine the kinetic energy of a race car in a circular track, as discussed in Example 5.3.

## Example 7.1: Kinetic Energy of a Race Car in a Circular Track

**Problem Statement:** As shown in Figure 7.2, a circular race track is designed with a bank angle $\theta$. The radius of the track is $\rho$, while the static friction coefficient $\mu_s$ under the dry condition is 0.8. Review the work done by various forces if the internal friction of the car is neglected by assuming that the car is one rigid body.

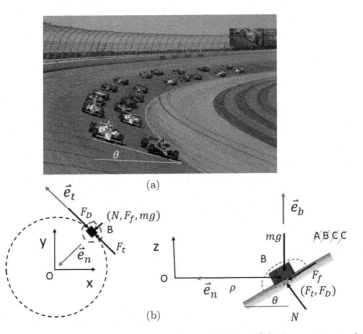

Figure 7.2: (a) A race track with a bank angle $\theta$; and (b) the motion and force analysis of a car on the race track.
*Source:* https://upload.wikimedia.org/wikipedia/commons/d/d4/Chicagoland_Speedway_ow.jpg.

**Solution:**

**Force Analysis:** The free-body diagram is shown in Figure 7.2(b). There is a normal force $N$, a frictional force $F_f$ along the bank direction, a tractive force $F_t$, pointing to the $\vec{e}_t$ direction, the wind drag $F_D$, in the opposite direction, and finally the car weight $mg$.

**Motion Analysis:** This is a 2D circular motion and we choose the $t - n$ description.

The velocity equation is

$$\vec{v} = v\vec{e}_t \tag{7.16}$$

**Governing Equations:** Applying the energy equation, the governing equation is

$$\frac{1}{2}mv_1^2$$

$$+ \int m\vec{g} \cdot d\vec{r}_g + \int \vec{N} \cdot d\vec{r}_N + \int \vec{F}_f \cdot d\vec{r}_f + \int \vec{F}_D \cdot d\vec{r}_D$$

$$+ \int \vec{F}_t \cdot d\vec{r}_t$$

$$= \frac{1}{2}mv_2^2 \tag{7.17}$$

**Solving for Unknowns:** From Equation (7.17), it is clear that if the total work done by the forces is not zero, the speed of the race car will change. Let's examine each of the work terms.

For the work done by $mg$, it is zero if the race car does not move up and down the banked track. For the work done by $N$, it is always zero because $\vec{N}$ and $\vec{v}$ (i.e., $d\vec{r}_N$)) are always perpendicular. For the work done by $F_f$, it is zero if the car does not move up and down the banked track. For the work done by $F_D$, it is not zero and is negative because $\vec{F}_D$ and $d\vec{r}_D = d\vec{r}$ are in the opposite directions. Finally, for the work done by $F_t$, it is not zero either, but is positive because $\vec{F}_t$ and $d\vec{r}_t = d\vec{r}$ are in the same direction. The rolling friction in line with the velocity is ignored in this case. Assuming that the car follows a circular path without moving up and down the banked track, Equation (7.17) can be simplified as

$$\frac{1}{2}mv_1^2 + \int \vec{F}_D \cdot d\vec{r}_D + \int \vec{F}_t \cdot d\vec{r}_t = \frac{1}{2}mv_2^2 \tag{7.18}$$

If the work done by the wind drag cannot be ignored, the car will need to provide a tractive force from the car engine to overcome the negative work consumed by the wind in order to keep $v_1 = v_2$. This is the experience we have. On a highway at a high cruising speed, if we let go of the gas peddle, the car will slow down.

If we ignore the wind drag, the car should be able to cruise at a constant speed without the engine. Of course, this is not quite true and it is because a car is not a rigid body and it has internal friction. The internal friction or the work done by the internal forces will be discussed later. In winter Olympics, a bobsled can maintain its speed without an engine because the wind drag is small (aerodynamic shape) and the sliding friction force ($F_s$) between the sled and the snow is low. Over a short distance, based on Equation (7.18) by including a sliding friction force, we have $v_1 \approx v_2$ because $\int \vec{F}_D \cdot d\vec{r} \approx 0$ and $\int \vec{F}_s \cdot d\vec{r} \approx 0$.

**Verification:** It is important that we conduct a force analysis with FBD and identify the work done by each force. We should always recognize the case when the force ($\vec{F}$) and the displacement vector ($d\vec{r}$) are perpendicular, it results in zero work.

## 7.2.1 *Work done by conservative forces*

In this section, let's consider cases when we can actually carry out the integration of the work done by a force, for which the force must be a function of position, including being a constant in both size and direction. This means that the size and direction of the force is only related to the position vector of the object it is acting on. If so, we can carry out the integration to determine the work as

$$W(r_1, r_2) = \int_{r_1}^{r_2} \vec{F}(\vec{r}) \cdot d\vec{r} \qquad (7.19)$$

Note that the work done is also a function of position, just like the force.

The forces with this property are very "nice" because we can calculate the work done by them. These forces have a special name: *Conservative Forces*. There are several specific examples of conservative forces important to engineering analysis. The first example is the gravitational force.

### 7.2.2    *Potential energy of gravitational forces*

For an object near the Earth's surface, the gravitational pull is $\vec{F} = M\vec{g}$, which is a constant both in terms of direction and size. As a constant, it is indeed a function of position because we can easily carry out the integration as in Equation (7.19).

As shown in Figure 7.3, an object of mass $M$ moves from position $\vec{r}_1$ to position $\vec{r}_2$. The work done by the gravitational force is

$$W(r_1, r_2) = \int_{r_1}^{r_2} \vec{F}(\vec{r}) \cdot d\vec{r} = \int_{y_1}^{y_2} M\,g\cos(180°)dy$$

$$= \int_{y_1}^{y_2} -Mg\,dy = Mgy_1 - Mgy_2 \qquad (7.20)$$

which is only a function of the initial and the final positions, independent from the path it travels from $\vec{r}_1$ to $\vec{r}_2$. From Equation (7.20), we can also define a new energy term, denoted as gravitational potential energy, $U_g = Mgy$. The work done by the gravitational force is the difference between the gravitational potential energy,

$$W(r_1, r_2) = Mgy_1 - Mgy_2 = U_{g1} - U_{g2} = -\Delta U_g \qquad (7.21)$$

From Equation (7.21), it becomes quite straightforward for us to determine the work done by the gravitational force. If an object moves from a lower position to a higher position, the work is negative,

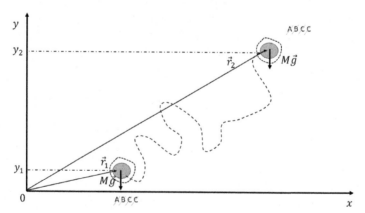

Figure 7.3:    An object moves through an arbitrary path from a new position $\vec{r}_1$ to a new position, $\vec{r}_2$. The work done by the gravitational force is independent of the path it has traveled.

while the work is positive if it moves from a high position to a lower position. In other words, a positive work done by the gravitational force will reduce the gravitational potential energy, i.e., $W_g = -\Delta U_g$. Finally, the datum to determine the vertical position is not unique. We can define any vertical position as the datum, i.e., $y = 0$. The resulting work done by the gravitational force will be the same. However, we must use the same datum throughout the calculation.

Let us also answer a simple question. Because the gravitational force near the Earth is conservative and constant, is every constant force a conservative force? For example, if a block is moving on a flat surface with a constant friction force, is this constant friction force a conservative force? The answer is NO because while the magnitude could be constant for this friction force, its direction will change when the block is moving toward a different direction. The magnitude can also change if it moves to a different surface or there is an external force acting on it.

### 7.2.3  *Potential energy of an elastic force*

Another example of the conservative forces is related to the spring, linear or nonlinear. We will discuss only the linear spring case. As shown in Figure 7.4, an ideal linear spring with an original length of $x_0$ has been stretched by $x$. At that point, the spring force on the free end is $\vec{F}_k$. If the spring is further stretched by an infinitesimal amount, $dx$, the spring force would be essentially the same, but there is an infinitesimal work done by $\vec{F}_k$ as

$$\delta W = \vec{F}_k \cdot d\vec{x} = -F_k \, dx \qquad (7.22)$$

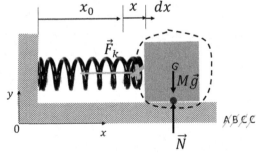

Figure 7.4:   A linear spring is stretched from its original length.

Here, we must adhere to the definition of the work and define $\delta W$ based on the FBD in Figure 7.4.

If the spring is stretched to $x_2$, then the work done by the spring force is

$$W(x_1, x_2) = \int_{x_1}^{x_2} -F_k \, dx$$

$$= \int_{x_1}^{x_2} -kx \, dx = \frac{1}{2}kx_1^2 - \frac{1}{2}kx_2^2 = U_{k1} - U_{k2} = -\Delta U_k$$

(7.23)

where $F_k = kx$ and $U_k = \frac{1}{2}kx^2$.

Here, we have the definition of a new potential energy, $U_k = \frac{1}{2}kx^2$. The work done by the spring to the block only depends on the potential energies at its initial stretch and the final stretch, similar to the gravitational force case. Also similar to the gravitational potential energy, a positive work by the spring force is to reduce the spring potential energy, i.e., $W_k = -\Delta U_k$. Note that the spring potential energy is positive either if the spring is compressed or stretched. The datum for calculating the spring potential energy is defined when there is no deformation in the spring. We cannot arbitrarily define a datum for the spring deformation. $U_k$ is also called the elastic potential energy because the spring is assumed to be elastic and does not suffer permanent deformation (plastic deformation).

Now, let us combine these two conservative forces in an example. As shown in Figure 7.5, object A is tied to a linear spring, with a spring constant $k$, on a frictionless slope and reaches an equilibrium with the spring compressed by $z_1$ from its original length $z_0$. We can easily find out $z_1$ by the equilibrium force equations, $z_1 = \frac{mg \sin \theta}{k}$.

Now, let us examine the potential energy condition at the equilibrium for the system of object A and the spring. Three forces are involved in Figure 7.5. The normal force $\vec{N}$ is not a conservative force, but it does not do any work in this case because it is always perpendicular to the motion direction of the block. With respect to datum $O$–$x$–$y$, the total potential energy is

$$U_{\text{total}} = U_{k1} + U_{g1} = \frac{1}{2}kz_1^2 + (z_0 - z_1)mg \sin \theta \qquad (7.24)$$

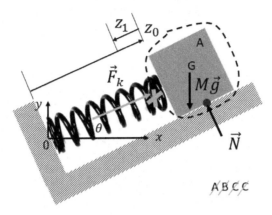

Figure 7.5: Potential energy analysis involves both gravity and spring.

### 7.2.4 Potential energy of non-viscous fluid pressure

In this example, we will trace an infinitesimal cubic of fluid volume along its path within the fluid. We recognize that the pressure field of a non-viscous (inviscid) fluid is a continuous function of position, i.e., $P = P(x, y, z)$, as shown in Figure 7.6. For this infinitesimal cubic of fluid, the pressure at each surface of cubic is defined accordingly. The pressure results in normal forces acting on each surface. It is also subjected to a gravitational field.

By applying the energy equation for this cubic volume of fluid, we obtain the following governing equation:

$$\frac{1}{2}\rho dV v_1^2 + \int - \left[ \left( \frac{\partial P}{\partial x}dx \right) dy\, dz\, \vec{i} + \left( \frac{\partial P}{\partial y}dy \right) dx\, dz\, \vec{j} \right.$$
$$\left. + \left( \frac{\partial P}{\partial z}dz \right) dx\, dy\, \vec{k} \right] \cdot d\vec{r} + \int (-\rho dV g)\vec{k} \cdot d\vec{r} = \frac{1}{2}\rho dV v_2^2 \quad (7.25)$$

where $d\vec{r} = dx\vec{i} + dy\vec{j} + dz\vec{k}$, $dV = dx\, dy\, dz$, and $\rho$ is the density of the fluid. In the above equation, we also assume that the fluid cubic is not rotating. Therefore, there is no rotational kinetic energy in the above equation. This assumption dictates that the fluid is non-rotational, denoted as Irrotational Flow in Fluid Mechanics.

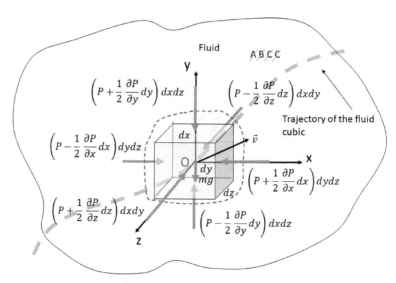

Figure 7.6:  Pressure field of a non-viscous fluid and the force and motion analysis of an infinitesimal cubic of fluid.

Equation (7.25) can be simplified as

$$\frac{1}{2}\rho v_1^2 + \int - \left(\frac{\partial P}{\partial x}dx + \frac{\partial P}{\partial y}dy + \frac{\partial P}{\partial z}dz\right) + \int(-\rho g)dz = \frac{1}{2}\rho v_2^2$$

$$\implies \frac{1}{2}\rho v_1^2 + \int -dP + \int(-\rho g)dz = \frac{1}{2}\rho v_2^2$$

$$\implies \frac{1}{2}\rho v_1^2 + P_1 + \rho g z_1 = \frac{1}{2}\rho v_2^2 + P_2 + \rho g z_2 = \text{constant} \qquad (7.26)$$

Equation (7.26) is the famous Bernoulli equation in Fluid Mechanics. Because we usually cannot identify an individual particle of a fluid and follow its path, we usually limit the application of the Bernoulli equation to a stream line of a steady flow. Under this circumstance, the stream line is the same as the trajectory of a fluid "particle." Along with other assumptions we made during the derivation, the flow also has to be inviscid and irrotational.

The potential energy due to fluid pressure is $P$, which is defined for a unit volume because we cannot identify individual fluid "particles," Accordingly, the potential energy due to gravity per unit volume of the fluid is $\rho g z$, instead of $mgz$ defined earlier.

### 7.2.5 *Energy equation with potential energy terms*

We can now present a more useful energy equation expression which incorporates the potential energies due to gravity and springs for an object with a mass, $m$.

$$T_1 + U_{g1} + U_{k1} + \int \sum \vec{F}_{\text{non}} \cdot d\vec{r} = T_2 + U_{g2} + U_{k2} \quad (7.27)$$

Let's revisit Example 3.3 and solve for it using the energy equation.

### Example 7.2: Spring Compression due to a Ball Dropping on it

What is the maximum compression of the spring when a ball of mass $m = 10$ kg is dropped from rest at a height of 1 meter? The neutral length of the spring is 0.1 meter. The spring constant is $k = 100,000$ N/m. In addition, at which height will the object achieve the highest speed?

**Solution:**

**Problem Statement:** The graphical representation of the problem is shown in Figure 7.7(a).

**Force Analysis:** The FBD of the spring when it is in contact with the ball is shown in Figure 7.7(b). As per the FBD, there are only two external forces involved and both are conservative forces. There are no non-conservative forces doing work on the ball.

**Motion Analysis:** Skipped.

**Governing Equations:** Based on Equation (7.27), we have

$$T_1 + U_{g1} + U_{k1} = T_2 + U_{g2} + U_{k2} \quad (7.28)$$

**Solving Equations:** We know that $T_1 = 0$, $U_{g1} = mgy_1 = 98.1\,\text{Nm}$, and $U_{k1} = 0$. At the maximum compression, $T_2 = 0$, $U_{g2} = mgy_2 = 98.1y_2$, and $U_{k2} = \frac{1}{2}k(h_0 - y_2)^2$. Introducing them to the governing

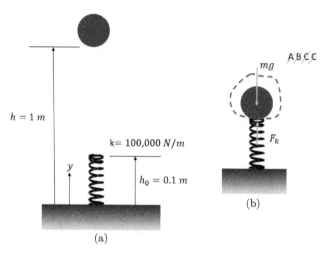

(a)

Figure 7.7: (a) A ball falling from a high position before striking the spring below and (b) the ball has sticken the spring with the spring compressed.

equation, we have

$$98.1 = 98.1 y_2 + \frac{1}{2}100{,}000(h_0 - y_2)^2 \qquad (7.29)$$

Solving the above equation for $y_2$, we found $y = 0.057\,m$. Therefore, the maximum compression of the spring is $0.043\,m$.

Compared with the solution method of Example 3.3, it is a lot simpler using the energy equation.

To answer the question at which height the speed of the object is maximum, we can apply the energy equation again,

$$T_1 + U_{g1} + U_{k1} = 98.1 = T_3 + U_{g3} + U_{k3}$$

$$= \frac{1}{2}mv_3^3 + mgy_3 + \frac{1}{2}k(h_0 - y_3)^2$$

$$\Longrightarrow T_3 = \frac{1}{2}mv_3^3 = 98.1 - mgy_3 - \frac{1}{2}k(h_0 - y_3)^2 \qquad (7.30)$$

When the speed is the maximum, $T_3$ is also at its maximum; therefore, taking a derivative of $T_3$ and letting it equal to zero, we can solve for $y_3$. The result is the same as obtained in Example 3.3. Again, the solution method is much simpler with the energy equation.

## 7.3 Multiple Rigid Bodies Systems

In this section, we extend the energy equation for a system of one rigid body to a system of multiple rigid bodies. As shown in Figure 7.8(a), the FBD of the loading system of a front loader is shown. The external forces related to this systems are $\vec{H}_x$, $\vec{H}_y$, $\vec{C}$, $\vec{R}_x$, $\vec{R}_y$, $M_B\vec{g}$, and $M_A\vec{g}$. Only $M_B\vec{g}$ and $M_A\vec{g}$ are conservative forces.

### 7.3.1 *External forces vs internal forces*

The loading system contains two large rigid bodies. The corresponding FBDs for each rigid body are shown in Figures 7.8(b) and 7.8(c). There are additional forces in these two FBDs, such as $\vec{B}_x$, $\vec{D}$, etc. These forces are not shown in Figure 7.8(a) because they are internal forces. FBD only shows external forces.

The internal forces, $\vec{B}_x$ and $\vec{B}_y$, are related to pin $B$. The force directions are opposite for rigid bodies $A$ and $B$. The displacement of pin $B$ on rigid body $A$ and $B$ are the same. Therefore, the work done by these two forces will cancel each other. However, for force $\vec{D}$,

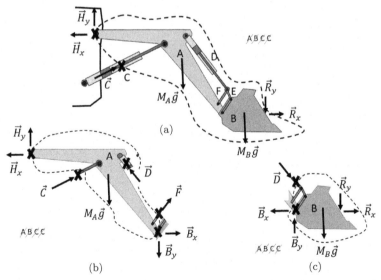

Figure 7.8: (a) The FBD of a front loader with multiple bodies, (b) the FBD for the arm only, and (c) the FBD for the loader only.

it is acting on the hydraulic cylinder rod attached to rigid body $B$. The opposite force of $\vec{D}$ is on the cylinder attached to rigid body $A$. The displacements for the opposing $\vec{D}$ are different. As a result, the work done by $\vec{D}$ on rigid body $A$ will not cancel the work done by $-\vec{D}$ on rigid body $B$. This situation exists no matter how we construct the FBD.

For the FBD of Figure 7.8(a), force $\vec{D}$ is an internal force. We recognize now that internal forces can still do work. We can write an energy equation for rigid body $A$ considering all external forces shown in Figure 7.8(b). We can do the same for rigid body $B$ with respect to Figure 7.8(c). We then add up both energy equations. Some of the work done by internal forces will cancel each other but some will not, such as $\vec{D}$. As a result, when we construct an overall energy equation for a system containing multiple bodies, we need to account for the work done by internal forces.

### 7.3.2 *Energy and work equation for a multiple-body system*

The energy equation for a multi-body system should be expressed as

$$
T_1 + U_{g1} + U_{k1} + \int \sum \vec{F}_{\text{non}} \cdot d\vec{r} + \int \sum \vec{f}_{\text{non}} \cdot d\vec{r}
$$
$$
= T_2 + U_{g2} + U_{k2} \tag{7.31}
$$

where $\vec{F}_{\text{non}}$ denotes non-conservative external forces, while $\vec{f}_{\text{non}}$ stands for non-conservative internal forces.

Unfortunately, the FBD does not show any internal forces; therefore, we must review these internal forces carefully when constructing the energy equation for multi-body systems. The next example discusses how to construct an energy equation correctly.

### Example 7.3: Energy Equations for Multi-body Systems

As shown in Figure 7.9, there is a car at a constant cruising speed and a person running in a treadmill. Write the energy equation for each of these cases.

**Problem Statement:** We will construct the energy equation based on an inertial reference on the ground.

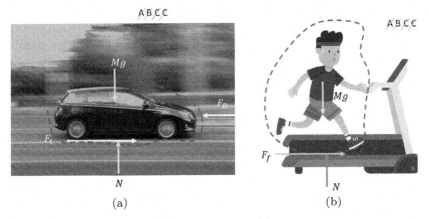

Figure 7.9: Examples of multi-body systems: (a) a car cruising at a constant speed; and (b) a person on a treadmill.

*Modified images from sources*: (a) https://images.pexels.com/photos/9235/road-traffic-car-business.jpg?cs=srgb&dl=car-driving-by-fast-9235.jpg&fm=jpg; and (b) https://pngimg.com/uploads/treadmill/treadmill_PNG45.png.

## Solution:

**Force analysis:** We construct FBDs for both cases. For the case of Figure 7.9(a), the external forces identified are $N$, $Mg$, $F_f$ (the tractive force), and $F_D$ (the wind drag).

If we consider the whole car as one rigid body and construct the energy equation without considering the internal forces, the equation will be

$$T_1 + \int \vec{F}_t \cdot d\vec{r}_1 - \int \vec{F}_D \cdot d\vec{r}_2 = T_2$$

$$\implies T + \int (\vec{F}_t - \vec{F}_D) \cdot \vec{v}\, dt = T$$

$$\implies 0 = 0 \tag{7.32}$$

where $F_t = F_D$ because the car is at a constant speed and we ignore the small rolling friction or include it to be a part of the wind drag. $F_t$ is moved to the mass center due to the rigid body assumption; therefore, $F_t$ and $F_D$ are seeing the same displacement, $d\vec{r} = \vec{v}\, dt$. We end up with a trivial equation, not very useful.

If we do not assume the car is a rigid body but a system of multiple rigid bodies, then we cannot move $F_t$ to the mass center. The work done by $F_t$ has to be based on the displacement vector, $d\vec{r}_1$, of the point where $F_t$ is acting. Because the car tire is assumed to be rolling without slipping, we know that the contact point is an instantaneous center and its velocity is zero; therefore, $d\vec{r}_1 = 0$. In other words, $\int \vec{F}_t \cdot d\vec{r}_1 = 0$. Based on this conclusion, Equation (7.32) is not even correct unless we ignore the wind drag entirely.

To obtain a correct energy equation, we need to consider the internal forces which do work. First, there are internal forces involved in the combustion engine which produce positive work. There are also internal friction forces which consume energy, doing negative work. We will produce a chart to detail all the forces involved and if they produce work or not. As shown in Table 7.1, only those forces doing work will be included in the energy equation. If a force is doing work and it is a conservative force, then it should appear as a potential energy.

Table 7.1 indicates that we only have to consider $F_D$, $f_e$, and $f_f$. The resulting energy equation is

$$T_1 - \int \vec{F}_D \cdot \vec{v}dt + \int \vec{f}_e \cdot d\vec{r}_3 - \int \vec{f}_f \cdot d\vec{r}_4 = T_2$$

$$\implies -\int \vec{F}_D \cdot \vec{v}dt + \int \vec{f}_e \cdot d\vec{r}_3 - \int \vec{f}_f \cdot d\vec{r}_4 = 0 \qquad (7.33)$$

Equation (7.33) indicates that the car engine produces positive work in order to overcome the negative work done by the wind drag and internal frictional forces. There is no net work to change the speed of the car; therefore, $T_1 = T_2$. This is different from the

Table 7.1: A force list to account for all forces, external and internal, for constructing the energy equation.

| External forces | Conservative? | Doing work? |
|---|---|---|
| Weight, $Mg$ | Yes | No |
| Tractive force, $F_t$ | No | No |
| Wind drag, $F_D$ | No | Yes |
| Normal force, $N$ | No | No |
| **Internal Forces** | | |
| Engine forces, $f_e$ | No | Yes |
| Internal friction, $f_f$ | No | Yes |

perspective of *Newton's Second Law*, for which the car has no acceleration because $F_t = F_D$, according to the FBD of Figure 7.9(a).

Because both $F_D$ and $f_f$ are proportional to the speed, the engine will be producing its maximum power when the vehicle reaches its maximum speed. Therefore, Equation (7.33) also indicates that the maximum speed a car can achieve is limited by its engine power. To demonstrate the awesome power of an engine, the best way is to take the car to a race track and find out the maximum speed the car can achieve. However, quite often, the engine power is demonstrated by a wrong way to burn the tire by flooring the gas pedal. The engine torque could exceed the maximum tractive force, causing the tire to spin and rub the road to produce burning and toxic smokes. This wrong practice does not showcase the power of the engine and is environmentally unfriendly.

We can now apply the same analysis to Figure 7.9(b). The human body apparently is not a rigid body. The corresponding FBD identifies three external forces, $F_f$, $N$, and $Mg$. The internal forces will include some chemical "forces" to produce energy, or to burn calories, forces which move body parts, and forces that consume energy. With respect to an inertial reference frame on the ground, the force $F_f$ will produce a negative work if it is pointing forward while the foot on the treadmill is moving backward. However, if we consider the whole body as a rigid body and move $F_f$ to the mass center, based on *Newton's Second Law*, the mass center of the body is not moving forward; therefore, the acceleration of the body mass center is zero, which renders $F_f = 0$. In other words, when running on a treadmill, with the belt driven by a motor, the shoes only experience some impact forces, $N$, when they engage the treadmill, and no need for $F_f$ to push the body forward or to move the treadmill. The corresponding energy equation will show that the chemical engine of the body burns calories to produce energy which is consumed by the work done by the internal forces of the body. The is how the workout of running on a treadmill is achieved. Let's consider a different workout mode. If a person is walking on the treadmill and the treadmill belt does not move unless the person is pushing it hard by feet through walking forcefully. In such a case, the person needs to hold on to the hand rail of the treadmill, so that the feet can push the treadmill belt. $F_f$ is not zero and will be doing a negative work, while the force on the hand rail does not do work and it is the same as $F_f$ but in the opposite direction, so that the body does not move forward. The workout mode now includes the chemical engine of the body burning calories to produce work which is consumed by internal forces and $F_f$.

In the above analysis, we do not need to reconcile both the energy equation and *Newton's Second Law*. So far, both laws are consistent with each other.

### 7.3.3 Who is right? Dr. Energy or Professor Newton?

Let's consider another example related to a car during braking.

**Example 7.4: Energy Equations for Car Braking**

As shown in Figure 7.10(a), there is a car at a speed, $v_0$, when the driver notices an obstacle ahead. The brake was first applied at $t_1$ and then it was pressed down hard at $t_2$ when the tires locked up. A tire mark was then produced until the car hit the obstacle and came to an abrupt stop at $t_3$. There are several questions to be answered.

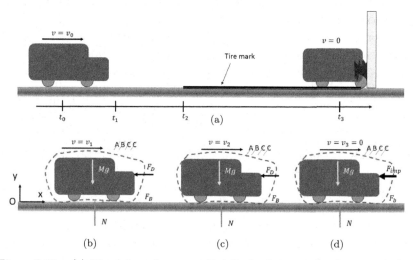

Figure 7.10: (a) The driver of a car applied the brake at $t_1$, the tires were locked up at $t_2$, and then the car came to an abrupt stop when it hit an obstacle; and (b)–(d) the corresponding FBDs at different time instants.

- Question (1): Which force slows down the car during $t_1 < t < t_2$?
- Question (2): When a police shows up and wants to estimate $v_0$ to determine if the driver was speeding, how should it be calculated?
- Question (3): Answer the first question but for $t_2 < t < t_3$.

**Solution:**

**Problem Statement:** Skipped.

**Solution:** We will first construct an FBD to establish *Newton's Second Law* to answer the first question. We will also construct the energy equation to answer the same question. After that, we will try to answer the second and the third questions.

**Force Analysis:** An FBD is constructed for $t_1 < t < t_2$ (Figure 7.10(b). Four external forces are identified, the normal force $N$, the braking force $F_B$, the wind drag $F_D$, and the weight of the car $Mg$. By applying *Newton's Second Law*, we have

$$-F_B - F_D = ma_x$$
$$\implies \quad a_x = -\frac{(F_B + F_D)}{m} \quad\quad (7.34)$$

From the above equation, it is clear that $F_B$ and $F_D$ are slowing down the car, according to *Newton's Second Law*. Let's ignore the wind drag because for braking performance, it is too risky to rely on the wind drag. We will assume no wind drag for the following analysis. We then conclude that

*It is $F_B$ which slows down the car according to Newton's Second Law.*

Let's construct an energy equation to answer the first question. As discussed in the previous example, the car is a multi-body system. We also have to consider the internal forces. During braking, the engine is idling; therefore, let's ignore the work done by the engine. We will also skip the internal frictions of the car. The main internal force which consumes energy is the braking force between the brake pads and the brake rotors, denoted as $f_B$. The energy consumed turns into heat to raise the pads and rotors to fairly high temperatures. For the external forces, the analysis is similar to that done in the previous example (Table (7.1)). However, we now have a braking force, $F_B$, applied by the ground to the tire. During $t_1 < t < t_2$, the tires are not locked up; therefore, they are rolling without slipping. The work done by $F_B$ has to be based on the displacement vector, $d\vec{r}_1$, of the point where $F_B$ is acting. Because the car tire is assumed to be rolling without slipping while braking, $d\vec{r}_1 = 0$. In other words, $\int \vec{F}_B \cdot d\vec{r}_1 = 0$. $F_B$ **does no work**. We can verify this statement by checking if there is a large amount of heat generated at the tire/ground interface during braking without lock-up. From experience, we know that the tire remains relatively cool, while the brake pad/rotor is very hot. Without considering the wind drag, the corresponding energy equation for $t_1$ and $t_2$ is

$$ T_1 - \int \vec{f}_B \cdot d\vec{r}_r = T_2 \tag{7.35} $$

where $\vec{r}_r$ is the relative displacement between the brake pads and the rotors. We know that $T_2 < T_1$ because the car is slowing down. The kinetic energy of the car is reduced because of $f_B$ and the kinetic energy loss becomes heat at the rotors and the pads, making them very hot. From Equation (7.35), we conclude that

*It is $f_B$, which slows down the car according to the energy equation.*

We have conflicting conclusions regarding which force slows down the car. The forces $F_B$ and $f_B$ are different. $F_B$ is an external force doing no work, while $f_B$ is an internal force, which is not shown in the FBD, but appears in the energy equation. According to *Newton's Second Law*, $f_B$ is NOT the force to slow down the car because it does not even appear in the FBD. On the other hand, according to the energy equation, $F_B$ does NOT slow down the car because it does not consume energy.

We must wonder **WHO IS RIGHT? Professor Newton or Dr. Energy? They apparently have different opinions.**

To resolve this seemingly conflicting conclusions presented so far, we need to construct another FBD, as shown in Figure 7.11, for the tire/wheel/rotor as an assembly. In this FBD, both $F_B$ and $f_B$ appear as external forces. Applying *Newton's Second Law* for the moments

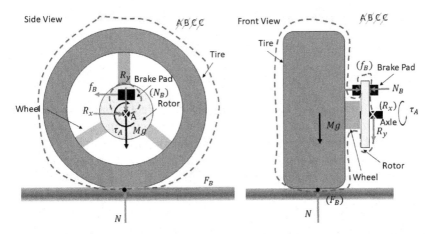

Figure 7.11: The front and side views of a tire/wheel/rotor as a free-body and the force analysis.

around point $A$, the axle, we have

$$-\tau_A + f_B \cdot d - F_B \cdot R = I_A \, \alpha_W \qquad (7.36)$$

where $\tau_A$ is the driving torque from the axle, $d$ is the distance from the brake pad to the axle center $A$, $R$ is the tire radius, while $I_A$ and $\alpha_W$ are the moment of inertia and the angular acceleration of the tire/wheel/rotor assembly, respectively.

During braking, $\tau_A$ can be considered as zero. We obtain,

$$f_B = \frac{R}{d} F_B + \frac{I_A}{d} \, \alpha_W \qquad (7.37)$$

It turns out that $f_B$ and $F_B$ are not independent, related by Equation (7.37). As a result, both Professor Newton and Dr. Energy are right. However, eventually, Professor Newton is right because the energy equation is derived from *Newton's Second Law*. It cannot be that Professor Newton is wrong while Dr. Energy is right.

Because $f_B$ depends on $F_B$, if $F_B$ is small, $f_B$ will be small. When a car is braking on ice, the friction coefficient is very low, resulting in a low $F_B$, i.e., a low $f_B$. The braking becomes ineffective based on *Newton's Second Law* and the energy equation.

During very hard braking, the Anti-Lock Braking System (ABS) cannot prevent the tire/wheel from being locked up to achieve $dr_r \neq 0$. As a result, $f_B$ no longer consumes any energy; however, because

$dr_1$ is no longer zero due to the tire skidding over the road surface, $F_B$ will now consume energy. During $t_2 < t < t_3$, the energy equation is

$$T_2 - \int \vec{F}_B \cdot d\vec{r}_1 = T_3 \qquad (7.38)$$

From Equations (7.34) and (7.38), Professor Newton and Dr. Energy both agree that $F_B$ slows down the car when the tire/wheel is locked up. This answers the third question.

To answer the second question, we first consider a typical calculation to determine the car speed before braking. A police will measure the length of the tire mark on the ground, $L_S$. The braking force is then determined as $F_B = \mu_d M g$, where $\mu_d$ is the sliding coefficient between the tire and the road.

The vehicle speed is then estimated as

$$v^* = \sqrt{2\,\mu_d\,g\,L_s} \qquad (7.39)$$

The above formula is widely used. It can derived by using Equation (3.31) with $a = -F_B/M = \mu_d g$. We can also use the energy equation presented in Equation (7.38).

However, is $v^*$ thus calculated accurate to match $v_0$? Let's construct the energy equation between $t_1$ and $t_3$.

$$\frac{1}{2}mv_0^2 - \int \vec{f}_B \cdot d\vec{r}_{1-2} - \int \vec{F}_B \cdot d\vec{r}_{2-3} - \int \vec{F}_{\text{imp}} \cdot d\vec{r}_3 = T_2 = 0$$

$$\Longrightarrow \frac{1}{2}mv_0^2 = \int \vec{f}_B \cdot d\vec{r}_{1-2} + \frac{1}{2}m(v^*)^2 + \int \vec{F}_{\text{imp}} \cdot d\vec{r}_3 \qquad (7.40)$$

From the above equation, it is obvious that $v_0$ could be substantially higher than the standard result of $v^*$. The standard estimate, $v^*$, does not include the speed reduction before the tires are locked up and the energy lost due to the impact. This example shows how important it is to consider the energy equation in its entirety with both external and internal forces.

## 7.4 Energy Consideration for the Impact Problem

Let's revisit the 1D, two-particle impact problem discussed in Section 6.2.1 and consider the energy aspect of the impact. With respect to Figure 6.5, we define Coefficient of Restitution (COR) $e$ as the ratio between the incoming and exiting speeds with respect to the mass center "wall" (Section 6.2.2). Based on the same inertial reference frame attached to the mass center of the two-particle system, the energy equation, corresponding to the times immediately before $(t_1)$ and after $(t_2)$ the impact are as follows:

$$\frac{1}{2}m_A(v_{A/G})_1^2 + \frac{1}{2}m_B(v_{B/G})_1^2 - \int \sum \vec{F}_{\text{ext}} \cdot d\vec{r}_e - \int \sum \vec{f}_{\text{imp}} \cdot d\vec{r}_i$$

$$= \frac{1}{2}m_A(v_{A/G})_2^2 + \frac{1}{2}m_B(v_{B/G})_2^2 \qquad (7.41)$$

where $\int \sum \vec{F}_{\text{ext}} \cdot d\vec{r}_e$ is the negative work done by external forces of the two-particle system, while $\int \sum \vec{f}_{\text{imp}} \cdot d\vec{r}_i$ is the negative work done by the impact force, which is an internal force of the two-particle system.

When formulating the linear momentum equation of the impact problem, we assured that the pseudo conservation of linear momentum is valid because $t_2 - t_1 \approx 0$ and no external forces are HUGE. However, we cannot automatically consider that the external forces do not do work in Equation (7.41) because we are not dealing with time. Let's reformulate the work done by the external forces as

$$\int \sum \vec{F}_{\text{ext}} \cdot d\vec{r}_e = \int_{t_1}^{t_2} \sum \vec{F}_{ext} \cdot \vec{v}_e \, dt = 0 \qquad (7.42)$$

where $v_e$ represents the speeds of $m_A$ and $m_B$ during impact. The work done by the external forces are essentially zero because both the external forces and speeds are not HUGE, their dot products are not HUGE either.

However, the internal force involved here is the impact force, which is considered as HUGE; therefore, the work done by the internal force, or the energy loss due to impact, is not zero, in general. Equation (7.41) becomes

$$\frac{1}{2}m_A(v_{A/G})_1^2 + \frac{1}{2}m_B(v_{B/G})_1^2 - \int \sum \vec{f}_{\text{imp}} \cdot d\vec{r}_i$$

$$= \frac{1}{2}m_A(v_{A/G})_2^2 + \frac{1}{2}m_B(v_{B/G})_2^2 \qquad (7.43)$$

Introducing Equation (6.46) into (7.43), we obtain

$$\frac{1}{2}m_A(v_{A/G})_1^2 + \frac{1}{2}m_B(v_{B/G})_1^2 - \int \sum \vec{f}_{\text{imp}} \cdot d\vec{r}_i$$

$$= e^2 \cdot \frac{1}{2}m_A(v_{A/G})_1^2 + e^2 \cdot \frac{1}{2}m_B(v_{B/G})_1^2$$

$$\implies (1 - e^2)\left[\frac{1}{2}m_A(v_{A/G})_1^2 + \frac{1}{2}m_B(v_{B/G})_1^2\right] = \int \sum \vec{f}_{\text{imp}} \cdot d\vec{r}_i$$

$$(7.44)$$

From the above equation, the total energy loss during impact can be determined by the COR. For elastic impact, $e = 1$, the energy loss is zero, while for plastic impact, $e = 0$, the entire kinetic energy is lost.

We must point out that the above statement is based on the mass center inertial reference frame. With different inertial reference frames, the above statement is not true. The entire kinetic energy may not be lost in a plastic impact if we do not use an inertial reference frame with respect to the mass center of the two-particle system.

We can apply the energy equation to each particle separately to obtain

$$\frac{1}{2}m_A(v_{A/G})_1^2 - \int \sum \vec{f}_{\text{imp}} \cdot d\vec{r}_A = \frac{1}{2}m_A(v_{A/G})_2^2 \qquad (7.45)$$

$$\frac{1}{2}m_B(v_{B/G})_1^2 - \int \sum \vec{f}_{\text{imp}} \cdot d\vec{r}_B = \frac{1}{2}m_B(v_{B/G})_2^2 \qquad (7.46)$$

Similarly, we can introduce Equation (6.46) into Equations (7.45) and (7.46) to obtain

$$(1 - e^2)\frac{1}{2}m_A(v_{A/G})_1^2 = \int \sum \vec{f}_{\text{imp}} \cdot d\vec{r}_A \qquad (7.47)$$

$$(1 - e^2)\frac{1}{2}m_B(v_{B/G})_1^2 = \int \sum \vec{f}_{\text{imp}} \cdot d\vec{r}_B \qquad (7.48)$$

When viewing $m_A$ and $m_B$ separately, the fraction of the energy lost to the initial energy is the same, as $1 - e^2$. From the above equations, we can conclude that the remaining kinetic energy is $e^2 T_1$, with $T_1$ as the initial kinetic energy.

What is the amount of energy lost with respect to different inertial reference frames? Are they the same or different? To answer this question, we will introduce Equations (6.39) and (6.40) into Equation (7.43), which yields

$$\frac{1}{2}m_A(v_{A/G})_1^2 + \frac{1}{2}m_B(v_{B/G})_1^2 - \int \sum \vec{f}_{\text{imp}} \cdot d\vec{r}_i$$

$$= \frac{1}{2}m_A(v_{A/G})_2^2 + \frac{1}{2}m_B(v_{B/G})_2^2$$

$$\implies \frac{1}{2}m_A[(v_{A/O})_1 - v_{G/O}]^2 + \frac{1}{2}m_B[(v_{B/O})_1 - v_{G/O}]^2$$

$$- \frac{1}{2}m_A[(v_{A/O})_2 - v_{G/O}]^2 + \frac{1}{2}m_B[(v_{A/O})_2 - v_{G/O}]^2$$

$$= \int \sum \vec{f}_{\text{imp}} \cdot d\vec{r}_i$$

$$\implies \frac{1}{2}m_A(v_{A/O})_1^2 + \frac{1}{2}m_B(v_{B/O})_1^2 - \frac{1}{2}m_A(v_{A/O})_2^2 - \frac{1}{2}m_B(v_{B/O})_2^2$$

$$- [m_A(v_{A/O})_1 + m_B(v_{B/O})_1]v_{G/O} + [m_A(v_{A/O})_2$$

$$+ m_B(v_{B/O})_2]v_{G/O}$$

$$= \int \sum \vec{f}_{\text{imp}} \cdot d\vec{r}_i \tag{7.49}$$

Due to conservation of linear momentum, Equation (7.49) becomes

$$\frac{1}{2}m_A(v_{A/O})_1^2 + \frac{1}{2}m_B(v_{B/O})_1^2 - \frac{1}{2}m_A(v_{A/O})_2^2 - \frac{1}{2}m_B(v_{B/O})_2^2$$

$$= \int \sum \vec{f}_{\text{imp}} \cdot d\vec{r}_i \tag{7.50}$$

Equation (7.50) indicates that the amount of energy loss will be the same for all inertial reference frames. However, the ratio of the energy loss to the initial kinetic energy will be different because the initial kinetic energy will be different with respect to different inertial reference frames. We again see the advantage of using the mass center inertial reference frame because $e^2 T_1$ is the remaining kinetic energy after impact, while $(1 - e^2)T_1$ is the amount of energy loss.

## 7.5   Concluding Remarks

In this chapter, we introduced the energy equation, derived from *Newton's Second Law*, for multi-body systems. We also introduced the potential energy for conservative forces, which are functions of positions. It is highly important that we also consider internal forces, unseen by an FBD, if they do work, in the energy equation. The energy equation is a scalar equation; therefore, we can only have one unknown if we use the energy equation alone to solve problems. Typically, if there are non-conservative forces, external or internal, which are doing work, the energy equation will contain multiple unknowns, making it unsolvable. In some special cases, if there are no non-conservative forces doing work, we then have conservation of mechanical energies. The Bernoulli equation is a well-known example of conservation of mechanical energies. Conservation of mechanical energy is also valid for elastic impact, while the entire kinetic energy is lost in a plastic impact if viewed based on the mass center inertial reference frame for a 1D, two-particle impact case. We again see the advantage of using the mass center inertial reference frame for the 1D, two-particle impact analysis, because $e^2 T_1$ is the remaining kinetic energy after impact, while $(1 - e^2) T_1$ is the amount of energy loss. The amount of energy loss is the same for all inertial reference frames.

Chapter 8

# Interesting Real-World Dynamic Problems

In the first seven chapters of this book, we have presented detailed motion analysis and fundamental laws of dynamics in various forms. We also demonstrated how to apply the motion analysis and these fundamental laws to solve many problems. In this chapter, we look at some interesting real-world problems. We should construct appropriate governing equations based on *Newton's Second Law*, the momentum equation, or the energy equation, or their combinations to solve problems. We do not know before hand which governing equations to use. We must apply these dynamics laws correctly and rigorously. We simply review the problem and decide the best course of action. Many of these seemingly puzzling problems become easy to understand if we simply apply the laws of dynamics properly.

## 8.1 Bullet/Block Experiments

The first problem we will consider is a famous experiment presented on youtube.[1]

---

[1]https://www.youtube.com/watch?v=vWVZ6APXM4w.

## Example 8.1: Which Block will Rise Higher?

**Problem Statement:** In this experiment, two identical blocks are placed on a fixture. A bullet is then fired into the block, as shown in Figure 8.1. In case A, the bullet is fired at the mass center of the block, while in case B, a bullet is fired off center. The blocks will rise. In case A, the block rises without spinning and reaches a maximum height, $h_A$. For case B, the block rises and spins at the same time, reaching a maximum height of $h_B$. We assume that the mass of the block, $M$, is much larger than that of the bullet. As a result, the moment of inertia of the block with the bullet stays the same as $I_G$.

In which case will the block rise higher? It is important that we do not rely on instinct to guess an answer or apply superficially dynamic laws to reach a conclusion. As shown in the above referenced youtube video, most of the guesses were wrong or citing the dynamic laws incorrectly.

**Force Analysis:** The corresponding FBDs for both cases are shown in Figure 8.2. We consider the bullet and the block as one system; therefore, the impact force when the bullet strikes the block is an

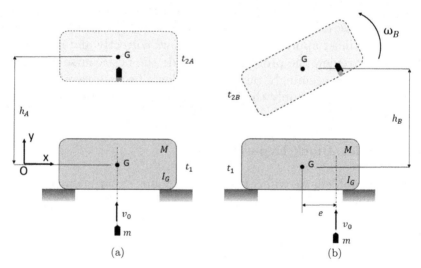

Figure 8.1:   Two blocks at rest are hit by an identical bullet: (a) the bullet hits the center of the block; and (b) the bullet hits the block at an off-center location.

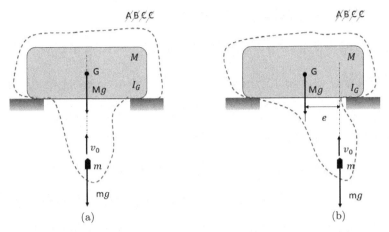

Figure 8.2: FBDs for both the bullet and the block as a system: (a) center hit case; and (b) off-center hit case.

internal force, not shown in the FBDs. The only external forces are related to the gravity.

**Motion Analysis:** The motion of the blocks after being hit by the bullet is a 2D rigid body motion. In the center-hit case, there is no rotational motion of the block when it rises. We define the time line as: at $t_1$, the bullet hits the block. At $t_{2A}$, the center-hit block reaches its maximum height, while at $t_{2B}$, the off-center-hit block reaches its maximum height. We do not know at this moment if $t_{2A} = t_{2B}$. In addition, $t_1^-$ is the instant right before the bullet hits, while $t_1^+$ is right after.

**Governing Equations:** We first consider the energy equation for the duration between $t_1^-$ and $t_1^+$. It is preferred to start with the energy equation because it is a scalar equation and is easier to construct.

The only external force is related to gravitation. The internal force to be considered is the impact force between the bullet and the block. This impact force is non-conservative and it dissipates energy due to the nature of plastic impact with the bullet embedded inside the block after the hit. We will ignore air resistance. At $t_1^+$, the mass center of block gains a velocity as $v_{GA}$ for the center-hit case, and $v_{GB}$ for the off-center-hit case. The rotational speed of the off-center-hit block at $t_1^+$ is $\omega_B$.

The corresponding energy equation for the center-hit case is

$$\frac{1}{2}mv_0^2 + U_{g1-} + \int \vec{f}_{impA} \cdot d\vec{r}_A = \frac{1}{2}(M+m)v_{GA}^2 + U_{g1+} \quad (8.1)$$

We know that at $t_1^+$ there is no vertical displacement of the mass center because the time is too short and the velocity gained during the short impact time is finite, NOT HUGE. Therefore, $U_{g1+} = U_{g1-}$. The above equation can be simplified to

$$\frac{1}{2}mv_0^2 - \int f_{impA}dr_A = \frac{1}{2}(M+m)v_{GA}^2 \quad (8.2)$$

Similarly, the energy equation for the off-center-hit case is

$$\frac{1}{2}mv_0^2 - \int f_{impB}dr_B = \frac{1}{2}(M+m)v_{GB}^2 + \frac{1}{2}I_G\omega_B^2 \quad (8.3)$$

where $\int f_{impA}dr_A$ and $\int f_{impB}dr_B$ are work done by the impact forces and they cause energy losses, thus a negative sign in the energy equations.

Here, we assume that the mass center of the off-center-hit block is at the same location and the moment of inertia is the same because $M \gg m$.

For Equation (8.2), there are two unknowns, $v_{GA}$ and $\int f_{impA}dr_A$. We cannot solve for them. For Equation (8.3), we have three unknowns, $v_{GA}$, $\omega_B$, and $\int f_{impB}dr_B$. We cannot solve for them either.

We need to consider additional equations. We will not consider *Newton's Second Law* because the impact force is unknown. We will first consider the linear momentum equation. For the center-hit case, the corresponding linear momentum equation is

$$\vec{P}_{A1-} + \int_{t_1^-}^{t_1^+} (M\vec{g} + m\vec{g})dt = \vec{P}_{A1+} \quad (8.4)$$

Because of the short impact time and $(M+m)g$ not being HUGE, we have a pseudo-conservation of linear momentum. Equation (8.4) can be simplified to

$$\vec{P}_{A1-} = \vec{P}_{A1+}$$

$$\implies mv_0 = (M+m)v_{GA} \quad (8.5)$$

For the off-center case, the linear momentum equation is the same,

$$mv_0 = (M + m)v_{GB} \tag{8.6}$$

From Equations (8.5) and (8.6), we have

$$v_{GA} = v_{GB} = v_G = \frac{m}{M + m}v_0 \tag{8.7}$$

Let's consider the angular momentum equation with respect to the mass center, $G$, for the system of the bullet and the block, for $t_1^-$ and $t_1^+$.

For the center-hit case,

$$\vec{H}_{A1-} + \int_{t_1^-}^{t_1^+} 0 \, dt = \vec{H}_{A1+}$$

$$\implies 0 = I_G \omega_A \tag{8.8}$$

We obtain that $\omega_A = 0$, confirming that the center hit block does not spin. For the second case,

$$\vec{H}_{B1-} + \int_{t_1^-}^{t_1^+} -mge \, dt = \vec{H}_{B1+}$$

$$\implies mv_0 e = I_G \omega_B \tag{8.9}$$

where $\int_{t_1^-}^{t_1^+} -mge \, dt = 0$ because $mge$ is NOT HUGE.

We obtain

$$\omega_1 = \omega_B = \frac{mv_0 e}{I_G} \tag{8.10}$$

To answer the question of which block will rise higher, we turn our attention to $t_1^+$ and $t_2$. The FBD during this time interval is basically the same, but there is no internal force to consider for the energy equation. We have a case of conservation of mechanical energy. The corresponding energy equation for the center-hit case is

$$\frac{1}{2}(M + m)v_G^2 + U_{gA1+} = U_{gA2}$$

$$\implies \frac{1}{2}(M + m)v_G^2 = (M + m)gh_A \tag{8.11}$$

We obtain

$$h_A = \frac{1}{2}\frac{v_G^2}{g} = \frac{1}{2g}\left(\frac{mv_0}{M+m}\right)^2 \tag{8.12}$$

For the off-center case, the energy equation is

$$\frac{1}{2}(M+m)v_G^2 + \frac{1}{2}I_G\omega_1^2 + U_{gB1+} = U_{gB2} + \frac{1}{2}I_G\omega_2^2$$

$$\implies \frac{1}{2}(M+m)v_G^2 + \frac{1}{2}I_G\omega_B^2 = (M+m)gh_B + \frac{1}{2}I_G\omega_{B2}^2 \tag{8.13}$$

where $\omega_{B2}$ is the angular velocity of the block at $t_{B2}$.

We cannot solve the above equation because we do not know $\omega_{B2}$. Let's apply the angular momentum equation for the same time interval with respect to the mass center. The only force is the gravitational force which passes through the mass center. As a result, there is no moment with respect to the mass center. We have a conservation of angular momentum for this time interval,

$$\vec{H}_{B1+} + \int_{t_1^+}^{t_2} 0\,dt = \vec{H}_{B2}$$

$$\implies I_G\omega_B = I_G\omega_{B2} \tag{8.14}$$

We obtain $\omega_{B2} = \omega_B$. Introducing this result to Equation (8.13), we obtain

$$h_B = \frac{1}{2}\frac{v_G^2}{g} = \frac{1}{2g}\left(\frac{mv_0}{M+m}\right)^2 \tag{8.15}$$

As a result, both blocks will rise to the same height, $h_A = h_B$. This result is confirmed from the high speed photography footage of the cited video, showing indeed both blocks rising to the same height.

**Verification and Extensions:** We should not stop here after answering the question of which block would rise higher.

Let's verify a few more things. Specifically, what are the energy losses for both cases? Are they the same? How about the impact forces?

Comparing Equations (8.2) and (8.3), and recognizing $v_{GA} = v_{GB}$, we obtain

$$\int f_{\text{imp}A} dr_A - \int f_{\text{imp}B} dr_B = \frac{1}{2} I_G \omega_B^2 \qquad (8.16)$$

This result indicates that the energy loss in the center-hit case is greater and the difference is the rotational kinetic energy of the off-center-hit block.

Furthermore, we can estimate the average impact force based on Equation (8.16),

$$\int f_{\text{imp}A} dr_A - \int f_{\text{imp}B} dr_B = \frac{1}{2} I_G \omega_B^2$$

$$\Longrightarrow \bar{f}_{\text{imp}A} r_A - \bar{f}_{\text{imp}B} r_B = \frac{1}{2} I_G \omega_B^2 \qquad (8.17)$$

where $r_A$ is the traveling distance of the bullet inside the block for the center-hit case, and $r_B$ for the off-center case. The average impact forces are $\bar{f}_{\text{imp}A}$ and $\bar{f}_{\text{imp}B}$.

The youtube video referenced above did an X-ray imaging and found that $r_A \approx r_B$. As a result, $\bar{f}_{\text{imp}A}$ must be larger than $\bar{f}_{\text{imp}B}$,

$$\bar{f}_{\text{imp}A} > \bar{f}_{\text{imp}B} \qquad (8.18)$$

In other words, the impact force in the off-center case is lower. This result is important and will be utilized in the next example.

In this example, we demonstrated how the governing equations of dynamics can be combined to solve a seemingly strange result. Most people would guess that the center-hit block will rise higher by incorrectly assuming conservation of energy of the entire process. We must avoid similar mistakes. Correct answers can be reached if we conduct the force and motion analyses thoroughly to reach correct governing equations.

## 8.2 Luckiest Outcome of a Potentially Deadly Car Accident

In this section, we consider the collision between a car and a person. We will develop a model to explain an actual accident which ended with the luckiest outcome. Is it pure luck?

## Example 8.2: Collision between a Car and a Person

**Problem Statement:** We will describe a car accident which actually happened from author's first hand knowledge. A car was moving at a low speed, about 25–30 mph, just under the speed limit. Suddenly, a person stepped in front of the car behind a parked car. The driver immediately hit the brake, but it was too late. The car hit the person straight on. The person tumbled over the engine hood and one of his feet slammed at the windshield and cracked it. The car came to a complete stop soon after and the person rolled down the engine hood to the ground. He then laid on the ground motionless. The impact was loud, hard, and scary. The driver got off the car and looked at the person on the ground, fearing that he had killed the person. People gathered around the car and the person. The police and ambulance were called.

The ambulance came and an emergency crew was whispering to the person on the ground. The person was then carefully moved to a stretcher into the ambulance. The ambulance sped away with its loud and screeching siren. The police talked to a few witnesses first before approaching the driver. The police said that there was no way the driver could avoid the impact and he was not at fault; i.e., no ticket. "You were just at the wrong place at the wrong time," said the police.

The driver was a bit traumatized by the experience and drove away very cautiously. Later in the afternoon, the driver contacted the hospital to check on the condition of the person but was told that the person had already been discharged.

Later in the evening, the driver called the person. The latter apologized to the driver for causing the shock and confirmed he was all Ok, except a bit sore in the leg, but with no broken bones. The person was over 70 years old at the time.

Isn't this the luckiest outcome of a potential deadly car accident? Is it pure luck? Can we explain this outcome with dynamics laws?

## Solution:

Let's compare the collisions of case (a) and case (b) in Figure 8.3, and relate them to Example 8.1. Let's adjust the vehicle speed with respect to its mass, so that the linear momentum of the vehicles are identical, as the bullet in Example 8.1. The person is hit by a sedan (case (a)) at the lower legs, while at the chest by a truck (case (b)).

(a)          (b)

Figure 8.3: A pedestrian hit by: (a) a sedan; and (b) a truck.

*Modified images from sources*: (1) https://libreshot.com/wp-content/uploads/ 2017/05/walking-with-mobile-phone.jpg; (2) https://pngimg.com/uploads/chevr olet/%D1%81hevrolet_PNG33.png; and (3) https://upload.wikimedia.org/wiki pedia/commons/3/33/Green_pickup_truck.png.

Case (b) is similar to the bullet/block example when the bullet hits at the center, while case (a) is similar to the off-center-hit case. From the discussion of Example 8.1, the impact force and the energy loss for the off-center-hit case are lower. As a result, we expect similar conditions for the sedan collision case. The induced rotation allows the person to roll over the engine hood. On the other hand, if the same person is hit by a truck or an SUV, it would hit him or her squarely in the chest. The person does not rotate as in the center-hit block case. The higher impact force and higher energy loss could lead to a more tragic ending. This is the reason why there is an engine hood height regulation in Europe. In conclusion, it is the luckiest outcome for this unfortunate accident of collision between the person and the car discussed above. However, because the car was a sedan with a low engine hood height and the person was tall, it led to this most fortunate outcome. It is not pure luck.

Let's consider extending the impact problem in reverse in the next example.

## 8.3 Toilet Paper Tearing

### Example 8.3: Tearing a Piece of Toilet Paper Off

**Problem Statement:** We all have the experience that when we pull the toilet paper slowly, the roll is simply unrolled and the toilet paper is not torn. However, if we pull the paper rapidly, we can snap a piece off. How do we explain this phenomenon with the dynamics laws?

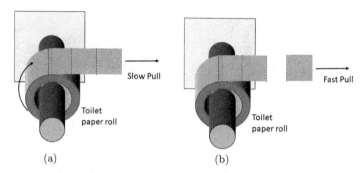

(a)             (b)

Figure 8.4: Two ways of pulling a toilet paper roll: (a) with a slow pull, the toilet roll is simply unrolled; and (b) with a fast pull, a piece of toilet paper is torn and the roll is not rotating.

**Force Analysis:** We will ignore the gravity in our analysis. For the slow pulling case, we construct two FBDs. The first FBD isolates a piece of toilet paper, which is attached to the roll. There are two forces involved: the slow pulling force, $P_{\text{slow}}$, which is small, and the constraint force, $F_2$, at the seam.

The FBD for the paper roll involves $F_2$, $F_3$, which is force at the roll shaft, and $\tau_z$, which is the friction torque of the roll over the supporting shaft.

**Motion Analysis:** Due to slow pulling, we can assume that the acceleration of the toilet paper is zero. The moment of inertia of the whole paper roll is $I_z$ and its angular acceleration is $\alpha$.

**Governing Equations:** We can construct the following two governing equations for the slow pulling case.

$$\sum F_x = ma_x \Longrightarrow P_{\text{slow}} - F_2 = 0 \qquad (8.19)$$

$$\sum \tau_z = I_Z \alpha \Longrightarrow \tau_z - F_2 r = I_z \alpha \qquad (8.20)$$

where $r$ is the radius of the roll. From the above equations, we have

$$P_{\text{slow}} = F_2 = \frac{1}{r}(\tau_z - I_z \alpha) \qquad (8.21)$$

From the above equation, we know that if $\tau_z$ is small, and with slow pulling, $\alpha$ is essentially zero, then $P_{\text{slow}} \approx \frac{\tau_z}{r}$, and it can be kept to be lower than $F_{\text{seam}}$, which is the force needed to tear the toilet paper. As a result, the toilet paper is simply unrolled without tearing.

If one holds the roll to keep it from rolling with one hand, $\tau_z$ can be made to be high enough to render $F_2 > F_{\text{seam}}$, then a piece of toilet paper can be torn off by the other hand.

However, if the person simply pulls the toilet paper by one hand rapidly, a piece of toilet paper can be torn off as shown in Figure 8.4(b). The corresponding FBDs are shown in Figure 8.5(b). The fast pulling force, $P_{\text{fast}}$, versus time is also plotted in Figure 8.5(b). In a snap, the fast pulling force lasts from $t_0$ to $t_2$, similar to a pulse. The corresponding momentum equations are

$$m\vec{v}_0 + \int_{t_0}^{t_2} (\vec{P}_{\text{fast}} - \vec{F}_2)dt = m\vec{v}_2 \qquad (8.22)$$

$$I_z\omega_0 + \int_{t_0}^{t_1} (F_2 r - \tau_z)dt = I_z\omega_1 \qquad (8.23)$$

where $m$ is the mass of the torn toilet paper piece. Note that during the time from $t_0$ to $t_1$, the toilet paper is still attached as shown in Figure 8.5(a). As $P_{\text{fast}}$ starts to increase, $F_2$ will also increase. However, $F_2$ cannot exceed $F_{\text{seam}}$. When $F_2 = F_{\text{seam}}$ at $t_1$, a piece of toilet paper is torn off and $F_2$ will drop to zero, as shown in Figure 8.5(b). Therefore, the angular momentum equation of Equation (8.23) only

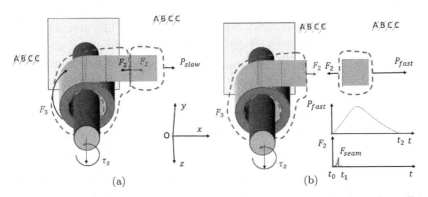

Figure 8.5:   FBDs for toilet paper pulling: (a) with a slow pull, the toilet roll is simply unrolled; and (b) with a fast pull, a piece of toilet paper is torn and the roll is not rotating.

needs to be integrated up to $t_1$. Because $F_{\text{seam}}$ and $\tau_z$ are still rather small and $t_1$ is very short, Equation (8.23) can be further simplified as a pseudo-conservation of angular momentum equation,

$$I_z \omega_0 = 0 = I_z \omega_1 \qquad (8.24)$$

Therefore, the toilet paper roll will stay in place without unrolling. From Equation (8.22), without solving it, we know $v_2$ is larger than zero as it is torn from the roll.

This example is also related to how a mechanic can remove lug bolts of a tire using an impact wrench, which is our next example.

## 8.4   Impact Wrenches

In Chapter 6, we presented the impact problems and discussed the nature of impact forces as impulse-like. In many applications, such impulse-like impact forces are useful. Common examples include hammering a nail, cutting down a tree with an axe, hitting a powerful tennis forehand, and slugging a home run.

The key in the above examples is not about the force exerted by the hand during impact but the speed, or, more precisely, the linear momentum of the hammer, the axe, the racket, the bat, etc. at the instant of impact.

One very useful power tool is the impact wrench. In Figure 8.6, the major components of an air-powered impact wrench are shown. The key components identified are wrench head (57), anvil (55), anvil dog (56), hammer dog (54), and camming ball (53). The wrench head, anvil and anvil dog are one rigid piece. The camming ball and hammer dog are coupled. The hammer dog is rotated by a power source, such as an air propeller or an electric motor. The camming ball will follow a camming arrangement when the hammer dog rotates, as shown in Figure 8.7(a). The hammer dog is free to rotate to gain a high speed. When the peak of the camming arrangement pushes the ball upward, the hammer dog also moves upward to hit the anvil dog, as shown in Figure 8.7(b). This causes an impact force, or an impact torque, to drive the wrench head. Immediately after the impact, the camming ball is pushed down by the spring. The hammer dog is free to rotate again. For each revolution of the hammer dog, one impact happens. An impact wrench will drive a bolt or a nail by a series of

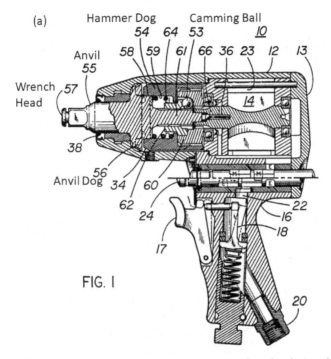

(a)

Hammer Dog    Camming Ball
54   64   53          _10_

Anvil    58 \ 59 / 6/ / 66  36  23   /2    /3
55

Wrench ,57
Head

_/4_

38

56
Anvil Dog  34 / 60

62
24

22
/6
/8

/7

FIG. I

20

Figure 8.6:   Impact wrench design and working principles: the design diagram of an air-powered impact wrench from US Patent 5083619A.
*Modified   images   from   sources*:   https://patents.google.com/patent/US5083 619A/en.

impacts, not by a steady but a lower torque. Because the torque is generated by the impact, not from the operator, the user does not need to apply a large torque, only large enough to hold the wrench in place for the motor to rotate with respect to the wrench. Let's examine how a technician can remove a tire in a tire shop compared with a driver removing a tire on the road side using a hand tool.

## Example 8.4: Why a Technician Can Remove a Tire with the Tire in the Air but a Driver Using a Hand Tool must Release the Bolts with the Tire on the Ground?

**Problem Statement:** As shown in Figure 8.8(a), if we want to remove a tire on the road side with a hand tool, the tire needs to be

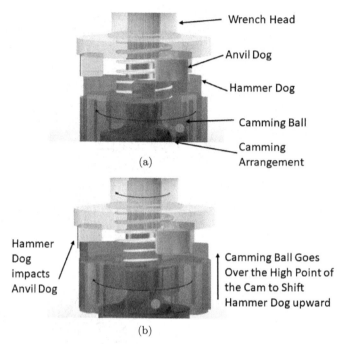

Figure 8.7:   Impact wrench design and working principles: (a) the impact wrench before impact occurs when the lower hammer dog is rotating; (b) a camming ball goes over a cam, which pushes the hammer dog upward to cause the hammer dog to impact the anvil dog. After impact, the spring pushes the hammer dog down (images from screen shots from a youtube video by Sioux Tools).

*Modified Images from Sources*: (a) and (b) https://www.youtube.com/watch? v=x-mpC8u6Cks.

on the ground. If not, as we apply a force to generate a torque, the tire will rotate, not being able to loosen the lug bolt. After loosening all the lug bolts with the tire on the ground, the tire is then lifted in the air to be removed. However, as shown in Figure 8.8(b), a tire shop technician can remove lug bolts with a tire in the air using an impact wrench. The tire will not rotate. Why? In fact, the answer is related to the toilet paper tearing example.

**Force Analysis:** The corresponding FBDs are shown in Figure 8.8. For the hand tool case, the tire needs to be in touch with the ground so that a friction force is needed to generate a counter moment to keep the tire from rotating. In this way, a brief but high torque can be produced on the bolt to loosen it. However, for the impact wrench

(a)

(b)

Figure 8.8: (a) Removing a tire with a hand tool requires that the tire be on the ground; and (b) a tire can be removed while in the air on a lift with an impact wrench.
*Modified Images from Sources*: (a) https://crawfordsautoservice.com/wp-content/uploads/2020/04/changing-a-tire-768x512.jpg; (b) and (c) http://www.automoblog.net/wp-content/uploads/2014/06/Roberto_LAM-600x400.jpg.

case, the FBD shown in Figure 8.8(b) does not explain how the bolt can be loosened. We need additional FBDs.

As shown in Figure 8.9, three FBDs are constructed. We only focus on the torques for these FBDs. The impact wrench generates a series of torque pulses, $\tau_1$, as shown. Each pulse lasts for a very short period of time. These torque pulses are applied to a lug bolt and they are resisted by the frictional torque, $\tau_2$, on the bolt threads. The applied torque, $\tau_1$, needs to be higher than the frictional torque, $\tau_2$, in order to loosen the bolt. The friction torque is due to static friction and it rises up to a peak value and drops to the lower sliding friction when the bolt starts to turn.

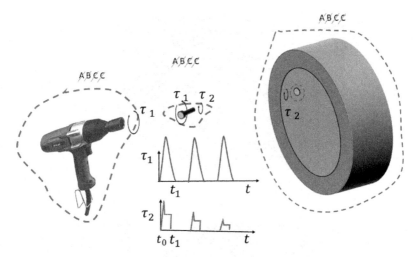

Figure 8.9: Free-body diagrams for the impact wrench, the lug bolt, and the tire/wheel assembly/.

*Source: Impact wrench image sources*: (a) https://images-na.ssl-images-amazon.com/images/I/41a%2BBOQPg4L.jpg.

**Governing Equations:** The angular momentum equation for the bolt is

$$I_b\omega_0 + \int_{t_0}^{t_1} (\tau_1 - \tau_2)dt = I_b\omega_1 \qquad (8.25)$$

where $I_b$ is the moment of inertia of the bolt around its longitudinal axis and the impact force lasts from $t_0$ to $t_1$ with $t_1 - t_0 \approx 0$. The rotational speed, $\omega_0$, at $t_0$ can be considered as zero, while $\omega_1$ is the gained rotational speed of the bolt at $t_1$. The value of $\omega_1$ is not zero when $\int_{t_0}^{t_1}(\tau_1 - \tau_2)dt > 0$. Even though $t_1 - t_0$ is short, because the peak magnitude of $\tau_1$ is HUGE due to its impulse-like nature and $\tau_2$ is limited in size, $\int_{t_0}^{t_1}(\tau_1 - \tau_2)dt$ will be larger than zero.

Once the bolt turns, the frictional torque gets smaller because the bolt is less tight. The bolt becomes more loosened and spins out of the tire wheel. Why is the tire not rotating with the impact wrench? To answer it, we need to construct the angular momentum equation for the tire,

$$I_t\omega_{t0} + \int_{t_0}^{t_1} \tau_2 dt = I_t\omega_{t1} \qquad (8.26)$$

where $I_t$ is the tire moment of inertia, $\omega_{t0} = 0$ is the tire angular speed at $t_0$, and $\omega_{t1}$ is the tire angular speed at $t_1$. Because $\tau_2$ is the friction torque and is limited in size, we have $\int_{t_0}^{t_1} \tau_2 dt \approx 0$, which leads to $\omega_{t1} = 0$.

Despite being quite different in appearance, the tire removal and the toilet paper tearing problem actually share the same dynamics principles.

## 8.5 Single-wheel Vehicles

In this section, we will discuss the working principle of a single-wheel vehicle. There are many examples of this type of vehicles, such as Segway, hovering disk, unicycle, or a one-wheel motorcycle, as shown in Figure 8.10. In the case of the Segway and hovering scooter, they have two wheels, but they can be considered as having only one wheel along the moving direction.

We will discuss how a single-wheel vehicle works, in comparison with regular cars with front and rear wheels.

### Example 8.5: How does a Single-Wheel Vehicle Work?

**Problem Statement:** As shown in Figure 8.11, an idealized single-wheel vehicle is shown. The wheel is fitted with a motor and the wheel accelerates toward the right. The radius of the wheel is $r$.

**Force Analysis:** Based on the FBD for the entire single-wheel vehicle, there are only three forces, weight $Mg$, normal $N$, and driving force $F_f$.

**Motion Analysis:** The disk is a rigid body with rolling-without-slipping motion, while the motor housing is a second rigid body with a different rotation.

**Governing Equations and Analysis:** We can construct *Newton's Second Law* based on an inertial reference frame for the motion of the mass center and with respect to the mass center for the rotational motion.

$$F_f = M a_{Gx} \tag{8.27}$$

(a)

(c)

(b)

(d)

Figure 8.10: Single-wheel vehicle examples: (a) Segway; and (b) hovering scooter; (c) unicycle; (d) a patented single-wheel motorcycle.
*Sources*: (a) http://2.bp.blogspot.com/_29_shKT4Elw/TKEEbQYGhiI/AAAA
AAAAK6s/2xEsSah88Ho/s1600/foo.jpg; (b) https://i2.cdscdn.com/pdt2/1/3/
6/1/700x700/ele2009823639136/rw/elenxs-r-skateboard-electrique-6-5-350w-mi
ni-gyr.jpg; (c) https://www.goodfreephotos.com/albums/vector-images/unicyc
le-vector-clipart.png; (d) https://upload.wikimedia.org/wikipedia/en/f/f8/Uno_
motorcycle.jpg.

$$N - Mg = Ma_{Gy} = 0 \qquad (8.28)$$

$$F_f \, r\vec{k} = \sum \dot{\vec{H}}_G = I_2\alpha_2\vec{k} - I_1\alpha_1\vec{k} \qquad (8.29)$$

where $M = m_1 + m_2$ is the total mass of the vehicle, including both the wheel and the motor, $\alpha_1$ is the angular acceleration of the wheel and $a_{Gx}$ is the acceleration of the mass center along the $x$ axis. $I_2$ is the moment of inertia of the motor housing which can rotate relative to the wheel and $\alpha_2$ is the angular acceleration of the motor housing. The directions of the angular accelerations are defined in Figure 8.11 accordingly. The acceleration of the mass center along the $y$ axis is zero.

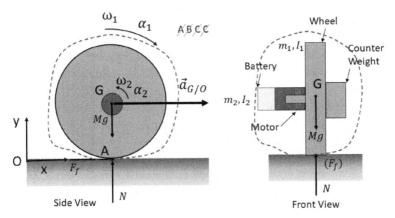

Figure 8.11: An idealized single wheel vehicle and the FBDs in both side and front views.

From Equations (8.27) and (8.29), we obtain

$$\alpha_2 = \frac{I_1 + Mr^2}{I_2}\alpha_1 \tag{8.30}$$

Equation (8.30) indicates that $\alpha_2 \gg \alpha_1$ because $I_1 + Mr^2 \gg I_2$. After some time from the initial rest condition, we also have

$$\omega_2 = \frac{I_1 + Mr^2}{I_2}\omega_1 \tag{8.31}$$

Let's consider the energy equation of the system of Figure 8.11. We have

$$T_1 + \Delta W_{\text{motor}} = T_2$$
$$= \frac{1}{2}Mv_G^2 + \frac{1}{2}I_1\omega_1^2 + \frac{1}{2}I_2\omega_2^2$$
$$= \frac{1}{2}[Mr^2 + I_1]\omega_1^2 + \frac{1}{2}I_2\omega_2^2 \tag{8.32}$$

where $v_G = r\omega_1$ due to the no-slip condition.

Introducing Equation (8.31) to (8.32), we have

$$\Delta W_{\text{motor}} = \frac{1}{2}[Mr^2 + I_1]\omega_1^2 + \frac{1}{2}\frac{(Mr^2 + I_1)^2}{I_2}\omega_1^2 \tag{8.33}$$

The second term is $\frac{(Mr^2+I_1)}{I_2}$ times that of the first term. In other words, most of the motor energy is used to turn the motor housing, instead of moving the disk forward. This single-wheel vehicle of Figure 8.11 may be able to move to the right, but its efficiency will be very poor.

This is similar to the situation of an open differential, discussed in Section 4.4.1 of Chapter 4, in which the energy of the engine simply spins the tire without traction.

We can also visualize this scenario by the case of using a cordless drill (not an impact wrench type) to drive a screw into a piece of wood. If you don't hold the cordless drill firmly, the drill will spin instead of driving the screw.

In order not to waste energy, we need to keep the motor housing from rotation. This is important because the rider will be on the motor housing and the rider obviously should not be spinning while riding.

We reconsider the single wheel vehicle problem as shown in Figure 8.12. We have two rigid bodies with different mass center locations.

In Figure 8.12, a foot stand is attached to the motor assembly and a vertical handle is attached for the passenger to have something to

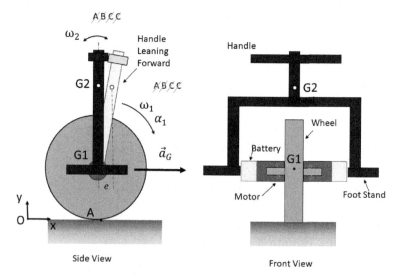

Figure 8.12: A practical single-wheel vehicle design with a handle which can lean forward or backward.

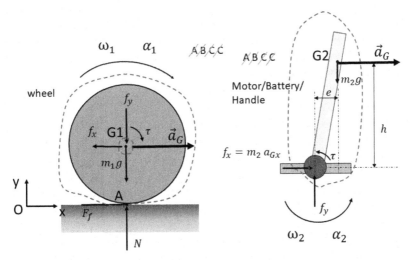

Figure 8.13: Two FBDs constructed for the wheel and the motor/battery/ handle, respectively.

hold on to. We also make the design symmetric with a motor on each side. The mass center of the wheel is at $G_1$, while the motor/batter/ stand/handle is at $G_2$. If a passenger steps over it, $G_2$ will be at a different location.

Apparently, we need $\omega_2 \approx 0$ and $\alpha_2 \approx 0$ for the passenger to ride on it. Let the mass center of the motor/batter/handle $G_2$ be stabilized at a distant $e$ horizontally from $G_1$. Because the vehicle system now contains two rigid bodies with different motions, it would be better to construct two separate FBDs for them, as shown in Figure 8.13.

Based on Figure 8.13 and applying *Newton's Second Law*, for the wheel we have

$$F_f - f_x = m_1 a_{Gx} \tag{8.34}$$

$$N - m_1 g - f_y = M a_{Gy} = 0 \tag{8.35}$$

$$F_f \, r - \tau = -I_{G1} \alpha_1 \tag{8.36}$$

For the motor/battery/handle, we have

$$f_x = m_2 a_{Gx} \tag{8.37}$$

$$f_y - m_2 g = M a_{Gy} = 0 \tag{8.38}$$

$$\tau + f_x h - f_y e = I_{G2} \alpha_2 = 0 \tag{8.39}$$

Note that the motor/battery/handle is stabilized in the orientation shown, accelerating to the right, i.e., $\alpha_2 = 0$ and $a_{G2x} = a_{Gx}$.

From the above equations, we found

$$\tau = [(m_1 + m_2)r^2 + I_{G1}]\alpha_1 \tag{8.40}$$

$$\alpha_1 = \frac{m_2 g e}{(m_1 + m_2)r^2 + I_{G1} + m_2 h r} \tag{8.41}$$

Equation (8.41) indicates that in order to have angular acceleration $\alpha_1$ to move the wheel to the right, $G_2$ needs to be at $e$ to the right of $G_1$ horizontally. If $e = 0$, $\alpha_1 = 0$. The torque needed to achieve this angular acceleration is determined by Equation (8.40).

To apply the brakes on the single-wheel vehicle, $G_2$ needs to be behind $G_1$.

If we examine the pictures of a Segway or a unicycle rider, as shown in Figure 8.14, we will notice the rider will be leans forward to accelerate to the right, while leaning backward to slow down, just as indicated by Equations (8.40) and (8.41). When cruising at a

(a)          (b)

Figure 8.14: A person riding a unicycle: (a) or a Segway; and (b) to accelerate forward with the body leaning forward as well.

*Sources*: (a) https://live.staticflickr.com/1093/1388397330_e5eb096a0f.jpg; and (b) https://www.movilidadhoy.com/wp-content/uploads/2018/04/segway-movili dadhoy_1-365x330.jpg.

A B C C

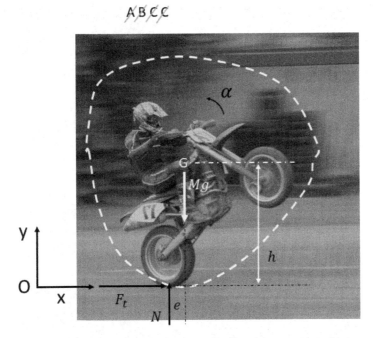

Figure 8.15: A person doing a dangerous wheelie with a motorcycle.
*Images modified from Source*: https://c1.staticflickr.com/1/72/155010932_
f82946340b_b.jpg.

constant speed, the body can stay upright. To ride a unicycle, one
has to practice so that the body can stabilize without falling forward
or backward. For other one-wheel vehicles, a control system is built in
to stabilize the handle and the rider, making it easy to ride without
much practice.

## Example 8.6: The Danger of Wheelies — Riding a Motorcycle on the Rear Tire Only.

Two-wheel vehicles, such as a motorcycle or a bicycle, have a lower
front end mass. As a result, with sufficient torque to achieve high
acceleration, the front tire of these vehicles can be raised into air
and a person can then ride them as a one-wheel vehicle. This is
called wheelie, which is very dangerous.

Figure 8.16:  (a)–(d) A terrible crash due to a wheelie.
*Images    modified    from    Sources*:    https://www.youtube.com/watch?v=Qp
02u0F3cO8.

As shown in Figure 8.15, the moment equation with respect to the mass center is

$$(-Ne + F_t h) = I\alpha \tag{8.42}$$

where $N$ is the normal force from the ground at the tire, $F_t$ is the tractive force, induced by the driving torque of the motorcycle, $e$ is the horizontal distance between the tire contact point to the mass center, $h$ is the height of the mass center from the ground, $I$ is the moment of inertia of the motorcycle/rider, and $\alpha$ is the angular acceleration of the motorcycle/rider. The moment of inertia of the rear tire is ignored.

The rider needs to control the driving torque, thus, the tractive $F_t$ so that $\alpha = 0$ to stabilize the motorcycle during the wheelie ride. However, any change in $F_t$ will result in changes in $e$ and $h$, making the ride inherently unstable and dangerous. When $F_t$ increases, $\alpha$ will increase counterclockwise to reduce $e$, which in turn increases $\alpha$ further, making it unstable.

As shown in Figure 8.16(a), a screen image captured from the referenced video, the rider had lost control. The motorcycle flipped so much, its seat slammed at the ground (Figure 8.16(b)). The rider likely let go of the accelerator causing $F_t$ to become zero, but the impact force $N_2$ flipped the motorcycle/rider in reversed direction, as shown in Figure 8.16(c). The front tire of the motorcycle then slammed to the ground (Figure 8.16(d)) and the rider was thrown to the ground, head first, leading likely to a major injury.

A wheelie should only be performed by stunt riders or properly trained persons, in a secured area, never on a common road or a major highway.

## 8.6    Concluding Remarks

In this chapter, we presented several interesting real-world examples for analysis. We utilize suitable governing equations for analysis. The key idea is that we should have the skill to construct and combine different equations, such as *Newton's Second Law*, the momentum equations, and the energy equation to solve the problem. We always need to construct proper FBDs for analysis. With patience and vigor, many seemingly puzzling problems can be explained clearly.

# Chapter 9

# Analysis of Vehicle Dynamics

In the previous chapters, we have presented many examples related to vehicles. For example, race track and car acceleration in Chapter 3, car suspensions and piston rods, planetary gear train systems, transmissions, differentials, and universal joints in Chapter 4, motorcycle acceleration and race car bank angle discussion in Chapter 5, car collision and impact problems in Chapter 6, energy view of car braking in Chapter 7, and passenger/car collision and one-wheel vehicle design in Chapter 8. In this chapter, we look into more vehicle dynamics issues related to traction limited acceleration, steering, roll-over, and four-wheel drive designs. This chapter and earlier discussions on vehicle dynamics would be an adequate preparation for readers to go deeper into more advanced topics related to vehicle dynamics.

## 9.1 Acceleration and Braking Performances

### 9.1.1 *Weight shifts during car acceleration or deceleration*

Similar to Example 5.7 for a motorcycle, we consider a car during acceleration or deceleration on a flat ground. The front end of a car typically is too heavy to be lifted off the ground entirely unlike the case for a motorcycle. In Figure 9.1(a), the car is in acceleration. The height of the car mass center from the ground is $h$, and it is at a distance $L_1$ from the front tire and $L_2$ from the rear tire.

459

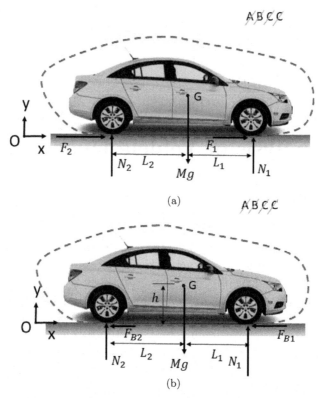

Figure 9.1: (a) A car during acceleration on a flat ground; and (b) a car during braking.

*Car Image Source*: https://pngimg.com/uploads/chevrolet/%D1%81lhevrolet_PNG33.png.

The corresponding FBDs for the acceleration and deceleration cases are shown in Figure 9.1. If not in acceleration or deceleration, the weight distributions on the front and rear tires are inversely proportional to the mass center distances, which could be easily derived from force equilibrium.

$$\frac{N_{10}}{N_{20}} = \frac{L_2}{L_1} \tag{9.1}$$

where $N_{10}$ and $N_{20}$ are dead weights on the front and rear tires, respectively. A typical sedan will have a weight distributed more toward the front due to the weight of the engine. Typical weight distribution for sedans is $N_{10}:N_{20} = 55:45$. For a sports car, it could be 50:50, and in some cases, it could be 35:65.

During acceleration, there is a weight shift to the rear tires, while the weight shift is reversed during braking. From *Newton's Second Law*, during acceleration, we have

$$F_1 + F_2 = Ma_x \qquad (9.2)$$

$$N_1 + N_2 - Mg = 0 \qquad (9.3)$$

$$\sum \tau_G = N_1 L_1 - N_2 L_2 + (F_1 + F_2)h = I_{Gz}\alpha = 0 \qquad (9.4)$$

where $F_1$ and $F_2$ are tractive forces on the front and rear tires, respectively, $N_1$ and $N_2$ are the normal forces acting on the front and rear tires, respectively, $M$ is the mass of the car, $I_{Gz}$ is the moment of inertia with respect to $G$ around the $z$-axis. The moments of inertia of tires and other rotating parts of the car are neglected and the entire car is treated as one rigid body.

We assume that the car is at a constant acceleration and reaches a steady state in terms of car body tilting, thus the angular acceleration, $\alpha = 0$.

For front-wheel-drive cars, $F_2 = 0$, for rear-wheel-drive cars, $F_1 = 0$, while for all-wheel-drive cars, $F_1 \neq 0$ and $F_2 \neq 0$. We will use rear-wheel-drive cars for illustration. There are four unknowns, $F_2$, $N_1$, $N_2$, and $a_x$ for a rear-wheel-drive car in acceleration on a flat ground, but there are only three governing equations. We can express $F_2$, $N_1$, and $N_2$ as functions of $a_x$. We obtain

$$N_1 = Mg\frac{L_2}{L} - M\frac{h}{L}a_x \qquad (9.5)$$

$$N_2 = Mg\frac{L_1}{L} + M\frac{h}{L}a_x \qquad (9.6)$$

$$F_2 = Ma_x \leq \mu_s N_2 \qquad (9.7)$$

where $L = L_1 + L_2$ is the wheel base and $\mu_s$ is the static friction coefficient between the tire and the ground. On a typical road, $\mu_s \approx 0.6$–$0.7$ in a dry condition and $\mu_s \approx 0.3$–$0.35$ in a wet condition. When $a_x = 0$, we obtain the same dead weight distribution as Equation (9.1).

During braking (Figure 9.1(b)), the weight shift is reversed by considering $a_x$ as negative. However, there are braking forces on both front and rear tires.

$$-F_{B1} - F_{B2} = -Ma_x \tag{9.8}$$

$$F_{B1} \leq \mu_s N_1 \tag{9.9}$$

$$F_{B2} \leq \mu_s N_2 \tag{9.10}$$

If the tires are locked up, they would be skidding over the road, the braking forces become

$$F_{B1} = \mu_d N_1 \tag{9.11}$$

$$F_{B2} = \mu_d N_2 \tag{9.12}$$

where $\mu_d$ is the sliding friction coefficient. Typically, $\mu_d < \mu_s$; therefore, to achieve the maximum braking performance and to avoid excessive tire wear, the tire lock-up should be avoided. Locked up tires could also lead to poor steering stability and potential major accidents. In modern cars, the Anti-Lock Braking System (ABS) is used to prevent the tire locked up situation. When raining, due to the phenomenon of hydro-planning, when the tire is sliding over a layer of water, the tires could lock up even equipped with the ABS. Tires with good trends are capable of dispelling water to avoid hydro-planning. Therefore, worn tires should be replaced for safety reasons.

### 9.1.2 *Maximum acceleration and deceleration limited by traction*

From Equation (9.7), if equipped with a sufficiently powerful engine, the maximum acceleration for a car is limited by the static friction. A higher normal force can afford a higher tractive force. Real-wheel-drive cars take advantage of the weight shift to the rear. For all-wheel-drive cars, the traction limited maximum acceleration is $\mu_s g$. For front-wheel-drive cars, it appears that they are at a disadvantage due to weight reduction in the front during acceleration; however, because front-wheel-drive cars typically have higher dead weight on the front, some weight shift to the rear does not impact their acceleration performance.

Based on Equations (9.6) and (9.7), for a rear-wheel-drive car with a sufficiently powerful engine, the maximum acceleration is found to be

$$(a_x)_{\max} = \mu_s \frac{L_1}{L - \mu_s h} g \qquad (9.13)$$

Equation (9.13) assumes the entire car is a rigid body. The actual maximum acceleration would be lower due to various factors, to be discussed later.

The maximum possible deceleration during braking, without considering air resistance, is $\mu_s g$, which is the same as the acceleration of an all-wheel-drive car.

### 9.1.3 *Engine power and maximum speed*

From the discussion in the previous section, one might wonder why there is an obsession with the engine power if maximum acceleration is the goal. There are engine tinkers who can boost a stock engine to a much high power with turbo-chargers, for example. The extra engine power will not increase the maximum possible acceleration performance. Instead, the engine would deliver excessive torque to cause the tire to spin and burn. Some people like to show off the awesome power of their engine by spinning and, thus, burning the tires. The smoke from burning tires is toxic. When the engine power is too much, a touch of the gas pedal could lead to spinning and burning tires, while achieving a lower acceleration.

What is a better way to show off the awesome engine power?

Let's try to answer this question based on the following energy equation,

$$\frac{1}{2} M v_1^2 + \int \sum \vec{F}_{\text{ext}} \cdot d\vec{r}_{\text{ext}} + \int \sum \vec{f}_{\text{int}} \cdot d\vec{r}_{\text{int}} = \frac{1}{2} M v_2^2 \qquad (9.14)$$

Let's consider the situation in which the car has achieved the maximum speed and maintains this maximum speed; thus, $v_1 = v_2 = v_{\max}$. Equation (9.14) becomes

$$\int \sum \vec{F}_{\text{ext}} \cdot d\vec{r}_{\text{ext}} + \int \sum \vec{f}_{\text{int}} \cdot d\vec{r}_{\text{int}} = 0 \qquad (9.15)$$

While cruising at the maximum speed, the related external forces are wind drag, gravity, rolling friction, tractive force, normal forces at the tire, etc. The internal forces include internal friction, engine forces at the piston (or the engine torque), etc. The forces related to the engine produce positive work, while all the frictional forces consume energy (negative work). The rolling friction and wind drag produce negative work. The positive work produced by the engine is consumed entirely by all the forces which are doing negative work. Let's assign an efficiency, $\eta$, account for the energy losses due to friction (internal and external). The effective engine power is then defined as $\eta P_e$. Assuming the wind drag, $F_D = \mu_D v^2$, as a quadratic function of speed, Equation (9.15) can be rewritten as

$$\int \eta P_e dt - \int \mu_D v_{\text{max}}^2 dr = 0$$

$$\implies \eta P_e dt = \mu_D v_{\text{max}}^2 dr$$

$$\implies \eta P_e = \mu_D v_{\text{max}}^2 \frac{dr}{dt} = \mu_D v_{\text{max}}^3 \tag{9.16}$$

The above equation indicates that the maximum speed is determined by the effective engine power, $\eta P_e$. Therefore, a better way to showcase the awesome engine power is to run the car on a race track to see how fast it can go instead of burning tires.

### 9.1.4 *Correlation of acceleration and speed*

Is there any correlation between the speed and the acceleration when a car is accelerating toward a higher speed? Let's consider a car as a particle in a straight line motion. We can convert Equation (3.25) into an energy form,

$$M v \, dv = M a \, ds = \sum F \, ds = \eta P_e \, dt$$

$$\implies M v \frac{dv}{dt} = M v a = \eta P_e$$

$$\implies a = \frac{\eta P_e}{M v} = \frac{\eta \tau_e \omega_e}{M v} \tag{9.17}$$

where $P_e = \tau_e \omega_e$, with $\tau_e$ as the engine torque and $\omega_e$ as the engine angular velocity. Equation (9.17) indicates that if we want to keep the same acceleration as the vehicle speed becomes higher, the needed power will be higher. At a constant gear ratio, $\omega_e$ and $v$ will cancel each other. Therefore, Equation (9.17) also indicates the acceleration is proportional to the engine torque, which is usually a function of the engine speed.

Figure 9.2 shows the engine performance of electric motors used in electric cars, a diesel engine, and a gasoline engine. For the electric driving motors (Figure 9.2(a)), the driving torque can stay constant until about 45 mph and then decreases as the speed increases. This indicates that a transmission is still needed in electric cars, but it could be just a single speed with one fixed gear ratio.

For diesel engines (Figure 9.2(b)), the torque is low when the engine speed is low. The torque reaches its maximum for this diesel

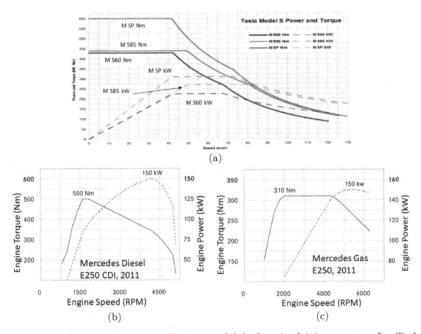

Figure 9.2:   The performance diagrams of (a) electric driving motors for Tesla Model S; and (b) the 2011 Mercedes diesel engine, E250 CDI; (c) the 2011 Mercedes gas engine, E250.

*Sources:* (a) https://electrek.co/2017/11/20/teslas-next-gen-roadster-technical-analysis/; (b) and (c) https://media.daimler.com/marsMediaSite/en/instance/ko/New-engines-for-the-E-class-E-class-more-efficient-than-ever.xhtml?oid=990 4566.

engine at about $1,600$ rpm and quickly decreases as the engine speed increases. For the gasoline engine (Figure 9.2(c)), the peak torque is maintained between 2,000 to about 4,500 rpm, and then decreases with higher engine speeds. Note that the diesel engine shown has a higher torque between 1,500 to 4,500 rpm than that of the gasoline engine in spite of the large variations. Due to the low torque at low speeds, diesel and gasoline engines need to couple with transmissions with multiple gear ratios in order to obtain an appropriate engine torque for different driving needs.

## 9.2 Acceleration Performance with Internal Rotating Mass Considered

In the previous section, we discussed the acceleration performance by assuming that the car is one rigid body. In this section, we will consider the effect due to the rotating components in the transmission, the drive shaft, the differential, the axles, and the tires. In Chapter 4, we presented the design of automatic transmissions using planetary gear trains. In this section, the analysis is based on manual transmission gear trains, but the results are general.

In Figure 9.3, a gear pair are shown. We assume that these gears are rigid bodies. Gear 1 is driven by torque, $\tau_1$, with an angular speed, $\omega_1$, and angular acceleration, $\alpha_1$. Its moment of inertia is $I_1$. The radius of gear 1 is $r_1$. At the engaging point of the gear pair, the driving force is $F$, in opposite directions for each gear. Correspondingly, gear 2 is subjected to a torque $\tau_2$. Gear 2 has a radius of $r_2$, moment of inertia $I_2$, angular speed $\omega_2$, and angular acceleration $\alpha_2$. In addition, we define the gear ratio, $N_{21} = r_2/r_1$. The angular speeds and accelerations are related as

$$N_{21} = \frac{r_2}{r_1} = \frac{\omega_1}{\omega_2} = \frac{\alpha_1}{\alpha_2} \tag{9.18}$$

The moment equation for the gears are

$$\tau_1 - Fr_1 = I_1\alpha_1 \tag{9.19}$$

$$-Fr_2 + \tau_2 = -I_2\alpha_2 \tag{9.20}$$

Eliminating $F$ from the above equations, we have

$$\tau_2 = \tau_1 N_{21} - (I_1 N_{21}^2 + I_2)\alpha_2 = \tau_1 N_{21} - I_{21}\alpha_2 \tag{9.21}$$

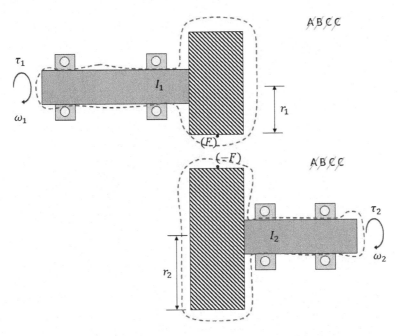

Figure 9.3: A pair of gears and their FBDs.

Equation (9.21) shows that $\tau_1$ is amplified by the gear ratio, $N_{21}$. However, the actual effective torque $\tau_2$ is reduced due to the angular acceleration of gears. The combined moment of inertia, $I_{21}$, as seen from the output side of the gear pair is $(I_1 N_{21}^2 + I_2)$. The moment of inertia of gear 1 is amplified by $N_{21}^2$, viewed from the output side.

The results of Equation (9.21) can be extended to a larger gear train, such as the one in Figure 9.4. Between gear 1 and gear 4, the gear ratio is $N_{41} = N_{21} \cdot N_{32} \cdot N_{43}$. The moment equation for the gear train is

$$\tau_4 = \tau_1 N_{41} - (((I_1 N_{21}^2 + I_2)N_{32}^2 + I_3)N_{43}^2 + I_4)\alpha_4$$

$$= \tau_1 N_{41} - I_{41}\alpha_4 \tag{9.22}$$

For a rear-wheel-drive car, the vehicle power train from the engine torque to the torque at the rear axle can be accordingly written as

$$\tau_w = \tau_e N_{we} - I_{we}\alpha_w \tag{9.23}$$

where $\tau_w$ is the torque by the axle to drive the wheel/tire, $\tau_e$ is the engine output torque, $N_{we}$ is the combined gear ratio of the

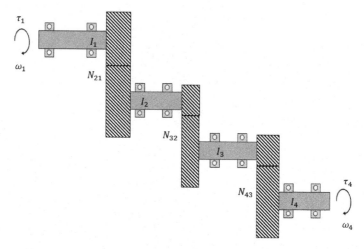

Figure 9.4: A gear train with three pairs of gears engaged.

power train, including the gear ratios of the transmission and the differential. The combined moment of inertia $I_{we}$ is due to those rotating components related to the transmission, the drive shaft, the differential, and the axle. Among them, according to Equation (9.22), the moment of inertia of the transmission is amplified the most.

The driving torque by the axle and the tractive force, $F_x$, at the tire can be related as

$$\tau_w - F_x R = I_w \alpha_w \tag{9.24}$$

where $\alpha_w$ is the wheel/tire angular acceleration, which is the same as the axle acceleration, and $I_w$ is the moment of inertia of the wheel/tire assembly. The radius of the tire is $R$.

Assuming that the tire is a perfect circle and rolls without slipping, $\alpha_w = a_x/R$, where $a_x$ is the acceleration of the vehicle. Combining Equations (9.23) and (9.24), we have

$$F_x = \frac{1}{R}\tau_e N_{we} - \frac{1}{R^2}(I_w + I_{we})a_x \tag{9.25}$$

Consider the mass center acceleration of the car and the external force, $F_x$, of Figure 9.1(a), we have

$$F_x = M a_x \tag{9.26}$$

Finally, introducing Equation (9.26) into (9.25), we obtain an equation to relate the engine torque to the vehicle acceleration as

$$\left[ M + \frac{1}{R^2}(I_w + I_{we}) \right] a_x = M_{eff} a_x = \frac{N_{we}}{R} \tau_e \qquad (9.27)$$

Equation (9.27) indicates that the engine torque is converted to a driving force via a factor $\frac{N_{we}}{R}$ through the power train and the wheel/tire assembly. However, this driving force has to deal with an effective vehicle mass $M_{eff}$ which is larger than $M$. The added mass $\frac{I_w + I_{we}}{R^2}$, denoted as rotating mass, is due to the rotating components of the power train.

When the transmission is set at first gear, for example, the gear ratio is higher, about 3 (see Section 4.3.4 of Chapter 4); therefore, the added rotating mass is higher. Because the driving force will be amplified with a higher gear ratio, the achieved acceleration will still be higher.

The ratio of $M_{eff}/M$ is denoted as the mass factor. For a small car, the mass factor at the first gear could be as high as 1.5 and at the high gear as 1.1.[1,2]

From Figure 9.2, it is also clear why we need to go through gear shifting during acceleration. Because the peak engine torque typically happens at a specific speed, for example, at about 1,500 rpm for the diesel engine (Mercedes E250 CDI) and for 2,000 to 5,000 rpm for the gas engine (Mercedes E250). To obtain suitable acceleration performance, a different gear ratio is needed during acceleration from stationary, so that the engine speed more or less is at the speed which produces the maximum engine torque.

Finally, from the analysis of the mass factor, we should be aware that the system inside a boundary of an FBD may not be a rigid body. This happens when the system within the FBD boundary is not a rigid body but a combination of many bodies with different kinematic properties. This issue was also addressed in Chapter 8 regarding the modeling of single-wheel vehicles.

---

[1]Gillespie, T.D. *Fundamentals of Vehicle Dynamics, Society of Automotive Engineers, Inc.*, ISBN 1-56091-199-9, p. 27, 1992.

[2]Taborek, J.J. *Mechanics of Vehicles*, Towmotor Corporation, Cleveland, Ohio, p. 93, 1957.

## 9.3 Body Roll During Acceleration and Traction Reduction

In addition to the mass factor which reduces the acceleration performance, there are other acceleration reduction factors related to the car body roll and the use of open differentials when the car is no longer considered as one rigid body.

Figure 9.5 is a simplified sketch to illustrate the main components of a rear-wheel drive vehicle with a solid drive axle. The engine/transmission and the car body are considered as a rigid body and together are commonly called the sprung mass because they are supported by the front and rear suspension systems. Here, we do not consider the shock absorbers. A drive shaft transmits the driving torque to a rear differential, which in turn distributes the driving torque to the rear tires through axles. During normal driving, the differential functions as an open differential. Most cars are equipped with open differentials.

The FBD for the rear solid drive axle and tires, suspensions, and drive shaft is shown in Figure 9.6. This FBD is simplified to only show the forces related to the torque around the $z$-axis. Furthermore, the forces due to dead weight of the car are removed as well. As shown in Figure 9.6, when a driving torque is applied, along the positive $z$-axis, the contact normal force at the rear right tire will increase by $W_2$, while that of the left tire will be reduced by the same amount.

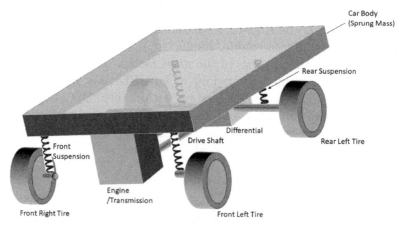

Figure 9.5: A simplified sketch for a rear-wheel drive vehicle with its car body, engine/transmission, solid drive axle and tires, and drive shaft.

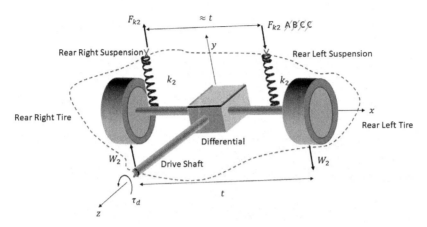

Figure 9.6:  The FBD for the rear solid drive axle, tires, and suspension.

Without considering the dead weight, this is similar to showing that $W_2$ points upward (positive $y$ direction) on the rear right tire, while it points downward on the rear left tire. The right and left designations are based on the driver sitting in the driver seat. It becomes more obvious later that the rear right suspension spring will be stretched, causing an addition tensile spring force, $F_{k2}$, pointing upward. The rear left spring will be compressed, resulting in a downward force by the same amount.

To show why the rear suspension springs would behave as described, we need to construct an FBD for the car body and engine/transmission, as shown in Figure 9.7. The engine torque acting on the car body/engine/transmission is in the reverse direction (around the negative $z$-axis), causing the car body to rotate clockwise when viewing from the front. Because the tires are in contact with the ground, the clock-wise rotation of the car body results in the right suspension springs being stretched and the left suspension springs, compressed. The resulting spring force directions are therefore correctly plotted in Figures 9.6 and 9.7.

The corresponding moment equations for Figures 9.6 and 9.7 are, with the car body at equilibrium,

$$\tau_d - W_2 t - F_{k2} t = 0 \qquad (9.28)$$

$$-\tau_d + F_{k1} t + F_{k2} t = 0 \qquad (9.29)$$

where the distance between the wheels is $t$. We also assume the distance between the springs as $t$ even though it could be smaller.

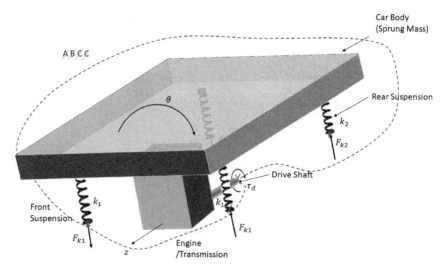

Figure 9.7: The FBD for the sprung mass.

When the car body rotates by $\theta$, the stretching and compression of the springs are $\theta t/2$. The suspension spring forces are

$$F_{k1} = k_1 \left( \frac{t}{2}\theta \right) \tag{9.30}$$

$$F_{k2} = k_2 \left( \frac{t}{2}\theta \right) \tag{9.31}$$

Equation (9.29) becomes

$$\tau_d = \frac{t^2}{2}(k_1 + k_2)\theta \tag{9.32}$$

From Equations (9.28) and (9.29), we also obtain,

$$W_2 = F_{k1} \tag{9.33}$$

From Equation (9.32), we obtain

$$\theta = \frac{2\tau_d}{t^2(k_1 + k_2)} \tag{9.34}$$

Finally, combining Equations (9.30), (9.33), and (9.34), we obtain

$$W_2 = \frac{k_1}{k_1 + k_2} \frac{\tau_d}{t} \tag{9.35}$$

Based on Equation (9.35), for a rear-wheel-drive vehicle with an open differential and solid drive axle during steady-state acceleration, the maximum tractive force is

$$F_x = 2\mu_s(W_{R0} - W_2) = F_{x0} - 2\mu_s W_2 = F_{x0} - 2\mu_s \frac{k_1}{k_1 + k_2} \frac{\tau_d}{t}$$

(9.36)

where $W_{R0}$ is the dead weight on each rear tire and $F_{x0} = 2\mu_s W_{R0}$.

According to Equation (9.36), due to the open differential, there is a reduction in tractive force as the normal force of the right tire is reduced by $W_2$.

To reduce the traction force loss due to the body roll, one should limit $\theta$, thus making $t(k_1 + k_2)$ large and $k_1$ small. We can accomplish this by having a stiffer suspension spring and a stronger shock absorber in the rear. We did not consider shock absorbers in the analysis above, but shocks contribute to the stiffness as well. Over time, when the springs and shock absorbers become weak, the acceleration performance will suffer.

A stiff $k_2$ may make the riding feel harsher. We can introduce an anti-roll bar to the rear, similar to a torsional spring, for roll stiffness without adding linear stiffness to the suspension. In this way, both acceleration and ride performances can be improved. Figure 9.8 shows the design of an anti-roll bar. Usually, the anti-roll bar is installed for both the front and the rear suspensions. However, we should keep the rear anti-roll bar stronger for a lower lateral weight shift due to the body roll. With the anti-roll bars installed, the weight shift equation becomes

$$W_2 = \frac{\frac{t^2}{2}k_1 + k_{\theta 1}}{\frac{t^2}{2}(k_1 + k_2) + (k_{\theta 1} + k_{\theta 2})} \frac{\tau_e N_{de}}{t} = \frac{K_{\phi f}}{K_\phi} \frac{\tau_e N_{de}}{t}$$

(9.37)

where $\tau_d = \tau_e N_{de}$ with $N_{de}$ as the gear ratio of the transmission. We neglect the effective moment of inertia of the transmission for simplicity, knowing it would introduce some errors. $k_{\theta 1}$ and $k_{\theta 2}$ are torsional stiffnesses of the front and rear anti-roll bars, respectively. $K_{\phi f}$ is the total front torsional stiffness and $K_\phi$ is the total torsional stiffness.

As shown in Figure 9.8, the mid section of the anti-roll bar is fixed to the car body, while the end points, $A$ and $B$, are free to move up and down together. Therefore, the anti-roll bar does not affect the vertical stiffness of the suspension. However, if point $A$ is moving up, while $B$ is moving down, as in the case of a body roll, then the anti-roll bar will be subjected to torsional loading. The anti-roll bar behaves like a torsional spring, adding to the roll stiffness.

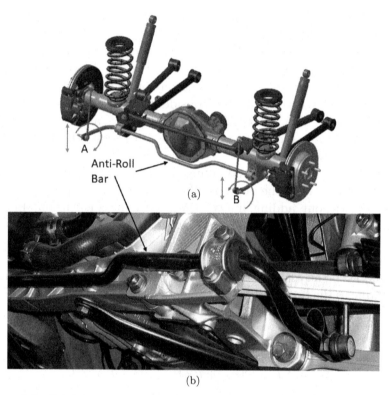

(a)

(b)

Figure 9.8: (a) An anti-roll bar design; and (b) an anti-roll bar in an actual car. The end points, A and B, of the anti-roll bar are free to move up and down, but resist moving in different directions.

*Sources*: (a) https://lh3.googleusercontent.com/-OYS3H6xrLyc/W1xTmR9nW UI/AAAAAAADlTk/osoCUkJDukMk08P-Gs1TPwRVopZKfaadgCHMYCw/s0/ 76768b8f2ad6add6bfa803bb9ea6b66bff02509c.gif; (b) https://upload.wikimedia. org/wikipedia/commons/thumb/3/3c/Stabilisator_(Porsche).jpg/1200px-Stabili sator_(Porsche).jpg.

By selecting proper roll stiffness for the front and rear, the body roll can be optimized for acceleration performance. The body roll also happens during cornering, to be discussed next.

Based on Equations (9.6) and (9.37), we can combine the effects of weight shift to the rear and the lateral weight shift during acceleration for a rear-wheel-drive car with a solid drive axle to determine the maximum tractive force

$$F_{xmax} = Ma_x = 2\mu_s \left( Mg\frac{L_1}{2L} + Ma_x\frac{h}{2L} - W_2 \right) \qquad (9.38)$$

where $W_2$ is determined by Equation (9.37), given the engine torque, $\tau_e$. Introducing Equations (9.37) and (9.27) into (9.38), we obtain

$$F_{xmax} = \frac{\mu_s Mg\frac{L_1}{L}}{1 - \mu_s\frac{h}{L} + 2\mu_s\frac{K_{\phi f}}{K_\phi}\frac{RM_{eff}}{tN_{wd}M}} \qquad (9.39)$$

where $\frac{N_{de}}{N_{we}} = \frac{1}{N_{ed}N_{we}} = \frac{1}{N_{wd}}$ and $N_{wd}$ is the gear ratio of the differential, or denoted as the final drive ratio.

If a locked differential or an open differential with independent rear suspension is used, $W_2$ should be removed from Equation (9.38). Equation (9.39) becomes

$$F_{xmax} = \frac{\mu_s Mg\frac{L_1}{L}}{1 - \mu_s\frac{h}{L}} \qquad (9.40)$$

For independent rear suspension, the differential is mounted on the car body. As a result, the drive shaft torque is countered by the torque from the differential mount without causing the body roll and the lateral weight shift; therefore, there is no reduction in the maximal tractive force even with an open differential.

## Example 9.1: Maximum Achievable Tractive Force

**Problem Statement:** Modified from an example from Gillespie (1992), the mass of a vehicle with a rear solid drive axle is 3950 lbm, $h = 21$ in, $L = 108$ in, $L_1 = 50.58$ in, $L_2 = 57.42$ in, $R = 13$ in, $t = 59$ in, $N_{wd} = 2.9$, $K_{\phi f} = 1{,}150$ in–lbf/deg, $K_\phi = 2{,}400$ in–lbf/deg, and $\mu_s = 0.62$. In addition, $M_{eff} = 1.5\,M$.

Determine the maximum tractive force and the maximum acceleration of this vehicle.

**Solution:** Based on Equation (9.39), we obtain

$$
F_{xmax} = \frac{\mu_s M g \frac{L_1}{L}}{1 - \mu_s \frac{h}{L} + 2\mu_s \frac{K_{\phi f}}{K_\phi} \frac{RM_{eff}}{tN_{wd}M}}
$$

$$
= \frac{0.62 \cdot 3950 \cdot \frac{50.48}{108}}{1 - 0.62 \cdot \frac{21}{108} + 2 \cdot 0.62 \cdot \frac{1150}{2400} \frac{13 \cdot 1.5}{59 \cdot 2.9}} = 1,209 \text{ lbf}
$$

The corresponding acceleration is 0.306 $g$.

The required engine torque to achieve this rather poor acceleration performance is, per Equation (9.27), 187 ft–lbf, assuming that the first gear ratio is 3.5 for the same final drive ratio, i.e., $N_{we} = 3.5 \cdot 2.9 = 10.15$.

If we can add a strong anti-roll bar to the rear to boost the overall torsional stiffness (also called roll stiffness) to $K_\phi = 2,800$ ft–lbf/deg, while keeping the front roll stiffness the same, the tractive force will improve to 1221 lbf and acceleration to 0.309 $g$, which may not appear to be significant for acceleration performance. However, anti-roll stiffness is also important for cornering stability, to be discussed in later sections. The required engine torque becomes 195 ft–lbf. These engine torque requirements can be easily met by modern engines and electric motors.

What if we use a much bigger engine which can produce a higher engine torque? For one, the vehicle can start via the second gear, which is quite common for modern cars. Starting with the second gear will reduce $M_{\text{eff}}$ and improve the acceleration performance.

If the engine produces excessive torque than needed for the maximum tractive force, the tires will spin, resulting in a poor acceleration performance and excessive tire wear.

Some people like to use larger tires for their cars, thinking that the acceleration performance might improve. According to Equation (9.39), it would actually reduce the acceleration performance if the car has a sold drive axle. A wider car, with a larger value of $t$, will help. The open-wheel cars for the Formula-1 racing have a larger value of $t$, which not only improves the acceleration, it also improves the cornering stability.

## 9.4   Braking System Design

As shown in Figure 9.1(b), when a car brakes, the weight shifts to the front. There is no body roll because the engine torque is essentially negligible during braking. The weight has now shifted to the front. The front braking and rear braking forces are

$$F_{B1} = \frac{\mu_1}{L}(MgL_2 + Ma_x h) \tag{9.41}$$

$$F_{B2} = \frac{\mu_2}{L}(MgL_1 - Ma_x h) \tag{9.42}$$

where $\mu_1$, $\mu_2 < \mu_s$ are the equivalent friction coefficient and are limited by the static friction coefficient, $\mu_s$. We can plot a chart with $F_{B1}$ versus $F_{B2}$. On this chart, we can plot a line for which $F_{B1}$ is at its maximum possible value while $F_{B2}$ is not. Equation (9.41) becomes

$$F_{B1\max} = \frac{\mu_s}{L}(MgL_2 + Ma_x h) \tag{9.43}$$

Combining Equations (9.42) and (9.43), we have

$$F_B = Ma_x = F_{B1\max} + F_{B2} \tag{9.44}$$

The deceleration is then found to be

$$a_x = F_B/M = \frac{\mu_s MgL_2 + F_{B2}L}{M(L - h\mu_s)} \tag{9.45}$$

Introducing Equation (9.45) to (9.43), we obtain

$$F_{B1\max} = \frac{\mu_s(MgL_2/L + F_{B2}h/L)}{1 - \mu_s h/L} \tag{9.46}$$

We can plot $F_{B1\max}$ as a function of $F_{B2}$ on the $F_{B1}-F_{B2}$ chart, denoted as the $(F_{B1\max} - F_{B2})$ line. Similarly, we can construct $F_{B2\max}$ as a function of $F_{B1}$ as

$$F_{B2\max} = \frac{\mu_s(MgL_1/L - F_{B1}h/L)}{1 + \mu_s h/L} \tag{9.47}$$

The corresponding line of the above equation is the $(F_{B2max} - F_{B1})$ line.

Finally, if maximum braking is achieved, we have

$$F_{B1} + F_{B2} = \mu_s M g \qquad (9.48)$$

We can plot Equation (9.48) as well, which is also a straight line. The resulting diagram is shown in Figure 9.9. All three lines, represented by Equations (9.46)–(9.48), respectively, intersect at point $C$ at which both braking forces are at their respective maximum and the vehicle has achieved the maximum braking.

When designing a brake system, we can control the hydraulic pressure applied to the brake pads and the rotor to control the brake force. Therefore, we can design a brake performance line as

$$F_{B1} = \beta F_{B2} \qquad (9.49)$$

As indicated in Figure 9.9, when $\beta > 1$, the line of $OA$ is followed. Both $F_{B1}$ and $F_{B2}$ will increase along line $OA$ when the brake pedal is pressed down harder and harder until it hits point $A$. At point $A$,

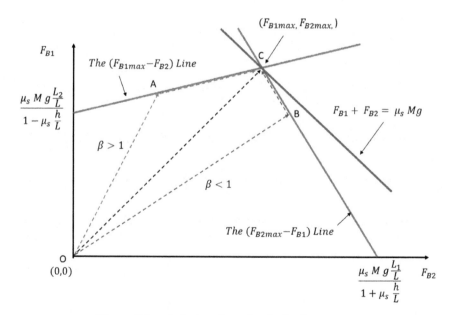

Figure 9.9: A design chart for the brake system.

the braking forces will follow the $(F_{B1\text{max}} - F_{B2})$ line until it reaches point $C$. Throughout this braking process, we assume that the anti-locking brake system will function properly to keep the front tires from locking up.

Similarly, if $\beta < 1$, we could be following the $OB$ line until it hits point $B$, upon which it follows the $(F_{B1} - F_{B2\text{max}})$ line until it reaches point $C$.

If we design a performance line corresponding to line $OC$, the braking forces will go from point $O$ to $C$.

Because a car might carry different weights due to the passengers, luggage, etc., it is difficult to design a performance line following $OC$ precisely. Between $OA$ and $OB$, performance line $OA$ is preferred because the rear tire lock-up should be avoided at all time (to be explained in later sections). Despite the excellent advances in anti-locking brake systems, once a tire hits its static limit, there is always a chance that it might skid. In addition, because the weight shifts to the front during braking, it is also natural to achieve higher front braking forces. This is the reason that the front brake rotors and pads are typically larger than those in the rear. Furthermore, the front brake pads and rotors also wear out sooner.

## 9.5 Lateral Weight Shift During Steady State Cornering

When a car makes a turn, i.e., in a cornering maneuver, there is a centripetal acceleration, pointing to the center of the trajectory arc. Let's consider the case of steady-state cornering for which the vehicle moves along a perfect circle at a constant speed, as shown in Figure 9.10(a). From the rear view, as shown in Figure 9.10(b), this centripetal acceleration points to the left. The tires on the left are called inside tires, while the ones on the right are outside tires. From the rear view, the car body will roll clockwise. For the steady-state cornering, the roll angle $\theta$ will be constant for a specific vehicle speed, $v$. From the car body suspension mechanism, we can identify a roll center and the mass center of the car body, $G_B$, will shift to $G'_B$, with respect to the roll center. The horizontal shift is then determined as $\Delta t = h_1 \theta$. Note that the roll center is similar to the instantaneous center of a linkage system, determined by the arrangement of the suspension system via the kinematic mechanism analysis.

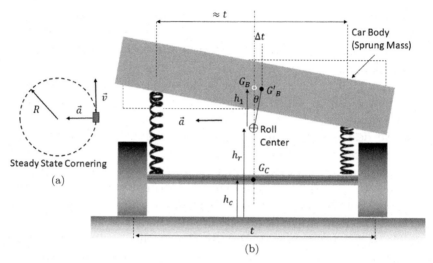

Figure 9.10: A car undergoing steady state cornering: (a) top view of the trajectory; and (b) rear view of the vehicle with a body roll.

We can construct two FBDs for the car undergoing steady-state cornering as shown in Figure 9.11. Applying *Newton's Second Law* to the car body, we have

$$F_{ki} + F_{ko} - m_B g = 0 \tag{9.50}$$

$$F_{Bli} + F_{Blo} = F_{Bl} = m_B a_n = m_B \frac{v^2}{R} \tag{9.51}$$

$$F_{ko}\left(\frac{t}{2} - \Delta t\right) - F_{ki}\left(\frac{t}{2} + \Delta t\right) - F_{Bl} h_1 = I_{GB} \alpha_B = 0 \tag{9.52}$$

For the chassis, the corresponding governing equations are

$$N_o + N_i - F_{ki} - F_{ko} - m_c g = 0 \tag{9.53}$$

$$-F_{Bli} - F_{Blo} + F_{li} + F_{lo} = -F_{Bl} + F_l = m_c a_n = m_c \frac{v^2}{R} \tag{9.54}$$

$$N_o \frac{t}{2} - N_i \frac{t}{2} - F_l h_c - F_{ko} \frac{t}{2} + F_{ki} \frac{t}{2} - F_{Bl}(h_r - h_c) = 0 \tag{9.55}$$

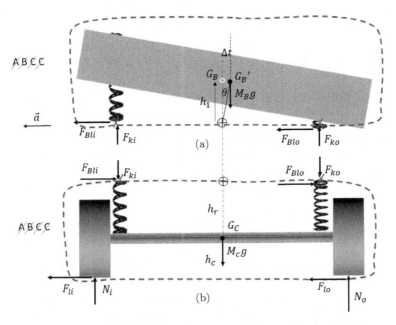

Figure 9.11: (a) FBD for the car body; (b) FBD for the chassis.

Let the lateral weight shift as $\Delta W$, the normal forces at the tires can be expressed as

$$N_O = \frac{mg}{2} + \Delta W \tag{9.56}$$

$$N_i = \frac{mg}{2} - \Delta W \tag{9.57}$$

Similarly, let the spring force change as $\Delta F_k$, the spring forces are

$$F_{kO} = \frac{m_B g}{2} + \Delta F_k \tag{9.58}$$

$$F_{ki} = \frac{m_B g}{2} - \Delta F_k \tag{9.59}$$

Furthermore, defining the roll stiffness as $K_\phi$ and from the roll angle, $\theta$, we have

$$\Delta F_k t = K_\phi \theta \tag{9.60}$$

Introducing Equations (9.56)–(9.60) into (9.50)–(9.52), we have

$$K_\phi \theta - m_B g \theta h_1 = m_B \frac{v^2}{R} h_1 \tag{9.61}$$

From Equations (9.54) and (9.55), we obtain

$$\Delta Wt - (m_B h_r + m_c h_c)\frac{v^2}{R} - K_\phi \theta = 0 \qquad (9.62)$$

From Equations (9.61) and (9.62), we obtain

$$\Delta W = (m_B h_r + m_c h_c)\frac{v^2}{R}\frac{1}{t} + K_\phi \frac{\theta}{t} \qquad (9.63)$$

$$\theta = \frac{m_B \frac{v^2}{R} h_1}{K_\phi - m_B g h_1} \qquad (9.64)$$

Finally, from the above two equations, we express the lateral weight shift during steady-state cornering as

$$\Delta W = \frac{K_\phi h_1}{t(K_\phi - m_B g h_1)}\frac{m_B v^2}{R} + \frac{m_B h_r + m_c h_c}{t}\frac{v^2}{R}$$
$$= \Delta W_\phi + \Delta W_h \qquad (9.65)$$

If we assume that $m_B \approx m$, $m_c \approx 0$, and $h_c \approx 0$, we get the same expression by Gillespie (1992) (see footnote 1) as

$$\Delta W^* = \frac{K_\phi h_1}{t(K_\phi - m g h_1)}\frac{m v^2}{R} + \frac{h_r}{t}\frac{m v^2}{R} \qquad (9.66)$$

Because $m_c$ typically is not small, it will cause errors by assuming $m_B \approx m$. However, we can safely assume that $h_c \approx h_r$ even when they are plotted as quite different from Figure 9.10. As a result, for easier calculation, Equation (9.65) can be converted to

$$\Delta W^+ = \frac{K_\phi h_1}{t(K_\phi - m_B g h_1)}\frac{m_B v^2}{R} + \frac{h_r}{t}\frac{m v^2}{R} \qquad (9.67)$$

Equation (9.65) indicates that the lateral force, $\frac{m v^2}{R}$, needed for cornering induces the lateral weight shift from the inside tires to the outside tires. This weight shift contains two parts. The first part, $\Delta W_\phi$, is related to the car body rolling with respect to the roll center. As a result, the mass center of the car body shifts toward the outside, causing a weight shift to the outside tires. The second part, $\Delta W_h$, is

related to the weight shift due to the dynamic equilibrium requirement, similar to the front to the rear weight shift during forward
acceleration. $\Delta W_h$ exists even if the car body is supported rigidly.

We can compare the expression of Equation (9.65) with an
approximation for the lateral weight shift without considering the
roll center and assuming the entire car is one rigid body. From the
moment equilibrium with respect to the mass center, the approximated weight shift is found to be

$$\Delta \hat{W} = \frac{mh}{t}\frac{v^2}{R} \tag{9.68}$$

Equations (9.65)–(9.67) can be simplified to Equation (9.68) by
assuming $K_\phi \longrightarrow \infty$ and $m_B \longrightarrow m$.

When deriving the above equations, we compress the front and
rear tires as one, illustrated in Figure 9.10. For the analysis of steering
stability (next section), it is of interest to determine the lateral weight
shift to the front and to the rear tires.

$$\Delta W = \Delta W_f + \Delta W_r \tag{9.69}$$

Experiments can be set up to verify Equation (9.65) and to determine (9.69). For analysis, it is rather difficult to find additional
equation to determine $\Delta W_f$ and $\Delta W_r$. However, we can try to estimate them. As indicated by Equation (9.65), the lateral weight shift
is proportional to the mass, roll stiffness, $K_\phi$, and the height of
the roll center. Combining this observation with Equation (9.69),
we assume

$$\Delta W_f = \frac{m_f K_{\phi f} h_{rf}}{m_f K_{\phi f} h_{rf} + m_r K_{\phi r} h_{rr}} \Delta W \tag{9.70}$$

$$\Delta W_r = \frac{m_r K_{\phi r} h_{rr}}{m_f K_{\phi f} h_{rf} + m_r K_{\phi r} h_{rr}} \Delta W \tag{9.71}$$

where $h_{rr}$ is the roll center height at the rear and $h_{rf}$ is the roll center
height at the front. The roll center along the longitudinal direction
of the vehicle often is not horizontal. At the front tires, it is usually
lower, as $h_{rf} < h_r$, while at the rear tires, it would be higher, as
$h_{rr} > h_r$.

In Gillespie (1992) (see footnote 1), by splitting the vehicle into a front rigid body $m_f$, and a rear rigid body $m_r$, the lateral weight shift distributions are expressed as

$$\Delta \hat{W}_f = \left( \frac{K_{\phi f} h_1}{t(K_\phi - mgh_1)} \right) \frac{mv^2}{R} + \frac{h_{rf}}{t} \frac{m_f v^2}{R} \tag{9.72}$$

$$\Delta \hat{W}_r = \left( \frac{K_{\phi r} h_1}{t(K_\phi - mgh_1)} \right) \frac{mv^2}{R} + \frac{h_{rr}}{t} \frac{m_r v^2}{R} \tag{9.73}$$

Although, $m_f$ and $m_r$ can be determined from the dead weight distribution for the above equations, $\Delta W_f + \Delta W_r \neq \Delta W$ based on Equation (9.66) unless we replace both $h_{rr}$ and $h_{rf}$ by $h_r$ in the two equations above.

Note that Equations (9.70)–(9.73) are all estimates. The dead weight distribution could be different with the driver, passengers, and luggage. The importance of these equations is to study how the lateral weight distribution could affect the steering stability, discussed in the next section.

In practice, a roll rate, $\dot{\theta}_a$, can be defined as

$$\dot{\theta}_a = \frac{\partial \theta}{\partial a_n} = \frac{m_B g h_1}{K_\phi - m_B g h_1} \; (\text{rad}/g) \tag{9.74}$$

The roll rate is usually in the range of $3-7°/g$ on typical passenger cars (see footnote 1).

Let's work out a numerical example to determine the lateral weight shift.

## Example 9.2: Lateral Weight Shifts During Steady-State Cornering

**Problem Statement:** Based on the vehicle data defined in Example 9.1, the car is undergoing a steady-state cornering at 60 mph along a circle with an 800 ft radius. The roll center height is $h_r = 6$ in. The front roll center height is $h_{rf} = 5.0$ in and the rear is $h_{rr} = 7.0$ in. The mass center height of the chassis is 5 in. The car body mass (sprung mass) is $m_B = 2,900$ lbm, while $m_c = 1,050$ lbm. The front mass is $m_f = 2,100$ lbm and the rear mass, $m_r = 1,850$ lbm. The mass center height is $h = 21$ in.

**Solution:** First, the centripetal acceleration is found to be

$$a_n = \frac{v^2}{R} = \frac{(60 \cdot 5280/(60 \cdot 60))^2}{800} = 9.68 \ (\text{ft/s}^2) = 0.30 \ (g) \quad (9.75)$$

The value of $h_1$ is found to be

$$h_1 = h - h_r = 21 - 6 = 15 \ (\text{in}) \quad (9.76)$$

The roll rate is found to be

$$\dot{\theta}_a = \frac{m_B g h_1}{K_\phi - m_B g h_1} = \frac{2900 \cdot (32.2/32.2) \cdot 15}{2400 \cdot (180/\pi) - 2900 \cdot (32.2/32.2) \cdot 15}$$

$$= 0.462 \ (\text{rad}/g) = 26.5 \ (\text{deg}/g) \quad (9.77)$$

The roll angle is found to be

$$\theta = \dot{\theta}_a \cdot a_n = 26.5 \cdot 0.3 = 8.0 \ (\text{deg}) \quad (9.78)$$

This angle is quite excessive due to the very high roll rate. If we add a stronger anti-roll bar to the rear as in Example 9.1 to make $k_\phi = 2,800$ ft–lbf/deg, the roll rate will drop to 21.3 deg/g and roll angle to 6.4°.

The lateral weight shift calculated via Equation (9.65) with the original roll stiffness is

$$\Delta W = \left(\frac{K_\phi h_1}{t(K_\phi - m_B g h_1)}\right)\frac{m_B v^2}{R} + \frac{m_B h_r + m_c h_c}{t}\frac{v^2}{R}$$

$$= \left(\frac{2400 \cdot (180/\pi) \cdot 15/12}{(59/12)(2400 \cdot (180/\pi) - 2900 \cdot (15/12))}\right)(2900 \cdot 0.3)$$

$$+ \frac{2900 \cdot (6/12) + 1050 \cdot (5/12)}{59/12}(0.3)$$

$$= 323.5 + 115.2 = 438.7 \ (\text{lbf}) \quad (9.79)$$

In comparison, based on Equation (9.68), we obtain

$$\Delta \hat{W} = \frac{mh}{t}\frac{v^2}{R} = \frac{3950 \cdot 21}{59} \cdot 0.3 = 421.8 \ (\text{lbf}) \quad (9.80)$$

which is lower by 3.8%.

The lateral weight shift based on Equation (9.66) is found to be 650 lbf, which is much higher, likely incorrect due to representing the sprung mass as the total mass in the calculation of $\Delta W_\phi$.

According to Equations (9.72) and (9.73), the lateral weight shift in the front and the rear are found to be $\Delta W_f = 187$ lbf and $\Delta W_r = 251$ lbf, respectively. Note that the rear lateral weight shift is higher and it is generally not desirable, as will be explained in the next section.

As shown in Figure 9.12, a substantial body roll is seen in the car in cornering. In addition, it also appears that the outside rear tire is in the air, totally unloaded, while the front outside tire is still in contact with the ground. This image indicates a situation that the rear tire lateral weight shift is much larger than the one in front. Having such high lateral weight shift in the rear could be dangerous.

In general, the body roll during cornering is not desirable but cannot be eliminated entirely unless active suspension systems are deployed.

Finally, for the analysis above, we did not consider the camber angle, tire deformation, tire steering angle, transient conditions, etc. These factors could affect the lateral weight shift and its distributions. The equations above are good starting points in initial vehicle design stages.

Figure 9.12: A car undergoing cornering with a substantial body roll.
*Sources*: https://live.staticflickr.com/65535/50316115821_aecbcb5bcf_b.jpg.

## 9.6 Steering Stability

As shown in Figure 9.13, a car undergoes steady-state cornering at a constant speed, $v_G$, following a circular path with a radius, $R$. The angular velocity of the vehicle as a rigid body is $\omega$ and at steady-state cornering, $\omega = v_G/R$.

During cornering, the mass center of the vehicle has a lateral acceleration, $a_G = v_G^2/R$, pointing to the center, $O$, of the circular path. At steady-state cornering, the moment equation with respect to the mass center is

$$F_2 L_2 - F_1 L_1 = I_G \alpha \qquad (9.81)$$

where $I_G$ is the moment of inertia of the vehicle with respect to the mass center and $\alpha$ is the angular acceleration of the vehicle. At steady-state cornering, $\alpha = 0$. If $\alpha > 0$, the vehicle will steer off the

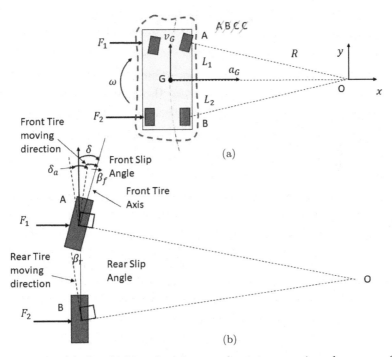

Figure 9.13: (a) A vehicle undergoing steady-state cornering along a perfect circular trajectory with a radius, $R$; and (b) the steering angles of the front and rear tires.

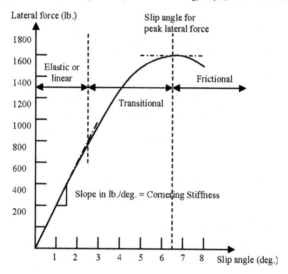

P215/60R15 Goodyear Eagle GT-S (shaved for racing) 31 psi, Load = 1800 lbf

Figure 9.14: The lateral force vs slip angle diagram of a shaved Goodyear GT-S tire (see footnote 3).

circular path outward, which is the condition denoted as understeering. On the other hand, if $\alpha < 0$, the vehicle will steer off the circular path inward, denoted as oversteering. If we have to choose between oversteering and understeering, the latter is preferred because oversteering is an unstable condition and the vehicle could spin uncontrollably, often reversing $180°$ or running into opposite lanes.

The lateral forces, $F_1$ and $F_2$, are related to the tire steering angles as shown in Figure 9.13(b). The slip angles are defined as the angle between the tire orientation axis and the tire traveling direction. A typical lateral force vs slip angle diagram is shown in Figure 9.14.[3] The lateral force typically is a nonlinear function of the slip angle. It is also a function of the normal force. At small slip angles, below $3°$, the relationship is nearly linear and a cornering stiffness can be defined.

---

[3]Koo, S., Tan, H., and Tomizuka, M. Nonlinear tire lateral force versus slip angle curve identification, *Proceedings of 2004 American Control Conference*, Boston, Massachusetts.

To keep the car from going off the road during cornering, the driver changes the front tire steering angle to change its slip angle so that $\alpha = 0$ for Equation (9.81). For a car designed to be neutral steering, the driver turns the steering wheel in an intuitive way, which results in an angle denoted as the Ackerman angle. Our human brain is hardwired to come up with this angle. Theoretically, the Ackerman angle in degrees is determined by the circular path radius, $R$, and the wheelbase, $L$,

$$\delta_a = 57.3 \frac{L}{R} \qquad (9.82)$$

The Ackerman angle is also the angle between the front and rear tire moving directions as shown in Figure 9.13(b). From geometry, we have the relationship between the steering angle $\delta$, the Ackerman angle $\delta_a$, the front slip angle $\beta_f$, and the rear slip angle $\beta_r$:

$$\delta = \delta_a + \beta_f - \beta_r \qquad (9.83)$$

From Equation (9.83), for a neutral steering vehicle, $\beta_f = \beta_r$ and thus, the steering angle $\delta = \delta_a$. The cornering maneuver for a neutral steering vehicle is natural for a driver. If a vehicle has a tendency of understeering, then the Ackerman angle is not sufficient. As indicated by Equation (9.81), if $\delta = \delta_a$, the resulting angular acceleration of the vehicle is $\alpha > 0$ because $F_2 L_2 > F_1 L_1$. The driver needs to add more steering angle so that $\beta_f > \beta_r$ and $\delta > \delta_a$. With a larger $F_1$, the angular acceleration can be made zero, i.e., $\alpha = 0$, for the vehicle to stay on the circular path.

On the other hand, for a vehicle with an oversteering tendency, if the driver keeps the steering angle the same as the Ackerman angle, it would result in $F_1 L_1 > F_2 L_2$, leading to $\alpha < 0$, thus oversteering. The driver needs to use a less steering angle than the Ackerman angle, making $\beta_f < \beta_r$ so that $\alpha = 0$.

Typically, a front-wheel-drive vehicle, because $L_1 < L_2$, has a tendency more toward understeering, while a rear-wheel-drive sports car has an oversteering tendency due to $L_1 >> L_2$. We can experience the understeering or oversteering tendency using a shopping cart as an example. If you put heavy weight over the front wheels of the cart (similar to a front-wheel-drive car), it would be hard to turn the shopping cart, indicating the tendency of understeering. On the other hand, if you put heavy weight over the rear wheels (similar to a super sports car), you can change direction of the cart very easily, indicating the oversteering tendency.

When a car has too much weight in the trunk, the steering tendency will move toward oversteering. When the rear tires are worn, a car will have the tendency of oversteering because the resulting $F_2$ will be lower when the road is wet due to hydroplaning. When hydroplaning happens, it is like the tire is sliding over the ground with a layer of water in between, greatly reducing the lateral force.

When undergoing a turn with the road being wet or icy, it is dangerous to hit the brake. By hitting the brake, there is a weight shift to the front, which increases $F_1$ and reduces $F_2$. Based on Equation (9.81), it can cause $\alpha < 0$, leading to oversteering. This can happen almost instantly especially on an icy road. The driver could lose the control of the car entirely within split seconds. Therefore, it is important that the driver should reduce the speed before entering a turn. During the turn, it is safer to simply coast without hitting the brake or accelerating.

On the other hand, for a super sports car with the mass center very near the rear tires (for example, by placing the engine at the rear), if the driver is accelerating to enter a sharp turn with the gas pedal pressed down, by merely lifting the gas pedal, the super sports car can oversteer. This can be explained based on Equation (9.81) as well. With the car accelerating, there is a weight shift to the rear which provides a higher $F_2$ and a lower $F_1$. This accelerating condition will promote an understeering tendency based on Equation (9.81). When the driver feels that by accelerating and with the understeering tendency, he or she could add a higher steering angle to make the turn, while letting go of the gas pedal by instinct, the car can suddenly oversteer. This sudden oversteering happens because the sudden reduction of the acceleration will shift the weight back to the front. Because the super sports cars also have a strong oversteering tendency combined with a higher steering angle, the car could suddenly oversteer. This is commonly called "lift-off" oversteering. Again, for safety, it is better to coast without accelerating when entering a turn.

As stated before, oversteering must be avoided because it can lead to very dangerous situations. If a person can only afford to buy two new tires, the new tires should be placed in the rear, no matter the car is front- or rear-wheel-drive.

## 9.7 Roll Over

As shown in Figure 9.15(a) when a car oversteers suddenly, the car could start to skid sideways, leading to a potential roll-over situation when the car flips over. Roll-over can also happen when a car goes in a sharp turn at high speeds, as shown in Figure 9.15(b). The FBD for the roll-over situation is shown in Figure 9.16.

The governing equations for the roll-over situation are

$$F_O + F_i = ma_{Gn} \tag{9.84}$$

$$N_o + N_i - mg = ma_{Gz} \tag{9.85}$$

$$(F_o + F_i)h + \frac{1}{2}tN_i - \frac{1}{2}tN_o = I_G\alpha \tag{9.86}$$

where $N_O = \frac{1}{2}mg + \Delta W$ and $N_i = \frac{1}{2}mg - \Delta W$.

Let's consider a threshold situation in which the outside tires just lift off the ground, totally unloaded, but the car maintains a steady

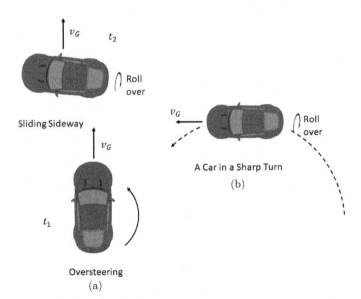

Figure 9.15: (a) A car moving sideways due to oversteering; and (b) a car could roll over when undergoing a sharp turn.
*Car Image Source*: https://pre00.deviantart.net/32fd/th/pre/f/2009/161/a/3/aston_martin_db9_by_bagera3005.png.

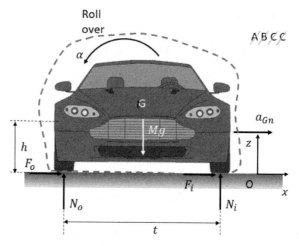

Figure 9.16:   FBD of a car in a roll-over situation.
*Car Image Source*: https://pre00.deviantart.net/32fd/th/pre/f/2009/161/a/3/ aston_martin_db9_by_bagera3005.png.

state without flipping over. As a result, we have

$$F_i = N_i = 0 \tag{9.87}$$

$$a_{Gz} = 0 \tag{9.88}$$

$$\Delta W = \frac{1}{2}\, mg \tag{9.89}$$

$$\alpha = 0 \tag{9.90}$$

The roll-over situation is related to the relative ratio of lateral acceleration, $a_{Gn}$, to the gravitational acceleration, $g$. We then define their ratio at threshold as the roll-over threshold. From Equations (9.84)–(9.90), the weight shift and the roll-over threshold, $\frac{a_{Gn}}{g}$ become

$$\Delta W = \frac{1}{2}mg = \frac{h}{t}a_{Gn} \tag{9.91}$$

$$(|\frac{a_{Gn}}{g}|)_{\text{threshold}} = \frac{t\Delta W}{hg} = \frac{t}{2h} \tag{9.92}$$

From Equation (9.92), the roll-over threshold is determined by the width between the tires and the height of the mass center. Therefore, a low and open-wheel sports car has a high roll-over threshold, while an SUV will have a lower threshold.

If we consider the roll of the car body (sprung mass) with respect to the roll center, Equation (9.92) can be modified by combining Equation (9.65) with $a_{Gn} = \frac{v^2}{R}$ and Equation (9.74). From Equation (9.65), we have

$$\Delta W = \frac{1}{2} mg = \frac{1}{t} \left[ \frac{m_B K_\phi h_1}{K_\phi - m_B g h_1} + m_B h_r + m_c h_c \right] a_{Gn} \qquad (9.93)$$

Introducing the roll rate (Equation (9.74)) to the above equation and with some lengthy derivations, we arrive at

$$\left( \left| \frac{a_{Gn}}{g} \right| \right)_{\text{threshold}} = \frac{t}{2h} \frac{m}{m_B} \frac{1}{1 + \dot{\theta}_a (1 - h_r/h) + \frac{m_c}{m_B} \frac{h_c}{h}} \qquad (9.94)$$

If we let $m \approx m_B$ and $m_c \approx 0$, Equation (9.94) becomes

$$\left( \left| \frac{a_{Gn}}{g} \right| \right)_{\text{threshold}} = \frac{t}{2h} \frac{1}{1 + \dot{\theta}_a (1 - h_r/h)} \qquad (9.95)$$

which is the same as the one by Gillespie (1992).[4]

From Equation (9.94) or (9.95), the roll-over threshold is further reduced when the mass center is shifted out of its original position when we consider the suspension of the car body.

If we only use Equation (9.92) to estimate the roll-over threshold, the threshold of most sedans ranges from 1.2–1.7, while SUVs and trucks could be as low as 0.4–1.1[5] When people "pimp" their trucks into "monster" trucks, the roll-over threshold is further reduced. From accident reports, the occurrences of roll-over is much higher for cars with lower roll-over thresholds, in general.

---

[4]Equation (9.6), p. 314, Gillespie, T.D., *Fundamentals of Vehicle Dynamics*, Society of Automotive Engineers, Inc., 1992.

[5]Gillespie, T.D. and Ervin, R.D. *Comparative Study of Vehicle Roll Stability*, The University of Michigan Transportation Research Institute, Report No. UMTRI-83-25, May 1983, p. 42.

## 9.8    Limited-Slip Differentials

In Section 4.4.1, the loading conditions of open differentials were presented. The key result is that, with an open differential, the torque is the same for both sides while the speeds are allowed to be different. The FBDs for describing this behavior of an open differential were presented in Figure 4.45.

On the other hand, for a locked differential, the torques can be different for both sides, but the speeds are locked to be the same.

There are several ingenious designs, such as the Torsen differential and limited-slip differentials, to allow a differential to act either as an open differential or locked differential depending on the driving situation. We will provide a more detailed analysis for the limited-slip differentials.

In the paper by Haas and Manwaring (1971), the development of early limited-slip differentials was explained.[6] While modern limited slip differentials or designs of the Torsen differential have more refinements and higher performance, the basic principles are similar.

As shown in Figures 4.44 and 4.45, also in Figure 9.17(a), the pinion gear of an open differential is allowed to rotate freely, making the torque to each axle the same, as are the tractive forces to the tires. Here, we need to consider the causality, the tractive force conditions at the tires will decide the torques transmitted through the differential. As shown in Figure 9.17(b), if the left side tire moves over a puddle of oil, the tractive force will be nearly zero, rendering the tractive force on the right side nearly zero as well.

If a differential can be locked, as shown in Figure 9.17(c), the pinion gear is not allowed to rotate and is locked with its shaft as one rigid body. As a result, the moment equation around the $y$-axis becomes

$$F_{Az}r_p - F_{Bz}r_p + \tau_{ay} = 0 \qquad (9.96)$$

---

[6]Haas, R.H. and Manwaring, R.C., *Development of a Limited Slip Differential*, 1971. ISSN: 0148-7191, e-ISSN: 2688-3627. DOI: https://doi.org/10.4271/710610. Published February 01, 1971 by SAE International in United States.

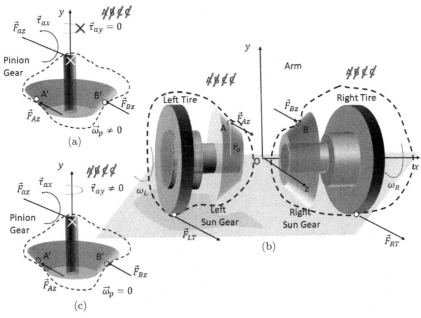

Figure 9.17: (a) The FBD for the pinion gear of an open differential; (b) the FBDs for the left and right axles with the left tire slipping; and (c) the FBD for the pinion gear of a locked differential.

We then obtain

$$F_{Bz} = F_{Az} + \frac{\tau_{ay}}{r_p} \tag{9.97}$$

$$\omega_A = \omega_B \tag{9.98}$$

From Equation (9.97), because $\tau_{ay} \neq 0$, $F_{Bz} \neq 0$ even if $F_{Az} \approx 0$. The torques in the right tires and the corresponding total driving torque are

$$\tau_R = F_{Bz}r_o = F_{Az}r_o + \frac{\tau_{ay}r_o}{r_p}$$

$$= \tau_L + \frac{\tau_{ay}r_o}{r_p} \tag{9.99}$$

$$\tau_d = \tau_R + \tau_L \tag{9.100}$$

One metric to evaluate a differential under different traction conditions is called the bias ratio, defined as the higher tractive torque divided by the lower one. For example, if the left tire slips, then the bias ratio is

$$\mu_B = \frac{\tau_R}{\tau_L} \tag{9.101}$$

A chart representing the bias ratio versus the total driving torque can be used to review the performance of a differential, as shown in Figure 9.18 (see footnote 6).

For an open differential, the bias ratio is 1, represented by the horizontal line of $\mu_B = 1$. With an open differential, we have

$$\tau_R = \tau_L = \frac{1}{2}\tau_d \tag{9.102}$$

Without slipping, an open differential can operate from a low value of $\tau_d$ to high values along this horizontal line. However, once the slipping occurs, the differential will be stuck at one point on the open differential horizontal line. For example, if one tire is slipping over snow and ice with $\tau_L = 100$ ft–lbf, then the differential will be at point $O1$ in Figure (9.18), with the total driving torque at 200 ft–lbf.

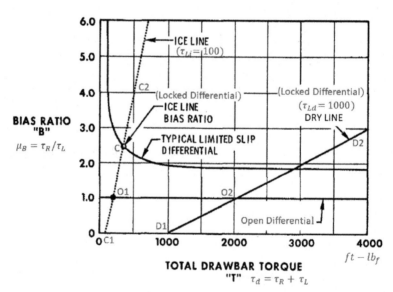

Figure 9.18:   The bias ratio diagram of a limited slip differential (see footnote 6). *Image Modified from*: Haas and Manwaring (1971).

If the left tire slips on some dry road, with $\tau_L = 1{,}000$ ft–lbf, it would be at point $O2$, with a total driving torque of 2,000 ft–lbf.

With a locked differential, the non-slipping tire can have a different tractive torque, as indicated by Equation (9.99). If the left tire is on ice and snow, slipping, while the right tire is on a dry surface, not slipping. Let's assume again that the torque on the left tire is a constant at $\tau_{Li} = 100$ ft–lbf. Because the right tire is not slipping, its torque can go from zero to the static limit, determined by the static friction coefficient and the normal load on the tire.

Equations (9.99) and (9.100) can be specified to describe this ice/snow condition with a locked differential as

$$\tau_d = \tau_R + \tau_{Li} = (\frac{\tau_R}{\tau_L} + 1)\tau_{Li} = (\mu_B + 1) \cdot 100 \qquad (9.103)$$

We can plot Equation (9.103) on the same $\mu_B$ vs $\tau_d$ diagram. Because Equation (9.103) is linear, the operating line is a straight line, denoted as the ICE LINE. When the total driving torque $\tau_d$ is 100 ft–lbf, this torque will be applied to the slipping left tire, making it spin, while the right tire will have zero torque, i.e., $\mu_B = 0$, identified as point $C1$ on Figure 9.18. When the total driving power increases to 500 ft–lbf, the left tire torque stays at 100 ft–lbf while the right tire torque becomes 400 ft–lbf, i.e., $\mu_B = 4$, identified as point $C2$ on Figure 9.18.

Similarly, if the left tire is slipping on a dry surface when $\tau_L = \tau_{Ld} = 1{,}000$ ft–lbf, the torque equation becomes

$$\tau_d = \tau_R + \tau_{Ld} = (\frac{\tau_R}{\tau_L} + 1)\tau_{Ld} = (\mu_B + 1) \cdot 1000 \qquad (9.104)$$

Equation (9.104) defines the DRY LINE of a locked differential on Figure 9.18. On the dry line, point $D1$ is when $\tau_L = \tau_{Ld} = 1{,}000$ ft–lbf, $\tau_R = 0$ ($\mu_B = 0$), and $\tau_d = 1{,}000$ ft–lbf. At $D2$, $\tau_L = \tau_{Ld} = 1{,}000$ ft–lbf, $\tau_R = 3{,}000$ ft–lbf ($\mu_B = 3$), and $\tau_d = 4{,}000$ ft–lbf. As discussed above, with an open differential, when one tire slips, it would be stuck at one point on the horizontal line represented by $\mu_B = 1$. With a locked differential, the differential can be at any point along its operating line, such as the ICE line or the Dry line. Note that, the ICE line has a high slope, indicating that the total torque to both tires would stay low, well below 1,000 ft–lbf in Figure 9.18. This means that even though the differential is locked, the total driving

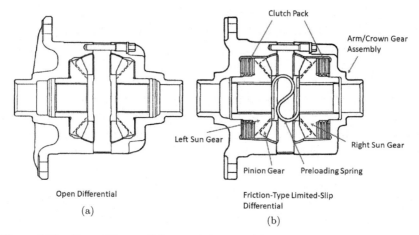

Figure 9.19: Two different differential designs: (a) An open differential; and (b) a friction type limited-slip differential.

*Image Modified from*: Haas and Manwaring (1971).

torque is still limited. Most engine power is wasted on spinning the slipping tire. However, it is far better than an open differential under the same situation.

A limited-slip differential was developed to offer an operation somewhat in between an open and a locked differential. A limited-slip differential can allow different speeds and different torques for the left and the right tires at the same time.

One of the more popular limited-slip differentials is the friction type limited-slip differential, as shown in Figures 9.19(b) and 9.20(a). In addition to the pinon gear, the friction type limited-slip differential offers a second coupling between the arm/crown gear and the sun gears using clutch packs. The clutch packs are engaged by the forces applied by the preloading spring and the separating forces between the pinion gear and the sun gears (Figure 9.20(b)). The separating force is the $x$ component of the driving force ($F_A$ or $F_B$). Because the driving force is higher at higher driving torque, the clutch packs are engaged more tightly when a larger driving torque is applied, allowing a high torque to be transmitted through the clutch packs.

When the clutch packs are not slipping, the limited-slip differential behaves like a locked differential. However, if one tire hits a slick surface, the clutch pack can slip, allowing the tire to spin faster, while applying a higher torque to the non-slipping tire.

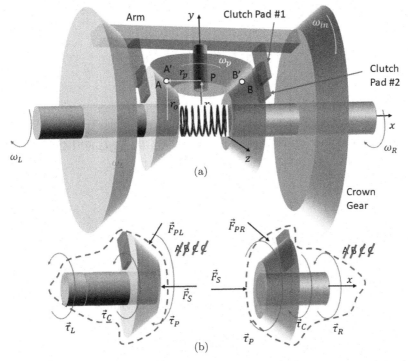

Figure 9.20: (a) The design of a friction type limited-slip differential; and (b) the FBDs for the left and right axles with the left tire slipping.

We can evaluate the performance of a limited-slip differential by the same $\mu_B$ versus $\tau_d$ diagram.

As shown in Figure 9.20(a), assuming that the left tire is slipping and the right tire not rotating, we have $\omega_L > \omega_{in} > \omega_R = 0$ and both clutch packs are slipping at the same rate. As a result, the clutch pack on the left tire is resisting the left tire rotation, while the clutch pack on the right is driving the right tire. Furthermore, because the spring force and the separating force acting on the clutch packs are the same, the coupling torque through the clutch packs, $\tau_C$, are the same as well. We can represent $\tau_C$ as a function of the spring preloading force and the driving torque,

$$\tau_C = f_1(\tau_d, F_S) \tag{9.105}$$

The moment equations of the axles of the limited-slip differential are

$$\tau_P + \tau_C - \tau_R = I_x \alpha_R \tag{9.106}$$

$$\tau_P - \tau_C - \tau_L = I_x \alpha_L \tag{9.107}$$

where $\tau_P$ is the torque transmitted through the free rotating pinion gear.

If we consider quasi-static state and let $\alpha_R \approx 0$ and $\alpha_L \approx 0$, we have

$$\tau_P + \tau_C = \tau_R \tag{9.108}$$

$$\tau_P - \tau_C = \tau_L \tag{9.109}$$

From the above equations, we also obtain

$$\tau_P = \frac{1}{2}\tau_d \tag{9.110}$$

The bias ratio of limited-slip differential becomes

$$\mu_B = \frac{\tau_d - \tau_L}{\tau_L} \tag{9.111}$$

When the left tire is lifted in the air, $\tau_L \approx 0$ and $\mu_B \longrightarrow \infty$. For different slip conditions at higher $\tau_L$, $\mu_B$ will decrease while $\tau_d$ will increase.

Equation (9.111) can be modified into a more convenient form,

$$\mu_B + 1 = \frac{\tau_d}{\tau_L} = \frac{\tau_d}{\tau_d - \tau_R} = \frac{\tau_d}{\tau_d - \tau_P - \tau_C} \tag{9.112}$$

Introducing Equations (9.110) and (9.105) into (9.112), we have

$$\mu_B + 1 = \frac{\tau_d}{\tau_P - \tau_C} = \frac{\tau_d}{\frac{1}{2}\tau_d - f_1(\tau_d, F_S)} \tag{9.113}$$

When $\tau_d$ becomes higher, the separating force by the pinion gear will be much larger than the spring preloading force. The torque transmitted by the clutch becomes mainly a function of the driving torque. Assuming that $\tau_C$ is a linear function of $\tau_d$, we have

$$\tau_c = f_1(\tau_d, F_S) \approx \Gamma \tau_d \tag{9.114}$$

where $\Gamma$ is a coefficient.

As a result, at a higher driving torque, Equation (9.113) is approximated as

$$\mu_B + 1 \approx \frac{1}{\frac{1}{2} - \Gamma} \tag{9.115}$$

Based on Equation (9.115), the operating line of the limited-slip differential becomes a constant, i.e., horizontal, at higher driving torque in Figure 9.18.

The overall operating line of a limited slip differential is shown in Figure 9.18, in which the operating line is a steeply decreasing line at lower values of the driving torque, becoming a horizontal line at higher values of the driving torque.

We can conduct experiments to determine the operating line of a limited-slip differential for different slipping conditions because $f_1$ in Equation (9.113) is typically unknown. The torques to the left and the rights are measured to determine the bias ratio and the total driving torque.

For the operating line shown in Figure 9.18 of a limited-slip differential, if one tire slips over snow and ice, the differential will operate at point $C$, with $\mu_B = 2.5$ and $\tau_d = 350$ ft–lbf, compared with 1.0 and 200 ft–lbf, respectively, for an open differential at point $C1$. For a locked differential, the total driving torque is 400 ft–lbf with $\mu_B = 4$ (point $C2$). Therefore, this limited slip differential is better than an open differential, but not as good as a locked differential. Some limited-slip differentials also offer a manual locking capability. The operating line of the limited-slip differential reaches an asymptotic horizontal line, $\mu_B \approx 1.8$, at higher driving torques. In other words, once the clutch packs start to slip, the right tire torque is about 1.8 times that of the left, an 80% improvement over an open differential.

Most locked differentials require the driver to stop the car and lock the differential by hand when one tire slips.

The Torsen differential discussed in Figure 4.46 can switch between the open mode and the locked mode depending on the situation automatically, which is quite ingenious.

## 9.9 Four-Wheel Drive Design

The design of the limited-slip differential can be modified to divide the total driving torque to the front and the rear tires, offering the

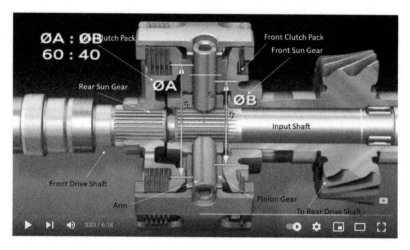

Figure 9.21:   The design of the Audi Quattro all-wheel-drive system.
*Source: Image Modified from*:   https://www.youtube.com/watch?v=9MlhY Hy4pwg.

all-wheel-drive capability. One of the most successful examples is the Quattro all-wheel-drive system offered by the car company, Audi. We will conduct a force analysis to show how this system works.

As shown in Figure 9.21, the Audi Quattro system in fact is very much like a friction type limited-slip differential with three main differences.

The first is that the two sun gears are at different sizes. The radius ratio of the rear sun gear to the front sun gear is 3:2.

The second difference is that the front clutch packs are bigger and with more pads.

The third main difference is that the clutch packs are not engaged until there is a speed difference between the front and the rear tires. This is likely achieved from the spinning of the pinion gear although Figure 9.21 does not show exactly how this is achieved.

The force analysis for the Audi Quattro system is shown in Figure 9.22. The FBD for the input shaft is illustrated by Figure 9.22(a), from which we have

$$\tau_e - \tau_a = I_a \alpha_e \approx 0$$

$$\implies \tau_e \approx \tau_a = F_a \cdot r_a \tag{9.116}$$

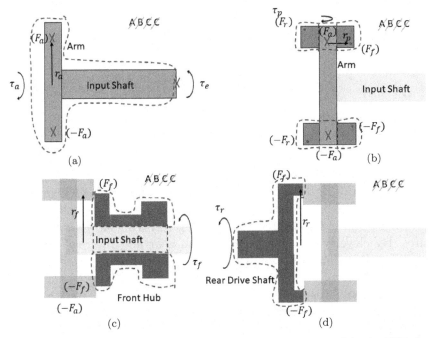

Figure 9.22: The force analysis for the Audi Quattro system: (a) The FBD for the input drive shaft; (b) the FBD for the pinion gear; (c) the FBD for the front sun gear and hub; and (d) the FBD for the rear sun gear and drive shaft.

where $r_a$ is the effective radius of the arm, $F_a$ is the radial force acting on the arm in order to transmit the driving torque, $\tau_a$, to the front and rear drive shafts, and $\tau_e$ is the torque from engine/transmission.

The FBD for the pinion gear is illustrated by Figure 9.22(b). The moment equation for the pinion gear along the axial direction is

$$-\tau_p + F_r r_p - F_f r_p = I_p \alpha_p \approx 0$$
$$\implies \quad F_r \approx F_f \tag{9.117}$$

where $\tau_p \approx 0$ for operating as a free rotating gear, similar to the setting of an open differential.

The FBD for the front sun gear and the hub to mesh to the front drive shaft is illustrated by Figure 9.22(c), from which we have

$$F_f r_f - \tau_f = I_f \alpha_f \approx 0$$
$$\implies \quad \tau_f \approx F_f r_f \tag{9.118}$$

The FBD for the rear sun gear and the rear drive shaft is illustrated by Figure 9.22(d), from which we have

$$F_r \, r_r - \tau_r = I_r \alpha_r \approx 0$$

$$\implies \quad \tau_r \approx F_r \, r_r \tag{9.119}$$

From Equations (9.118) and (9.119), we obtain

$$\frac{\tau_r}{\tau_f} = \frac{r_r}{r_f} = \frac{3}{2} \tag{9.120}$$

Equation (9.120) confirms that during normal driving where there is no speed difference between the front and the rear tire, 60% of the torque or power is delivered to the rear tires, while 40% to the front tires. By cleverly meshing the front sun gear and the rear sun gear at different radial positions, this 60:40 torque/power distribution is accomplished.

If the front tires hit a slick patch and start to spin faster than the rear tires, the Quattro system will engage both the clutch packs. The Quattro system now operates as a limited-slip differential, resulting in more torque being delivered to the rear tires, up to 85:15, as needed.

On the other hand, if the rear tires start to slip and spin faster, the reverse will happen. Because the clutch packs for the front are smaller, only up to 70% of total torque will be delivered to the front tires from the 40% initial setting. The rear tires now receive only 30% of the total torque.

The Quattro system design makes this torque/power distribution automatic and almost instantly without intervention by the driver.

## 9.10 Concluding Remarks

In this chapter, we presented detailed dynamic analyses for different motor vehicle-related problems. These problems together with many other examples presented in Chapters 1–8, constitute a fairly complete treatment of vehicle dynamics. In particular, all the analyses are presented with detailed FBDs and governing equations. Readers with proper background of engineering dynamics should be able to understand the essence of these practical vehicle dynamics problems.

## Chapter 10

# Beyond Rigid Body Dynamics

In Chapter 4, the rigid body assumption was explained. The question is what if the object we are analyzing is NOT a rigid body? Can the laws and governing equations we presented still be used for analysis and problem solving? To answer this question, let's revisit the force analysis and motion analysis under the rigid body assumption. From there, we will proceed to discussing fluid systems and elastic solid systems. The objective is to extend the basic dynamics principles to non-rigid body systems.

## 10.1    Force Analysis for Non-Rigid Bodies

As discussed in Section 2.3 of Chapter 2, when the rigid body assumption is made, the force vector can move along its line of action without altering its effect on the rigid body. In addition, a moment, or a force couple, is a free vector and it can be placed anywhere on the rigid body without altering its effect.

However, if the body is NOT rigid, we no longer have the luxury to move the force and the moment. The force vector and the moment vector are now fixed vectors and they cannot be moved out of their points of application. The same force acting at different points of a non-rigid body will result in different deformations or motions.

However, if the system reaches its static equilibrium, the force and moment vectors can be moved just like they are on a rigid body.

This is because, at static equilibrium, the deformation of the system is no longer changing as if it is rigid. Because we are dealing with dynamics problems, we often do not have the static equilibrium. Therefore, we must not move forces and moments without proper consideration of the resulting consequences.

Figure 2.11 illustrates why the force and moment vectors cannot be moved out of their points of applications.

## 10.2  Motion Analysis for Non-Rigid Bodies

As discussed in Section 4.1, Chapter 4, the rotational properties for a rigid body are universal. The motion analysis for a rigid body is relatively simple because if we know the motion of one point on the rigid body and we know the rotational properties, then we know the motion of every single point, as defined by Equations (4.5) and (4.6).

With the rigid body assumption, a 2D rigid body has three degrees of freedom and a 3D rigid body has six degrees of freedom.

However, if a body is not rigid, it has infinite degrees of freedom. There are no universal rotational properties. Knowing the motion of one point on a non-rigid body is not sufficient to predict the motion of other points unless they are very close to the known point. Under this circumstance, we can only derive governing equations for a very small portion of the system, such as an infinitesimal element of the entire system. Luckily, if the entire system follows some specific patterns, we often can extend the governing equations for the infinitesimal element to the entire system.

Another issue with non-rigid bodies is that it is often difficult to track individual particles of the system. This is particularly true for fluid flows. The difficulty of tracking specific particles poses a fundamental problem for dynamics analysis. Recall that after we isolate a subsystem with an FBD for force analysis, *Newton's Second Law* is applied for the subsystem inside the boundary to obtain the governing equations. We cannot alternate this isolated subsystem by adding or removing some particles from it without changing the governing equations. This is obviously difficult if we cannot track the subsystem for which the FBD is constructed. Therefore, we must consider a new way to conduct analysis.

## 10.3  Extension of Free-Body Diagrams

We introduce a new way to isolate a subsystem for dynamic analysis in this section. As shown in Figure 10.1, at $t_0$, we isolate a system, consisting of $m_2$, $m_3$, $m_4$, and $m_5$ within an FBD boundary. At $t_1$, a short $\Delta t$ later, because all these particles are in motion, we need to move this FBD boundary accordingly to keep them inside the FBD boundary. We also need to update the force conditions at the boundary at $t_1$. This is quite cumbersome and often hard to do.

Let's introduce a new way of isolating a subsystem. Instead of tracking specific particles, let's define a space volume at $t_0$ whose boundary happens to coincide with the FBD boundary defined earlier. This space volume is denoted as the Control Volume, or CV. The CV does not move with the particles. As a result, at $t_1$, $m_5$ moves out of the CV, while $m_1$ moves into the CV. For the isolated subsystem of $m_2, \ldots, m_5$, *Newton's Second Law* dictates that

$$\sum \vec{F} = \frac{d}{dt}\left(\sum_{i=2}^{5} m_i \vec{v}_i\right) = \frac{d}{dt}(\vec{P}_{FB}) = \dot{\vec{P}}_{FB} \qquad (10.1)$$

where $\vec{P}_{FB}$ is the total linear momentum inside the free-body boundary related to $m_2, \ldots, m_5$.

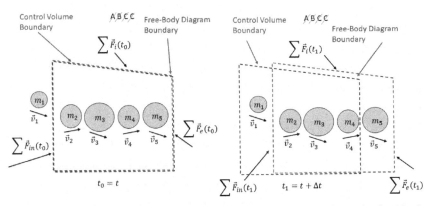

Figure 10.1:  The concept of control volume versus the boundary of a free-body diagram.

We can also calculate the rate change of the linear momentum within the control volume. Note that at $t_0$, the total linear momentum within the CV is $\sum_{i=2}^{5} m_i \vec{v}_i(t)$, while at $t_1$, it is $\sum_{i=1}^{4} m_i \vec{v}_i(t + \Delta t)$. As a result, we have

$$\sum \dot{\vec{P}}_{CV} = \lim_{\Delta t \longrightarrow 0} \frac{1}{\Delta t} \left( \sum_{i=1}^{4} m_i \vec{v}_i(t + \Delta t) - \sum_{i=2}^{5} m_i \vec{v}_i(t) \right)$$

$$= \lim_{\Delta t \longrightarrow 0} \frac{1}{\Delta t} \left[ \sum_{i=2}^{5} (m_i \vec{v}_i(t + \Delta t) - m_i \vec{v}_i(t)) \right.$$

$$\left. + m_1 \vec{v}_1(t + \Delta t) - m_5 \vec{v}_5(t + \Delta t) \right]$$

$$= \lim_{\Delta t \longrightarrow 0} \frac{1}{\Delta t} (\Delta \vec{P}_{FB} + \Delta \vec{P}_{in} - \Delta \vec{P}_{out})$$

$$= \dot{\vec{P}}_{FB} + \dot{\vec{P}}_{in} - \dot{\vec{P}}_{out} \tag{10.2}$$

where, without loss of generality, the linear momentum entering the CV by $m_1$ is denoted as $\vec{P}_{in}$, as the linear momentum influx into the CV, while the linear momentum exiting the CV by $m_5$ is denoted as $\vec{P}_{out}$, as the linear momentum outflow exiting the CV.

Introducing Equation (10.1) into (10.2) and rearranging it accordingly, we have

$$\sum \vec{F} = \dot{\vec{P}}_{CV} - \dot{\vec{P}}_{in} + \dot{\vec{P}}_{out} \tag{10.3}$$

Equation (10.3) is a new way of applying *Newton's Second Law* based on a CV, instead of a specific free-body isolated by the FBD boundary. The right-hand side of Equation (10.3) can be interpreted as

*In order to obtain correct $\dot{\vec{P}}_{FB}$ for applying Newton's Second Law, we need to subtract from $\dot{\vec{P}}_{CV}$ the extra linear momentum influx that enters the CV ( $-\dot{\vec{P}}_{in}$) and bring back the outflow, the one that exits ($+\dot{\vec{P}}_{out}$).*

This new method is very useful for dealing with systems associated with fluid flows or chains, for example, for which it is difficult to trace specific particles.

Based on Equation (10.3), the angular moment equation for an FBD/CV is

$$\sum \vec{\tau}_O = \dot{\vec{H}}_{CV/O} - \dot{\vec{H}}_{\text{in}/O} + \dot{\vec{H}}_{\text{out}/O} \qquad (10.4)$$

where $O$ is a convenient stationary point for determining the moment and the angular momentum.

Similar to the selection of the FBD boundary, the selection of the CV needs some careful considerations. In fact, we actually create an identical boundary for the FBD and for the CV. We use the boundary for the FBD for the force analysis, while using the boundary for the CV to determine $\dot{\vec{P}}_{CV}$, $\dot{\vec{P}}_{\text{in}}$, and $\dot{\vec{P}}_{\text{out}}$. We don't have to draw two identical boundaries as shown in Figure 10.1, but understand that the boundary has a dual use. Just like the boundary of the free-body, a CV can be moving or accelerating. Furthermore, it could be shrinking or expanding. We simply have to determine the correct influx and outflow across the CV.

We must emphasize that the calculation of $\vec{P}_{\text{in}}$ and $\vec{P}_{\text{out}}$ must be based on the mass of the particles across the boundary, and their velocities are with respect to an inertial reference frame. The general expressions for the linear momentum influx and outflow across the CV boundary during a $\Delta t$ time period are

$$\Delta \vec{P}_{\text{in}} = \int_{t'}^{t'+\Delta t} \int_S \rho \vec{v}_{i/S} \cdot (-\vec{e}_{Sn}) \, \vec{v}_{i/O} \, dS \, dt \qquad (10.5)$$

$$\Delta \vec{P}_{\text{out}} = \int_{t'}^{t'+\Delta t} \int_S \rho \vec{v}_{j/S} \cdot \vec{e}_{Sn} \, \vec{v}_{j/O} \, dS \, dt \qquad (10.6)$$

where $\vec{e}_{Sn}$ is the unit directional vector of the outward normal direction of an infinitesimal surface $dS$ of the CV. A negative sign is introduced in Equation (10.5) to make the dot multiplication positive, so that it can be used correctly for Equation (10.3). The velocity, $\vec{v}_{i/O}$ is the velocity of the influx mass flow observed by an inertial reference frame at $O$, while $\vec{v}_{j/O}$ is the velocity of the mass outflow with respect to the same inertial reference frame at $O$.

From Equations (10.5) and (10.6), the rate of linear momentum influx and outflow are

$$\dot{\vec{P}}_{\text{in}} = \int_S \rho \vec{v}_{i/S} \cdot (-\vec{e}_{Sn}) \, \vec{v}_{i/O} \, dS \qquad (10.7)$$

$$\dot{\vec{P}}_{\text{out}} = \int_S \rho \vec{v}_{j/S} \cdot \vec{e}_{Sn} \, \vec{v}_{j/O} \, dS \qquad (10.8)$$

If the influx and outflow velocities are the same across the boundary, Equations (10.7) and (10.8) can be further simplified to

$$\dot{\vec{P}}_{\text{in}} = \int_S \rho \vec{v}_{i/S} \cdot (-\vec{e}_{Sn}) \, dS \, \vec{v}_{i/O}$$

$$= \dot{m}_{\text{in}} \vec{v}_{i/O} \qquad (10.9)$$

$$\dot{\vec{P}}_{\text{out}} = \int_S \rho \vec{v}_{j/S} \cdot \vec{e}_{Sn} \, dS \, \vec{v}_{j/O}$$

$$= \dot{m}_{\text{out}} \vec{v}_{j/O} \qquad (10.10)$$

where $\dot{m}_{\text{in}}$ and $\dot{m}_{\text{out}}$ are the entering and exiting mass flow rates, respectively, defined accordingly in the above equations.

Equations (10.9) and (10.10) are fairly easy to use for many applications when the influx and outflow are uniform across the CV boundary.

## 10.4 Mass Flow Systems

We will consider a few examples involving mass flows in this section.

### Example 10.1: Forces Exerted by a Water Jet

**Problem Statement:** As shown in Figure 10.2, a water jet with a jet speed of $v_{W/O} = 2$ m/s impinges upon a cart with a blade, which diverts the water jet upward. The density of water is $\rho_W = 1,000$ kg/m$^3$ and the water jet has a constant cross-section area of $A_W = 0.002$ m$^2$. At the instant of time, the cart is observed to move horizontally at a constant speed, $v_{A/O} = 1$ m/s. How much force does the water jet exert on the cart? Neglect the gravity and the friction between the water jet and the blade.

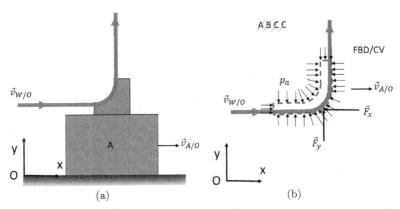

Figure 10.2: (a) A water jet impinges horizontally upon a cart with a blade to divert the water jet upward; and (b) the FBD and CV for the water jet.

**Force Analysis:** First, we draw a boundary of FBD/CV as shown in Figure 10.2(b). Going through the ABCC method, we have distributed contact forces due to the atmosphere pressure, $p_a$, and the contact forces between the water jet and the blade, represented as two concentrated forces, $F_x$ and $F_y$. We neglect the gravity and the frictional force between the water jet and the blade. Note that the net force due to the atmospheric pressure is zero; therefore, $p_a$ is excluded from the subsequent analysis.

**Motion Analysis:** Because the cart moves at a constant speed based on the problem statement, the selected CV also moves at a constant speed. We will define an inertial reference attached to the CV as $A$-$x$-$y$. With respect to the CV boundary, we will determine the rates of linear momentum changes within the CV, $\dot{\vec{P}}_{CV}$, entering the CV, $\dot{\vec{P}}_{\text{in}}$, and exiting the CV, $\dot{\vec{P}}_{\text{out}}$.

From the continuity principle of fluid flow, we know the jet speed is the same at every cross-section of the water jet.

Note that the CV moves to the right at 1 m/s. In order to calculate the influx and outflow, we need to determine the relative velocity of the water jet with respect to the CV. It is found that

$$\vec{v}_{\text{in}/S} = \vec{v}_{W/A} = \vec{v}_{W/O} - \vec{v}_{A/O} = 1\vec{i} \ (\text{m/s}) \tag{10.11}$$

$$\vec{v}_{\text{out}/S} = \vec{v}_{\text{out}/A} = v_{\text{in}/S} \ \vec{j} = 1\vec{j} \ (\text{m/s}) \tag{10.12}$$

The mass flow rate of the water jet entering the CV is

$$\dot{m}_{in} = \rho_W A_W v_{in/S} = 2 \ (kg/s) \tag{10.13}$$

$$\dot{m}_{out} = \dot{m}_{in} \ (kg/s) \tag{10.14}$$

Because the CV moves at a constant speed and we can attach an inertial reference to the CV, the velocity terms for determining the linear momentum terms are the same velocity terms determined above.

$$\vec{v}_{in/A} = \vec{v}_{in/S} \tag{10.15}$$

$$\vec{v}_{out/A} = \vec{v}_{out/S} \tag{10.16}$$

**Governing Equations:** Based on Equation (10.3), we have

$$\sum \vec{F} = -F_x \vec{i} + F_y \vec{j} = \dot{\vec{P}}_{CV} - \dot{\vec{P}}_{in} + \dot{\vec{P}}_{out}$$

$$= \frac{d}{dt}(\vec{P}_{CV}) - \dot{m}_{in} \ \vec{v}_{in/A} + \dot{m}_{out} \ \vec{v}_{out/A}$$

$$= -2\vec{i} + 2\vec{j} \ (kg \ m/s^2) \tag{10.17}$$

where $\dot{\vec{P}}_{CV} = 0$ because the cart is moving at a constant speed and there is no change of the linear momentum inside the CV.

Solving Equation (10.17), we obtain

$$F_x = 2 \ (N) \tag{10.18}$$

$$F_y = 2 \ (N) \tag{10.19}$$

The directions of the forces are as illustrated in Figure 10.2(b).

**Verification:** In this example, the CV happens to move at a constant velocity. If not, we will need to determine $P_{CV}$ and determine its time derivative. We will also need to establish an inertial reference frame not attached to the CV.

### Example 10.2: Rocket Propulsion Analysis

**Problem Statement:** As shown in Figure 10.3, we want to determine the acceleration of a rocket. The mass of the rocket structure and payload is $m_1$, while the mass of the remaining fuel is

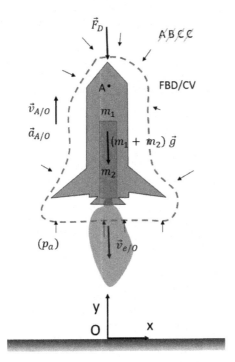

Figure 10.3:   The FBD/CV analysis to determine the acceleration of a rocket.

$m_2 = m_{20} - \dot{m}_e t$, with $\dot{m}_e$ as the fuel consumption rate and $t$, is the time since the firing of the rocket. Typically, $\dot{m}_e$ is kept constant. At this instant of time, the rocket has an upward velocity $\vec{v}_{A/O}$ with respect to an inertial reference frame on the ground. The exhaust coming out of the rocket tail has a velocity of $\vec{v}_{e/O}$. The air drag experienced by the rocket is $F_D$, which is due to the air pressure distribution surrounding the rocket. Determine the acceleration of the rocket.

**Force Analysis:** An FBD is drawn around the entire rocket as shown in Figure 10.3. There are no applied forces, the body force is $(m_1 + m_2)g$, the contact forces are due to the rocket in contact with the air, and there are no constraint forces. The distributed contact forces with the air are summed to a resultant air drag, $F_D$. Note that there is no so-called thrust force at the bottom of the rocket. The static pressure across the frame is essentially similar to that of the surrounding atmosphere.

**Motion Analysis:** We define a CV overlapped with the FBD boundary as shown in Figure 10.3. This CV, as a result, is accelerating with the rocket and has velocity of $\vec{v}_{A/O}$. The flame mass flow exiting the CV lower boundary should be determined by the relative velocity of $\vec{v}_{e/A}$. We have the following relative velocity relationship,

$$\vec{v}_{e/O} = \vec{v}_{A/O} + \vec{v}_{e/A} = (v_{A/O} - v_{e/A})\vec{j} \qquad (10.20)$$

where $\vec{v}_{A/O} = v_{A/O}\vec{j}$ and $\vec{v}_{e/A} = -v_{e/A}\vec{j}$.

By keeping $\dot{m}_e$ a constant, $v_{e/A}$ is also a constant when the flame density, $\rho_f$, and the exit area, $A_f$, are constant. The exiting mass flow rate is determined as

$$\dot{m}_{\text{out}} = \dot{m}_e = \rho_f A_f v_{e/A} = \rho_f A_f |v_{e/O} - v_{A/O}| \qquad (10.21)$$

Note that an absolute operator is applied in the above equation to ensure $\dot{m}_{\text{out}}$ as positive by definition.

Equation (10.3) for the rocket with respect to $O$–$x$–$y$ becomes

$$-F_D\vec{j} - (m_1 + m_2)g\,\vec{j} = \dot{\vec{P}}_{CV} - \dot{\vec{P}}_{\text{in}} + \dot{\vec{P}}_{\text{out}}$$

$$= \frac{d}{dt}((m_1 + m_2)\vec{v}_{A/O}) + \dot{m}_{\text{out}}\vec{v}_{e/O}$$

$$= (\dot{m}_1 + \dot{m}_2)\vec{v}_{A/O} + (m_1 + m_2)\dot{\vec{v}}_{A/O}$$

$$\quad + \dot{m}_e(v_{A/O} - v_{e/A})\vec{j}$$

$$= (-\dot{m}_e)\vec{v}_{A/O} + (m_1 + m_{20} - \dot{m}_e t)a_{A/O}\vec{j}$$

$$\quad + \dot{m}_e(v_{A/O} - v_{e/A})\vec{j}$$

$$= (m_1 + m_{20} - \dot{m}_e t)a_{A/O}\,\vec{j} - \dot{m}_e v_{e/A}\,\vec{j} \qquad (10.22)$$

where $\dot{m}_{\text{in}} = 0$ and $\dot{v}_{e/A} = 0$.

The acceleration of the rocket is then obtained as

$$a_{A/O} = \frac{1}{m_1 + m_{20} - \dot{m}_e t}(\dot{m}_e v_{e/A} - F_D) - g \qquad (10.23)$$

The acceleration of the rocket comes from the term $\dot{m}_e v_{e/A}$. Often, we consider this term as the thrust force of the rocket. The thrust force will be in the FBD if we define the FBD boundary inside the flame nozzle where the pressure is much higher. However, inside the nozzle the exiting velocity with respect to the nozzle is unknown and likely small with a higher density. As a result, we merely add an extra variable to Equation (10.22), making it hard to solve. With the boundary shown in Figure 10.3, there is no thrust force in the FBD.

## Example 10.3: Chain Pulling Analysis

**Problem Statement:** As shown in Figure 10.4(a), a chain is pulled upward at a constant speed, $v$. We want to determine the force, $F_1$, which is needed when the vertical length of the chain is $y$. For Figure 10.4(b), the chain is lowered down at the same speed, $v$, we need to know force, $F_2$, needed to hold the chain while lowering. The density of the chain per unit length is $\rho$ and the total length of the chain is $L$.

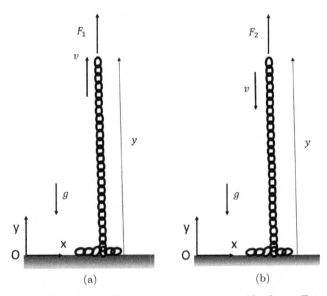

Figure 10.4: (a) A chain is pulled up at a constant speed by force, $F_1$; and (b) the same chain is lowered at the same speed with force $F_2$ to resist the gravitational force.

**Force Analysis:** For the case of Figure 10.4(a), we draw several CV/FBDs to solve the problem in different ways, as shown in Figure 10.5. The first CV/FBD (Figure 10.5(a)) is for the entire vertical portion of the chain. A separate CV/FBD is defined for the chain elements on the ground (Figure 10.5(b)). The third FBD (Figure 10.5(c)) is for one infinitesimal chain element which just got pulled up off the ground. Although a real chain is made of finite-sized rings or beads, we will model the chain as a continuous and flexible body. The fourth FBD (Figure 10.5(d)) is for the entire chain.

**Governing Equations and Solutions:** For the CV/FBD of Figure 10.5(a), we define the CV big enough on the top so that no chain element exits the CV. The influx mass flow is

$$\dot{m}_{\text{in}} = \rho v \tag{10.24}$$

we have the governing equation as

$$F_1 - \rho y g - F_s = \frac{d}{dt}(\rho y v) - \dot{m}_{\text{in}} v$$

$$= \rho \frac{dy}{dt} v + \rho \frac{dv}{dt} y - \rho v^2 = 0 \tag{10.25}$$

There are two unknowns, $F_1$ and $F_s$, in Equations (10.25). We cannot solve for them. The FBD of Figure 10.5(b), it does not help because it contains two more unknowns, $F_s$ and $N$.

With the FBD of Figure 10.5(c), not a CV, we follow an infinitesimal element with a mass $dm = \rho \cdot dy$. At $t$, this little element was on the ground and at $t+dt$, it is now off the ground with a velocity $v$. The only forces involved are $F_s$ and the gravity assuming the chain pile on the ground does not impede the rise of the chain element. Invoking the linear momentum equation for the infinitesimal element, we have

$$(\vec{P}_{dy})_t + \int_t^{t+dt} (\vec{F}_s - \rho g \, dy \, \vec{j}) dt' = (\vec{P}_{dy})_{t+dt} \tag{10.26}$$

where $(\vec{P}_{dy})_t = 0$ because it has a zero velocity, while $(\vec{P}_{dy})_{t+dt} = \rho v \, dy \, \vec{j}$. The gained linear momentum is due to $F_s$ because the time is too short for the tiny gravitational force, $\rho g dy$, to contribute to

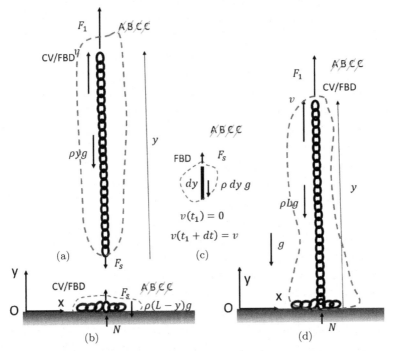

Figure 10.5: (a) The FBD for the vertical chain; (b) the FBD for the pile on the ground; (c) the FBD for one infinitesimal chain piece just off the ground; and (d) the FBD for the entire chain including the pile on the ground.

any linear momentum. Also during this short time, we can treat $F_s$ as a constant; therefore, we have

$$\int_t^{t+dt} (F_s - \rho g \, dy) dt' = F_s \, dt = (P_{dy})_{t+dt} = \rho v \, dy \qquad (10.27)$$

Dividing $dt$ to the right-hand side of the above equation, we obtain $F_s$ as

$$F_s = \rho v^2 \qquad (10.28)$$

Note that $v = dy/dt$. Introducing Equation (10.28) into (10.25), we obtain

$$F_1 = \rho y g + F_s = \rho g y + \rho v^2 \qquad (10.29)$$

Let's use the CV/FBD of the pile on the ground Figure 10.5(b)). The governing equation becomes

$$(F_s - \rho(L - y)g + N)\vec{j} = \dot{\vec{P}}_{cv} - \dot{\vec{P}}_{in} + \dot{\vec{P}}_{out}$$

$$= \dot{\vec{P}}_{cv} + \dot{m}_{out}\vec{v} = 0 + \rho v^2 \vec{j} = \rho v^2 \vec{j} \qquad (10.30)$$

where $\dot{\vec{P}}_{cv} = 0$ because inside the CV, there are just chain elements with zero velocity.

With $F_s = \rho v^2$, we found out

$$N = \rho(L - y)g \qquad (10.31)$$

The force, $N$, is equal to the dead weight of the chain pile on the ground. In other words, the ground does not push the chain element to rise. It is all due to $F_1$.

Finally, let's examine the CV/FBD for the entire chain. The governing equation is

$$(F_1 - \rho Lg + N)\vec{j} = (F_1 - \rho yg)\vec{j} = \dot{\vec{P}}_{cv} - \dot{\vec{P}}_{in} + \dot{\vec{P}}_{out}$$

$$= \dot{\vec{P}}_{cv} + 0 + 0 = (\rho \dot{y} v)\vec{j} = \rho v^2 \vec{j} \qquad (10.32)$$

We obtain the same result for $F_1$.

Let's consider the lowering chain case of Figure 10.3(b). The corresponding CV/FBDs are shown in Figure 10.6.

Based on the CV/FBD of Figure 10.6(a), there are no influx or outflow mass flows. The governing equation becomes

$$F_2 - \rho Lg + N = \frac{d}{dt}(\rho y(-v)) = \rho v^2 \qquad (10.33)$$

Note that $\frac{dy}{dt} = -v$. We need to determine $N$ in order to determine $F_2$. We will split $N$ into two components, one related to the weight of the chain already on the ground ($N_0 = \rho(L - y)g$) and the impact force ($N_W$) applied by the ground to the chain element to reduce its speed from $v$ to 0.

Let's consider the FBD for an infinitesimal piece of the chain right before it hits the ground, as shown in Figure 10.6(b). The only forces acting on this little piece are the tensile force $F_W$ and the little gravitational force $\rho g \cdot dy$. An impact force $N_W$ is applied by the ground to make the infinitesimal chain element have a zero velocity at $t_1 + dt$.

The linear momentum equation for the infinitesimal element is

$$-\rho \cdot dy \cdot v + \int_{t_1}^{t_1+dt'} (F_W - \rho g\, dy + N_W)dt = 0$$

$$\implies -\rho \cdot dy \cdot v + N_W\, dt = 0 \qquad (10.34)$$

where $F_W$ and $\rho g\, dy$ are NOT HUGE to contribute any linear momentum changes due to the short impact time.

As a result, we have

$$N_W = \rho v^2 \qquad (10.35)$$

Introducing Equation (10.35) into (10.33) and with $N = N_0 + N_W$, We finally have

$$F_2 = \rho y g \qquad (10.36)$$

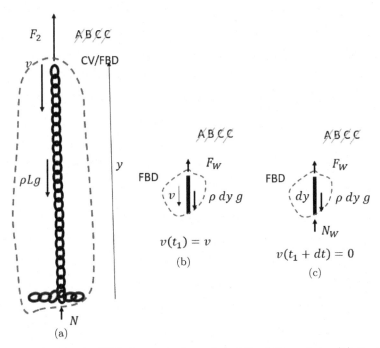

Figure 10.6: (a) The FBD for vertical portion of the falling chain; (b) the FBD for the infinitesimal chain piece right before it hits the ground; and (c) the FBD for the infinitesimal piece the moment it hits the ground.

## Example 10.4: Steady-State Chain Fountain Analysis

**Problem Statement:** In 2013, Mould[1] demonstrated a long chain would hoist itself up, like a cobra, when it slips out of a cup and falls to the ground some distance below. This is called the chain fountain. A video presented by Yokoyama[2] demonstrates a chain fountain from beginning to finish. In Figure 10.7(a), a screen shot of the above video is shown, in which the chain fountain reaches its maximum height, 16 cm, with the chain falling to 123 cm below the cup. Throughout the video, it is shown that the chain fountain gradually rises, reaching its maximum height, maintaining this height in a quasi-steady-state, but never quite steady, until the last piece of the chain falls out of the cup and to the ground. The chain speed was found to be about 4.2 m/s.

In Figure 10.7(b), we model the chain fountain with an idealized shape at steady state. The maximum vertical height of the rising chain before the chain forms the curve is defined as $y_1$. Similarly, the maximum vertical height of the falling chain right below the curve to the lower ground is $y_2$. The arc length of the curve is $l_c$, which not necessarily is a semi-circle. We also define the mass center of the curved chain portion as point $C$. Based on this idealized steady-state chain fountain, we attempt to provide a model of the chain fountain based on basic laws of dynamics and vigorous force analysis. From there, we will discuss a few issues related to a real chain fountain not included in the idealized model above, such as the rather wrinkly look of the rising chain, compared with that of the falling chain which is relatively straight.

**Solutions:**

**Force Analysis:** Similar to Example (10.3), we construct several FBDs for different parts of the chain fountain (Figures 10.8). In Figure 10.8(a), an FBD/CV is constructed for the entire chain fountain.

---

[1]Mould, S. Self siphoning beads, 2013. http://stevemould.com/siphoning-beads/.

[2]Yokoyama, H. Reexamining the Chain Fountain, 2018. arXiv:1810.13008v1, arXiv.org, https://arxiv.org/ftp/arxiv/papers/1810/1810.13008.pdf.

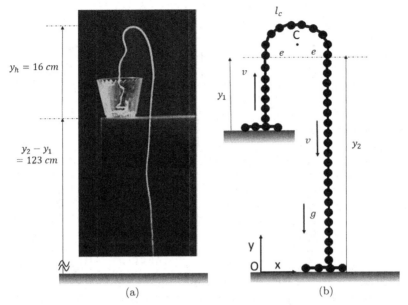

Figure 10.7: (a) An image of a chain fountain from a video by Yokoyama[6]; and (b) an idealized chain fountain at steady state.

*Source*: (a) Modified from https://www.dropbox.com/s/tevep4lpwggj95s/Fig %202%20Mould%20Chain%20Fountain.wmv?dl=0.

The total gravitational force of the chain fountain is $\rho L g$, not including the chain elements inside the cup or on the lower ground, with $\rho$ as the density of the chain and $L = y_1 + l_c + y_2$. The contact force by the upper ground (or cup) is $N_1$. This force is often described as the kick force in the literature. The contact force on the lower ground is $N_2$. These two contact forces, $N_1$ and $N_2$, do not include the weights of the beads in the cup and on the lower ground because they are balanced by static forces from the ground and are not relevant to the chain fountain.

Based on Figure 10.8(a), the governing equation for a steady-state chain fountain is

$$(N_1 - \rho L g + N_2)\vec{j} = \dot{\vec{P}}_{cvI} - \dot{\vec{P}}_{\text{in}I} + \dot{\vec{P}}_{\text{out}I}$$

$$= 0 + 0 + 0 = 0 \qquad (10.37)$$

With the CV as defined in Figure 10.8(a), there is no influx or outflow of linear momentum, while inside the CV, there is no change

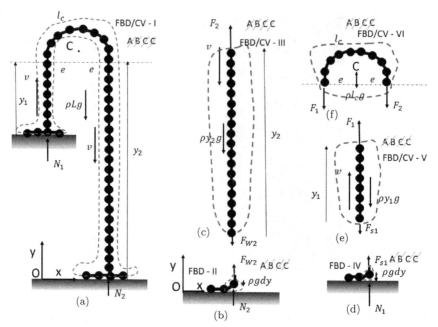

Figure 10.8: (a) The FBD for vertical portion of the falling chain; (b) the FBD for the infinitesimal chain piece right before it hits the ground; (c) the FBD for the infinitesimal piece the moment it hits the ground; (d) the FBD fountain chain on the ground; (e) the FBD for the vertical portion of the falling chain; and (f) the FBD for the arc on the top.

of the linear momentum at the steady state. From Equation (10.37), we have

$$N_1 = \rho L g - N_2 \qquad (10.38)$$

For the FBD defined in Figure 10.8(b) for an infinitesimal chain piece just off the ground, it is observed that at time $t$, it has a speed $v$ and then at $t + dt$, its impact with the ground reduces its speed to zero by the impact force $N_2$. The linear momentum equation for $t$ and $t + dt$ is

$$(\vec{P}_{dyII})_t + \int_t^{t+dt} (\vec{N}_2 + \vec{F}_{W2} - \rho g \cdot dy \vec{k}) dt' = (\vec{P}_{dyII})_{t+dt} = 0$$

$$\implies -\rho \cdot dy \cdot v + \int_t^{t+dt} N_2 \, dt' = 0 \qquad (10.39)$$

where the contribution of $\rho g \cdot dy$ and $F_{W2}$ are excluded because they are NOT HUGE and the integration interval is too short. Assuming $N_2$ is a uniform pulse, we have

$$N_2 = \rho \left( \frac{dy}{dt} \right) v = \rho v^2 \qquad (10.40)$$

The result of Equation (10.40) is the same classical result of a falling chain at a constant speed, as derived in Example (10.3).

Introducing Equation (10.40) into (10.38), we obtain $N_1$ as

$$N_1 = \rho L g - \rho v^2 \qquad (10.41)$$

There is a debate in the literature on whether $N_1$ exists or not. Biggins and Warner[3] proposed its existence while Yokoyama (see footnote 2) presented cases that $N_1 = 0$.

Based on Equation (10.41), $N_1 = 0$ when $\rho L g = \rho v^2$, yielding $v = \sqrt{Lg}$, which is not correct in most cases. However, as will be shown later, typically, $v < \sqrt{Lg}$; therefore, $N_1 \neq 0$ in most cases.

Note that $N_1 = 0$ in the chain upward pulling case in Example 10.3. However, because likely $N_1 \neq 0$ in the chain fountain, the pulling up portion of the chain fountain is different from a simple upward pulling straight chain case.

With respect to the FBD/CV of Figure 10.8(c), the mass flow rate in and out of the CV is

$$\dot{m}_{\text{in}} = \dot{m}_{\text{out}} = \rho v \qquad (10.42)$$

The governing equation becomes

$$(F_2 - \rho y_2 g - F_{W2})\vec{j} = \dot{\vec{P}}_{cvIII} - \dot{\vec{P}}_{\text{in}III} + \dot{\vec{P}}_{\text{out}III}$$
$$= 0 - (\rho v(-v)) + (\rho v(-v)) = 0 \qquad (10.43)$$

We then obtain

$$F_2 = \rho y_2 g + F_{W2} \approx \rho y_2 g \qquad (10.44)$$

where $F_{W2}$ is the tension of the chain just off the lower ground. This tension likely can be assumed to be zero.

---

[3]Biggins, J.S. and Warner, M. Understanding the chain fountain, *Proc. R. Soc. A* **470**(20130689) (2014). http://dx.doi.org/10.1098/rspa.2013.0689.

The force, $F_2$, is the same as Equation (10.36) for the falling chain at a constant speed case of Example 10.3.

With respect to the FBD of Figure 10.8(d), the linear momentum equation for one infinitesimal chain piece between $t$ and $t + dt$ is

$$(\vec{P}_{dyIV})_t + \int_t^{t+dt'} (\vec{N}_1 + \vec{F}_{s1} - \rho g \cdot dy\vec{k})dt' = (\vec{P}_{dyIV})_{t+dt}$$

$$\implies 0 + \int_t^{t+dt'} (N_1 + F_{s1})dt' = \rho \cdot v \cdot dy$$

$$\implies N_1 + F_{s1} = \rho v^2 \tag{10.45}$$

where the linear momentum contribution of $\rho g dy$ is too small and is excluded. We cannot exclude $N_1$ and $F_{s1}$ at this moment.

From the FBD/CV of Figure 10.8(e), we have

$$(F_1 - \rho y_1 g - F_{s1})\vec{j} = \dot{\vec{P}}_{cvV} - \dot{\vec{P}}_{inV} + \dot{\vec{P}}_{outV}$$

$$= 0 - (\rho v(v)) + (\rho v(v)) = 0 \tag{10.46}$$

We obtain

$$F_1 = \rho y_1 g + F_{s1} \tag{10.47}$$

With respect to Figure 10.8(f), the governing equation related to the linear momentum is

$$(-F_1 - \rho l_c g - F_2)\vec{j} = \dot{\vec{P}}_{cvVI} - \dot{\vec{P}}_{inVI} + \dot{\vec{P}}_{outVI}$$

$$= 0 - \rho v(v\vec{j}) + \rho v(-v\vec{j}) = -2\rho v^2 \vec{j} \tag{10.48}$$

which yields, with the introduction of Equation (10.44),

$$F_1 = 2\rho v^2 - F_2 - \rho l_c g$$

$$= 2\rho v^2 - \rho(l_c + y_2)g \tag{10.49}$$

With respect to the same FBD/CV of Figure 10.8(f) in terms of the angular momentum and moments with respect to a stationary point $C$, we have

$$F_1 e\vec{k} - F_2 e\vec{k} = \dot{\vec{H}}_{cvVI/C} - \dot{\vec{H}}_{\text{in}VI/C} + \dot{\vec{H}}_{\text{out}VI/C}$$

$$= 0 - (-\rho v^2)e\vec{k} + (-\rho v^2)e\vec{k} = 0 \qquad (10.50)$$

The above equation leads to

$$F_1 = F_2 \qquad (10.51)$$

With Equations (10.49) and (10.51), we obtain that

$$v = \sqrt{(y_2 + \frac{l_c}{2})g} \qquad (10.52)$$

Subsequently, we also obtain

$$F_{s1} = \rho(y_2 - y_1)g \qquad (10.53)$$

$$N_1 = \rho(y_1 + \frac{l_c}{2})g \qquad (10.54)$$

We now have a complete solution of an idealized steady-state chain fountain, as shown in Figure 10.9. The entire chain is under tension. At different locations of the chain, the tensions are different and the tension is zero ($F_{W2} = 0$) right before it hits the ground.

If we follow one infinitesimal chain element and account for its energy equation between when it just lifts off the cup (at $t = t_c$) and right before it hits the ground (at $t = t_g$), we have

$$T_{t_c} + \rho\, dy(y_2 - y_1)g + \int \sum \vec{F}_{\text{non}} \cdot d\vec{r}_e + \int \sum \vec{f}_{\text{non}} \cdot d\vec{r}_i = T_{t_g}$$

$$(10.55)$$

where $T_{t_c} = 0$ is the kinetic energy of $\rho\, dy$ before it gains the velocity due to $N_1$ and $F_{s1}$, and $T_{t_g} = \frac{1}{2}\rho \cdot dy \cdot v^2$ before it hits the ground. $\vec{F}_{\text{non}}$ are non-conservative external forces acting on the chain fountain, which are mainly $N_1$ and the tension of the chain when it travels along the fountain. The force, $N_2$, is not considered because $t_g$ is right before it hits the ground. The term, $\int (\sum \vec{F}_{\text{non}} \cdot d\vec{r}_e)$, is the work done by the non-conservative external forces. $\vec{f}_{\text{non}}$ is the non-conservative internal forces related to the chain fountain. When we assume the

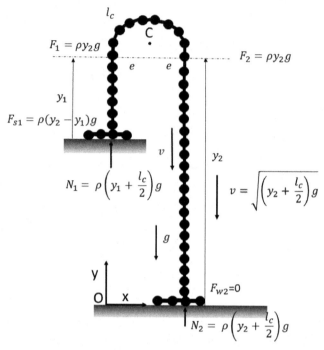

Figure 10.9: The forces and the velocity of an idealized chain fountain at steady state.

idealized chain fountain with infinitesimal chain element, $dy$, the net work done by the internal forces is zero because each $dy$ element is considered as a rigid body.

Rewriting Equation (10.55) in a simpler form, we have

$$\rho \cdot dy \cdot (y_2 - y_1)g + \Delta W_{dy} = \frac{1}{2}\rho \cdot dy \cdot v^2$$

$$= \frac{1}{2}\rho \cdot dy \cdot \left(y_2 + \frac{l_c}{2}\right)g \quad (10.56)$$

We obtain the dissipating work $\Delta W_{dy}$ as

$$\Delta W_{dy} = -\rho \cdot dy \cdot g \left(\frac{1}{2}y_2 - y_1 - \frac{l_c}{4}\right) \quad (10.57)$$

If we simply drop a $dy$ element of chain at rest by a distance of $y_2 - y_1$, the speed before hitting the ground is

$$\hat{v} = \sqrt{2(y_2 - y_1)g} \quad (10.58)$$

The kinetic energy loss ratio is

$$\eta = \frac{|v^2 - \hat{v}^2|}{\hat{v}^2} = \frac{1}{2} - \frac{y_1 + l_c/2}{2(y_2 - y_1)} \tag{10.59}$$

The speed ratio can be defined as the chain speed over the free-falling speed, as

$$\epsilon = \frac{v}{\hat{v}} = \sqrt{\frac{y_2 + l_c/2}{2(y_2 - y_1)}} \tag{10.60}$$

Let's verify the above results with the actual chain fountain of Figure 10.6(a). We estimate that $l_c = 7.62$ cm and $y_1 = 13.57$ cm, $y_2 - y_1 = 123$ cm, and $y_2 + l_c/2 = 140.4$ cm. The chain speed obtained from Equation (10.52) is $v = 3.7$ m/s, compared with the measured speed of 4.2 m/s, which is about 12% error. However, it should be noted that the actual chain speed will change throughout the entire fountain chain process. Therefore, it is not clear if the measured speed actually corresponds to the heights reported by Yokoyama.

The free-falling speed is calculated to be $\hat{v} = 4.97$ m/s. The corresponding speed ratio for calculated speed is $\epsilon = 0.744$, while the measured speed ratio is 0.845, which is quite high.

The measured kinetic energy loss ratio is 28.6%, which appears to be low, calling into question the measured speed above. The calculated kinetic energy loss ratio is 44.3%.

The experiments done by Pantaleone[4] for three different chains indicate that the measured speed ratios are $0.70 \pm 0.03$, $0.69 \pm 0.03$, and $0.71 \pm 0.04$ m/s, compared with the predicted speed ratios as $0.748 \pm 0.003$, $0.748 \pm 0.003$, and $0.755 \pm 0.003$ m/s based on a similar speed prediction of Equation (10.52). The measured kinetic energy loss ratio is 51%, which mostly is due to the internal forces of the actual bead element.

### 10.4.1  *Practical considerations of the chain fountain*

A discussion of the deviation of the idealized model versus the actual chain fountain is in order.

---

[4]Pantaleone, J. A quantitative analysis of the chain fountain, *American Journal of Physics* **85**(2017) 414. Doi: 10.1119/1.4980071.

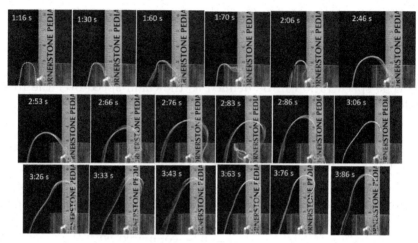

Figure 10.10:    Screen shots of a chain fountain video with a bead chain of 10.1 m. The chain fountain starts at 1.16 s and reaches a quasi-steady state at about 3.86 s.

### 10.4.1.1   *Velocity factor*

In the idealized model, the speed of the chain is assumed to be always tangential to the chain loop. In the straight portion of the chain, the velocity will be either vertically up or down, while in the curved portion, the velocity will be tangential to the curve. However, the video images of the chain fountain indicate that the chain element could have a lateral velocity component. Figures 10.10 and 10.11 are video images of a fountain chain experiment conducted by the author. The overall length of the chain is 10.1 m as shown in Figure 10.10. From these images, it is clear that the chain element could have a lateral velocity component. This lateral motion most likely is due to the chain pile being spread inside the cup and the new chain element picking up the velocity not always being aligned vertically. However, it is also clear that the lateral velocity component could appear in both the rising and falling portions of the chain fountain. Therefore, there is a chain lateral vibration.

### 10.4.1.2   *Asymmetric fountain profile*

Even at a relatively steady state situation, as shown in Figure 10.11 from 3.76 s to 4.30 s, along with the image of Figure 10.7(a), the rising portion of the chain fountain is not straight, but curvy, similar to a

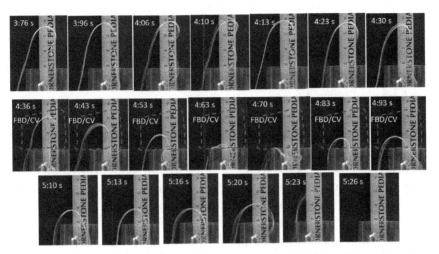

Figure 10.11: Screen shots of the same chain fountain of Figure 10.10. The chain fountain starts to collapses at 4.30 s and completes at 5.26 s.

slacking string not under sufficient tension to keep it straight. As a result, the forces $F_1$ and $F_2$ may no longer be the same as indicated by Equation (10.51).

### 10.4.1.3 Transient state of chain fountain

As shown in Figure 10.23 for images from 4.36 to 4.93 s, the upper portion of the chain fountain collapses and then regrows, after previously being at a fairly steady-state condition. If we define a fixed FBD/CV for the upper portion of the chain fountain as indicated in Figure 10.23, the linear momentum inside the CV is no longer a constant. Equation 10.47 should be rewritten as

$$-F_1\vec{j} - \rho l_c g \vec{j} - F_2 \vec{j} = \dot{\vec{P}}_{cvVI} - \dot{\vec{P}}_{\text{in}VI} + \dot{\vec{P}}_{\text{out}VI}$$
$$\implies F_1 + \rho l_c g + F_2 = -\dot{P}_{cvVI} + (\rho v_1^2) + \rho v_2^2 \qquad (10.61)$$

where $v_1$ is the chain speed entering the CV, while $v_2$ is the one leaving the CV. Here, $v_1 \neq v_2$. In the above equation, we assume that $F_2$ is relatively steady, predicted by Equation (10.44), because it is related to the falling portion of the chain. Similarly, we can assume that $v_2 \approx v$. As a result, based on Equation (10.61), when the fountain chain collapses (from 4.36 to 4.70 s), $\dot{P}_{cvVI} < 0$ and

$F_1$ and $l_c$ get smaller. To maintain Equation (10.61), $v_1$ needs to be smaller. The end result is the collapsing chain fountain. From 4.7 to 4.93 s, the opposite happens.

#### 10.4.1.4 *Changes of $N_1$*

When the chain fountain is collapsing or growing, $v_1$ will change accordingly. Based on Equation (10.45), when a little chain element gains linear momentum, it is due to $N_1$ and $F_{s1}$. At transient states, we need to replace $v$ as $v_1$ for Equation (10.45). When the chain fountain is collapsing, with $v_1$ smaller, so are $N_1$ and $F_{s1}$. The opposite happens when the chain fountain is growing.

#### 10.4.1.5 *Energy Consideration*

From Equations (10.55)–(10.60), we derived the energy loss of the chain fountain. The energy loss ratio, compared with a free-falling case, is about 44.3%, compared with actually measured value of 51%. The extra loss is due to the negative work done by non-conservative external and internal forces for a real chain. In a real chain, there are frictional losses between the finite-sized chain elements, which are not considered for the ideal chain with infinitesimal chain elements. In addition, because of the lateral velocity component of the chain fountain, the kinetic energy associated with the lateral velocity is considered as waste. In conclusion, a chain fountain is a highly interesting phenomenon. Many journal papers have been published on this topic, but most of these papers presented their equations without proper motion and force analysis. The idealized steady-state model presented here is based on fundamental governing equations of dynamics and the results match the experimental results very well. The idealized model can also be extended to explain the behaviors of a real chain fountain deviating from the idealized steady-state model.

### 10.5    Kinematic Analysis for Elastic Bodies

As discussed early in this chapter, an elastic body will have infinite degrees of freedom, unlike a rigid body. However, because the particles of an elastic body will still follow some material properties to stay as a body, it is possible to approximate the entire elastic body

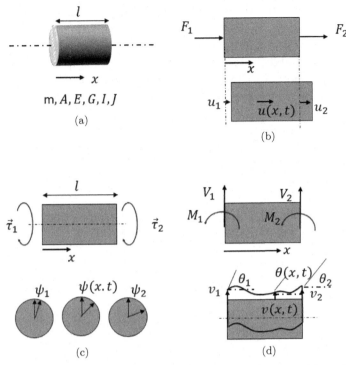

Figure 10.12: The finite element formulation: (a) a short beam; (b) axial loading and axial displacement; (c) torsional loading and displacement; and (d) transverse loading and displacement.

with finite degrees of freedom. Without loss of generality, we will use a short elastic cylindrical beam as an example to discuss the kinematic analysis for elastic bodies.

One most effective way of representing the motion of an elastic body is the Finite Element Method (FEM). The underlying principle of FEM is to break an entire elastic body to many finite-sized elements. Each of these elements will have an idealized geometry. For example, a long beam with different cross-sections can be broken down to many short beams with a finite size (not infinitesimal), as shown in Figure 10.12(a). For a finite-sized short beam, we can determine the motion and loading at the boundaries. We then try to determine the motion of any point on the beam based on the motions at the boundaries. In Figure 10.12(b), the short beam is subjected to axial loading, $F_1$ and $F_2$, at both ends. We also observe that the axial displacement on the left and right ends are $u_1$ and $u_2$, respectively, at time $t$. We need to determine the axial displacement $u(x,t)$ at $x$. If the beam is short enough, $u(x,t)$ likely will be related to $u_1$ and $u_2$. It is reasonable to assume that the relationship is linear and it does not change with time. If this assumption does not hold, we can shorten the beam further so that a linear relationship becomes valid. We therefore represent $u(x,t)$ as

$$u(x,t) = [\phi_{a1}(x) \ \phi_{a2}(x)] \begin{bmatrix} u_1(t) \\ u_2(t) \end{bmatrix} = \phi_{a1}(x)u_1(t) + \phi_{a2}(x)u_2(t)$$

$$(10.62)$$

The functions, $\phi_{a1}$ and $\phi_{a2}$, are called shape functions for the axial loading case and are assumed to be linear,

$$\phi_{a1}(x) = ax + b \qquad (10.63)$$

$$\phi_{a2}(x) = cx + d \qquad (10.64)$$

Because $u(0,t) = u_1(t)$ and $u(l,t) = u_2(t)$, the shape functions are found to be

$$\phi_{a1}(x) = 1 - \frac{x}{l} = 1 - \xi \qquad (10.65)$$

$$\phi_{a2}(x) = \frac{x}{l} = \xi \qquad (10.66)$$

where $\xi = \frac{x}{l}$ is a dimensionless variable.

Similarly, if the short beam is subjected to torsional loading, the torsional displacement, $\psi(x,t)$, at $x$ is found to be

$$\psi(x,t) = [\phi_{t1}(x) \ \phi_{t2}(x)] \begin{bmatrix} \psi_1(t) \\ \psi_2(t) \end{bmatrix}$$

$$= \phi_{t1}(x)\psi_1(t) + \phi_{t2}(x)\psi_2(t) \qquad (10.67)$$

$$\phi_{t1}(x) = 1 - \frac{x}{l} = 1 - \xi \qquad (10.68)$$

$$\phi_{t2}(x) = \frac{x}{l} = \xi \qquad (10.69)$$

where $\phi_{t1}$ and $\phi_{t2}$ are shape functions for the torsional loading case.

If the beam is subjected to the transverse loading and bending moment, as shown in Figure (10.12d), the resulting displacements involve both the vertical displacement and the angular displacement. Both of these two displacements can be caused by both the transverse loading and the bending moment. The total vertical displacement, $v(x, t)$, with the respect to time $t$ and at position $x$ can be observed directly. The function $v(x, t)$ represents the deformed beam profile. The total angular displacement of the beam is the gradient of the deformed profile, i.e., $\frac{\partial v(x, t)}{\partial x}$. We only need to determine $v(x, t)$ and take a derivative with respect to $x$ to determine the gradient.

When we try to represent $v(x, t)$ with the displacements at both ends, we cannot rely on the vertical displacements only. We have to consider the angular displacements at both ends as well because two very different deformed beam profile could have the same vertical displacements at both ends.

The total vertical displacements at both ends are $v_1 = v(0, t)$ and $v_2 = v(l, t)$, respectively. However, we typically do not want to choose the total angular displacement at both ends. Instead, it is preferred to choose only the portion of the angular displacement directly caused by the bending moments. Lets $\theta$ as the angular displacement, often denoted as the rotation of the beam, solely caused by the bending moment. Two bending angular displacements at both ends are $\theta_1 = \theta(0, t)$ and $\theta_2 = \theta(l, t)$, respectively.

With appropriate shape function, the transverse displacement of any point of the short beam becomes.

$$v(x, t) = [\phi_{b1}(x) \ \phi_{b2}(x) \ \phi_{b3}(x) \ \phi_{b4}(x)] \begin{bmatrix} v_1(t) \\ \theta_1(t) \\ v_2(t) \\ \theta_2 t) \end{bmatrix}$$

$$= \phi_{b1}(x)v_1(t) + \phi_{b2}(x)\theta_1(t) + \phi_{b3}(x)v_2(t) + \phi_{b4}(x)\theta_2(t) \quad (10.70)$$

where $\phi_{b1}(x)$, $\phi_{b2}(x)$, $\phi_{b3}(x)$, and $\phi_{b4}(x)$ are shape functions for the transverse loading case.

These shape functions are not linear functions and they are determined based on static equilibrium conditions, to be discussed later.

When a short beam is subjected to combined axial, torsional, transverse, and bending loading, the displacement at the $x$ position on the short beam is expressed as

$$
\begin{bmatrix} u(x,t) \\ \psi(x,t) \\ v(x,t) \end{bmatrix} = \begin{bmatrix} \phi_{a1} & \phi_{a2} & 0 & 0 & 0 & 0 & 0 & 0 \\ 0 & 0 & \phi_{t1} & \phi_{t2} & 0 & 0 & 0 & 0 \\ 0 & 0 & 0 & 0 & \phi_{b1} & \phi_{b2} & \phi_{b3} & \phi_{b4} \end{bmatrix} \begin{bmatrix} u_1(t) \\ u_2(t) \\ \psi_1(t) \\ \psi_2(t) \\ v_1(t) \\ \theta_1(t) \\ v_2(t) \\ \theta_2(t) \end{bmatrix}
\tag{10.71}
$$

Equations (10.62), (10.67), (10.70), and (10.71) are formulations to represent the infinite degrees of freedom of an elastic beam by finite degrees of freedom.

## 10.6    Energy for Elastic Bodies

A rigid body can only carry kinetic energy related to its mass center velocity and its rotation, as discussed in Chapter 7, but no elastic potential energy. An elastic body can carry both.

Before we discuss the kinetic and elastic energies stored in an elastic body, we will introduce a more general expression for Equation (10.71) as

$$
\begin{bmatrix} u(x,t) \\ \psi(x,t) \\ v(x,t) \end{bmatrix} = \begin{bmatrix} \phi_1 & \phi_2 & 0 & 0 & 0 & 0 & 0 & 0 \\ 0 & 0 & \phi_3 & \phi_4 & 0 & 0 & 0 & 0 \\ 0 & 0 & 0 & 0 & \phi_5 & \phi_6 & \phi_7 & \phi_8 \end{bmatrix} \begin{bmatrix} q_1(t) \\ q_2(t) \\ q_3(t) \\ q_4(t) \\ q_5(t) \\ q_6(t) \\ q_7(t) \\ q_8(t) \end{bmatrix} = \phi q
\tag{10.72}
$$

where $\phi$ is the shape function matrix and $q$ is the general displacement vector. The components of $\phi$ and $q$ are defined according to Equation (10.71).

In addition, we can define additional shape functions for the rotational displacement due to the bending moment.

$$\theta(x,t) = \begin{bmatrix} \phi_9 & \phi_{10} & \phi_{11} & \phi_{12} \end{bmatrix} \begin{bmatrix} q_5(t) \\ q_6(t) \\ q_7(t) \\ q_8(t) \end{bmatrix} = \phi_\theta q_v \qquad (10.73)$$

Similarly, the transverse displacement is

$$v(x,t) = \begin{bmatrix} \phi_5 & \phi_6 & \phi_7 & \phi_8 \end{bmatrix} \begin{bmatrix} q_5(t) \\ q_6(t) \\ q_7(t) \\ q_8(t) \end{bmatrix} = \phi_v q_v \qquad (10.74)$$

The shape functions are usually determined from the static equilibrium of an infinitesimal element as shown in Figure 10.13. When we isolate an infinitesimal element from the finite-sized element of Figure 10.12, we assume that there are no external forces acting on the infinitesimal element other than the constraint forces from the neighboring materials. To determine the shape functions, we need to delve into the elastic beam theory.

## 10.7 A Little Elastic Beam Theory

With respect to Figure 10.13, when an infinitesimal elastic beam element is deformed elastically, the corresponding forces are functions of the deformations. The definition of positive direction of forces could be confusing. We should always define the positive direction of a normal force aligned with the normal direction of a surface. For example, a positive $x$ surface is the surface whose normal direction is pointing to the same positive $x$ direction. Therefore, on this positive $x$ surface, a force pointing to the positive $x$ is considered positive. On the other hand, for a normal force on a negative $x$ surface, its positive direction points to the negative $x$ direction. However, when we combine forces, we can assign vectors based on a reference frame. The definition of the positive direction is only for the convenience of labeling the forces.

As shown in Figure 10.13(a), $B$ surface is the positive $x$ surface, while $A$ surface is a negative $x$ surface. For the axial loading Figure 10.13(b), $F + dF$ points to the positive direction as it is on the positive $x$ surface. Similarly, for the torsional loading (Figure 10.13(c)), $\tau + d\tau$ points to the positive direction based on the right-hand rule.

For the transverse loading case, it is more complicated. As shown in Figure 10.13(d), the shear force $V + dV$, pointing down, is defined as positive, while $M + dM$, which is counterclockwise, is positive. While $M + dM$ is in line with the positive angular displacement, $V + dV$ is the opposite to the positive direction of the transverse displacement $v$. We define them in this way because it is more convenient to establish the force equilibrium equations.

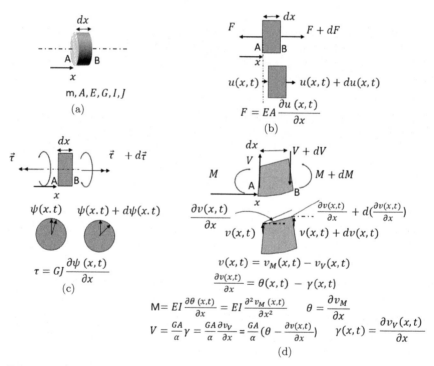

Figure 10.13: The forces and deformations of an infinitesimal beam element: (a) an infinitesimal element without loading; (b) axial displacement due to axial loading; (c) torsional displacement due to torsional loading; and (d) transverse displacements due to transverse force and bending moment.

The axial, torsional, and bending deformations can be related to their respective loadings as

$$F = EA\frac{\partial u}{\partial x} \tag{10.75}$$

$$\tau = GJ\frac{\partial \psi}{\partial x} \tag{10.76}$$

$$M = EI\frac{\partial^2 v_M}{\partial^2 x} \tag{10.77}$$

$$V = \frac{GA}{\alpha}\gamma = \frac{GA}{\alpha}\frac{\partial v_V}{\partial x} \tag{10.78}$$

where $v_M$ is the transverse displacement due to the bending moment $M$ and $v_V$ is the transverse displacement due to the shear force $V$. The shear strain is $\gamma$. The properties of the beam element are $E$, Young's modulus, $G$, the shear modulus, $A$, the cross-section area, $I$, the moment of inertia, $J$, the torsional moment of inertia, $\nu$, the Poisson ratios, and $m$, the density per unit length. The shear modulus can be derived from Young's modulus and Poisson's ratio as $G = E/2(1 + \nu)$. In addition, $\alpha$ is a function of the beam cross-section. For a solid circular beam, the value of $\alpha$ is about 1.08; for a solid rectangular beam, 1.15; for a thin-walled circular tube, 1.95; and for a thin-walled square tube, 2.35. These values are determined via integration with respect to the shear flux. Readers can refer to a textbook of solid mechanics for details.

The total transverse displacement, based on the definitions of Figure 10.13(d), is

$$v(x,t) = v_M(x,t) - v_V(x,t) \tag{10.79}$$

With the direction defined in Figure 10.13(d), the moment, $M$, results in higher vertical displacement, while the shear force, $V$, reduces it; i.e., the negative sign in Equation (10.79).

The total angular displacement of the beam is the derivative of the transverse displacement,

$$\frac{\partial v}{\partial x} = \frac{\partial v_M}{\partial x} - \frac{\partial v_V}{\partial x} = \theta - \gamma \tag{10.80}$$

where $\theta$ is the angular displacement solely due to the bending moment and $\gamma$ is the shear strain due to the shear force. It is easier to determine $\gamma$ based on the total angular displacement and the rotation,

$$\gamma = \theta - \frac{\partial v}{\partial x} \tag{10.81}$$

The shear force is then represented as

$$V = \frac{GA}{\alpha}\gamma = \frac{GA}{\alpha}\frac{\partial[v_M - v]}{\partial x} \tag{10.82}$$

To derive the shape functions, we will establish the force equilibrium equations as

$$dF = 0 \implies \frac{\partial}{\partial x}[EA\frac{\partial u}{\partial x}] = 0 \tag{10.83}$$

$$d\tau = 0 \implies \frac{\partial}{\partial x}\left[GJ\frac{\partial \psi}{\partial x}\right] = 0 \tag{10.84}$$

$$dV = 0 \implies \frac{\partial}{\partial x}\left[\frac{GA}{\alpha}\frac{\partial[v_M - v]}{\partial x}\right] = 0 \tag{10.85}$$

$$dM = Vdx \implies \frac{\partial}{\partial x}\left[EI\frac{\partial^2 v_M}{\partial x^2}\right] = \frac{\partial}{\partial x}\left[EI\frac{\partial \theta}{\partial x}\right] = V \tag{10.86}$$

Combining Equations (10.77)–(10.81) and integrating Equations (10.85) and (10.86), we can obtain the general form of the shape functions for the transverse displacement, $v$, and rotation, $\theta$, as

$$V = c_0 \tag{10.87}$$

$$M = c_0 x + c_1 \tag{10.88}$$

$$\theta = \frac{1}{EI}(\frac{c_0}{2}x^2 + c_1 x + c_2) \tag{10.89}$$

$$v = \frac{1}{EI}\frac{c_0}{6}x^3 + \frac{1}{EI}\frac{c_1}{2}x^2 + \frac{1}{EI}c_2 x - \frac{1}{GA/\alpha}c_0 x + c_3 \tag{10.90}$$

The coefficients, $c_0$, $c_1$, $c_2$, and $c_3$, are determined based on boundary conditions defined for each shape function. For example, to determine $\phi_5$ and $\phi_9$, $q_5 = v_1 = 1$, $q_6 = \theta_1 = 0$, $q_7 = v_2 = 0$, and $q_8 = \theta_2 = 0$. Introducing these boundary conditions into Equations

(10.89) and (10.90), we obtain

$$c_3 = 1 \tag{10.91}$$

$$c_2 = 0 \tag{10.92}$$

$$c_0 = \frac{12EI}{l^3} \frac{1}{1 + \frac{12EI}{l^2 GA/\alpha}} \tag{10.93}$$

$$c_1 = -\frac{6EI}{l^2} \frac{1}{1 + \frac{12EI}{l^2 GA/\alpha}} \tag{10.94}$$

We can proceed to solve for all coefficients for each shape function. Defining the dimensionless variable, $\xi = \frac{x}{L}$, and letting $\Phi = \frac{12EI}{l^2 GA/\alpha}$, we obtain

$$\phi_1 = 1 - \xi \tag{10.95}$$

$$\phi_2 = \xi \tag{10.96}$$

$$\phi_3 = 1 - \xi \tag{10.97}$$

$$\phi_4 = \xi \tag{10.98}$$

$$\phi_5 = \frac{1}{1 + \Phi}(1 - 3\xi^2 + 2\xi^3 + \Phi(1 - \xi)) \tag{10.99}$$

$$\phi_6 = \frac{l}{1 + \Phi}(\xi - 2\xi^2 + \xi^3 + \frac{1}{2}\Phi(\xi - \xi^2)) \tag{10.100}$$

$$\phi_7 = \frac{1}{1 + \Phi}(3\xi^2 - 2\xi^3 + \Phi\xi) \tag{10.101}$$

$$\phi_8 = \frac{l}{1 + \Phi}(-\xi^2 + \xi^3 + \frac{1}{2}\Phi(-\xi + \xi^2)) \tag{10.102}$$

In addition, we can define the shape functions for the rotational displacement due to the bending moment, $\theta = [\phi_9 \ \phi_{10} \ \phi_{11} \ \phi_{12}]q_v = \phi_M q_v$, as

$$\phi_9 = \frac{6}{l(1 + \Phi)}(-\xi + \xi^2) \tag{10.103}$$

$$\phi_{10} = \frac{1}{1 + \Phi}(1 - 4\xi + 3\xi^2 + \Phi(1 - \xi)) \tag{10.104}$$

$$\phi_{11} = \frac{6}{l(1 + \Phi)}(\xi - \xi^2) \tag{10.105}$$

$$\phi_{12} = \frac{1}{1 + \Phi}(-2\xi + 3\xi^2 + \Phi\xi) \tag{10.106}$$

The above shape functions are defined for the Timoshenko beam with the shear deformation included. If $\Phi = 0$, the shape functions become those well-known ones for the Euler–Bernoulli beam, in which the transverse displacement is only due to the bending moment,

$$\phi_{M5}(x) = 1 - 3\xi^2 + 3\xi^3 \tag{10.107}$$

$$\phi_{M6}(x) = l\xi - 2l\xi^2 + l\xi^3 \tag{10.108}$$

$$\phi_{M7}(x) = 3\xi^2 - 2\xi^3 \tag{10.109}$$

$$\phi_{M8}(x) = -l\xi^2 + l\xi^3 \tag{10.110}$$

## 10.8   Some More Math Tools

Let us learn a few new mathematical tools. In basic calculus, the derivative is taken with respect to a scalar, such as $dt$ in $\dot{x} = \frac{dx}{dt}$. We can take a derivative of a vector with respect to a scalar too, such as, $\frac{d\vec{r}}{dt} = \frac{dx}{dt}\vec{i} + \frac{dy}{dt}\vec{j} + \frac{dz}{dt}\vec{k}$ if the unit directional vectors are fixed in space. Can we take a derivative with respect to a vector? To answer this question, we introduce some new notations so that we can deal with any number of dimensions, not just $x, y$, and $z$.

For a vector in an $n$-dimensional space, the unit vectors for each dimension are defined as $\vec{e}_i$, $i = 1, 2, \ldots, n$. The position vector of a point in this $n$-dimensional space is

$$\vec{q} = q_1\vec{e}_1 + q_2\vec{e}_2 + \cdots + q_n\vec{e}_n \tag{10.111}$$

Mathematically, we can introduce a row vector and a column vector so that

$$\vec{q} = q_1\vec{e}_1 + q_2\vec{e}_2 + \cdots + q_n\vec{e}_n = \vec{e}\boldsymbol{q} \tag{10.112}$$

where $\vec{e} = [\vec{e}_1\vec{e}_2 \ldots \vec{e}_n]$ and $\boldsymbol{q} = \begin{bmatrix} q_1 \\ q_2 \\ \vdots \\ q_n \end{bmatrix}$.

It becomes troublesome to write down $\vec{e}$ every time. We will just use $\boldsymbol{q}$, which represents $\vec{q}$, while keeping it as a column vector. We will further simplify it and use $q$ for this column vector instead of $\boldsymbol{q}$.

Now, let us take a derivative of a scalar function, $U$, with respect to $q$. This is called directional derivative and is defined as

$$\frac{\partial U}{\partial q} = \begin{bmatrix} \dfrac{\partial U}{\partial q_1} & \dfrac{\partial U}{\partial q_2} & \cdots & \dfrac{\partial U}{\partial q_n} \end{bmatrix} \tag{10.113}$$

where an operator $\frac{\partial}{\partial q} = [\frac{\partial}{\partial q_1} \ \frac{\partial}{\partial q_2} \ \cdots \ \frac{\partial}{\partial q_n}]$ is defined as a row vector. Therefore, if we take a directional derivative of a scalar, we end up with a row vector using the directional derivative operator. A directional derivative is a partial derivative. To get the total derivative, we need to use the total derivative operator,

$$\frac{\partial}{\partial q} dq = \begin{bmatrix} \dfrac{\partial}{\partial q_1} & \dfrac{\partial}{\partial q_2} & \cdots & \dfrac{\partial}{\partial q_n} \end{bmatrix} \begin{bmatrix} dq_1 \\ dq_2 \\ \vdots \\ dq_n \end{bmatrix}$$

$$= dq_1 \frac{\partial}{\partial q_1} + dq_2 \frac{\partial}{\partial q_2} + \cdots + dq_n \frac{\partial}{\partial q_n} \tag{10.114}$$

Applying the total derivative vector to $U$, we have

$$\frac{\partial U}{\partial q} dq = \begin{bmatrix} \dfrac{\partial U}{\partial q_1} & \dfrac{\partial U}{\partial q_2} & \cdots & \dfrac{\partial U}{\partial q_n} \end{bmatrix} \begin{bmatrix} dq_1 \\ dq_2 \\ \vdots \\ dq_n \end{bmatrix}$$

$$= \frac{\partial U}{\partial q_1} dq_1 + \frac{\partial U}{\partial q_2} dq_2 + \cdots + \frac{\partial U}{\partial q_n} dq_n \tag{10.115}$$

The total derivative of a scalar is still a scalar, as indicated by Equation (10.115).

To take second directional derivatives, such as $\frac{\partial^2 U}{\partial q_1 \partial q_2}$, we will need a new operator,

$$\left(\frac{\partial}{\partial q}\right)' \left(\frac{\partial}{\partial q}\right) = \begin{bmatrix} \dfrac{\partial}{\partial q_1} \\[6pt] \dfrac{\partial}{\partial q_2} \\[6pt] \vdots \\[6pt] \dfrac{\partial}{\partial q_n} \end{bmatrix} \begin{bmatrix} \dfrac{\partial}{\partial q_1} & \dfrac{\partial}{\partial q_2} & \cdots & \dfrac{\partial}{\partial q_n} \end{bmatrix}$$

$$= \begin{bmatrix} \dfrac{\partial^2}{\partial q_1 \partial q_1} & \cdots & \dfrac{\partial^2}{\partial q_1 \partial q_n} \\[6pt] \vdots & \ddots & \vdots \\[6pt] \dfrac{\partial^2}{\partial q_n \partial q_1} & \cdots & \dfrac{\partial^2}{\partial q_n \partial q_n} \end{bmatrix} = \left(\frac{\partial^2}{\partial q \partial q}\right) \qquad (10.116)$$

where $'$ is the transpose operator to turn a column vector to a row vector or vice versa.

The second directional derivative of a scalar becomes

$$\left(\frac{\partial}{\partial q}\right)' \left(\frac{\partial U}{\partial q}\right) = \left(\frac{\partial^2 U}{\partial q \partial q}\right) = \begin{bmatrix} \dfrac{\partial}{\partial q_1} \\[6pt] \dfrac{\partial}{\partial q_2} \\[6pt] \vdots \\[6pt] \dfrac{\partial}{\partial q_n} \end{bmatrix} \begin{bmatrix} \dfrac{\partial U}{\partial q_1} & \dfrac{\partial U}{\partial q_2} & \cdots & \dfrac{\partial U}{\partial q_n} \end{bmatrix}$$

$$= \begin{bmatrix} \dfrac{\partial^2 U}{\partial q_1 \partial q_1} & \cdots & \dfrac{\partial^2 U}{\partial q_1 \partial q_n} \\[6pt] \vdots & \ddots & \vdots \\[6pt] \dfrac{\partial^2 U}{\partial q_n \partial q_1} & \cdots & \dfrac{\partial^2 U}{\partial q_n \partial q_n} \end{bmatrix} \qquad (10.117)$$

Note that the second directional derivative of a scalar is an $n$-by-$n$ matrix.

## 10.9    Kinetic Energy of a Short Elastic Beam

It can be shown that the total kinetic energy of a short beam with axial, torsional, and transverse displacements is the sum of each of these displacements (proof skipped). Let's consider the displacement vector of an elastic body,

$$u = \phi q \tag{10.118}$$

The velocity vector becomes

$$\dot{u} = \phi \dot{q} \tag{10.119}$$

If the dimension of $u$ is 3, while the dimension of $q$ is 8, then $\phi$ is 3 by 8. Consider the mass per unit length of the elastic body is $m$, the total kinetic energy is

$$
\begin{aligned}
T &= \frac{1}{2} \int \dot{u}^2 m \, dx \\
&= \frac{1}{2} \int (\phi \dot{q})' \, \phi \dot{q} \, m \, dx \\
&= \frac{1}{2} \dot{q}' m \int \phi' \phi \, dx \, \dot{q} \\
&= \frac{1}{2} \dot{q}' M \dot{q} \tag{10.120}
\end{aligned}
$$

where $M = m \int \phi' \phi \, dx$ is the equivalent mass matrix for the elastic body corresponding to $q$.

For the axial displacement of a short beam element, the equivalent mass matrix can be found by carrying out the integration of the shape function,

$$M_a = \frac{ml}{6} \begin{bmatrix} 2 & 1 \\ 1 & 2 \end{bmatrix} \tag{10.121}$$

Similarly, the equivalent mass matrix for the torsional displacement is

$$M_t = \frac{Jl}{6} \begin{bmatrix} 2 & 1 \\ 1 & 2 \end{bmatrix} \tag{10.122}$$

$J$ is the moment of inertial per unit length.

The equivalent mass matrix for the transverse displacements is

$$M_b = [M_b]_1 + [M_b]_2 + [M_b]_3 \tag{10.123}$$

where

$$[M_b]_1 = \frac{ml}{420(1+\Phi)^2} \begin{bmatrix} 156 & 22l & 54 & -13l \\ 22l & 4l^2 & 13l & -3l^2 \\ 54 & 13l & 156 & -22l \\ -13l & -3l^2 & -22l & 4l^2 \end{bmatrix} \tag{10.124}$$

$$[M_b]_2 = \frac{\Phi\, ml}{420(1+\Phi)^2} \begin{bmatrix} 294 & 38.5l & 126 & -31.5l \\ 38.5l & 7l^2 & 31.5l & -7l^2 \\ 126 & 31.5l & 294 & -38.5l \\ -31.5l & -7l^2 & -38.5l & 7l^2 \end{bmatrix} \tag{10.125}$$

$$[M_b]_3 = \frac{(\Phi)^2 \; ml}{420(1+\Phi)^2} \begin{bmatrix} 140 & 17.5l & 70 & -17.5l \\ 17.5l & 3.5l^2 & 17.5l & -3.5l^2 \\ 70 & 17.5l & 140 & -17.5l \\ -17.5l & -3.5l^2 & -17.5l & 3.5l^2 \end{bmatrix} \quad (10.126)$$

In the above mass matrix, we do not consider the mass matrix related to the moment of inertia for spinning beam elements. Readers can refer to the paper by Lin *et al.* (2003)[5] for the mass matrices related to a spindle shaft at high speed rotations.

The combined equivalent mass matrix for all three displacements is

$$M = \begin{bmatrix} M_a & 0 & 0 \\ 0 & M_t & 0 \\ 0 & 0 & M_b \end{bmatrix} \quad (10.127)$$

The 0s in the matrix above should carry appropriate dimensions. The total kinetic energy due to all three modes of displacements becomes

$$T = \frac{1}{2} \dot{q}' M \dot{q} \quad (10.128)$$

## 10.10 Elastic Energy Stored in an Elastic Beam

For a idealized linear spring, the elastic potential energy stored by the spring is proportional to its deformation. For many practical problems, we cannot consider every object as a rigid body. For example, a cantilever beam, such as the diving platform of a swimming pool, must be considered as a flexible body. In this section, we will consider the elastic energy stored in an elastic beam. We can use the general displacement vector for deriving the elastic energy.

First, let's consider the work done by the external force to an elastic body, as shown in Figure 10.12. The infinitesimal work done

---

[5]Lin, CW, Tu, JF, and Kamman, J. An integrated thermo-mechanical-dynamic model to characterize motorized machine tool spindles during very high speed rotation, *Int J Mach Tool Manu* **43**(2003) 1035–1050. https://doi.org/10.1016/S0890-6955(03)00091-9.

by the axial loadings, $F_1$ and $F_2$, is

$$\delta W_a = F_1 \, du_1 + F_2 \, du_2 \tag{10.129}$$

Note that the displacements, $u_1$ and $u_2$, cannot go from zero to its current value instantly. Therefore, to determine the total stored elastic energy, we must integrate Equation (10.129).

Similarly, for the torsional and transverse loadings, we have

$$\delta W_t = \tau_1 \, d\phi_1 + \tau_2 \, d\phi_2 \tag{10.130}$$

$$\delta W_b = V_1 \, dv_1 + M_1 \, d\theta_1 + V_2 \, dv_2 + M_2 \, d\theta_2 \tag{10.131}$$

Note that $M_1$ and $M_2$ are bending moments acting on the beam element, as defined in Figure (10.13d), not to be confused with the mass matrix.

Similar to Equation (10.72), corresponding to the general displacement vector $q$, we will define a general force vector $Q$ as

$$Q = \begin{bmatrix} F_1 \\ F_2 \\ \tau_1 \\ \tau_2 \\ V_1 \\ M_1 \\ V_2 \\ M_2 \end{bmatrix} = \begin{bmatrix} Q_1(t) \\ Q_2(t) \\ Q_3(t) \\ Q_4(t) \\ Q_5(t) \\ Q_6(t) \\ Q_7(t) \\ Q_8(t) \end{bmatrix} \tag{10.132}$$

The total infinitesimal work done by $Q$ is

$$\delta W_{\text{total}} = Q dq \tag{10.133}$$

To obtain the total work done by $Q$ when the displacement goes from $q(t_0) = 0$ to $q(t_1)$, we can integrate Equation (10.133). However, in general, $Q$ is not a function of $q$; therefore, we cannot carry out the integration. In other words, these external loading forces are not conservative.

On the other hand, if we focus on an infinitesimal elastic element, as shown in Figure 10.13, we can consider only the conservative forces and they are functions of $q$.

### 10.10.1  *Elastic potential energy*

From the above definition, we can define the elastic potential energy stored in the $dx$ element as

$$
dU_a = \frac{1}{2}Fdu = \frac{1}{2}EA\frac{\partial u}{\partial x}\cdot\frac{\partial u}{\partial x}dx
$$

$$
= \frac{1}{2}EA\left(\frac{\partial\phi_u}{\partial x}q_u\right)'\left(\frac{\partial\phi_u}{\partial x}q_u\right)dx
$$

$$
= \frac{1}{2}q_u'\left[EA\left(\frac{\partial\phi_u}{\partial x}\right)'\left(\frac{\partial\phi_u}{\partial x}\right)dx\right]q_u \qquad (10.134)
$$

$$
dU_t = \frac{1}{2}\tau d\psi = \frac{1}{2}GJ\frac{\partial\psi}{\partial x}\cdot\frac{\partial\psi}{\partial x}dx
$$

$$
= \frac{1}{2}GJ\left(\frac{\partial\phi_t}{\partial x}q_t\right)'\left(\frac{\partial\phi_t}{\partial x}q_t\right)dx
$$

$$
= \frac{1}{2}q_t'\left[GJ\left(\frac{\partial\phi_t}{\partial x}\right)'\left(\frac{\partial\phi_t}{\partial x}\right)dx\right]q_t \qquad (10.135)
$$

The elastic energy for the transverse displacement and the shear strain needs more careful consideration. The elastic energy due to the bending moment is

$$
dU_M = \frac{1}{2}Md\theta = \frac{1}{2}EI\frac{\partial\theta}{\partial x}\cdot\frac{\partial\theta}{\partial x}dx
$$

$$
= \frac{1}{2}EI\left(\frac{\partial\phi_\theta}{\partial x}q_v\right)'\left(\frac{\partial\phi_\theta}{\partial x}q_v\right)dx
$$

$$
= \frac{1}{2}q_v'\left[EI\left(\frac{\partial\phi_\theta}{\partial x}\right)'\left(\frac{\partial\phi_\theta}{\partial x}\right)dx\right]q_v \qquad (10.136)
$$

The elastic energy due to shearing is

$$
dU_V = \frac{1}{2}Vdv_V = \frac{1}{2}\frac{GA}{\alpha}\gamma\cdot\gamma dx
$$

$$
= \frac{1}{2}\frac{GA}{\alpha}\left(\theta-\frac{\partial v}{\partial x}\right)\cdot\left(\theta-\frac{\partial v}{\partial x}\right)dx
$$

$$= \frac{1}{2}\frac{GA}{\alpha}\left[\left(\phi_\theta - \frac{\partial\phi_v}{\partial x}\right)q_v\right]'\left(\phi_\theta - \frac{\partial\phi_v}{\partial x}\right)q_v\, dx$$

$$= \frac{1}{2}q_v'\left[\frac{GA}{\alpha}\left(\phi_\theta - \frac{\partial\phi_v}{\partial x}\right)'\left(\phi_\theta - \frac{\partial\phi_v}{\partial x}\right)dx\right]q_v \quad (10.137)$$

The total elastic energy due to transverse loading becomes

$$dU_b = dU_M + dU_V \quad (10.138)$$

The total elastic energy stored in a finite-sized beam with the length $l$ due to transverse deformations is

$$U_b = \int_0^l dU_b$$

$$= \frac{1}{2}q_v'\int_0^l EI\left(\frac{\partial\phi_\theta}{\partial x}\right)'\left(\frac{\partial\phi_\theta}{\partial x}\right)dx\, q_v$$

$$+\frac{1}{2}q_v'\int_0^l \frac{GA}{\alpha}\left(\phi_\theta - \frac{\partial\phi_v}{\partial x}\right)'\left(\phi_\theta - \frac{\partial\phi_v}{\partial x}\right)dx\, q_v$$

$$= \frac{1}{2}q_v' K_b q_v \quad (10.139)$$

The stiffness matrix $K_b$ due to transverse loading is obtained by carrying out the integration with the shape functions defined in Equations (10.99)–(10.106). We obtain

$$K_b = \frac{EI}{l^3(1+\Phi)}\begin{bmatrix} 12 & 6l & -12 & 6l \\ 6l & (4+\Phi)l^2 & -6l & (2-\phi)l^2 \\ -12 & -6l & 12 & -6l \\ 6l & (2-\phi)l^2 & -6l & (4+\Phi)l^2 \end{bmatrix} \quad (10.140)$$

If we do not consider the energy stored due to shearing, the stiffness matrix becomes the well-known one for the Euler–Bernoulli beam,

$$K_b = \frac{EI}{l^3}\begin{bmatrix} 12 & 6l & -12 & 6l \\ 6l & 4l^2 & -6l & 2l^2 \\ -12 & -6l & 12 & -6l \\ 6l & 2l^2 & -6l & 4l^2 \end{bmatrix} \quad (10.141)$$

Note that for a Euler-Bernoulli beam, we have $\theta = \partial v/\partial x = v'$.

The stiffness matrices for the axial loading and the torsional loading can be found similarly, as

$$K_a = \frac{EA}{l} \begin{bmatrix} 1 & -1 \\ -1 & 1 \end{bmatrix} \tag{10.142}$$

$$K_t = \frac{GJ}{l} \begin{bmatrix} 1 & -1 \\ -1 & 1 \end{bmatrix} \tag{10.143}$$

The combined stiffness matrix becomes

$$M = \begin{bmatrix} K_a & 0 & 0 \\ 0 & K_t & 0 \\ 0 & 0 & K_b \end{bmatrix} \tag{10.144}$$

### 10.10.2 *Castigliano's Theorems*

Knowing how to calculate the elastic energy of a flexible body can be very useful for static and dynamic analysis. In this section, we will discuss Castigliano's Theorem which relates force and deformation through the elastic energy.

Carlo Alberto Castigliano (1847–1884) was an Italian railroad engineer, mathematician, and physicist. In 1879, he published his now well-known First and Second Castigliano's theorems.

As discussed earlier, for a linear elastic structure, the force and elastic deformation are linked via a linear relationship. For a general force vector, $Q$, and the corresponding general displacement, $q$, Castigliano's First Theorem[6] states that

*If the strain energy of an elastic structure can be expressed as a function of generalised displacement q, then the partial derivative of the strain energy with respect to generalised displacement gives the generalised force Q.*

$$Q = \frac{\partial U}{\partial q} \tag{10.145}$$

*where U is the elastic energy or strain energy stored in the structure.*

---

[6]Stephen P. Timoshenko, *History of Strength of Materials*, Dovers Publications Inc., New York, USA, 2003. ISBN: 9780486611877.

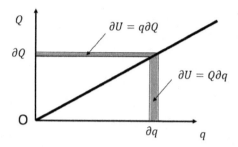

Figure 10.14:   Graphical representation of Castigliano's Theorems.

Castigliano's Second Theorem states

*If the strain energy of a linearly elastic structure can be expressed as a function of generalised force, $Q$, then the partial derivative of the strain energy with respect to generalised force gives the generalised displacement, $q$, in the direction of $Q$.*

$$q = \frac{\partial U}{\partial Q} \tag{10.146}$$

Graphically, we can see clearly how both Castigliano's Theorems exist for linear elastic energy (Figure 10.14).

According to Figure 10.14, $Q$ and $q$ are related by a linear function, $Q = kq$. The elastic energy stored from the unloaded condition can be integrated either as $U = \int Q\,dq$ or $U = \int q\,dQ$. In other words, $\partial U = Q\partial q$ or $\partial U = q\partial Q$, which yield both of Castigliano's Theorems.

The general force and displacement can be generalized to $n$-dimension. Let's apply Castigliano's Theorem to a Euler–Bernoulli elastic beam subjected to transverse loadings Figure 10.12(d). The general force vector, $Q$, and the general displacement vector, $q$, are related as

$$Q = \begin{bmatrix} Q_1 \\ Q_2 \\ Q_3 \\ Q_4 \end{bmatrix} = \begin{bmatrix} F_1 \\ M_1 \\ F_2 \\ M_2 \end{bmatrix} = Kq = K \begin{bmatrix} q_1 \\ q_2 \\ q_3 \\ q_4 \end{bmatrix} = K \begin{bmatrix} v_1 \\ \theta_1 \\ v_2 \\ \theta_2 \end{bmatrix} \tag{10.147}$$

From Equation (10.136), we know that the elastic energy can be calculated from the shape function, $\phi_\theta$ (Equations (10.107)–(10.110)) as

$$U_M = \frac{1}{2}q_v' \int_0^l EI \left(\frac{\partial \phi_\theta}{\partial x}\right)' \left(\frac{\partial \phi_\theta}{\partial x}\right) dx \; q_v = \frac{1}{2}q_v' K q_v \qquad (10.148)$$

where $K$ is given by Equation (10.141).

Taking partial derivative of $U_M$ with respect to $q$, we obtain Castigliano's First Theorem. Let's express $U_M$ explicitly as a function of $q$ by carrying out the multiplication of Equation (10.148),

$$U_M = \frac{EI}{2l^3}(12q_1^2 + 12lq_1q_2 - 24q_1q_3$$
$$+ 12lq_1q_4 + 4l^2q_2^2 - 12lq_2q_3$$
$$- 12lq_3q_4 + 12q_3^2 + 4l^2q_4^2 + 4l^2q_2q_4) \qquad (10.149)$$

Taking the directional derivative of $U_M$ with respect to $q$, we obtain

$$\frac{\partial U_M}{\partial q} = \left[\frac{\partial U_M}{\partial q_1} \; \frac{\partial U_M}{\partial q_2} \; \frac{\partial U_M}{\partial q_3} \; \frac{\partial U_M}{\partial q_4}\right] \qquad (10.150)$$

where

$$\frac{\partial U_M}{\partial q_1} = \frac{EI}{2l^3}(24q_1 + 12lq_2 - 24q_3 + 12q_4) \qquad (10.151)$$

$$\frac{\partial U_M}{\partial q_2} = \frac{EI}{2l^3}(12lq_1 + 8l^2q_2 - 12lq_3 + 4l^2q_4) \qquad (10.152)$$

$$\frac{\partial U_M}{\partial q_3} = \frac{EI}{2l^3}(-24q_1 - 12lq_2 + 24q_3 - 12lq_4) \qquad (10.153)$$

$$\frac{\partial U_M}{\partial q_4} = \frac{EI}{2l^3}(12lq_1 - 12lq_3 + 8l^2q_4 + 4l^2q_2) \qquad (10.154)$$

Now, let's consider some special situations when $q^{*1} = [1\ 0\ 0\ 0]^T$, $q^{*2} = [0\ 1\ 0\ 0]^T$, $q^{*3} = [0\ 0\ 1\ 0]^T$, and $q^{*4} = [0\ 0\ 0\ 1]^T$, and determine the forces required to achieve these specific displacements. Also recall that under these specific displacements, the shape of the beam will be those described by the shape functions. For $q^{*1} = [1\ 0\ 0\ 0]^T$, we obtain

$$Q_1|_{q*1} = F_1|_{q*1} = \frac{\partial U_M}{\partial q_1}|_{q*1} = \frac{12EI}{l^3} \tag{10.155}$$

$$Q_2|_{q*1} = M_1|_{q*1} = \frac{\partial U_M}{\partial q_2}|_{q*1} = \frac{6EI}{l^2} \tag{10.156}$$

$$Q_3|_{q*1} = F_2|_{q*1} = \frac{\partial U_M}{\partial q_3}|_{q*1} = \frac{-12EI}{l^3} \tag{10.157}$$

$$Q_1|_{q*1} = M_2|_{q*1} = \frac{\partial U_M}{\partial q_4}|_{q*1} = \frac{6EI}{l^2} \tag{10.158}$$

Positive values indicate either an upward force or a counterclockwise moment.

We can proceed to determine all other cases and plot the results as shown in Figure 10.15. These loadings required to achieve each specific displacement represent the columns of the stiffness matrix. For example,

$$K_{11} = Q_1|_{q*1} = F_1|_{q*1} = \frac{\partial U_M}{\partial q_1}|_{q*1} = \frac{12EI}{l^3} \tag{10.159}$$

$$K_{21} = Q_2|_{q*1} = M_1|_{q*1} = \frac{\partial U_M}{\partial q_2}|_{q*1} = \frac{6EI}{l^2} \tag{10.160}$$

$$K_{31} = Q_3|_{q*1} = F_2|_{q*1} = \frac{\partial U_M}{\partial q_3}|_{q*1} = \frac{-12EI}{l^3} \tag{10.161}$$

$$K_{41} = Q_4|_{q*1} = M_2|_{q*1} = \frac{\partial U_M}{\partial q_4}|_{q*1} = \frac{6EI}{l^2} \tag{10.162}$$

The results match the stiffness matrix of Equation (10.141). We can also express the elastic energy as a function of $Q$ as

$$U = \frac{1}{2}q'Kq = \frac{1}{2}(K^{-1}Q)'K(K^{-1}Q) = \frac{1}{2}Q'(K^{-1})'Q \tag{10.163}$$

The inverse of the stiffness matrix is the flexibility matrix

$$a = K^{-1} \tag{10.164}$$

However, sometimes, the stiffness matrix might not have a full rank, thus, without an inverse. The stiffness matrix of

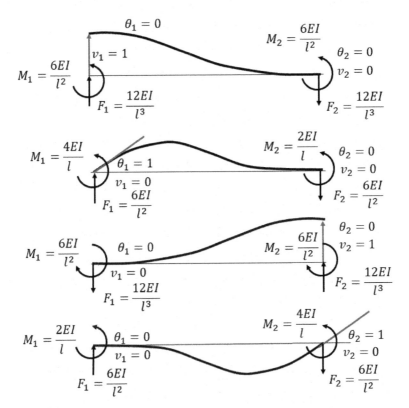

Figure 10.15: Four shape functions of beam transverse displacement.

Equation (10.141) is such a case because the beam can move in space in a rigid body motion. Let's consider a cantilever beam clamped on the left side to hold down the beam. The boundary condition renders $q_1 = 0$ and $q_2 = 0$. The force and displacement equation is then reduced to

$$\begin{bmatrix} Q_3 \\ Q_4 \end{bmatrix} = \begin{bmatrix} F_2 \\ M_2 \end{bmatrix} = K_C \begin{bmatrix} q_3 \\ q_4 \end{bmatrix} = \frac{EI}{l^3} \begin{bmatrix} 12 & -6l \\ -6l & 4l^2 \end{bmatrix} \begin{bmatrix} v_2 \\ \theta_2 \end{bmatrix} \quad (10.165)$$

Based on the flexibility matrix, we have

$$\begin{bmatrix} q_3 \\ q_4 \end{bmatrix} = K^{-1} \begin{bmatrix} F_2 \\ M_2 \end{bmatrix} = \begin{bmatrix} \dfrac{l^3}{3EI} & \dfrac{l^2}{2EI} \\ \dfrac{l^2}{2EI} & \dfrac{l}{EI} \end{bmatrix} \begin{bmatrix} F_2 \\ M_2 \end{bmatrix} \quad (10.166)$$

The elastic energy, expressed in $F_2$ and $M_2$, is

$$U_C = \frac{l^3}{6EI}F_2^2 + \frac{l^2}{2EI}F_2 M_2 + \frac{l}{2EI}M_2^2 \qquad (10.167)$$

Applying Castigliano's Second Theorem, we have

$$\frac{\partial U_C}{\partial Q} = \begin{bmatrix} \dfrac{\partial U_C}{\partial F_2} & \dfrac{\partial U_C}{\partial M_2} \end{bmatrix} \qquad (10.168)$$

where

$$\frac{\partial U_C}{\partial F_2} = \frac{l^3}{3EI}F_2 + \frac{l^2}{2EI}M_2 \qquad (10.169)$$

$$\frac{\partial U_C}{\partial M_2} = \frac{l^2}{2EI}F_2 + \frac{l}{EI}M_2 \qquad (10.170)$$

From the above equations, we obtain the well-known results of the deflections of a cantilever beam,

$$v_2 = \frac{l^3}{3EI}F_2 + \frac{l^2}{2EI}M_2 \qquad (10.171)$$

$$\theta_2 = \frac{l^2}{2EI}F_2 + \frac{l}{EI}M_2 \qquad (10.172)$$

When only $F_2$ is applied with $M_2 = 0$, the vertical deflection of a cantilever beam is $\frac{l^3}{3EI}F_2$ and the angular deformation is $\frac{l^2}{2EI}F_2$. On the other hand, if $F_2 = 0$, we have $\frac{l^2}{2EI}M_2$ and $\frac{l}{EI}M_2$, for vertical and angular deformation, respectively.

From Equations (10.171) and (10.172), we can work out the values of $F_2$ and $M_2$ needed to make $v_2 = 1$ and $\theta_2 = 0$. The results will match the third shape function listed in Figure 10.15.

We will apply Castigliano's First Theorem for a few more examples.

### Example 10.5: Stiffness Matrix of a Flexible 2D Structure

**Problem Statement:** A flexible structure is constructed with two flexible beams as shown in Figure 10.16.[7] The deformations of the

---

[7]Thomson, W. and Dahleh, M.D., *Theory of Vibration with Applications*, 5th edition, Prentice Hall, Inc., 1998. ISBN: 0-13-651068-X.

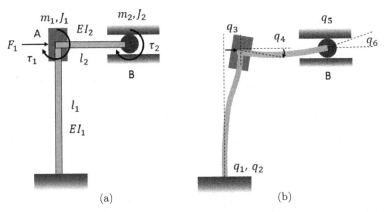

Figure 10.16: The deformation of a flexible structure for Example 10.5: (a) general forces; and (b) general displacements.

flexible structure are shown in Figure 10.16(b). The corner, $A$, maintains its 90° angle, while point $B$ is confined inside a groove. Six transverse deformations are identified for both flexible beams, but due to the constraints, we have $q_1 = q_2 = q_5 = 0$. Construct a stiffness matrix for this structure.

**Solution:** We will invoke Castigliano's First Theorem to solve this problem. First, the total elastic energy of the structure, based on Equation (10.149), is

$$U = \frac{EI_1}{2l_1^3}(-12l_1 q_3 q_4 + 12q_3^2 + 4l_1^2 q_4^2)$$

$$+ \frac{EI_2}{2l_2^3}(4l_2^2 q_4^2 + 4l_2^2 q_6^2 - 4l_2^2 q_4 q_6) \qquad (10.173)$$

Note that we need to add a negative sign for $q_3$ and $q_4$ because it is defined in the negative directions in Figure 10.16 compared with those in Figure 10.15.

Applying Castigliano's First Theorem, we have

$$\frac{\partial U}{\partial q} = \left[\frac{\partial U}{\partial q_3} \; \frac{\partial U}{\partial q_4} \; \frac{\partial U}{\partial q_6}\right] \qquad (10.174)$$

where

$$\frac{\partial U}{\partial q_3} = \frac{EI_1}{2l_1^3}(-12l_1 q_4 + 24q_3) \tag{10.175}$$

$$\frac{\partial U}{\partial q_4} = \frac{EI_1}{2l_1^3}(-12l_1 q_3 + 8l_1^2 q_4) + \frac{EI_2}{2l_2^3}(8l_2 q_4 - 4l_2^2 q_6) \tag{10.176}$$

$$\frac{\partial U}{\partial q_6} = \frac{EI_2}{2l_2^3}(8l_2^2 q_6 - 4l_2^2 q_4) \tag{10.177}$$

Accordingly, with $q^{*1} = [1\ 0\ 0]^T$, the first column of the stiffness matrix is found to be

$$k_{11} = \frac{\partial U}{\partial q_3}\Big|_{q^{*1}} = \frac{12EI_1}{l_1^3} \tag{10.178}$$

$$k_{21} = \frac{\partial U}{\partial q_4}\Big|_{q^{*1}} = \frac{-6EI_1}{l_1^2} \tag{10.179}$$

$$k_{31} = \frac{\partial U}{\partial q_6}\Big|_{q^{*1}} = 0 \tag{10.180}$$

Similarly, the entire stiffness matrix is found to be

$$K = \begin{bmatrix} \dfrac{12EI_1}{l_1^3} & \dfrac{-6EI_1}{2l_1^2} & 0 \\[3mm] \dfrac{-6EI_1}{2l_1^2} & \dfrac{4EI_1}{l_1} + \dfrac{4EI_2}{l_2} & \dfrac{-2EI_2}{l_2} \\[3mm] 0 & \dfrac{-2EI_2}{l_2} & \dfrac{4EI_2}{l_2} \end{bmatrix} \tag{10.181}$$

## Example 10.6: Deflections and Reactions of Double-clamped Beam

**Problem Statement:** We will revisit the well known result of a doubly clamped beam, as shown in Figure 10.17.

Because a doubly clamped been is statically indeterminant, solving for the deflection and the reaction forces/moments at the supports require extra equations based on the superposition principle. For this example, we will utilize the energy method and Castigliano's theorem for solutions.

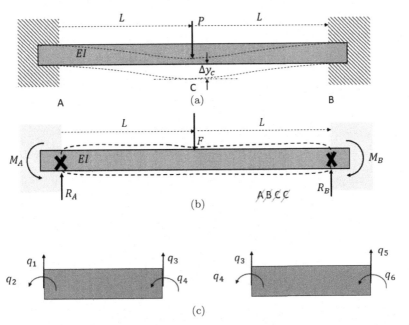

Figure 10.17: The deflection and reaction forces/moments of a doubly clamped beam based on Castigliano's theorem. (a) The doubly clamped beam under a center force; (b) the FBD; and (c) the two sections with their general displacement variables.

**Solution:** First, we break down the beam to two sections and define the general displacement variables, $q$, as shown in Figure 10.17(c). The total elastic energy of the entire beam based on Equation (10.149) is

$$U = \frac{EI}{2l^3}(12q_1^2 + 12lq_1q_2 - 24q_1q_3$$

$$+12lq_1q_4 + 4l^2q_2^2 - 12lq_2q_3$$

$$-12lq_3q_4 + 12q_3^2 + 4l^2q_4^2 + 4l^2q_2q_4$$

$$+12q_3^2 + 12lq_3q_4 - 24q_3q_5$$

$$+12lq_3q_6 + 4l^2q_4^2 - 12lq_4q_5$$

$$-12lq_5q_6 + 12q_5^2 + 4l^2q_6^2 + 4l^2q_4q_6) \tag{10.182}$$

Applying Castigliano's theorem, we have

$$Q' = \frac{\partial U}{\partial q} = \left[\frac{\partial U}{\partial q_1} \; \frac{\partial U}{\partial q_2} \; \frac{\partial U}{\partial q_3} \; \frac{\partial U}{\partial q_4} \; \frac{\partial U}{\partial q_5} \; \frac{\partial U}{\partial q_6}\right] \tag{10.183}$$

We work out the partial derivatives and then apply the boundary conditions, $q^* = [0\ 0\ q_3\ 0\ 0\ 0]$, to obtain

$$Q_1 = R_A = \frac{\partial U}{\partial q_1}\Big|_{q^*} = \frac{-12EI}{l_1^3}q_3 \tag{10.184}$$

$$Q_2 = M_A = \frac{\partial U}{\partial q_2}\Big|_{q^*} = \frac{-6EI}{l_1^2}q_3 \tag{10.185}$$

$$Q_3 = -P = \frac{\partial U}{\partial q_3}\Big|_{q^*} = \frac{24EI}{l_1^3}q_3 \tag{10.186}$$

$$Q_4 = 0 = \frac{\partial U}{\partial q_4}\Big|_{q^*} \tag{10.187}$$

$$Q_5 = R_B = \frac{\partial U}{\partial q_5}\Big|_{q^*} = \frac{-12EI}{l_1^3}q_3 \tag{10.188}$$

$$Q_6 = M_B = \frac{\partial U}{\partial q_6}\Big|_{q^*} = \frac{6EI}{l_1^2}q_3 \tag{10.189}$$

From Equation (10.186), we obtain

$$q_3 = \frac{-Pl^3}{24EI} \tag{10.190}$$

From Equation (10.190), we determine that

$$Q_1 = R_A = \frac{P}{2} \tag{10.191}$$

$$Q_2 = M_A = \frac{Pl}{4} \tag{10.192}$$

$$Q_5 = R_B = \frac{P}{2} \tag{10.193}$$

$$Q_6 = M_B = \frac{-Pl}{4} \tag{10.194}$$

The above results match the well-known results, but the calculation is mainly mathematical derivation without the FBD analysis.

## 10.11   Curved Beam Theory

As shown in Figure 10.18(a), we would like to know the displacement of the curved beam when a force is applied at the center. We also

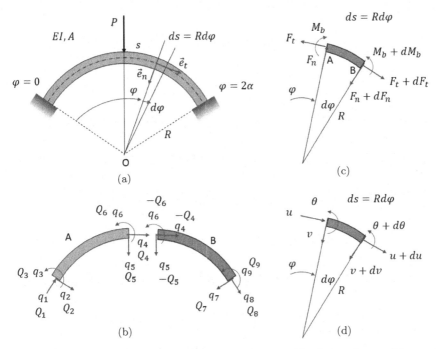

Figure 10.18: The loading and deformation of a double-clamped curved beam: (a) the configuration; (b) the whole beam is broken into two beams with their general displacements and forces; (c) the FBD for an infinitesimal curved beam element; and (d) the corresponding displacements.

would like to determine the reaction forces and moment at the supports. Just like in the previous example of a double-clamped straight beam, this is a statically indeterminate case. Due to the curved beam shape, the analysis becomes much more complicated.

We could follow a similar strategy as in the previous example to break the curved beam into two pieces, as shown in Figure 10.18(b). We will need to define the beam theory for a curved beam before we can apply Castigliano's theorem.

As shown in Figure 10.18(c), the loading on an infinitesimal curved beam element is presented. We define two directional unit vectors:, $\vec{e}_t$ is the tangential unit vector and $\vec{e}_n$ is the normal unit vector, pointing to the center of the curved beam as shown in Figure 10.18(a). The tangential direction is consistent with the angular position defined by $\varphi$.

The positive force directions are defined based on surface $B$, which is a positive surface with respect to $\vec{e}_t$. The tangential force, $F_t$, on $B$ is positive when it is pointing to the $\vec{e}_t$ direction. The normal force, $F_n$, is positive when it is pointing to the $\vec{e}_n$ direction. The tangential and normal displacements are defined the same way. However, note that the positive moment direction on surface $B$ is defined in the counterclockwise direction as shown in Figure 10.18(c), so is the angular displacement, $\theta$, shown in Figure 10.18(d). These definitions are similar to those defined for a straight beam (Figure 10.13) but different in the definition of the normal force versus the shear force, as well as the normal displacement versus the transverse displacement.

### 10.11.1  *Shape functions of a curved beam*

Based on Figure 10.18, we will present a curved beam theory and define its shape function similar to Section 10.6.1 for the straight beam.

Based on the static equilibrium shown in Figure 10.19, we can establish the governing equations for the curved beam. Figure 10.19(b) presents the force equilibrium in the tangential and normal directions. The corresponding force equilibrium equations are

$$\sum (F_i)_t = 0 \Longrightarrow -F_t + (F_t + dF_t)\cos(d\varphi) - (F_n + dF_n)\sin(d\varphi) = 0$$

$$\Longrightarrow dF_t - F_n d\varphi = 0$$

$$\Longrightarrow \frac{dF_t}{d\varphi} = F_n \tag{10.195}$$

$$\sum (F_i)_n = 0 \Longrightarrow (F_n + dF_n)\cos(d\varphi) - F_n + (F_t + dF_t)\sin(d\varphi) = 0$$

$$\Longrightarrow dF_n + F_t d\varphi = 0$$

$$\Longrightarrow \frac{dF_n}{d\varphi} = -F_t \tag{10.196}$$

The moment equilibrium with respect to point $A$ is shown in Figure 10.18(c) and we have

$$\sum (M_i)_A = 0 \Longrightarrow -M_b + (M_b + dM_b) - (F_n + dF_n)ds = 0$$

$$\Longrightarrow dM_b - F_n ds = dM_b - F_n R d\varphi = 0$$

$$\Longrightarrow \frac{dM_b}{d\varphi} = RF_n \tag{10.197}$$

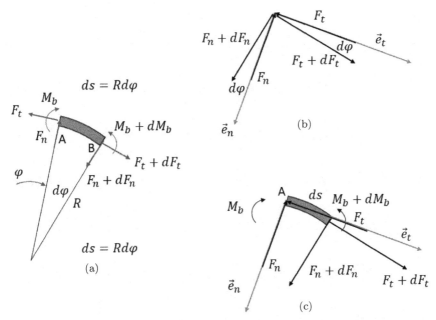

Figure 10.19: The force analysis of an infinitesimal curved beam: (a) the FBD; (b) the force equilibrium; and (c) the moment equilibrium with respect to point $A$.

Relating the angular rotation and the bending moment, together with the elastic beam properties, we have

$$\frac{d\theta}{d\varphi} = \frac{R}{EI}M_b \qquad (10.198)$$

The hoop strain of the curved beam, $\epsilon_t$, is affected both by the tangential and normal displacements.

$$\epsilon_t = \epsilon_u + \epsilon_v \qquad (10.199)$$

It is obvious that $\epsilon_u = \frac{du}{ds}$, but $\epsilon_v$ requires some explanation. If the curved beam has only a normal displacement, $v$, its radius will be reduced to $R - v$. The circumferential length is then shortened to $2\pi(R - v)$. This results in a compressive hoop strain as

$$\epsilon_v = \frac{2\pi(R - v) - 2\pi R}{2\pi R} = -\frac{v}{R} \qquad (10.200)$$

The hoop strain is due to the tangential force; therefore,

$$F_t = EA\epsilon_t = EA\left(\frac{du}{ds} - \frac{v}{R}\right)$$

$$\implies \frac{du}{d\varphi} = v + \frac{R}{EA}F_t \tag{10.201}$$

The total normal displacement is due to both the normal force and the tangential force as well as the bending moment,

$$v = v_{F_n} - v_{F_t} - v_{M_b} \tag{10.202}$$

Taking a derivative of Equation (10.202), we have

$$\frac{dv}{ds} = \frac{dv_{F_n}}{ds} - \frac{dv_{F_t}}{ds} - \frac{dv_{M_b}}{ds} \tag{10.203}$$

If we ignore the shear strain due to $F_n$, i.e., considering the Euler–Bernoulli beam instead of the Timoshenko beam, Equation (10.203) is simplified to

$$\frac{dv}{ds} = -\frac{dv_{F_t}}{ds} - \frac{dv_{M_b}}{ds} \tag{10.204}$$

The angular displacement due to the tangential displacement is

$$\frac{dv_{F_t}}{ds} = \frac{ud\varphi}{ds} = \frac{u}{R} \tag{10.205}$$

The angular displacement due to the bending moment is $\theta$. Equation (10.204) thus becomes

$$\frac{dv}{ds} = -\frac{u}{R} - \theta$$

$$\implies \frac{dv}{d\varphi} = -u - R\theta \tag{10.206}$$

We can now integrate Equations (10.195)–(10.198), (10.201), and (10.206) to obtain $F_t$, $F_n$, $M_b$, $\theta$, $u$, and $v$ as functions of $\varphi$.
First for $F_t$ and $F_n$, we obtain

$$F_t = c_1 \cos\varphi + c_2 \sin\varphi \tag{10.207}$$

$$F_n = -c_1 \sin\varphi + c_2 \cos\varphi \tag{10.208}$$

Subsequently, we have

$$M_b = Rc_1 \cos\varphi + Rc_2 \sin\varphi + c_3 \tag{10.209}$$

$$\theta = \frac{R^2}{EI}c_1 \sin\varphi - \frac{R^2}{EI}c_2 \cos\varphi + \frac{R}{EI}c_3\varphi + c_4 \tag{10.210}$$

The solutions for $v$ and $u$ are more complicated and they involve the procedures to solve an ordinary differential equation with the homogeneous and particular solutions. The results are

$$v = a_1 \cos\varphi + a_2 \sin\varphi + \frac{1}{2}\left(\frac{R}{EA} + \frac{R^3}{EI}\right)c_2\varphi\cos\varphi$$

$$-\frac{1}{2}\left(\frac{R}{EA} + \frac{R^3}{EI}\right)c_1\varphi\sin\varphi - \frac{R^2}{EI}c_3 \tag{10.211}$$

$$u = \left[-a_2 - \frac{1}{2}\left(\frac{R}{EA} - \frac{R^3}{EI}\right)c_2\right]\cos\varphi$$

$$+\left[a_1 + \frac{1}{2}\left(\frac{R}{EA} - \frac{R^3}{EI}\right)c_1\right]\sin\varphi$$

$$+\frac{1}{2}\left(\frac{R}{EA} + \frac{R^3}{EI}\right)c_1\varphi\cos\varphi$$

$$+\frac{1}{2}\left(\frac{R}{EA} + \frac{R^3}{EI}\right)c_2\varphi\sin\varphi$$

$$-\frac{R^2}{EI}c_3\varphi - Rc_4 \tag{10.212}$$

For the functions of Equations (10.207)–(10.212), we have six coefficients, $c_1$, $c_2$, $c_3$, $c_4$, $a_1$, and $a_2$, to be determined by boundary conditions. We will examine a few special cases.

### 10.11.2  *Three special curved beam cases*

Let's consider special cases of a quarter-circular cantilever beam loaded by a normal force, a moment, or a tangential force at the free end, respectively, as shown in Figure 10.20. The boundary conditions for each case are defined in the figures. For loading with $P$ only, the boundary conditions at the fixed end, $\varphi = 0$, are $u(0) =$

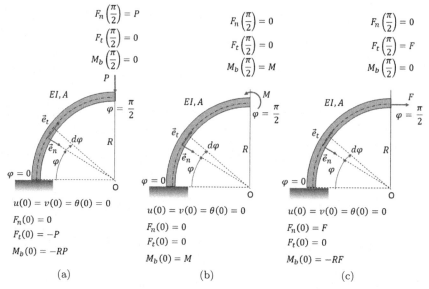

Figure 10.20: A quarter-circular cantilever beam at different loading conditions at the free end: (a) a normal force only; (b) a moment only; and (c) a tangential force only.

$v(0) = \theta(0) = 0$, while at the free end, they are $F_t(\pi/2) = -P$, $F_n(\pi/2) = 0$, and $M_b(\pi/2) = 0$. Conducting an FBD analysis for the entire quarter beam, the constraint forces and moment at the fixed end can be found. The corresponding boundary forces and moment are $F_n(0) = 0$, $F_t(0) = -P$, and $M(0) = -RP$.

Similarly, for loading with $M$ only, the boundary conditions at the fixed end, $\varphi = 0$, are $u(0) = v(0) = \theta(0) = 0$, while at the free end, they are $F_t(\pi/2) = 0$, $F_n(\pi/2) = 0$, and $M_b(\pi/2) = M$. Conducting an FBD analysis for the entire quarter beam, the constraint forces and moment at the fixed end can be found. The corresponding boundary forces and moment are $F_n(0) = 0$, $F_t(0) = 0$, and $M(0) = M$.

Finally, for loading with $F$ only, the boundary conditions at the fixed end, $\varphi = 0$, are $u(0) = v(0) = \theta(0) = 0$, while at the free end, they are $F_t(\pi/2) = F$, $F_n(\pi/2) = 0$, and $M_b(\pi/2) = 0$. Conducting an FBD analysis for the entire quarter beam, the constraint forces and moment at the fixed end can be found. The corresponding boundary forces and moment are $F_n(0) = F$, $F_t(0) = 0$, and $M(0) = -RF$.

### 10.11.2.1 *Normal force only case*

Introducing these boundary conditions to Equations (10.207)–(10.212) leads to

$$F_t(0) = c_1 = -P \tag{10.213}$$

$$F_n(0) = c_2 = 0 \tag{10.214}$$

$$M_b(0) = -RP + c_3 = -RP \implies c_3 = 0 \tag{10.215}$$

$$\theta(0) = c_4 = 0 \tag{10.216}$$

$$v(0) = a_1 - \frac{R^2}{EI}c_3 = 0 \implies a_1 = 0 \tag{10.217}$$

$$u(0) = -a_2 = 0 \tag{10.218}$$

The tangential and normal displacements at the free end can now be found as

$$u\left(\frac{\pi}{2}\right) = \frac{1}{2}\left(-\frac{R}{EA} + \frac{R^3}{EI}\right)P \tag{10.219}$$

$$v\left(\frac{\pi}{2}\right) = \frac{\pi}{4}\left(\frac{R}{EA} + \frac{R^3}{EI}\right)P \tag{10.220}$$

The rotation at the free end is

$$\theta\left(\frac{\pi}{2}\right) = -\frac{R^2}{EI}P \tag{10.221}$$

For a curved beam with $A = 0.01$ m$^2$, $E = 2 \times 10^{10}$ N/m$^2$, $I = 8.333 \times 10^{-6}$ m$^4$, $R = 1$ m, and $P = 100$ N, the tangential, normal, and rotational displacements at the free end are found to be

$$u\left(\frac{\pi}{2}\right) = 299.8 \times 10^{-6}\,\text{m} \approx 0.3\,\text{mm} \tag{10.222}$$

$$v\left(\frac{\pi}{2}\right) = 471.7 \times 10^{-6}\,\text{m} \approx 0.47\,\text{mm} \tag{10.223}$$

$$\theta\left(\frac{\pi}{2}\right) = -0.00060\,\text{rad} \approx -0.034° \tag{10.224}$$

### 10.11.2.2 *Moment only case*

Introducing these boundary conditions to Equations (10.207)–(10.212) leads to

$$F_t(0) = c_1 = 0 \tag{10.225}$$

$$F_n(0) = c_2 = 0 \tag{10.226}$$

$$M_b(0) = c_3 = M \tag{10.227}$$

$$\theta(0) = c_4 = 0 \tag{10.228}$$

$$v(0) = a_1 - \frac{R^2}{EI}c_3 = 0 \implies a_1 = \frac{R^2}{EI}M \tag{10.229}$$

$$u(0) = -a_2 = 0 \tag{10.230}$$

The tangential and normal displacements at the free end can now be found as

$$u\left(\frac{\pi}{2}\right) = \frac{(2 - \pi)R^2}{2EI}M \tag{10.231}$$

$$v\left(\frac{\pi}{2}\right) = -\frac{R^2}{EI}M \tag{10.232}$$

The rotation at the free end is

$$\theta\left(\frac{\pi}{2}\right) = \frac{\pi R}{2EI}M \tag{10.233}$$

### 10.11.2.3 *Tangential force only case*

Introducing these boundary conditions to Equations (10.207)–(10.212) leads to

$$F_t(0) = c_1 = 0 \tag{10.234}$$

$$F_n(0) = c_2 = F \tag{10.235}$$

$$M_b(0) = c_3 = -RF \tag{10.236}$$

$$\theta(0) = -\frac{R^2}{EI} + c_4 = 0 \implies c_4 = \frac{R^2}{EI}F \tag{10.237}$$

$$v(0) = a_1 - \frac{R^2}{EI}c_3 = 0 \implies a_1 = -\frac{R^3}{EI}F \tag{10.238}$$

$$u(0) = -a_2 - \frac{1}{2}\left(\frac{R}{EA} - \frac{R^3}{EI}\right)c_2 - Rc_4 = 0$$

$$\implies a_2 = -\frac{1}{2}\left(\frac{R}{EA} + \frac{R^3}{EI}\right)F \tag{10.239}$$

The tangential and normal displacements at the free end can now be found as

$$u\left(\frac{\pi}{2}\right) = \left[\frac{\pi R}{4EA} + \frac{(3\pi - 8)R^3}{4EI}\right]F \tag{10.240}$$

$$v\left(\frac{\pi}{2}\right) = -\frac{RF}{2EA} + \frac{R^3}{2EI}F \tag{10.241}$$

The rotation at the free end is

$$\theta\left(\frac{\pi}{2}\right) = \frac{2-\pi}{2EI}R^2F \tag{10.242}$$

### 10.11.3  *A pinched ring case*

Let's consider a pinched ring case, as shown in Figure 10.21(a), subjected to a compressive force, $2P$.

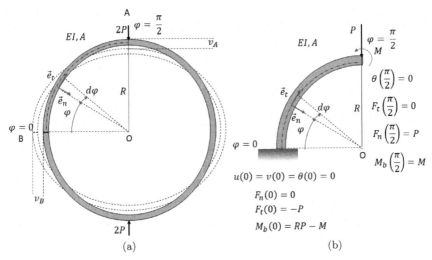

(a)                   (b)

Figure 10.21:  A pinched-ring case: (a) the original condition of a pinched ring; and (b) the equivalent quarter-ring case.

When the ring is pinched, it would deform into an oval shape. The pinched ring case can be represented as a fixed quarter ring case with a force $P$ and a moment $M$ acting at the free end Figure 10.21(b) so that the tangential displacement at the free end is the same as the normal displacement at $\varphi = 0$ for the ring, i.e., $u(\pi/2) = v_B$, while the normal displacement at the free end is the same as that of the ring at $\varphi = \pi/2$, i.e., $v(\pi/2) = v_A$. In the meanwhile, $\theta(0) = \theta(\pi/2) = 0$.

With an FBD analysis for Figure 10.21(b), we also determine that $F_n(0) = 0$, $F_t(0) = -P$, and $M_b(0) = RP - M$.

Introducing the boundary conditions for $F_t$, $F_n$, $\theta$, $v$, and $u$ for $\varphi = 0$, we obtain

$$F_t(0) = c_1 = -P \tag{10.243}$$

$$F_n(0) = c_2 = 0 \tag{10.244}$$

$$\theta(0) = c_4 = 0 \tag{10.245}$$

$$v(0) = a_1 - \frac{R^2}{EI}c_3 = 0 \tag{10.246}$$

$$u(0) = -a_2 = 0 \tag{10.247}$$

For the boundary conditions at $\varphi = \pi/2$, we have

$$M_b\left(\frac{\pi}{2}\right) = c_3 = M \tag{10.248}$$

$$\theta\left(\frac{\pi}{2}\right) = \frac{-R^2}{EI}P + \frac{\pi R}{2EI}c_3 = 0 \implies c_3 = \frac{2}{\pi}RP \tag{10.249}$$

The tangential and normal displacements at the free end can now be found as

$$u\left(\frac{\pi}{2}\right) = -\left[\frac{(\pi - 4)R^3}{2\pi EI} + \frac{R}{2EA}\right]P \tag{10.250}$$

$$v\left(\frac{\pi}{2}\right) = \frac{\pi RP}{4EA} + \frac{(\pi^2 - 8)R^3}{4\pi EI}P \tag{10.251}$$

The applied moment, $M$, at the free end is a function of $P$ in order to achieve $\theta(\pi/2) = 0$ and we have from Equation (10.248),

$$M = \frac{2}{\pi}RP \tag{10.252}$$

We can obtain the above result differently by using the principle of superposition based on the results of quarter-ring cases. Because $\theta(\pi/2) = 0$, we have

$$\theta(\pi/2) = \theta_P(\pi/2) + \theta_M(\pi/2) = -\frac{R^2 P}{EI} + \frac{\pi R M}{2EI} = 0 \quad (10.253)$$

Subsequently, we obtain the same result of Equation (10.252). The tangential displacement is found to be

$$u(\pi/2) = u_P(\pi/2) + u_M(\pi/2)$$
$$= \frac{1}{2}\left(-\frac{R}{EA} + \frac{R^3}{EI}\right)P + \frac{2-\pi}{2EI}R^2 M \quad (10.254)$$

Introducing Equation (10.252) into the above equation, we obtain the same result of Equation (10.250).

Similarly, with $v(\pi/2) = v_P(\pi/2) + v_M(\pi/2)$, we obtain the same result of Equation (10.251).

For a ring with $A = 0.01$ m$^2$, $E = 2 \times 10^{10}$ N/m$^2$, $I = 8.333 \times 10^{-6}$ m$^4$, $R = 1$ m, and $2\,P = 200$ N, the tangential and normal displacements at the free end are found to be

$$u\left(\frac{\pi}{2}\right) = 81.7 \times 10^{-6} \text{ m} \approx 0.082 \text{ mm} \quad (10.255)$$

$$v\left(\frac{\pi}{2}\right) = 89.7 \times 10^{-6} \text{ m} \approx 0.090 \text{ mm} \quad (10.256)$$

In comparison, a ring is three to four times stiffer than a fixed quarter beam of the same properties. From the quarter-ring perspective, this is because moment $M$ can counteract force $P$.

### 10.11.4 A fixed–fixed half-circular beam case

We will consider a third special case with a half ring fixed at both ends as shown in Figure 10.22(a). A more general fixed end case is shown in Figure 10.19(a) with $\varphi = 2\alpha$. Here, we consider the special case with $\alpha = \frac{\pi}{2}$. We will use the principle of superposition to solve this problem. For the quarter ring to behave like the fixed–fixed half-circular beam, we must require that

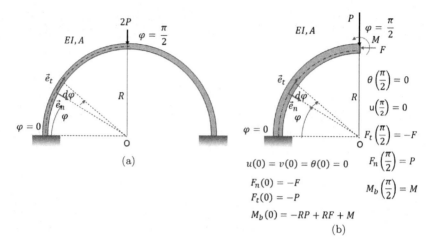

Figure 10.22: A fixed-half ring case: (a) the original condition of a fixed half ring; (b) the equivalent quarter-ring case.

$$\theta\left(\frac{\pi}{2}\right) = \theta_P\left(\frac{\pi}{2}\right) + \theta_M(\frac{\pi}{2}) + \theta_F\left(\frac{\pi}{2}\right) = 0 \qquad (10.257)$$

$$u\left(\frac{\pi}{2}\right) = u_P\left(\frac{\pi}{2}\right) + u_M\left(\frac{\pi}{2}\right) + u_F\left(\frac{\pi}{2}\right) = 0 \qquad (10.258)$$

Solving the above two equations, we obtain that

$$M = 4\left(\frac{(\pi-1)EIR + (\pi-3)EAR^3}{\pi^2 EI + (\pi^2-8)EAR^2}\right)P \qquad (10.259)$$

$$F = 2\left(\frac{\pi EI + (\pi-4)EAR^2}{\pi^2 EI + (\pi^2-8)EAR^2}\right)P \qquad (10.260)$$

Finally, we obtain the normal displacement as

$$v\left(\frac{\pi}{2}\right) = u_P\left(\frac{\pi}{2}\right) + u_M\left(\frac{\pi}{2}\right) + u_F\left(\frac{\pi}{2}\right)$$

$$= \frac{\pi}{4}\left(\frac{R}{EA} + \frac{R^3}{EI}\right)P - \frac{R^2}{EI}M - \frac{R}{2EA}F$$

$$+ \frac{R^3}{2EI}F \qquad (10.261)$$

where $M$ and $F$ are from Equations (10.259) and (10.260), respectively.

A closed-form of the normal displacement of Equation (10.261) is not readily available in the literature, but many numerical results exist. For comparison, we will ignore the effects of the axial deformation and only consider the deformation caused by the bending moment, contributed by $P$, $M$, and $F$. This approximation is generally true when $EI << EAR^2$. With only a force $P$, instead of $2P$, acting on the half-circle beam, Equation (10.261) can then be simplified as

$$v\left(\frac{\pi}{2}\right) = \frac{\pi^3 - 20\pi + 32}{8(\pi^2 - 8)}\frac{R^3}{EI}P \approx \frac{1}{85.8}\frac{R^3}{EI}P \qquad (10.262)$$

We can compare the result of Equation (10.262) to the numerical results in the literature,[8] as shown in Figure 10.23. The results of Equation (10.262) essentially match the linear region of the dimensionless numerical chart presented.

Based on Equation (10.261) for the same beam property with a half-circular beam with $A = 0.01$ m$^2$, $E = 2 \times 10^{10}$ N/m$^2$, $I = 8.333 \times 10^{-6}$ m$^4$, $R = 1$ m, and $2P = 200$ N, normal displacement at $\varphi = \pi/2$ is found to be

$$v\left(\frac{\pi}{2}\right) = 15.17 \times 10^{-6} \text{ m} \approx 0.015 \text{ mm} \qquad (10.263)$$

In comparison, a clamped–clamped half-circular beam is about six times stiffer than a pinched ring. The much enhanced stiffness is due to the restriction of tangential displacement.

Based on Equation (10.190), a clamped–clamped straight beam with the same beam property and a total length of $2R$ with a center loading of $2P$ will have a deflection of 0.050 mm. As a result, a clamped half-circular beam will be nearly three times stiffer. This is the shape reinforcement and the reason for many arches in buildings.

### 10.11.5 *Elastic energy stored in a curved beam*

We will derive equations for the elastic energy stored in a curved beam for a general case with respect to Figure 10.19(a). We will only consider the elastic energy stored due to bending. As discussed in

---

[8]Noor, A.K., Greene, W.H., and Hartley, S.J., Nonlinear finite element analysis of curved beams, *Computer Methods in Applied Mechanics and Engineering* **12**(3) (1977), 289–307. https://doi.org/10.1016/0045-7825(77)90018-4.

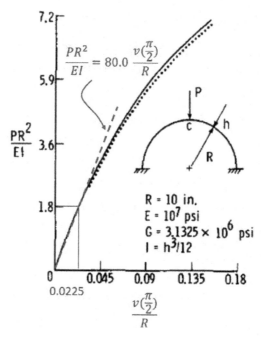

Figure 10.23: The dimensionless numerical results of a clamped half-circle beam subjected to $P$. The chart is modified from the result of Noor *et al.* (1977) (see footnote 8).

Section 10.6.4.1, if we know the shape functions, then based on Equation (10.136), we can work out the stored energy and corresponding stiffness matrix with respect to a general displacement vector, $q$, defined in Figure 10.18(b).

To determine the shape functions, we use the general shape functions for curved beams based on Equations (10.210)–(10.212) and introduce boundary conditions to solve for $c_1$, $c_2$, $c_3$, $c_4$, $a_1$, and $a_2$ for different specific shapes.

For example, for the first shape, $q^{(1)} = [1\ 0\ 0\ 0\ 0\ 0]$, we have boundary conditions as

$$u(0) = 1 = [-a_2 - \frac{1}{2}(\frac{R}{EA} - \frac{R^3}{EI})c_2] - Rc_4 \qquad (10.264)$$

$$v(0) = 0 = a_1 - \frac{R^2}{EI}c_3 \qquad (10.265)$$

$$\theta(0) = 0 = -\frac{R^2}{EI}c_2 + c_4 \qquad (10.266)$$

$$u(\varphi) = 0 = \left[ -a_2 - \frac{1}{2} \left( \frac{R}{EA} - \frac{R^3}{EI} \right) c_2 \right] \cos \varphi$$

$$+ \left[ a_1 + \frac{1}{2} \left( \frac{R}{EA} - \frac{R^3}{EI} \right) c_1 \right] \sin \varphi$$

$$+ \frac{1}{2} \left( \frac{R}{EA} + \frac{R^3}{EI} \right) c_1 \varphi \cos \varphi$$

$$+ \frac{1}{2} \left( \frac{R}{EA} + \frac{R^3}{EI} \right) c_2 \varphi \sin \varphi - \frac{R^2}{EI} c_3 \varphi - R c_4 \quad (10.267)$$

$$v(\varphi) = 0 = a_1 \cos \varphi + a_2 \sin \varphi$$

$$+ \frac{1}{2} \left( \frac{R}{EA} + \frac{R^3}{EI} \right) c_2 \varphi \cos \varphi$$

$$- \frac{1}{2} \left( \frac{R}{EA} + \frac{R^3}{EI} \right) c_1 \varphi \sin \varphi - \frac{R^2}{EI} c_3 \quad (10.268)$$

$$\theta(\varphi) = 0 = \frac{R^2}{EI} c_1 \sin \varphi - \frac{R^2}{EI} c_2 \cos \varphi + \frac{R}{EI} c_3 \varphi + c_4 \quad (10.269)$$

We solve Equations (10.264)–(10.269) for $c_1$, $c_2$, $c_3$, $c_4$, $a_1$, and $a_2$. Introducing these coefficients back to Equations (10.210)–(10.212), we obtain the shape functions $\phi_{\theta 1}$, $\phi_{u1}$, and $\phi_{v1}$. The shape function $\phi_{\theta 1}$ is the one we need to determine the elastic energy stored in the curved beam if only the work done by the bending moment is considered, as in the case of a Euler-Bernoulli beam. We will need to use symbolic equation solvers for this task.

We can also use numerical solutions for specific beam configurations. By defining the actual values of $R$, $EI$, $EA$, and $\varphi$, we can solve for the shape functions numerically. We proceed to define the shape functions of $\phi_{\theta 2}$–$\phi_{\theta 6}$. After that, we determine the stiffness matrix based on Equation (10.148) with $x$ changed to $ds = Rd\varphi$.

The elastic energy stored can be obtained as

$$U_M = \int_0^\phi dU_M = \frac{1}{2} q' \frac{EI}{R} \int_0^\phi \left( \frac{\partial \phi_{\theta i}}{\partial \varphi} \right)' \left( \frac{\partial \phi_{\theta j}}{\partial \varphi} \right) d\varphi \, q$$

$$= \frac{1}{2} q' K \, q \quad (10.270)$$

Based on the general shape function of $\theta$ defined by Equation (10.210), each component of the stiffness matrix can be defined as

$$K_{ii} = \frac{R}{EI}[R^2 c_{1i}^2(0.5\varphi + 0.5\cos(\varphi)\sin(\varphi))$$

$$-2Rc_{2i}c_{3i}(\cos(\varphi) - 1) + 2Rc_{3i}c_{1i}\sin(\varphi)$$

$$+R^2 c_{2i}^2(0.5\varphi - 0.5\cos(\varphi)\sin(\varphi))$$

$$-R^2 c_{1i}c_{2i}(\cos(\varphi)^2 - 1) + c_{3i}^2\varphi] \qquad (10.271)$$

$$K_{ij} = \frac{R}{EI}[R^2 c_{1i}c_{1j}(0.5\varphi + 0.5\cos(\varphi)\sin(\varphi))$$

$$-R(c_{2i}c_{3j} + c_{2j}c_{3i})(\cos(\varphi) - 1)$$

$$+R(c_{3i}c_{1j} + c_{1i}c_{3j})\sin(\varphi)$$

$$+R^2 c_{2i}c_{2j}(0.5\varphi - 0.5\cos(\varphi)\sin(\varphi))$$

$$+R^2(c_{1i}c_{2i} + c_{1i}c_{2j})(-0.5\cos(\varphi)^2 + 0.5)$$

$$+c_{3i}c_{3j}\varphi] \qquad (10.272)$$

where $c_{1i}$ with $i = 1, \ldots, 6$ are the coefficients of $c1$ for the $i$-$th$ shape function, solved numerically earlier. The coefficients of $c_{2i}$ and $c_{3i}$ are defined similarly.

We can validate the stiffness matrix by comparing it to the earlier results. The results of Equations (10.222)–(10.224) can be obtained as

$$\begin{bmatrix} q_4 \\ q_5 \\ q_6 \end{bmatrix} = \begin{bmatrix} u(\pi/2) \\ v(\pi/2) \\ \theta(\pi/2) \end{bmatrix} = \begin{bmatrix} K_{44} & K_{45} & K_{46} \\ K_{54} & K_{55} & K_{56} \\ K_{64} & K_{65} & K_{66} \end{bmatrix}^{-1} \begin{bmatrix} 0 \\ P \\ 0 \end{bmatrix} \qquad (10.273)$$

The results essentially match. Similarly, the results of Equations (10.255) and (10.256) also match. For the result of Equation (10.263), it is actually easier to obtain the results using the stiffness matrix. For the case of a fixed half ring, the deformation of the equivalent quarter ring matches the shape function when $q^T = [0\ 0\ 0\ 0\ q_5\ 0]$. As a result,

$$
\begin{bmatrix} Q_1 \\ Q_2 \\ Q_3 \\ Q_4 \\ Q_5 \\ Q_6 \end{bmatrix}_5 = \begin{bmatrix} Q_1 \\ Q_2 \\ Q_3 \\ Q_4 \\ P \\ Q_6 \end{bmatrix}_5 = K \begin{bmatrix} 0 \\ 0 \\ 0 \\ 0 \\ q_5 \\ 0 \end{bmatrix} = \begin{bmatrix} K_{15} \\ K_{25} \\ K_{35} \\ K_{45} \\ K_{55} \\ K_{65} \end{bmatrix} q_5 \qquad (10.274)
$$

From the above equation, we determine

$$
v\left(\frac{\pi}{2}\right) = q_5 = 100/K_{55} = 16.4 \ \mu\text{m} \qquad (10.275)
$$

$$
F = Q_4 = K_{45}q_5 = -90.4 \ \text{N} \qquad (10.276)
$$

$$
M = Q_6 = K_{65}q_5 = 30.8 \ \text{Nm} \qquad (10.277)
$$

Using Equations (10.259)–(10.261), we obtain $v(\pi/2) = 15.1 \ \mu\text{m}$, $F = -91.1$ N and $M = 30.5$ Nm. The difference is mainly due to the fact that the stiffness matrix method does not consider the energy stored by tangential deformation; thus the half-ring appears to be less stiff, showing about 8.6% reduction.

With the stiffness matrix defined, we can proceed to determine the elastic energy stored in a general curved beam. Finally, we can invoke the Castigliano theorem to determine the deformation of a general case of Figure 10.18(a). However, we do not have to work out the expression of the elastic energy because by applying the Castigliano theorem, we have

$$
Q = \frac{\partial U_M}{\partial q} = \frac{\partial(\frac{1}{2}q'Kq)}{\partial q} = Kq \qquad (10.278)
$$

We can simply combine the stiffness matrices to calculate the forces and deformations.

Let's finally solve the problem of Figure 10.18(a) with $\alpha = \pi/3$. This curved beam fixed at both ends is broken into two shorter beams as shown in Figure 10.18(b). The curved beam properties are as defined before.

For each shorter beam, we have

$$
\begin{bmatrix} Q_1 \\ Q_2 \\ Q_3 \\ Q_4 \\ Q_5 \\ Q_6 \end{bmatrix}^{(1)} = \begin{bmatrix} Q_1 \\ Q_2 \\ Q_3 \\ Q_4 \\ Q_5 \\ Q_6 \end{bmatrix} = K^{(1)} \begin{bmatrix} q_1 \\ q_2 \\ q_3 \\ q_4 \\ q_5 \\ q_6 \end{bmatrix}^{(1)} = K^{(1)} \begin{bmatrix} q_1 \\ q_2 \\ q_3 \\ q_4 \\ q_5 \\ q_6 \end{bmatrix} \tag{10.279}
$$

$$
\begin{bmatrix} Q_1 \\ Q_2 \\ Q_3 \\ Q_4 \\ Q_5 \\ Q_6 \end{bmatrix}^{(2)} = \begin{bmatrix} Q_4 \\ Q_5 \\ Q_6 \\ Q_7 \\ Q_8 \\ Q_9 \end{bmatrix} = K^{(2)} \begin{bmatrix} q_1 \\ q_2 \\ q_3 \\ q_4 \\ q_5 \\ q_6 \end{bmatrix}^{(2)} = K^{(2)} \begin{bmatrix} q_4 \\ q_5 \\ q_6 \\ q_7 \\ q_8 \\ q_9 \end{bmatrix} \tag{10.280}
$$

We introduce two conversion matrices defined as

$$
A_1 = \begin{bmatrix} 1 & 0 & 0 & 0 & 0 & 0 & 0 & 0 & 0 \\ 0 & 1 & 0 & 0 & 0 & 0 & 0 & 0 & 0 \\ 0 & 0 & 1 & 0 & 0 & 0 & 0 & 0 & 0 \\ 0 & 0 & 0 & 1 & 0 & 0 & 0 & 0 & 0 \\ 0 & 0 & 0 & 0 & 1 & 0 & 0 & 0 & 0 \\ 0 & 0 & 0 & 0 & 0 & 1 & 0 & 0 & 0 \end{bmatrix} \tag{10.281}
$$

$$
A_2 = \begin{bmatrix} 0 & 0 & 0 & 1 & 0 & 0 & 0 & 0 & 0 \\ 0 & 0 & 0 & 0 & 1 & 0 & 0 & 0 & 0 \\ 0 & 0 & 0 & 0 & 0 & 1 & 0 & 0 & 0 \\ 0 & 0 & 0 & 0 & 0 & 0 & 1 & 0 & 0 \\ 0 & 0 & 0 & 0 & 0 & 0 & 0 & 1 & 0 \\ 0 & 0 & 0 & 0 & 0 & 0 & 0 & 0 & 1 \end{bmatrix} \tag{10.282}
$$

The combined relationship can be derived as

$$
Q = \begin{bmatrix} Q_1 \\ Q_2 \\ Q_3 \\ Q_4 \\ Q_5 \\ Q_6 \\ Q_7 \\ Q_8 \\ Q_9 \end{bmatrix} = A_1' Q^{(1)} + A_2' Q^{(2)} = A_1' K^{(1)} q^{(1)} + A_2' K^{(2)} q^{(2)}
$$

$$
= A_1' K^{(1)} A_1 q + A_2' K^{(2)} A_2 q = K \begin{bmatrix} q_1 \\ q_2 \\ q_3 \\ q_4 \\ q_5 \\ q_6 \\ q_7 \\ q_8 \\ q_9 \end{bmatrix} \tag{10.283}
$$

$$
K = A_1' K^{(1)} A_1 + A_2' K^{(2)} A_2 \tag{10.284}
$$

For the case of Figure 10.18(a) with $\alpha = \pi/3$, we also know that

$$
q_1 = q_2 = q_3 = q_4 = q_6 = q_7 = q_8 = q_9 = 0 \tag{10.285}
$$

Therefore, just like Equation (10.274), we have

$$
\begin{bmatrix} Q_1 \\ Q_2 \\ Q_3 \\ Q_4 \\ Q_5 \\ Q_6 \\ Q_7 \\ Q_8 \\ Q_9 \end{bmatrix} = \begin{bmatrix} Q_1 \\ Q_2 \\ Q_3 \\ Q_4 \\ 2P \\ Q_6 \\ Q_7 \\ Q_8 \\ Q_9 \end{bmatrix} = K \begin{bmatrix} 0 \\ 0 \\ 0 \\ 0 \\ q_5 \\ 0 \\ 0 \\ 0 \\ 0 \end{bmatrix} = \begin{bmatrix} K_{15} \\ K_{25} \\ K_{35} \\ K_{45} \\ K_{55} \\ K_{65} \\ K_{75} \\ K_{85} \\ K_{95} \end{bmatrix} q_5 \tag{10.286}
$$

The vertical displacement at the center of the beam is then found to be $v(\pi/3) = 7.8\ \mu$m and the reaction forces and moments are $F_t(0) = 160.8$ N, $F_n(0) = 78.4$ N, and $M_b(0) = -9.6$ Nm. This shorter curved beam is stiffer than the half-ring beam with the same beam properties.

We will now go back to straight beams for the following sections.

## 10.12   Equivalent General Forces

When we derive the elastic energy, the general forces are defined specifically at the same locations as the general displacement. What if there are other loadings not at the locations of the general displacements? Can we still account for their effect using the same elastic energy concept?

As shown in Figure 10.24, there is an external force, $P$, acting at $x = a$, a moment, $\tau$, acting at $x = b$, and a distributed force, $w$, acting over $c < x < c + d$. In order to use the formulations we derived in

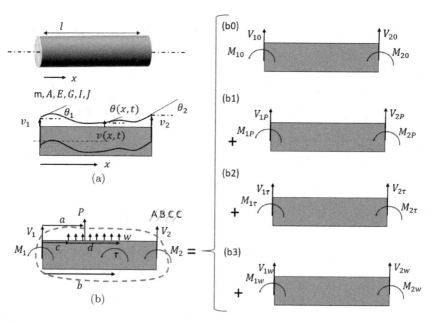

Figure 10.24:   Conversion of external loading on a beam element to equivalent general forces. (a) Displacements of a beam element and (b) the loading conditions and their equivalent general forces.

the previous sections, we need to convert these external loadings to equivalent general forces at the boundaries. As shown in Figure 10.24(b), the general force vector, $[V_{10}\ M_{10}\ V_{20}\ M_{20}]^T$, is the force vector when there are no external loadings. The external force, $P$, is replaced by $[V_{1P}\ M_{1P}\ V_{2P}\ M_{2P}]^T$, the moment, $\tau$, by $[V_{1\tau}\ M_{1\tau}\ V_{2\tau}\ M_{2\tau}]^T$, and the distributed force, $w$, by $[V_{1w}\ M_{1w}\ V_{2w}\ M_{2w}]^T$. As a result, the general force vector becomes

$$Q = \begin{bmatrix} V_1 \\ M_1 \\ V_2 \\ M_2 \end{bmatrix} = Q_0 + Q_P + Q_\tau + Q_w$$

$$= \begin{bmatrix} V_{10} \\ M_{10} \\ V_{20} \\ M_{20} \end{bmatrix} + \begin{bmatrix} V_{1P} \\ M_{1P} \\ V_{2P} \\ M_{2P} \end{bmatrix} + \begin{bmatrix} V_{1\tau} \\ M_{1\tau} \\ V_{2\tau} \\ M_{2\tau} \end{bmatrix} + \begin{bmatrix} V_{1w} \\ M_{1w} \\ V_{2w} \\ M_{2w} \end{bmatrix} \qquad (10.287)$$

To determine these equivalent general force vectors, we assume that they are equivalent if they produce the same elastic energy. For the external loading, $P$, we have

$$\delta U_P = P\delta v(a) = P\phi(a)\delta q = Q_P \delta q \qquad (10.288)$$

From the above equation and the shape functions of Equations (10.107)–(10.110), we have

$$Q_P = P\phi(a) = P \begin{bmatrix} \phi_{v1}(a) \\ \phi_{v2}(a) \\ \phi_{v3}(a) \\ \phi_{v4}(a) \end{bmatrix} = \begin{bmatrix} P(1 - 3(\frac{a}{l})^2 + 2(\frac{a}{l})^3) \\ P(l(\frac{a}{l}) - 2l(\frac{a}{l})^2 + l(\frac{a}{l})^3) \\ P(3(\frac{a}{l})^2 - 2(\frac{a}{l})^3) \\ P(-l(\frac{a}{l})^2 + l(\frac{a}{l})^3) \end{bmatrix} \qquad (10.289)$$

If $P$ is acting at the middle of the beam, i.e., $a = \frac{l}{2}$, the corresponding $Q_P$ becomes

$$Q_P\left(a = \frac{l}{2}\right) = \begin{bmatrix} \dfrac{P}{2} \\[6pt] \dfrac{Pl}{8} \\[6pt] \dfrac{P}{2} \\[6pt] -\dfrac{Pl}{8} \end{bmatrix} \tag{10.290}$$

Similarly, for the external moment $\tau$,

$$\delta U_\tau = \tau \delta\theta = \tau\delta\left(\frac{dv}{dx}\right) = \tau\delta\left(\frac{\partial\phi}{\partial x}q\right) = \tau\frac{\partial\phi}{\partial x}\delta q = Q_\tau \delta q \tag{10.291}$$

Introducing the location of $\tau$, we have

$$Q_\tau = \tau\frac{\partial\phi(b)}{\partial x} = \begin{bmatrix} \tau(\frac{1}{l})(-6(\frac{b}{l}) + 6(\frac{b}{l})^2) \\[8pt] \tau(\frac{1}{l})(l - 4l(\frac{b}{l}) + 3l(\frac{b}{l})^2) \\[8pt] \tau(\frac{1}{l})(6(\frac{b}{l}) - 6(\frac{b}{l})^2) \\[8pt] \tau(\frac{1}{l})(-2l(\frac{b}{l}) + 3l(\frac{b}{l})^2) \end{bmatrix} \tag{10.292}$$

For the distributed loading, $w$, we have

$$\delta U_w = \int_c^{c+d} w\, dx\, \delta v(x) = \int_c^{c+d} w\phi(x)dx\, \delta q = Q_w \delta q \tag{10.293}$$

and

$$Q_w = \int_c^{c+d} w\phi(x)dx = w \begin{bmatrix} l(\xi - \xi^3 + \frac{1}{2}\xi^4) \\[8pt] l(\frac{l}{2}\xi^2 - \frac{2l}{3}\xi^3 + \frac{l}{4}\xi^4) \\[8pt] l(\xi^3 - \frac{1}{2}\xi^4) \\[8pt] l(-\frac{l}{3}\xi^3 + \frac{l}{4}\xi^4) \end{bmatrix}_c^{c+d} \tag{10.294}$$

If $c = 0$ and $d = l$, we have

$$Q_w = \int_0^l w\phi(x)dx = \begin{bmatrix} \dfrac{l}{2}w \\[2mm] \dfrac{l^2}{12}w \\[2mm] \dfrac{l}{2}w \\[2mm] -\dfrac{l^2}{12}w \end{bmatrix} \qquad (10.295)$$

Finally, what if the beam element is connected to another component, such as a spring, as shown in Figure 10.25?

By conducting a free-body diagram analysis, the spring is replaced by a spring force, $F_k$, and then this spring force is replaced by an equivalent general force vector, $Q_k$, via the following relationship,

$$\delta U_k = -F_k\delta v(e,t) = -kv(e,t)\delta v(e,t)$$
$$= -kv(e,t)\phi(e)\delta q = [-k\phi(e)'\phi(e)q]'\delta q = Q_k'\delta q \qquad (10.296)$$

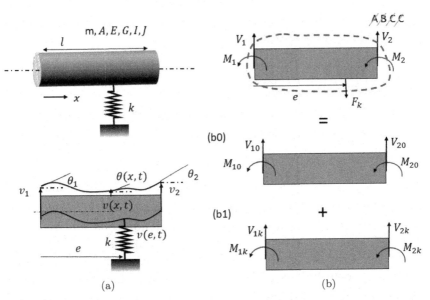

Figure 10.25: Conversion of a spring element to an equivalent general force.

and

$$
Q_k = -k
\begin{bmatrix}
\phi_1(e)\phi_1(e) & \phi_1(e)\phi_2(e) & \phi_1(e)\phi_3(e) & \phi_1(e)\phi_4(e) \\
\phi_2(e)\phi_1(e) & \phi_2(e)\phi_2(e) & \phi_2(e)\phi_3(e) & \phi_2(e)\phi_4(e) \\
\phi_3(e)\phi_1(e) & \phi_3(e)\phi_2(e) & \phi_3(e)\phi_3(e) & \phi_3(e)\phi_4(e) \\
\phi_4(e)\phi_1(e) & \phi_4(e)\phi_2(e) & \phi_4(e)\phi_3(e) & \phi_4(e)\phi_4(e)
\end{bmatrix}
\begin{bmatrix}
v_1 \\ \theta_1 \\ v_2 \\ \theta_2
\end{bmatrix}
$$

$$
= -k\bar{K}\, q \tag{10.297}
$$

The equivalent general force vector, $Q_k$, is a function of $q$ and it should be combined with the stiffness matrix of the beam of Equation (10.141) as

$$
K_{\text{all}} = K_b + k\bar{K} \tag{10.298}
$$

If $e = 0$, with the spring attached to the left edge, $\bar{K}$ becomes

$$
\bar{K} =
\begin{bmatrix}
1 & 0 & 0 & 0 \\
0 & 0 & 0 & 0 \\
0 & 0 & 0 & 0 \\
0 & 0 & 0 & 0
\end{bmatrix}
\tag{10.299}
$$

The external spring constant is simply added to the $k_{11}$ element of the stiffness matrix of Equation (10.141).

## 10.13 Equation of Motion for General Dynamic Systems

To derive the equation of motion for a general dynamic system, it is more convenient to go with the energy method. The general energy equation is

$$
d(T + U) = dT + dU = \delta W = Q^T dq \tag{10.300}
$$

where $T$ is the total kinetic energy of the system, $U$, the total potential energy, and $W$ the work done by all non-conservative forces involved with the system, including external and internal forces.

In the next section, we will present the celebrated Lagrange method to convert the general energy equation expression to the one related to the general force vector and the general displacement vector.

## 10.13.1 *Lagrange's equations*

The total kinetic energy is a function of the general velocity vector and the general displacement vector,

$$T = T(q, \dot{q}) \tag{10.301}$$

We have

$$dT = \frac{\partial T}{\partial q} dg + \frac{\partial T}{\partial \dot{q}} d\dot{q} \tag{10.302}$$

The total potential energy is a function of the general displacement only,

$$U = U(q) \tag{10.303}$$

We have

$$dU = \frac{\partial U}{\partial q} dg \tag{10.304}$$

Introducing the mass matrix and stiffness matrix to represent the total kinetic energy and the total potential energy, we found

$$\frac{\partial T}{\partial \dot{q}} = \frac{\partial}{\partial \dot{q}} \left[ \frac{1}{2} \dot{q}' M q \right]$$
$$= \dot{q}' M \tag{10.305}$$

Multiplying the above term with $\dot{q}$, we have

$$\frac{\partial T}{\partial \dot{q}} \dot{q} = \dot{q}' M \dot{q} = 2T \tag{10.306}$$

Taking the total derivative of $2T$, we have

$$d(2T) = 2dT = d \left( \frac{\partial T}{\partial \dot{q}} \dot{q} \right)$$
$$= d \left( \frac{\partial T}{\partial \dot{q}} \right) \dot{q} + \frac{\partial T}{\partial \dot{q}} d\dot{q} \tag{10.307}$$

Subtracting Equation (10.302) from (10.307), we obtain

$$dT = d\left(\frac{\partial T}{\partial \dot{q}}\right)\dot{q} - \frac{\partial T}{\partial q}dq$$

$$= d\left(\frac{\partial T}{\partial \dot{q}}\right)\frac{dq}{dt} - \frac{\partial T}{\partial q}dq$$

$$= \frac{d}{dt}\left(\frac{\partial T}{\partial \dot{q}}\right)dq - \frac{\partial T}{\partial q}dq$$

$$= \left[\frac{d}{dt}\left(\frac{\partial T}{\partial \dot{q}}\right) - \frac{\partial T}{\partial q}\right]dq \qquad (10.308)$$

The above manipulation allows us to eliminate the term related to $d\dot{q}$.

Finally, introducing Equations (10.304) and (10.308) into (10.300), we obtain the expression of the well-known Lagrange equation,

$$\frac{d}{dt}\left(\frac{\partial T}{\partial \dot{q}}\right) - \frac{\partial T}{\partial q} + \frac{\partial U}{\partial q} = Q' \qquad (10.309)$$

However, the expression of Lagrange's equation with the column vector $Q$, instead of $Q^T$ is more convenient to use. The explicit expression of Lagrange's equation in column vector form becomes

$$\frac{d}{dt}\begin{bmatrix}\dfrac{\partial T}{\partial \dot{q}_1}\\[2mm]\dfrac{\partial T}{\partial \dot{q}_2}\\[2mm]\vdots\\[2mm]\dfrac{\partial T}{\partial \dot{q}_n}\end{bmatrix} - \begin{bmatrix}\dfrac{\partial T}{\partial q_1}\\[2mm]\dfrac{\partial T}{\partial q_2}\\[2mm]\vdots\\[2mm]\dfrac{\partial T}{\partial q_n}\end{bmatrix} + \begin{bmatrix}\dfrac{\partial U}{\partial q_1}\\[2mm]\dfrac{\partial U}{\partial q_2}\\[2mm]\vdots\\[2mm]\dfrac{\partial U}{\partial q_n}\end{bmatrix} = \begin{bmatrix}Q_1\\Q_2\\\vdots\\Q_n\end{bmatrix} \qquad (10.310)$$

Let's revisit the structure of Figure 10.16 and derive its equation of motion using Lagrange's method.

The beam elements are considered massless. The total kinetic energy is

$$T = \frac{1}{2}(m_1 + m_2)\dot{q}_3^2 + \frac{1}{2}J_1\dot{q}_3^2 + \frac{1}{2}J_2\dot{q}_6^2 \qquad (10.311)$$

The total potential energy was already derived by Equation (10.173). The general force vector is $Q = [F_1 \; \tau_1 \; \tau_2]^T$ and the general displacement vector is $Q = [q_3 \; q_4 \; q_6]^T$. Calculating the terms in Equation (10.310), we found

$$\frac{\partial T}{\partial \dot{q}_3} = (m_1 + m_2)q_3 \tag{10.312}$$

$$\frac{\partial T}{\partial \dot{q}_4} = J_1 q_4 \tag{10.313}$$

$$\frac{\partial T}{\partial \dot{q}_6} = J_2 q_6 \tag{10.314}$$

$$\frac{\partial T}{\partial q_3} = \frac{\partial T}{\partial q_4} = \frac{\partial T}{\partial q_6} = 0 \tag{10.315}$$

$$\frac{\partial U}{\partial q_3} = \frac{EI_1}{2l_1^3}(-12l_1 q_4 + 24q_3) \tag{10.316}$$

$$\frac{\partial U}{\partial q_4} = \frac{EI_1}{2l_1^3}(-12l_1 q_3 + 8l_1^2 q_4) + \frac{EI_2}{2l_2^3}(8l_2^2 q_4 + 4l_2^2 q_6) \tag{10.317}$$

$$\frac{\partial U}{\partial q_6} = \frac{EI_2}{2l_2^3}(8l_2^2 q_6 + 4l_2^2 q_4) \tag{10.318}$$

Introducing all the terms in the above equation to Equation (10.310), we have

$$\frac{d}{dt}\begin{bmatrix} (m_1 + m_2)\dot{q}_3 \\ J_1 \dot{q}_4 \\ J_2 \dot{q}_6 \end{bmatrix} - \begin{bmatrix} 0 \\ 0 \\ 0 \end{bmatrix}$$

$$+ \begin{bmatrix} \dfrac{EI_1}{2l_1^3}(-12l_1 q_4 + 24q_3) \\ \dfrac{EI_1}{2l_1^3}(-12l_1 q_3 + 8l_1^2 q_4) + \dfrac{EI_2}{2l_2^3}(8l_2^2 q_4 + 4l_2^2 q_6) \\ \dfrac{EI_2}{2l_2^3}(8l_2^2 q_6 + 4l_2^2 q_4) \end{bmatrix} = \begin{bmatrix} F_1 \\ \tau_1 \\ \tau_2 \end{bmatrix} \tag{10.319}$$

Rearranging the above equation accordingly, we obtain

$$M\ddot{q} + Kq = Q \tag{10.320}$$

where

$$M = \begin{bmatrix} m_1 + m_2 & 0 & 0 \\ 0 & J_1 & 0 \\ 0 & 0 & J_2 \end{bmatrix} \tag{10.321}$$

$$K = \begin{bmatrix} \dfrac{12EI_1}{l_1^3} & \dfrac{-6EI_1}{2l_1^2} & 0 \\[2ex] \dfrac{-6EI_1}{2l_1^2} & \dfrac{4EI_1}{l_1} + \dfrac{4EI_2}{l_2} & \dfrac{-2EI_2}{l_2} \\[2ex] 0 & \dfrac{-2EI_2}{l_2} & \dfrac{4EI_2}{l_2} \end{bmatrix} \tag{10.322}$$

$$Q = \begin{bmatrix} F_1 \\ \tau_1 \\ \tau_2 \end{bmatrix} \tag{10.323}$$

Equation (10.320) is the general form of a linear dynamic system without damping. With damping, the general expression becomes

$$M\ddot{q} + B\dot{q} + Kq = Q \tag{10.324}$$

With today's computing power, the solution to the above equation is no longer an issue. In the next section, we present some Matlab codes for solving the general linear Equation of Motion.

## 10.14 Introduction to the Finite Element Method for Dynamic Analysis

In this section, we will present an example to discuss the general methodology of the Finite Element Method (FEM) for dynamic systems. As shown in Figure 10.26, a cantilever beam has a mass per unit length, $m = 100$ lbm/ft and has four segments, each of equal length, $l = 3$ in. The bending stiffness of the beam are $EI_1 = 5 \times 10^6$ lbf–in$^2$

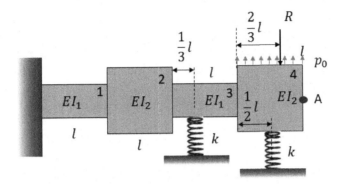

Figure 10.26: A composite cantilever beam subjected to different loadings and connected to springs.

and $EI_2 = 2EI_1$. Two springs are connected to the beam at locations as shown and the spring constant is $k = 2{,}400$ lbf/ft. A force $R$ is applied at the fourth segment at $2/3l$ off the front edge, while a uniformly distributed force is also applied.

We will present the FEM to determine the motion of point $A$ under three loading conditions:

1. Plot the displacement of point $A$ as a function of time when $p_0 = 0$ and

$$R(t) = 500 \text{ lbf}, \quad 0 \le t \le 0.01 \text{ s}$$
$$R(t) = 0, \quad t > 0.01 \text{ s}$$

2. Plot the displacement of point $A$ as a function of time when $p_0 = 0$ and

$$R(t) = 500 \sin(2\pi \cdot 10 \cdot t) \text{ lbf}$$

3. Plot the displacement of point $A$ as a function of time when $R = 0$ and

$$p_0(t) = 100 \text{ lbf}, \quad t \ge 0 \text{ s}$$
$$p_0(t) = 0, \quad t < 0 \text{ s}$$

We will model the transverse displacement of each segment based on the general displacement and general force vectors defined in Figure 10.13(d). We will be considering only the Euler–Bernoulli beam theory; therefore the shape functions are defined by Equations (10.107)–(10.110).

The mass matrix can be simplified from Equation (10.124) with $\Phi = 0$ as

$$
M_b = \frac{\mathrm{ml}}{420}
\begin{bmatrix}
156 & 22l & 54 & -13l \\
22l & 4l^2 & 13l & -3l^2 \\
54 & 13l & 156 & -22l \\
-13l & -3l^2 & -22l & 4l^2
\end{bmatrix}
\tag{10.325}
$$

The mass matrix is the same for all segments.

The stiffness matrix is defined by Equation (10.141). The first and third segments have the same stiffness matrix, while the second and the fourth segments are the same.

$$
K_{b1} = K_{b3} = \frac{EI_1}{l^3}
\begin{bmatrix}
12 & 6l & -12 & 6l \\
6l & 4l^2 & -6l & 2l^2 \\
-12 & -6l & 12 & -6l \\
6l & 2l^2 & -6l & 4l^2
\end{bmatrix}
\tag{10.326}
$$

$$
K_{b2} = K_{b4} = 2K_{b1}
\tag{10.327}
$$

### 10.14.1 Element #1 model

We now consider each element separately to establish its equation of motion, mainly the $M$ and $K$ matrices. For element #1, as shown in Figure 10.27(a), the general displacement and force vectors are

$$
q^{(1)} =
\begin{bmatrix}
v_1 \\
\theta_1 \\
v_2 \\
\theta_2
\end{bmatrix}^{(1)}
\tag{10.328}
$$

$$Q^{(1)} = \begin{bmatrix} V_{10} \\ M_{10} \\ V_{20} \\ M_{20} \end{bmatrix}^{(1)} \qquad (10.329)$$

where the superscript (1) indicates element #1.

Note that $V_{10}$ and $M_{10}$ are the reaction force and moment at the fixed end of the overall beam, respectively, while $V_{20}$ and $M_{20}$ are internal reaction force and moment which would be canceled when Elements #1 and #2 are combined.

The stiffness matrix is defined by Equation (10.327),

$$K^{(1)} = K_{b1} \qquad (10.330)$$

### 10.14.2 Element #2 model

For element #2, as shown in Figure 10.27(b), the general displacement and force vectors are

$$q^{(2)} = \begin{bmatrix} v_1 \\ \theta_1 \\ v_2 \\ \theta_2 \end{bmatrix}^{(2)} \qquad (10.331)$$

$$Q^{(2)} = \begin{bmatrix} V_{10} \\ M_{10} \\ V_{20} \\ M_{20} \end{bmatrix}^{(2)} \qquad (10.332)$$

Note that for element #2, $V_{10}$, $M_{10}$, $V_{20}$, and $M_{20}$ are all internal reaction forces and moments which would be canceled when Elements #1, #2, and #3 are combined.

The stiffness matrix is

$$K^{(2)} = K_{b2} = 2K_{b1} \qquad (10.333)$$

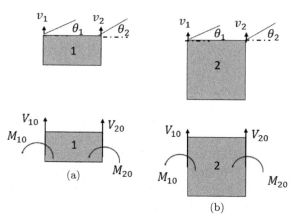

Figure 10.27: Models for beam elements: (a) Element #1; and (b) Element #2.

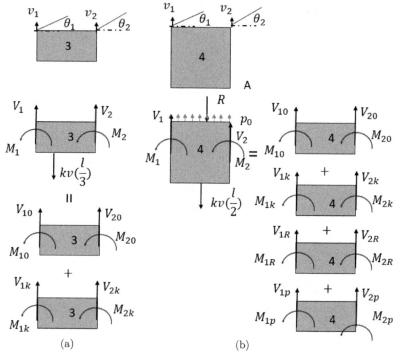

Figure 10.28: Models for beam elements: (a) Element #3; (b) Element #4.

### 10.14.3   *Element #3 model*

For element #3, the general displacement vector is similarly defined as

$$
q^{(3)} = \begin{bmatrix} v_1 \\ \theta_1 \\ v_2 \\ \theta_2 \end{bmatrix}^{(3)}
\tag{10.334}
$$

The general force vector is more complicated due to the presence of the spring. The general force vector becomes

$$
Q^{(3)} = \begin{bmatrix} V_{10} + V_{1k} \\ M_{10} + M_{1k} \\ V_{20} + V_{2k} \\ M_{20} + M_{2k} \end{bmatrix}^{(2)}
\tag{10.335}
$$

The internal reaction forces and moments will be canceled when element #3 is combined with other elements, but those related to the spring will become an equivalent stiffness matrix, as expressed by Equations (10.296)–(10.298). The combined stiffness matrix with the spring becomes

$$
K^{(3)} = K_{b1} + K_k
$$

$$
= K_{b1} + k \begin{bmatrix}
\phi_1(e)\phi_1(e) & \phi_1(e)\phi_2(e) & \phi_1(e)\phi_3(e) & \phi_1(e)\phi_4(e) \\
\phi_2(e)\phi_1(e) & \phi_2(e)\phi_2(e) & \phi_2(e)\phi_3(e) & \phi_2(e)\phi_4(e) \\
\phi_3(e)\phi_1(e) & \phi_3(e)\phi_2(e) & \phi_3(e)\phi_3(e) & \phi_3(e)\phi_4(e) \\
\phi_4(e)\phi_1(e) & \phi_4(e)\phi_2(e) & \phi_4(e)\phi_3(e) & \phi_4(e)\phi_4(e)
\end{bmatrix}
\tag{10.336}
$$

where $e = l/3$.

### 10.14.4 *Element #4 model*

For element #4, the general displacement vector is

$$q^{(4)} = \begin{bmatrix} v_1 \\ \theta_1 \\ v_2 \\ \theta_2 \end{bmatrix}^{(4)} \tag{10.337}$$

The general force vector needs to account for the attached spring, the distributed force, and the concentrated force. It becomes

$$Q^{(4)} = \begin{bmatrix} V_{10} + V_{1k} + V_{1R} + V_{1p} \\ M_{10} + M_{1k} + M_{1R} + M_{1p} \\ V_{20} + V_{2k} + V_{2R} + V_{2p} \\ M_{20} + M_{2k} + V_{2R} + V_{2p} \end{bmatrix}^{(4)} \tag{10.338}$$

The internal reaction forces and moments will be canceled when element #4 is combined with other elements, but those related to the spring will become an equivalent stiffness matrix as before. The equivalent forces related to $R$ and $p$ will remain. The stiffness matrix with the spring becomes

$$K^{(4)} = K_{b2} + K_k$$

$$= 2K_{b1} + k \begin{bmatrix} \phi_1(e)\phi_1(e) & \phi_1(e)\phi_2(e) & \phi_1(e)\phi_3(e) & \phi_1(e)\phi_4(e) \\ \phi_2(e)\phi_1(e) & \phi_2(e)\phi_2(e) & \phi_2(e)\phi_3(e) & \phi_2(e)\phi_4(e) \\ \phi_3(e)\phi_1(e) & \phi_3(e)\phi_2(e) & \phi_3(e)\phi_3(e) & \phi_3(e)\phi_4(e) \\ \phi_4(e)\phi_1(e) & \phi_4(e)\phi_2(e) & \phi_4(e)\phi_3(e) & \phi_4(e)\phi_4(e) \end{bmatrix}$$

$$\tag{10.339}$$

where $e = 2l/3$.

Invoking Equations (10.289) and (10.293), we have

$$
\begin{bmatrix} V_{1R} \\ M_{1R} \\ V_{2R} \\ V_{2R} \end{bmatrix}^{(4)} = R \begin{bmatrix} \phi_1(\frac{2}{3}) \\ \phi_2(\frac{2}{3}) \\ \phi_3(\frac{2}{3}) \\ \phi_4(\frac{2}{3}) \end{bmatrix} = R \begin{bmatrix} -\dfrac{7}{27} \\ -\dfrac{2}{27}l \\ -\dfrac{20}{27} \\ \dfrac{4}{27}l \end{bmatrix} \tag{10.340}
$$

and

$$
\begin{bmatrix} V_{1p} \\ M_{1p} \\ V_{2p} \\ V_{2p} \end{bmatrix}^{(4)} = p \begin{bmatrix} \int_0^l \phi_1(x)dx \\ \int_0^l \phi_2(x)dx \\ \int_0^l \phi_3(x)dx \\ \int_0^l \phi_4(x)dx \end{bmatrix} = p \begin{bmatrix} \dfrac{1}{2}l \\ \dfrac{1}{12}l^2 \\ \dfrac{1}{2}l \\ -\dfrac{1}{12}l^2 \end{bmatrix} \tag{10.341}
$$

## 10.14.5 *Global equation of motion*

The global displacement and force vectors are

$$
q = \begin{bmatrix} q_1 \\ q_2 \\ q_3 \\ q_4 \\ q_5 \\ q_6 \\ q_7 \\ q_8 \\ q_9 \\ q_{10} \end{bmatrix} \tag{10.342}
$$

$$Q = \begin{bmatrix} Q_1 \\ Q_2 \\ Q_3 \\ Q_4 \\ Q_5 \\ Q_6 \\ Q_7 \\ Q_8 \\ Q_9 \\ Q_{10} \end{bmatrix} \qquad (10.343)$$

The conversion matrices to combine element equations of motion are

$$A_1 = \begin{bmatrix} 1 & 0 & 0 & 0 & 0 & 0 & 0 & 0 & 0 & 0 \\ 0 & 1 & 0 & 0 & 0 & 0 & 0 & 0 & 0 & 0 \\ 0 & 0 & 1 & 0 & 0 & 0 & 0 & 0 & 0 & 0 \\ 0 & 0 & 0 & 1 & 0 & 0 & 0 & 0 & 0 & 0 \end{bmatrix} \qquad (10.344)$$

$$A_2 = \begin{bmatrix} 0 & 0 & 1 & 0 & 0 & 0 & 0 & 0 & 0 & 0 \\ 0 & 0 & 0 & 1 & 0 & 0 & 0 & 0 & 0 & 0 \\ 0 & 0 & 0 & 0 & 1 & 0 & 0 & 0 & 0 & 0 \\ 0 & 0 & 0 & 0 & 0 & 1 & 0 & 0 & 0 & 0 \end{bmatrix} \qquad (10.345)$$

$$A_3 = \begin{bmatrix} 0 & 0 & 0 & 0 & 1 & 0 & 0 & 0 & 0 & 0 \\ 0 & 0 & 0 & 0 & 0 & 1 & 0 & 0 & 0 & 0 \\ 0 & 0 & 0 & 0 & 0 & 0 & 1 & 0 & 0 & 0 \\ 0 & 0 & 0 & 0 & 0 & 0 & 0 & 1 & 0 & 0 \end{bmatrix} \qquad (10.346)$$

$$A_4 = \begin{bmatrix} 0 & 0 & 0 & 0 & 0 & 0 & 1 & 0 & 0 & 0 \\ 0 & 0 & 0 & 0 & 0 & 0 & 0 & 1 & 0 & 0 \\ 0 & 0 & 0 & 0 & 0 & 0 & 0 & 0 & 1 & 0 \\ 0 & 0 & 0 & 0 & 0 & 0 & 0 & 0 & 0 & 1 \end{bmatrix} \qquad (10.347)$$

The global stiffness matrix is combined as

$$K = A_1' K^{(1)} A_1 + A_2' K^{(2)} A_2$$
$$+ A_3' K^{(3)} A_3 + A_4' K^{(4)} A_4 \qquad (10.348)$$

The global mass matrix is

$$M = A_1' M^{(1)} A_1 + A_2' M^{(2)} A_2$$
$$+ A_3' M^{(3)} A_3 + A_4' M^{(4)} A_4 \qquad (10.349)$$

The global general force vector is

$$Q = A_1' Q^{(1)} + A_2' Q^{(2)} + A_3' Q^{(3)} + A_4' Q^{(4)}$$

$$= \begin{bmatrix} F_{10} \\ F_{10} \\ 0 \\ 0 \\ 0 \\ 0 \\ -\dfrac{7}{27}R + \dfrac{l}{2}p \\ -\dfrac{2l}{27}R + \dfrac{l^2}{12}p \\ -\dfrac{20}{27}R + \dfrac{l}{2}p \\ \dfrac{4l}{27}R - \dfrac{l^2}{12}p \end{bmatrix} \qquad (10.350)$$

### 10.14.6  *Model reduction with boundary conditions*

From the boundary conditions, we know $q_1 = q_2 = 0$; therefore, the global matrices should be reduced from $10 \times 10$ to $8 \times 8$. The reduced equation of motion becomes

$$\hat{M}\ddot{\hat{q}} + \hat{K}\hat{q} = \hat{Q} \qquad (10.351)$$

where

$$
\hat{Q} = \begin{bmatrix} 0 \\ 0 \\ 0 \\ 0 \\ -\dfrac{7}{27}R + \dfrac{l}{2}p \\ -\dfrac{2l}{27}R + \dfrac{l^2}{12}p \\ -\dfrac{20}{27}R + \dfrac{l}{2}p \\ \dfrac{4l}{27}R - \dfrac{l^2}{12}p \end{bmatrix} \tag{10.352}
$$

$$
\hat{K} = \begin{bmatrix} K_{33} & K_{34} & K_{35} & K_{36} & K_{37} & K_{38} & K_{39} & K_{310} \\ K_{43} & K_{44} & K_{45} & K_{46} & K_{47} & K_{48} & K_{49} & K_{410} \\ K_{53} & K_{54} & K_{55} & K_{56} & K_{57} & K_{58} & K_{59} & K_{510} \\ K_{63} & K_{64} & K_{65} & K_{66} & K_{67} & K_{68} & K_{69} & K_{610} \\ K_{73} & K_{74} & K_{75} & K_{76} & K_{77} & K_{78} & K_{79} & K_{710} \\ K_{83} & K_{84} & K_{85} & K_{86} & K_{87} & K_{88} & K_{89} & K_{810} \\ K_{93} & K_{94} & K_{95} & K_{96} & K_{97} & K_{98} & K_{99} & K_{910} \\ K_{103} & K_{104} & K_{105} & K_{106} & K_{107} & K_{108} & K_{109} & K_{1010} \end{bmatrix} \tag{10.353}
$$

$$
\hat{M} = \begin{bmatrix} M_{33} & M_{34} & M_{35} & M_{36} & M_{37} & M_{38} & M_{39} & M_{310} \\ M_{43} & M_{44} & M_{45} & M_{46} & M_{47} & M_{48} & M_{49} & M_{410} \\ M_{53} & M_{54} & M_{55} & M_{56} & M_{57} & M_{58} & M_{59} & M_{510} \\ M_{63} & M_{64} & M_{65} & M_{66} & M_{67} & M_{68} & M_{69} & M_{610} \\ M_{73} & M_{74} & M_{75} & M_{76} & M_{77} & M_{78} & M_{79} & M_{710} \\ M_{83} & M_{84} & M_{85} & M_{86} & M_{87} & M_{88} & M_{89} & M_{810} \\ M_{93} & M_{94} & M_{95} & M_{96} & M_{97} & M_{98} & M_{99} & M_{910} \\ M_{103} & M_{104} & M_{105} & M_{106} & M_{107} & M_{108} & M_{109} & M_{1010} \end{bmatrix} \tag{10.354}
$$

In addition, we have two equations related to the boundary conditions,

$$F_{10} = \begin{bmatrix} M_{13} & M_{14} & M_{15} & M_{16} & M_{17} & M_{18} & M_{19} & M_{110} \end{bmatrix} \begin{bmatrix} \ddot{q}_3 \\ \ddot{q}_4 \\ \ddot{q}_5 \\ \ddot{q}_6 \\ \ddot{q}_7 \\ \ddot{q}_8 \\ \ddot{q}_9 \\ \ddot{q}_{10} \end{bmatrix}$$

$$+ \begin{bmatrix} K_{13} & K_{14} & K_{15} & K_{16} & K_{17} & K_{18} & K_{19} & K_{110} \end{bmatrix} \begin{bmatrix} q_3 \\ q_4 \\ q_5 \\ q_6 \\ q_7 \\ q_8 \\ q_9 \\ q_{10} \end{bmatrix} \qquad (10.355)$$

$$M_{10} = \begin{bmatrix} M_{23} & M_{24} & M_{25} & M_{26} & M_{27} & M_{28} & M_{29} & M_{210} \end{bmatrix} \begin{bmatrix} \ddot{q}_3 \\ \ddot{q}_4 \\ \ddot{q}_5 \\ \ddot{q}_6 \\ \ddot{q}_7 \\ \ddot{q}_8 \\ \ddot{q}_9 \\ \ddot{q}_{10} \end{bmatrix}$$

$$+ \begin{bmatrix} K_{23} & K_{24} & K_{25} & K_{26} & K_{27} & K_{28} & K_{29} & K_{210} \end{bmatrix} \begin{bmatrix} q_3 \\ q_4 \\ q_5 \\ q_6 \\ q_7 \\ q_8 \\ q_9 \\ q_{10} \end{bmatrix} \qquad (10.356)$$

We can solve Equation (10.351) for $\hat{q}$ and work out $\hat{\ddot{q}}$. Together we obtain the reaction force and moment at the fixed end using Equations (10.355) and (10.356). In the next section, we will present a MATLAB code to solve Equation (10.351).

## 10.15   Solutions to Linear Equations of Motion

It is more convenient to represent a general linear equation of motion similar to Equation (10.351) to a form, denoted as the State-Space representation. We will consider a general equation of motion with damping as

$$M\ddot{q} + C\dot{q} + Kq = Q \qquad (10.357)$$

We will define

$$x_1 = q \qquad (10.358)$$

$$x_2 = \dot{q} \qquad (10.359)$$

$$x = \begin{bmatrix} x_1 \\ x_2 \end{bmatrix} = \begin{bmatrix} q \\ \dot{q} \end{bmatrix} \qquad (10.360)$$

With the above definitions, Equation (10.357) can be represent as

$$M\dot{x}_2 + Cx_2 + Kx_1 = Q \qquad (10.361)$$

With some arrangement, we now have a set of first-order differential equations,

$$\dot{x}_1 = x_2 \tag{10.362}$$

$$\dot{x}_2 = -M^{-1}Cx_2 - M^{-1}Kx_1 + M^{-1}Q \tag{10.363}$$

The above equations can be re-organized in matrix form as

$$\dot{x} = \begin{bmatrix} \dot{x}_1 \\ \dot{x}_2 \end{bmatrix} = \begin{bmatrix} 0 & I \\ -M^{-1}K & -M^{-1}C \end{bmatrix} \begin{bmatrix} x_1 \\ x_2 \end{bmatrix} + \begin{bmatrix} 0 \\ M^{-1} \end{bmatrix} Q \tag{10.364}$$

where $I$ is an identity matrix of an appropriate dimension and $0$ is a null matrix of an appropriate dimension.

The above equation can be expressed in a state-space expression as

$$\dot{x} = AAx + BBu \tag{10.365}$$

where $u = Q$ and matrices of $AA$ and $BB$ are defined accordingly based on Equation (10.364).

Often, we are only interested in the motion of a certain point of the system. We can define an output function as

$$y = CCx + DDu \tag{10.366}$$

For example, if we are only interested in the displacement of point $A$ for the example of Figure 10.26, the matrices of $CC$ and $DD$ become

$$CC = [0\ 0\ 0\ 0\ 0\ 0\ 1\ 0\ 0\ 0\ 0\ 0\ 0\ 0\ 0\ 0] \tag{10.367}$$

$$DD = 0 \tag{10.368}$$

For the example of Figure 10.26, we only consider single input with $u = p$ or $u = R$. If $u = R$, the related matrices are defined as

$$AA = \begin{bmatrix} 0_{8\times8} & I_{8\times8} \\ -M^{-1}K & 0_{8\times8} \end{bmatrix} \tag{10.369}$$

$$BB = \begin{bmatrix} \mathbf{0}_{8\times1} \\ 0 \\ 0 \\ 0 \\ 0 \\ -\dfrac{7}{27} \\ -\dfrac{2l}{27} \\ -\dfrac{20}{27} \\ \dfrac{4l}{27} \end{bmatrix} \tag{10.370}$$

$$CC = \begin{bmatrix} 0\ 0\ 0\ 0\ 0\ 0\ 1\ 0\ 0\ 0\ 0\ 0\ 0\ 0\ 0\ 0 \end{bmatrix} \tag{10.371}$$

$$DD = 0 \tag{10.372}$$

With the above definitions, the following Matlab script can solve the equation of motion readily with $R$ as a short pulse of 0.00001 and 500 N in size.

```
time= 0:0.00001:1;
time= time';
force= 0 * (time+1) ./ (time+1);
force(1)= 500;
system = ss(AA, BB, CC, DD);
[YY, TT] = lsim(sys, force, time);
```

## 10.16   Concluding Remarks

In this chapter, we took a substantial step to go beyond rigid body dynamics. We introduced the control volume concept to deal with fluid and other mass flow systems. We introduced elastic energy to deal with elastic bodies. We then discussed the motion and deformation of elastic straight and curved beams. Finally, we presented the Finite Element Method for dynamic analysis. This chapter provides a stepping stone for more advanced materials in dynamics and vibration systems.

Chapter 11

# Difficult Dynamics Problems
# Better Solved

In this chapter, we will solve classical problems in Dynamics using the methodologies discussed in Chapters 1–10. These problems are based on the review problems from *Dynamics* by Meriam, 2nd Edition.[1]

**Problem 11.1:** A jet airliner with a mass of 115 tons develops a constant jet thrust of 90 kN for each of its four engines. Determine the length $s$ of runway required for take-off at a speed of 270 km/h. Neglect air resistance.

**Solution:**
**Motion analysis:** First, we recognize the entire airliner is consider as one particle undergoing a straight line motion. With this observation, the equations listed in Section 3.4 could be used.

**Force analysis:** The corresponding FBD is shown in Figure 11.1. We are only interested in the force analysis along the $x$-axis. From the design point of view, we need to consider the worst case scenario for which the maximum acceleration occurs. As a result, we will neglect the air drag, $F_D$, the frictional forces, $F_1$ and $F_2$, and consider the engine thrust force, $F_e$, is at its maximum throughout the acceleration process on the runway. The thrust force of the engine is considered as the Applied Force based on the ABCC method.

---

[1]Meriam, J.L. *Dynamics*, 2nd Edition, 1975. John Wiley & Sons, Inc., pp. 442–451, ISBN 0-471-59607-8.

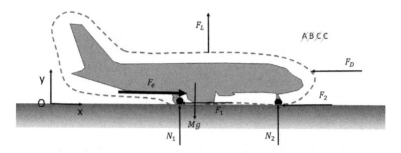

Figure 11.1:   FBD for Problem 1.

**Governing equations:** From Netwon's Second Law along the $x$-axis, we have

$$F_e = M a_{max} \qquad (11.1a)$$

where $F_e = 4 \times 90\,\text{kN}$ and $M = 115 \times 10^3\,\text{kg}$.

We obtain the maximum acceleration, $a_{max} = 3.13\,\text{m/s}^2$.

Assuming the airliner is at this maximum acceleration throughout the take-off process, we have

$$\int_0^s a_{max}\, dx = \int_0^{v_f} v\, dv \qquad (11.1b)$$

where $v$ is the speed, $x$ is the traveling distance, and $s$ is the runway length. The initial speed is zero, while the speed at the end of the runway is $v_f = 270\,\text{km/h}$. Solving the above equation, we obtain that the runway distance needs to be

$$s = 898.4\,\text{m} \qquad (11.1c)$$

**Verification:** For the runway design, we should round the above design to 900 m. We should also verify if this length of runway is sufficient for all airplanes. Some safety factors should be considered so that if an airplane does not achieve its maximum acceleration, the runway is still long enough for the airplane to achieve the needed take-off speed.

**Problem 11.2:** Consider a time-variant vector $\hat{V} = 10\,\hat{i}\,(\text{m/s})$ at this instant, described by a rotating reference frame $\vec{\Omega} = \Omega \hat{k}\,(\text{rad/s})$. The total time derivative of $\hat{V}$ is $\frac{d\hat{V}}{dt}$. It is found that $\frac{d\hat{V}}{dt}$ is 120° with respect to $\hat{V}$ and its magnitude is $6\,\text{m/s}^2$ at this instant. Determine the frame rotation, $\Omega$, and the partial time derivative of $\hat{V}$, i.e., $\frac{\partial \hat{v}}{\partial t}$ (Figure 11.2).

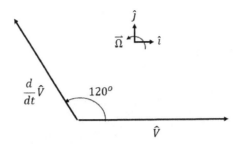

Figure 11.2:   Figure for Problem 2.

**Solution:** This problem is about the total time derivative of a vector described by a rotating reference frame. Per Section 3.10.4, the total time derivative of a vector described by a rotation reference is

$$\frac{d}{dt}\hat{V} = \vec{\Omega} \times \hat{V} + \frac{\partial}{\partial t}\hat{V}$$

$$= \Omega V \hat{j} + \frac{\partial \hat{v}}{\partial t}\hat{i} = 10\Omega\hat{j} + \frac{\partial \hat{v}}{\partial t}\hat{i} \qquad (11.2a)$$

We can also represent the above equation graphically. From there, we obtain the following two equations,

$$\left|\frac{d}{dt}\hat{V}\right| = 6 = \sqrt{\left(\frac{\partial \hat{v}}{\partial t}\right)^2 + (10\Omega)^2} \qquad (11.2b)$$

$$\tan(60°) = |10\Omega| / \left|\frac{\partial \hat{v}}{\partial t}\right| \qquad (11.2c)$$

Solving the above two equations and recognizing the direction of the vectors, we obtain

$$\frac{\partial \hat{v}}{\partial t} = -3 \text{ m/s} \qquad (11.2d)$$

$$\Omega = 0.520 \text{ rad/s} \qquad (11.2e)$$

**Problem 11.3:** The jet transport has a mass of 90 tons and its four engines develop a total thrust $T$ of 240 kN at its take-off speed of 280 km/h. Immediately after take-off the direction of its actual velocity $v$ and acceleration $a$ make an angle of 7° with the horizontal, whereas the aircraft axis and thrust are directed 10° above the horizontal. Neglect air resistance (drag) in the direction opposite to the velocity and compute the acceleration $a = \dot{v}$ (Figure 11.3).

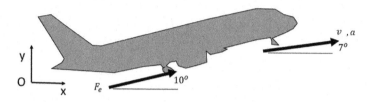

Figure 11.3: Figure for Problem 3.

**Solution:**

**Motion analysis:** We will still model the entire aircraft as one particle in a 2D motion. The inertial reference frame is attached to the ground, $x$ for the horizontal direction and $y$ for the vertical direction.

**Force analysis:** The corresponding FBD is constructed using the ABCC method, as shown in Figure 11.4. We include both the drag $F_D$ and lift $F_D$. $F_L$ will be neglected in calculation initially.

**Governing equations:** We will use *Newton's Second Law* to solve the problem because the force information is given and we would like to find out the acceleration. We will use the inertial reference frame, $O$–$x$–$y$. The aircraft experiences two forces from the air: the wind drag and the lift. The wind drag is in the direction of $v$, while the lift is perpendicular to $v$. By modeling the airplane as a particle, we only have two equations along the $x$ and $y$ axes.

$$-F_D \cos(7°) - F_L \sin(7°) + F_e \cos(10°) = Ma_x \quad (11.3a)$$

$$-Mg + F_L \cos(7°) + F_e \sin(10°) - F_D \sin(7°) = Ma_y \quad (11.3b)$$

In addition, the total acceleration is 7° from the horizontal; therefore

$$\tan(7°) = \frac{a_y}{a_x} \quad (11.3c)$$

We can also construct the above equations based on the orientation of the aircraft; however, the $x$–$y$ reference frame is a good decision as well.

**Solving for Unknowns:** Before we solve the above equations, let's count the unknowns. There are four unknowns, $F_D$, $F_L$, $a_x$, and $a_y$, but we only have three equations. By neglecting the drag $F_D$,

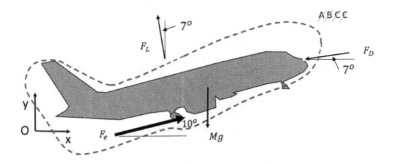

Figure 11.4: FBD for Problem 3.

we can solve for $F_L$, $a_x$, and $a_y$. However, we should recognize that the obtained acceleration will be overestimated.

With $F_D = 0$, solving for the remaining unknown, we obtain

$$a_x = 1.467 \text{ m/s}^2 \qquad (11.3\text{d})$$

$$a_y = 0.1788 \text{ m/s}^2 \qquad (11.3\text{e})$$

$$a = \sqrt{a_x^2 + a_y^2} = 1.468 \text{ m/s}^2 \qquad (11.3\text{f})$$

$$F_L = 863.8 \text{ kN} \qquad (11.3\text{g})$$

**Verification:** We can confirm that the angle of the acceleration is indeed 7°. The above solutions assume that the wind drag $F_D$ is zero. In practice, this is hardly the case. We can conduct experiments to determine the wind drag, which is often modeled as

$$F_D = \frac{1}{2}C_D A v^2 \qquad (11.3\text{h})$$

where $C_D$ is the wind drag coefficient and $A$ is the equivalent cross-section area of the aircraft. The value of $C_D$ can be a function of the angle between the aircraft orientation and $v$.

For this example, let's simply examine the effect of $F_D$ as a percentage of $F_L$. If $F_D = 0.05 \ F_L$, 5% of the lift, the acceleration is found to be $0.988 \text{ m/s}^2$, which is a typical acceleration of a large aircraft during take-off.

**Problem 11.4:** The end of the coil of rope of total length $L$ is released from rest at $x = 0$ and falls with increasing vertical velocity

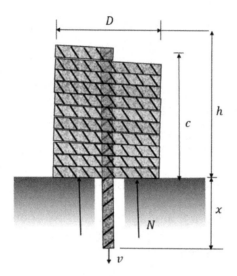

Figure 11.5:  Figure for Problem 4.

as the rope peels off the coils and passes through the center of the coil in a smooth and continuous flow. Neglect all friction and derive the expression for the velocity $v$ of the rope as the upper end passes through the opening at $x = 0$. The mass of the rope per unit length is $\rho$ and the initial height of the coil with $x = 0$ is $h$ (Figure 11.5).

**Solution:**

**Motion analysis:** We recognize that the entire rope is separated into two portions. One portion is at rest, while the other portion is falling vertically at an increasing speed. This problem bears similarities to the rope and chain fountain problems discussed in Examples 10.3 and 10.4. The difference is that this case is not a steady-state case.

**Force analysis:** We construct several control volumes/FBDs for analysis as shown in Figure 11.6. We may not need all of them for obtaining the answer, but we can uncover additional information on the problem.

**Governing equations:** As in the past, we have three sets of governing equations to consider. We should never assume that we know right away which one to choose. Among them, if we choose *Newton's Second Law*, we can receive the information about acceleration and forces. To obtain the speed information, we will have to conduct integration by time over the acceleration.

For the linear momentum equation, we will deal with velocity directly, but because there is an integration over time, we will need to know the forces involved as functions of time. It still seems complicated if we choose this route. Because we are dealing with straight line motion, the angular momentum equation is irrelevant.

Finally, the energy equation might be useful, but we need to evaluate all forces involved using the force accounting strategy.

With respect to the control volume/FBD of the entire rope (Figure 11.6(a)), the external forces involved are $N$ and $mg$. The non-conservative force $N$ is not doing work because the rope it acts on is not moving. The conservative force $mg$ is doing work and must be considered in the energy equation.

The internal forces involved are the tension of the rope $T$ and some small friction in the rope sections of the coil. Because the rope is not stretching, $T$ does not do work. Per the problem statement, we ignore the energy consumption due to any internal friction.

From the above analysis, we conclude that the conservation of mechanical energy holds for the entire rope system; therefore,

$$T_1 + U_{g1} = T_2 + U_{g2} \qquad (11.4a)$$

We know that $T_1 = 0$ and $T_2 = \frac{1}{2}mv^2$ when the entire rope is falling as a straight line with the top of the rope aligned with the resting surface, where $m$ is the total mass of the rope.

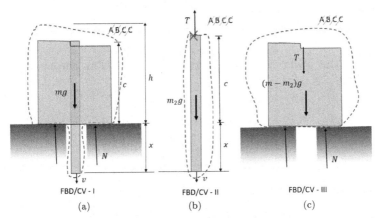

Figure 11.6: Control Volumes and FBDs for Problem 4: (a) entire rope; (b) vertical portion only; and (c) stationary coil only.

The mass center of the rope coil before falling is $\frac{1}{2}h$ above the surface and when it becomes a straight line with the top edge aligned with the surface, the mass center is $\frac{1}{2}L$ below the surface. Let's choose the datum at the surface. As a result, Equation (11.4a) becomes

$$mg\left(\frac{1}{2}h\right) = \frac{1}{2}mv^2 + mg\left(-\frac{1}{2}L\right) \tag{11.4b}$$

where $m = \rho L$.

*Solving for Unknowns*: There is only one unknown involved in Equation (11.4b) and the solution is

$$v = \sqrt{g(L + h)} \tag{11.4c}$$

**Verification and Extension:** This is a seemingly easy problem, but what if we would like to know the velocity of the rope during the uncoiling process? How about the rope tension? For the first question, we can use the energy equation again. Let's establish a few additional conditions.

The number of coil layer $N$ should be

$$N = \frac{L}{\pi D} \tag{11.4d}$$

where $D$ is the diameter of the coil.

The diameter of the rope $d$ becomes

$$d = \frac{h\pi D}{L} \tag{11.4e}$$

When the lower edge of the rope is at $x$ below the surface, the coil height $c$ is

$$c = h - \left(\frac{x + c}{\pi D}\right)d$$

$$\Longrightarrow c = \frac{h(L - x)}{L + h} \tag{11.4f}$$

Therefore, the mass center of the coil and rope above the surface when its height is $c$ is $\frac{c}{2}$. The mass center of the rope portion below the surface is $-\frac{x}{2}$. Applying the energy equation again, we have

$$\rho L g \left(\frac{1}{2}h\right) = \frac{1}{2}\rho(c+x)v^2 + \rho(L-x)g\left(\frac{1}{2}c\right) - \rho x g \left(\frac{1}{2}x\right)$$

(11.4g)

Finally, we have

$$v(x) = \sqrt{g(h+x)} \tag{11.4h}$$

With $x = L$, we obtain $v(L) = \sqrt{g(L+h)}$ as before. When $x = -h$, before the rope starts falling, $v = 0$.

We can take a time derivative of the velocity to obtain the acceleration as

$$a(x) = \frac{d}{dt}v(x) = \frac{1}{2}\frac{1}{\sqrt{g(h+x)}}g\dot{x} = \frac{1}{2}g \tag{11.4i}$$

It turns out that the rope will fall at a constant acceleration of $g/2$, half that of the free-falling case. The reason is that the coil will have a tension, pulling up the rope to reduce the downward acceleration by gravity.

Let's try to determine the rope tension at the top of the coil. Based on the FBD/CV of the falling rope portion (Figure 11.6(b)), we have

$$T - \rho(x+c)g = \frac{dP_{cv}}{dt} - \frac{dP_{in}}{dt} + \frac{dP_{out}}{dt}$$

$$= \frac{d}{dt}(\rho(x+c)(-v)) - \frac{\rho dx(-v)}{dt}$$

$$= -\rho v^2 - \rho v\dot{c} - \rho(x+c)a + \rho v^2$$

$$= -\rho v\dot{c} - \rho(x+c)a \tag{11.4j}$$

The tension at the top of the vertical rope is

$$T = \rho(x+c)g - \rho v\dot{c} - \rho(x+c)a$$

$$= \frac{\rho g}{2}\left(\frac{Lh + Lx + 2hx + 2h^2}{L+h}\right), \quad x < L \tag{11.4k}$$

However, Equations (11.4j) and (11.4k) cannot be applied when $x = L$ because there will be no $\frac{dP_{in}}{dt}$. At $x = L$, $c = 0$, and $\frac{d}{dt}(\rho(x + c)(-v)) = -\rho La$, the correct expression for the tension when $x = L$ should be

$$T(x = L) = \rho(x + c)g - \rho(x + c)a|_{x=L} = \frac{1}{2}\rho Lg \qquad (11.4l)$$

The tension when $x = L$ immediately becomes zero and the entire rope becomes free falling with an acceleration $g$.

For the rope coil on the surface, we use a control volume/FBD as shown in Figure 11.6(c) to obtain

$$N - T - \rho(L - x - c)g = \frac{dP_{cv}}{dt} - \frac{dP_{in}}{dt} + \frac{dP_{out}}{dt} = \frac{dP_{out}}{dt}$$

$$= \rho v(-v) = -\rho v^2 \qquad (11.4m)$$

The normal force at the surface becomes

$$N = T(x < L) + \rho(L - x - c)g - \rho v^2$$

$$= \frac{1}{2}\rho g\left(\frac{-3Lx - Lh + 2L^2}{L + h}\right), \quad x < L \qquad (11.4n)$$

Similarly, Equation (11.4n) cannot be applied when $x = L$ because $\frac{dP_{out}}{dt} = 0$ at that moment. The normal force $N$ at $x = L$ and $c = 0$ should be

$$N(x = L) = T(x = L) + \rho(L - x - c)g|_{x=L}$$

$$= \frac{1}{2}\rho Lg \qquad (11.4o)$$

At the instant of $x = L$, the tension on the top of the rope is due to force $N$ of the surface. Immediately after, the rope and the surface lose their contact, making $T = N = 0$. The rope becomes free falling at $g$.

In summary, we combined the energy equation and *Newton's Second Law* to obtain the information of rope velocity, acceleration, rope tension, and surface normal force.

**Problem 11.5:** The two spacecrafts $A$ and $B$ are to preform a docking operation in space. Spacecraft $A$ moves with a constant velocity of 27,000 km/h relative to the Earth. When the separation distance is 15 km, spacecraft $B$, which has a mass of 800 kg, has a velocity of 27,150 km/h. If the retro rockets on $B$ are fired at this point, determine the constant thrust $F$ which is required to bring $A$ and $B$ together with zero relative velocity. What total impulse $P$ is required? Assume collinear straight line motion during the interval (Figure 11.7).

**Solution:**

**Motion Analysis:** From the problem statement, we are dealing with the straight line motion. Basically, we need to determine the constant thrust $F$ and how long the retro rocket should be fired for the needed $P = \int_0^t F dt$ so that after traveling 15 km after the rocket being fired, spacecraft $B$ will have a velocity of 27,000 km/h.

**Force Analysis:** By considering the spacecrafts are in the straight line motion, the only force to be considered is the retro thrust force $F$ acting on spacecraft $B$. Because $F$ is a constant, we know that $B$ is at a constant deceleration during the docking interval. We also assume that the loss of fuel during the rocket firing is small; therefore, the total mass of the spacecraft is the same.

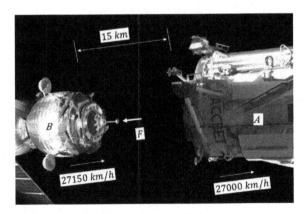

Figure 11.7: Figure for Problem 5.

*Images Modified from Source*: https://upload.wikimedia.org/wikipedia/commons/thumb/d/d3/Soyuz_TMA-03M_docking.jpg/640px-Soyuz_TMA-03M_docking.jpg.

**Governing Equations:** Basically, we are dealing with the straight line motion of a particle at a constant acceleration (deceleration). Those equations of Section 3.4 can be used. First, let's convert the units. The initial velocity of $B$ is $v_0 = 27,150$ km/h $= 7,541.667$ m/s. The final velocity is $v_f = 27,500$ km/h $= 75,00.000$ m/s. The distance between the aircrafts is $d = 15$ km $= 15,000$ m.

One equation which is useful is

$$\int_0^d a \, ds = \int_{v_0}^{v_f} v \, dv \qquad (11.5\text{a})$$

To use the above equation, the value of $d$ is the traveling distance of aircraft $B$ during docking. If we attach an inertial reference frame to spacecraft $A$, then the traveling distance is 15 km. If we use an inertial reference frame attached to the Earth, then the traveling distance of $B$ is unknown. With this consideration, we will use an inertial reference frame at $A$. The relative velocity of $B$ with respect to $A$ can be determined by the following equation.

$$\vec{v}_{B/A} = \vec{v}_{B/O} - \vec{v}_{A/O} \qquad (11.5\text{b})$$

where $O$ is the observer attached to the Earth. The initial and final relative speeds of $B$ with respect to $A$ are

$$(v_{B/A})_0 = 41.667 \, \text{m/s} \qquad (11.5\text{c})$$

$$(v_{B/A})_f = 0.000 \, \text{m/s} \qquad (11.5\text{d})$$

Solving the Equation (11.5a), we have

$$a = -0.0589 \, \text{m/s}^2 \qquad (11.5\text{e})$$

Subsequently, the thrust force needed is

$$F = m|a| = 800 \cdot 0.0589 = 46.28 \, (\text{N}) \qquad (11.5\text{f})$$

The linear momentum equation for $B$ is

$$m_B(v_{B/A})_0 - P = m_B(v_{B/A})_f \qquad (11.5\text{g})$$

We obtain

$$P = 33,330 \, \text{kg m/s} \qquad (11.5\text{h})$$

Finally, we can also determine how long the retro rocket should be firing.

$$P = 33330 = \int_0^T F \, dt = 46.28 \, T \qquad (11.5\text{i})$$

The firing time is $T = 720.2\,\text{s} = 12.00\,\text{min}$.

**Verification:** The actual docking maneuver of aircrafts is much more complicated than this problem indicates. One can search videos for the actual docking maneuvers.[2] One interesting number is that the docking relative speed when an aircraft is within about 150 m to International Space Station is 0.1 m/s. At such speed, it would take 25 minutes to dock from 150 m away.

**Problem 11.6:** The path of a non-rotating spacecraft $A$ as it nears a planet $P$ lies in the plane of rotation of the planet about the Sun. The planet has a velocity $v_p = 126{,}000$ km/h in its near-circular orbit around the sun with a radius $R$. If the planet appears to have a velocity $v_r = 22{,}200$ km/h away from the sun as seen from the spacecraft as it passes from the shaded to the lighted side of the planet, calculate the magnitude of the velocity $v_A$ of the spacecraft as measured in the solar reference system.

**Solution:**

**Motion analysis:** First, we establish an inertial reference frame $x$–$y$ attached to the sun, assuming that the sun is stationary. For this instant, we also establish a radial direction $\vec{e}_r$ and transverse direction $\vec{e}_\theta$, as defined in Figure 11.8. Note that $\vec{e}_r$ and $\vec{e}_\theta$ will change direction when the planet moves to a new position. However, for this instant of time, we can establish a specific relationship between $\vec{e}_r$ and $\vec{e}_\theta$ and the inertial reference frame $x$–$y$.

Per the problem statement, the spacecraft is moving along a straight line, while the planet is along a perfect circular orbit.

---

[2] https://www.youtube.com/watch?v=M2_NeFbFcSw.

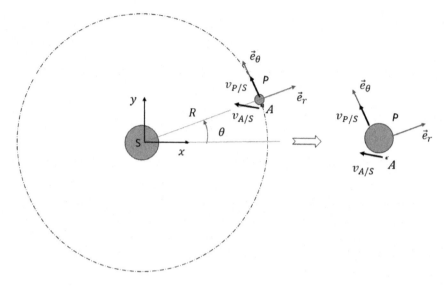

Figure 11.8: Figure for Problem 6.

The observers inside the spacecraft are inertial observers. For this instant, they will use the same $\vec{e}_r$ and $\vec{e}_\theta$.

Accordingly, the velocities of the planet and the spacecraft are defined as

$$\vec{v}_{P/S} = v_t \vec{e}_\theta = 126{,}000 \vec{e}_\theta \ (\text{km/h}) \tag{11.6a}$$

$$\vec{v}_{P/A} = v_r \vec{e}_r = 22{,}200 \vec{e}_r \ (\text{km/h}) \tag{11.6b}$$

Because the axes of the reference frames attached to the sun and to the spacecraft are the same, we have a simple relative velocity equation as

$$\vec{v}_{P/S} = \vec{v}_{A/S} + \vec{v}_{P/A} \tag{11.6c}$$

Introducing Equations (11.6a) and (11.6b) into (11.6c), we have

$$\vec{v}_{A/S} = -22{,}200 \vec{e}_r + 126{,}000 \vec{e}_\theta \tag{11.6d}$$

The speed of the spacecraft is found to be

$$v_{A/S} = \sqrt{126{,}000^2 + 22{,}200^2} = 127{,}940 \ (\text{km/h}) \tag{11.6e}$$

**Verification:** We need to be careful to observe that Equation (11.6c) is valid only when the observers attached to the Sun and to the spacecraft are using the same reference frame axes. Equation (11.6c) is still valid even through the $r-\theta$ reference frame is rotating, as long as both observers at the sun and the aircraft are using the same reference frame. If we want to determine the relative motion of the spacecraft to an observer attached to the planet (assuming that the planet is not spinning), we need to consider the rotation of $\vec{e}_r$ and $\vec{e}_\theta$. The relative velocity equation is

$$\vec{v}_{A/S} = \vec{v}_{P/S} + \vec{\Omega} \times \hat{r}_{A/P} + \hat{v}_{A/P} \qquad (11.6f)$$

where $\Omega = v_{P/S}/R$ is the frame rotation and $\hat{r}_{A/P}$ is the position vector of the spacecraft with respect to the planet. Note that $\hat{v}_{A/P} \neq -\vec{v}_{P/A}$.

**Problem 11.7:** Determine the acceleration $a$ of the supporting surface required to keep the center $G$ of the circular cylinder in a fixed position during motion. Friction is sufficient to prevent any slipping between the cylinder and its support.

**Solution:**

**Motion Analysis:** We will conduct rigid body motion analysis. With respect to the inertial reference $O-x-y$ of Figure 11.9(a), aligned with the inclined surface, plate $A$ is sliding along the inclined surface with a velocity and acceleration as

$$\vec{v}_{A/O} = v\vec{i} \qquad (11.7a)$$

$$\vec{a}_{A/O} = a\vec{i} \qquad (11.7b)$$

The cylinder is in contact with plate $A$ at point $C$ ($C'$ on the cylinder). The cylinder is rolling without slipping with respect to the

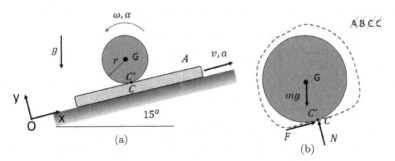

Figure 11.9: Figures for Problem 7: (a) a rolling-without-slipping cylinder over an inclined plate; and (b) FBD for the cylinder.

plate. Therefore,

$$(\vec{v}_{C'/O})_t = v\vec{i} \tag{11.7c}$$

$$(\vec{a}_{C'/O})_t = \vec{a}_{A/O} = a\vec{i} \tag{11.7d}$$

where the subscript $t$ indicates the tangential direction.

The cylinder is observed by $O-x-y$ to have an angular velocity and angular acceleration as

$$\vec{\omega} = \omega\vec{k} \tag{11.7e}$$

$$\vec{\alpha} = \alpha\vec{k} \tag{11.7f}$$

With $G$ fixed in position, the velocity and acceleration of point $G$ can be expressed as

$$\begin{aligned}\vec{v}_{G/O} = 0 &= \vec{v}_{C'/O} + \vec{v}_{G/C'}\\ &= \vec{v}_{C'/O} + \vec{\omega} \times \vec{r}_{G/C'} \end{aligned}\tag{11.7g}$$

$$\begin{aligned}\vec{a}_{G/O} = 0 &= \vec{a}_{C'/O} + \vec{a}_{G/C'}\\ &= \vec{a}_{C'/O} + \vec{\omega} \times (\vec{\omega} \times \vec{r}_{G/C'}) + \vec{\alpha} \times \vec{r}_{G/C'} \end{aligned}\tag{11.7h}$$

From the above equations, we obtain

$$v = r\omega \tag{11.7i}$$

$$a = r\alpha \tag{11.7j}$$

**Force Analysis:** The FBD for the cylinder is shown in Figure 11.9(b).

**Governing Equations:**

Per *Newton's Second Law*, we have

$$\sum F_x = F - mg\sin(15°) = m\vec{a}_{Gx} = 0 \tag{11.7k}$$

$$\sum F_y = N - mg\cos(15°) = m\vec{a}_{Gy} = 0 \tag{11.7l}$$

$$\sum \tau_{Gz} = Fr = I_G\alpha = \left(\frac{1}{2}mr^2\right)\alpha \tag{11.7m}$$

**Solving for Unknowns:** We can solve for $F$, $N$, and $\alpha$ using Equations (11.7k)–(11.7m). The results are

$$F = mg\sin(15°) \tag{11.7n}$$

$$N = mg\cos(15°) \tag{11.7o}$$

$$\alpha = \frac{Fr}{I_G} = \frac{2}{r}g\sin(15°) \tag{11.7p}$$

Finally, we have

$$a = r\alpha = 2g\sin(15°) = 5.078\,(\text{m/s}^2) \tag{11.7q}$$

**Verification:** The ratio of $F$ and $N$ is $\tan(15°) = 0.268$. Therefore, as long as the static friction coefficient between the cylinder and the plate is larger than $tan(15°)$, the rolling-without-slipping condition can be achieved.

**Problem 11.8:** The circular disk of mass $m$ and radius of gyration $k$ is attached to the torsional spring, which is secured to the fixed shaft. The disk is otherwise free to rotate on the shaft. The spring exerts a torque $\tau = C\theta$ on the disk, where $C$ is a constant and where $\theta$ is the angle of rotation of the disk in radians from the neutral position of the spring. If the disk is released from rest with an angular displacement of $\theta_0$, determine the maximum power output $P$ of the spring and the angle $\theta_m$ at which this condition occurs (Figure 11.10).

**Solution:**

**Motion Analysis:** We have a rigid disk rotating around a fixed point. Therefore, we only need to know the angular motion properties, $\omega$ and $\alpha$.

**Force Analysis:** A proper FBD analysis (not shown) will indicate that the only force doing work is the conservative spring torque, $\tau$.

**Governing Equations:** From the above analysis, we will establish the energy equation which is conserved

$$T_1 + U_{k1} = 0 + \frac{1}{2}C\theta_0^2 = T_2 + U_{k2} \tag{11.8a}$$

where $T_2 = \frac{1}{2}I_G\omega_2^2 = \frac{1}{2}mk^2\omega_2^2$ and $U_{k2} = \frac{1}{2}C\theta_2^2$. The angular velocity $\omega_2$ and angular position $\theta_2$ are two unknowns to be solved. We need additional equations.

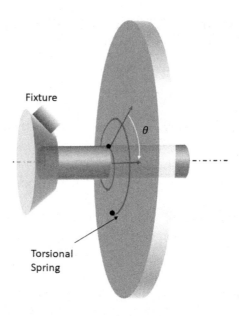

Figure 11.10:   Figure for Problem 8.

The torque of the spring on the disk is in the reverse direction of the angular displacement,

$$\tau = -C\theta \tag{11.8b}$$

The power output of the torsional spring is

$$P = |\tau\omega| = C\theta\omega \tag{11.8c}$$

To achieve that largest power output, the derivative of the power should be zero. Therefore,

$$\frac{d}{dt}P = C\dot{\theta}\omega + C\theta\dot{\omega} = C\omega_2^2 + C\theta_2\alpha_2 = 0 \tag{11.8d}$$

where $\theta_2$, $\omega_2$, and $\alpha_2$ are the angular displacement, velocity, and acceleration, respectively, when the output power is at its maximum.

*Newton's Second Law* for the disk when the power is the maximum is

$$\tau_2 = -C\theta_2 = I_G\alpha_2 = mk^2\alpha_2 \tag{11.8e}$$

**Solving for Unknowns:** From Equations (11.8d) and (11.8e), we have

$$\omega_2^2 = \frac{C}{I_G}\theta_2^2 \qquad (11.8\text{f})$$

Introducing Equation (11.8f) into (11.8a), we have

$$\frac{1}{2}C\theta_0^2 = \frac{1}{2}I_G\omega_2^2 + \frac{1}{2}C\theta_2^2 = C\theta_2^2 \qquad (11.8\text{g})$$

We obtain

$$\theta_2 = \frac{\theta_0}{\sqrt{2}} \qquad (11.8\text{h})$$

The maximum power output $P_{max}$ is found to be

$$P_{max} = C\theta_2\omega_2 = C\sqrt{\frac{C}{I_G}}\theta_2^2 = \frac{C}{2k}\sqrt{\frac{C}{m}}\theta_0^2 \qquad (11.8\text{i})$$

**Problem 11.9:** A simple pendulum consists of a wire and a mass $m$. The pendulum is released at rest when $\theta = \theta_0$. Determine the tension of the wire as a function of $\theta$.

**Solution:**

**Motion analysis:** We establish an inertial reference frame $O-x-y$ as shown in Figure 11.11(a). The counterclockwise angular displacement is defined as positive, consistent with the right-hand rule. The mass $A$ is in a perfect circular motion with a radius $r$.

At an instant when mass $A$ is swinging downward from $\theta_0$, its velocity and acceleration are determined as

$$\vec{v}_{A/O} = -r\omega\vec{e}_\theta \qquad (11.9\text{a})$$

$$\vec{a}_{A/O} = -r\omega^2\vec{e}_r - r\alpha\vec{e}_\theta \qquad (11.9\text{b})$$

**Force Analysis:** The FBD for mass $A$ is shown in Figure 11.11(b).

**Governing Equations:** First, because the problem statement asks for the wire tension, we naturally would apply *Newton's Second Law*. We have

$$\sum F_\theta = -mg\sin\theta = m(-r\alpha) \qquad (11.9\text{c})$$

$$\sum F_r = -T + mg\cos(\theta) = m(-r\omega^2) \qquad (11.9\text{d})$$

**Solving for Unknowns:** For Equations (11.9c) and (11.9d), there are three unknowns, $T$, $\omega$, and $\alpha$, but we only have two equations.

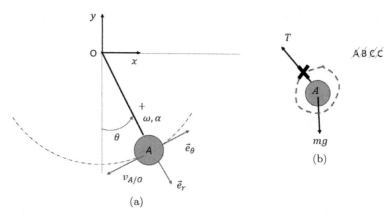

Figure 11.11:  Figures for Problem 9: (a) simple pendulum; and (b) FBD for the mass.

We need to find one additional equation without introducing new unknowns.

From the FBD and the motion analysis, we observe that the wire tension $T$ does not do work because it is also normal to the velocity of mass $A$. As a result, we have an energy conserved system. With respect to $O-x-y$, the energy equation becomes

$$T_0 + U_{g0} = -mgr\cos\theta_0$$

$$= T_1 + U_{g1} = \frac{1}{2}mv^2 - mgr\cos\theta \qquad (11.9\text{e})$$

The velocity becomes

$$v = \sqrt{-mgr\cos\theta_0 + mgr\cos\theta} \qquad (11.9\text{f})$$

We can now solve for $T$ and $\alpha$. The tension of the wire is found to be

$$T = mg(3\cos\theta - 2\cos\theta_0) \qquad (11.9\text{g})$$

$$\alpha = -\frac{g}{r}\sin\theta \qquad (11.9\text{h})$$

**Problem 11.10:** A test chamber $A$ is used for studying motion sickness. It is capable of oscillating about a horizontal axis around $O$ according to $\theta = \theta_0\sin(2\pi f_1 t)$ (rad), and at the same time the chamber has a linear motion $\hat{y} = \hat{y}_0\sin(2\pi f_2 t)$ (m) relative to the frame.

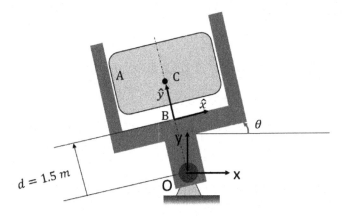

Figure 11.12: Figure for Problem 10.

For a certain series of tests the amplitudes are set at $\theta_0 = \pi/4\,\text{rad}$ and $\hat{y}_0 = 150\,\text{mm}$, while the frequencies are $f_1 = 0.25\,\text{Hz}$ and $f_2 = 0.5\,\text{Hz}$. Determine the vector expression for the acceleration of point $C$ in the chamber at the instant when $t = 2\,\text{s}$ (Figure 11.12).

**Solution:**

**Motion Analysis:** We need to establish different reference frames to observe the motion of point $C$ properly. The inertial reference frame is $O-x-y$. We also attach a rotating reference frame to the frame at point $B$. Frame $B-\hat{x}-\hat{y}$ has a frame rotation $\vec{\Omega} = \dot{\theta}\vec{k}$.

The motion of $C$ observed by $B-\hat{x}-\hat{y}$ is along a straight line,

$$\hat{v}_{C/B} = \dot{\hat{y}}\,\hat{j} = 0.3\pi f_2 \cos(2\pi f_2 t)\,\hat{j} \qquad (11.10a)$$

$$\hat{a}_{C/B} = \ddot{\hat{y}}\,\hat{j} = -0.6\pi^2 f_2^2 \sin(2\pi f_2 t)\,\hat{j} \qquad (11.10b)$$

The motion of $B$ observed by $O-x-y$ is a perfectly circular motion,

$$\vec{v}_{B/O} = \vec{\omega} \times \vec{r}_{B/O} \qquad (11.10c)$$

$$\vec{a}_{B/O} = \vec{\omega} \times (\vec{\omega} \times \vec{r}_{B/O}) + \vec{\alpha} \times \vec{r}_{B/O} \qquad (11.10d)$$

where

$$\vec{\omega} = \dot{\theta}\vec{k} = 0.5\pi^2 f_1 \cos(2\pi f_1 t)\vec{k} \qquad (11.10e)$$

$$\vec{\alpha} = \dot{\vec{\omega}} = -\pi^3 f_1^2 \sin(2\pi f_1 t)\vec{k} \qquad (11.10f)$$

Point $B$ appears to be a perfect circular motion to $O$ because they are on the same rigid body.

The motion of $C$ with respect to $O-x-y$ can now be constructed as

$$\vec{v}_{C/O} = \vec{v}_{B/O} + \vec{\Omega} \times \hat{r}_{C/B} + \hat{v}_{C/B} \qquad (11.10\text{g})$$

$$\vec{a}_{C/O} = \vec{a}_{B/O} + \dot{\vec{\Omega}} \times \hat{r}_{C/B}$$

$$+ \vec{\Omega} \times (\vec{\Omega} \times \hat{r}_{C/B}) + 2\vec{\Omega} \times \hat{v}_{C/B} + \hat{a}_{C/B} \qquad (11.10\text{h})$$

The unit direction vectors between two observation frames are

$$\hat{i} = \cos\theta \vec{i} + \sin\theta \vec{j} \qquad (11.10\text{i})$$

$$\hat{j} = -\sin\theta \vec{i} + \cos\theta \vec{j} \qquad (11.10\text{j})$$

$$\hat{k} = \vec{k} \qquad (11.10\text{k})$$

We further obtain

$$\vec{v}_{C/O} = \omega \vec{k} \times (-d\sin\theta \vec{i} + d\cos\theta \vec{j}) + \dot{\theta}\vec{k} \times \hat{y}\hat{j} + \dot{\hat{y}}\hat{j}$$

$$= -d\omega\cos\theta \vec{i} - d\omega\sin\theta \vec{j} - \dot{\theta}\hat{y}\hat{i} + \dot{\hat{y}}\hat{j}$$

$$= (-d\omega\cos\theta - \omega\hat{y}\cos\theta - \dot{\hat{y}}\sin\theta)\vec{i}$$

$$+ (-d\omega\sin\theta - \omega\hat{y}\sin\theta + \dot{\hat{y}}\cos\theta)\vec{j} \qquad (11.10\text{l})$$

$$\vec{a}_{C/O} = (d\omega^2\sin\theta - d\alpha\cos\theta - (\alpha\hat{y} + 2\omega\dot{\hat{y}})\cos\theta$$

$$- (-\hat{y}\omega^2 + \ddot{\hat{y}})\sin\theta)\vec{i} + (-d\omega^2\cos\theta - d\alpha\sin\theta$$

$$- (\alpha\hat{y} + 2\omega\dot{\hat{y}})\sin\theta + (-\hat{y}\omega^2 + \ddot{\hat{y}})\cos\theta)\vec{j}$$

$$(11.10\text{m})$$

At $t = 2\,s$, we obtain

$$\vec{a}_{C/O} = 1.1627\vec{i} - 3.6528\vec{j}\,(\text{m/s}^2) \qquad (11.10\text{n})$$

**Verification:** This is a rather simple problem but with tedious derivations. One must define the reference frames clearly and relate their unit vectors correctly. In the end, all expressions should be with respect to an inertial reference frame if we want to apply *Newton's Second Law* correctly.

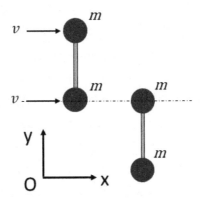

Figure 11.13: Figures for Problem 11.

**Problem 11.11:** The upper pair of connected spheres each of mass $m$ moves initially with a velocity $v$ before colliding with the lower pair of identical spheres initially at rest. The coefficient of restitution between the two colliding spheres is $e$. Prove that the angular velocities of the two links which connect the spheres have the same magnitude immediately after impact. The length of the link is $d$ (Figure 11.13).

**Solution:**

**Motion Analysis:**
According to Figures 11.14(a)–11.14(d), we have an eccentric impact problem. At the impact point, we will assume a virtual mass center velocity $v_G$, as described in Section 6.4. We do not know how to determine $v_G$ because this is an eccentric impact, unlike a 1D two-particle impact problem.

Based on the inertial reference frame attached to the virtual $G$, based on Figure 11.14(d), the approaching and exiting velocities for balls 2 and 3 with respect to the $G$ wall, based on the restitution coefficient $e$, are

$$(v_{2/G})_1 = v - v_G \tag{11.11a}$$

$$(v_{3/G})_1 = -v_G \tag{11.11b}$$

$$(v_{2/G})_2 = -e(v - v_G) \tag{11.11c}$$

$$(v_{3/G})_2 = ev_G \tag{11.11d}$$

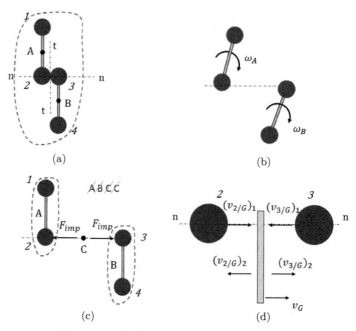

Figure 11.14: (a) Impact normal and tangential directions and FBD; (b) after impact rotations; (c) FBDs during the impact; and (d) motion analysis for eccentric impact along the normal direction.

The velocities with respect to $O-x-y$ are

$$(v_{2/O})_1 = v \tag{11.11e}$$

$$(v_{3/O})_1 = 0 \tag{11.11f}$$

$$(v_{2/O})_2 = -e(v - v_G) + v_G = -ev + (1+e)v_G \tag{11.11g}$$

$$(v_{3/O})_2 = ev_G + v_G = (1+e)v_G \tag{11.11h}$$

In addition, we have

$$(v_{1/O})_1 = v \tag{11.11i}$$

$$(v_{4/O})_1 = 0 \tag{11.11j}$$

The velocities for balls 1 and 4 immediately after impact are unknown and they are labeled as $(v_{1/O})_2$ and $(v_{4/O})_2$, respectively. Immediately after impact, all four balls still have velocities only in the x direction.

Finally, we can define the angular momentum for each system with respect to its own mass center as $\vec{H}_A$ nad $\vec{H}_B$. After impact, we have

$$(H_A)_2 = I\omega_A \qquad (11.11\text{k})$$

$$(H_B)_2 = I\omega_B \qquad (11.11\text{l})$$

where $I$ is the moment of inertia with respect to the center. These motion terms will be used for additional analysis.

**Force Analysis:** The FBDs for the impact are shown in Figures 11.14(a) and 11.14(c). The impact force $F_{imp}$ acts along the normal direction during the impact. However, for the entire system (Figures 11.14(a)), there are no external forces.

**Governing Equations:** To prove that the rotations of $A$ and $B$ will be the same after impact, we can invoke the angular momentum equation for $A$ and $B$, separately, based on Figure 11.14(c),

$$(\vec{H}_A)_1 - \int F_{imp}\frac{d}{2}\vec{k}dt = (\vec{H}_A)_2 = -I\omega_A\vec{k} \qquad (11.11\text{m})$$

$$(\vec{H}_B)_1 - \int F_{imp}\frac{d}{2}\vec{k}dt = (\vec{H}_B)_2 = -I\omega_B\vec{k} \qquad (11.11\text{n})$$

**Solving for unknowns:**
Subtracting Equation (11.11n) from (11.11m), we have

$$(\vec{H}_A)_1 - (\vec{H}_B)_1 = (\vec{H}_A)_2 - (\vec{H}_B)_2$$

$$\Longrightarrow 0 = (\vec{H}_A)_2 - (\vec{H}_B)_2 = -I(\omega_A - \omega_B) \qquad (11.11\text{o})$$

where $(\vec{H}_A)_1 = (\vec{H}_B)_1 = 0$.

Equation (11.11o) proves that, immediately after impact, $\omega_A = \omega_B$ and it is true for any value of the coefficient of restitution.

**Verification:** We can try to solve for the velocities after impact and the velocity for the virtual mass center $G$. To do so, we will invoke the linear momentum equation for the entire system (Figure 11.14(a)), which is conserved,

$$(\vec{P})_1 = (\vec{P})_2$$

$$\implies m(\vec{v}_{1/O})_1 + m(\vec{v}_{2/O})_1 + m(\vec{v}_{3/O})_1 + m(\vec{v}_{4/O})_1 = 2m v \vec{i}$$

$$= m(\vec{v}_{1/O})_2 + m(\vec{v}_{2/O})_2 + m(\vec{v}_{3/O})_2 + m(\vec{v}_{4/O})_2$$

$$= m(\vec{v}_{1/O})_2 + (-ev + (1+e)v_G)\vec{i}$$

$$+ (1+e)v_G \vec{i} + m(\vec{v}_{4/O})_2 \tag{11.11p}$$

There are three unknowns in Equation (11.11p). We need to invoke more governing equations. Let's consider the angular momentum for the entire system with respect to a stationary point $C$ (Figure 11.14(c)). Fortunately, we also have a conservation of angular momentum with respect to point $C$.

$$(\vec{H}_C)_1 = (\vec{H}_C)_2$$

$$\implies -md \cdot (v_{1/O})_1 = -md \cdot (v_{1/O})_2 + md \cdot (v_{4/O})_2 \tag{11.11q}$$

Now, we have two equations, (11.11p) and (11.11q), with three unknowns, $(v_{1/O})_2$, $(v_{4/O})_2$, and $v_G$. We will express $(v_{1/O})_2$ and $(v_{4/O})_2$ as functions of $v_G$ by solving the above-mentioned two equations,

$$(v_{1/O)})_2 = \frac{3+e}{2}v - (1+e)v_G \tag{11.11r}$$

$$(v_{4/O)})_2 = \frac{1+e}{2}v - (1+e)v_G \tag{11.11s}$$

Since we know that $\omega_A = \omega_B$, we have

$$\omega_A = [(v_{1/O})_2 - (v_{2/O})_2]/d$$

$$= \omega_B = [(v_{3/O})_2 - (v_{4/O})_2]/d \tag{11.11t}$$

Introducing all the velocity terms after impact into Equation (11.11t), we obtain

$$v_G = \frac{v}{2} \tag{11.11u}$$

Subsequently, we have

$$(v_{1/O})_2 = v \tag{11.11v}$$

$$(v_{2/O})_2 = \frac{1-e}{2}v \tag{11.11w}$$

$$(v_{3/O})_2 = \frac{1+e}{2}v \tag{11.11x}$$

$$(v_{4/O})_2 = 0 \tag{11.11y}$$

$$\omega_A = \omega_B = \frac{1+e}{2d}v \tag{11.11z}$$

In a special case when $e = 1$, elastic eccentric impact, we have

$$(v_{1/O})_2 = v \tag{11.11aa}$$

$$(v_{2/O})_2 = 0 \tag{11.11ab}$$

$$(v_{3/O})_2 = v \tag{11.11ac}$$

$$(v_{4/O})_2 = 0 \tag{11.11ad}$$

$$\omega_A = \omega_B = \frac{v}{d} \tag{11.11ae}$$

**Problem 11.12:** The tracker is used to hoist a steel casing of mass $m$ out of a dry well with the cable and pulley arrangement as shown in Figure 11.15. When $x = 0$, end $A$ and pulley $B$ coincide at $C$. Write the expression for the tension $T$ in the cable below $B$ as a function of $x$ if the tractor moves forward with a constant velocity $v$. The radius of the pulleys may be neglected, and the casing is free from the sides of the well.

**Solution:**

**Motion Analysis:** The motion of $B$ is vertical, while the motion of $A$ is horizontal. From the pulley setup, we know that the total length $L$ of the wire is

$$L = 3h = 2(h - y) + \sqrt{h^2 + x^2} \tag{11.12a}$$

By taking derivatives of $L$, we have

$$0 = \frac{d}{dt}L = -2\dot{y} + \frac{1}{2}\frac{2x}{\sqrt{h^2 + x^2}}\dot{x}$$

$$\implies \dot{y} = \frac{xv}{2\sqrt{h^2 + x^2}} \tag{11.12b}$$

$$\ddot{y} = \frac{h^2v^2}{2(\sqrt{h^2 + x^2})^3} \tag{11.12c}$$

**Force Analysis:** The FBD for the casing is simple, as shown in Figure 11.15.

**Governing Equations:**
Invoking *Newton's Second Law*, we have

$$T - mg = m\ddot{y}$$

$$\implies T = mg + \frac{mh^2v^2}{2(\sqrt{h^2 + x^2})^3} \tag{11.12d}$$

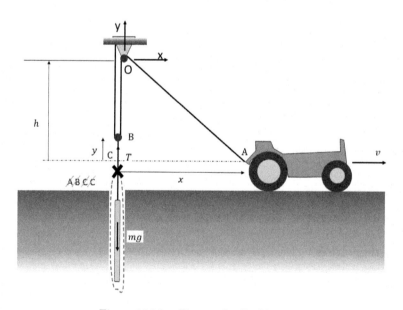

Figure 11.15: Figures for Problem 12.

**Problem 11.13:** The slotted disk is rotating about a vertical axis through $O$ and starts from rest with a constant angular acceleration $\omega$. The small slider $A$ has a mass $m$ and is free to move a slight amount before one or the other of the wires becomes tight (Figure 11.16).

(a) Find the initial acceleration $a_A$ of $A$ relative to the slot before being restrained by the wire.
(b) Determine the tension $T$ in the tight wire prior to the acquisition of any appreciable angular velocity.
(c) Determine the angular velocity during the acceleration period when tension is transferred from one wire to the other.

**Solution:**

**Motion Analysis:** We need to establish an inertial reference $O-x-y$ and a rotating reference frame $D-\hat{x}-\hat{y}$ as shown in Figure 11.17(a).

Point $D$ is at the center of the slot and the rotating frame is attached to the disk; therefore, the frame rotation is defined as

$$\vec{\Omega} = \omega\vec{k} \tag{11.13a}$$

$$\dot{\vec{\Omega}} = \dot{\omega}\vec{k} \tag{11.13b}$$

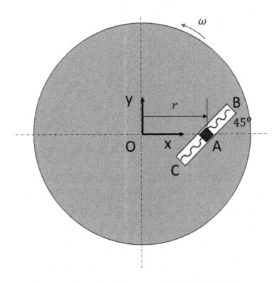

Figure 11.16: Figures for Problem 13.

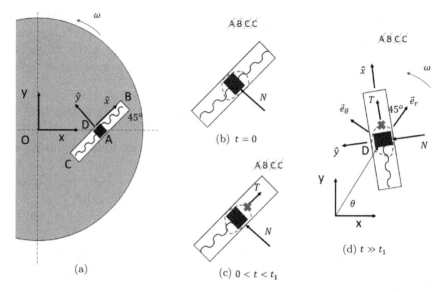

Figure 11.17: (a) Relative motion analysis; and (b)–(d) FBDs for analysis.

The relative motion equations can now be established as

$$\vec{v}_{A/O} = \vec{v}_{D/O} + \vec{\Omega} \times \hat{r}_{A/D} + \hat{v}_{A/D}$$

$$= \vec{\omega} \times \vec{r}_{D/O} + \vec{\Omega} \times \hat{r}_{A/D} + \hat{v}_{A/D} \qquad (11.13c)$$

$$\vec{a}_{A/O} = \vec{a}_{D/O} + \vec{\Omega} \times (\vec{\Omega} \times \hat{r}_{A/D})$$

$$+ \dot{\vec{\Omega}} \times \hat{r}_{A/D} + 2\vec{\Omega} \times \hat{v}_{A/D} + \hat{a}_{A/D}$$

$$= \vec{\omega} \times (\vec{\omega} \times \vec{r}_{D/O}) + \dot{\vec{\omega}} \times \vec{r}_{D/O}$$

$$+ \vec{\Omega} \times (\vec{\Omega} \times \hat{r}_{A/D})$$

$$+ \dot{\vec{\Omega}} \times \hat{r}_{A/D} + 2\vec{\Omega} \times \hat{v}_{A/D} + \hat{a}_{A/D} \qquad (11.13d)$$

As the system starts from rest with $\dot{\omega}$ as a constant, we know that at $t = 0$, $\omega(t = 0) = 0$. Similarly, mass $A$ does not have time to gain any velocity yet, i.e., $\hat{v}_{A/D}(t = 0) = 0$ because its acceleration is finite, not $\infty$. In addition, $\hat{r}_{A/D}(t = 0) = 0$ because mass $A$ has not moved yet.

Introducing the above conditions to Equation (11.13c), we obtain

$$\vec{v}_{A/O}(t=0) = \hat{v}_{A/D}(t=0) = 0 \tag{11.13e}$$

$$\vec{a}_{A/O}(t=0) = \dot{\vec{\omega}} \times \vec{r}_{D/O} + \hat{a}_{A/D}$$

$$= \dot{\omega} r \vec{j} + \ddot{x} \hat{i}$$

$$= \left( \frac{\sqrt{2}}{2} \ddot{x} \right) \vec{i} + \left( \frac{\sqrt{2}}{2} \ddot{x} + \dot{\omega} r \right) \vec{j} \tag{11.13f}$$

where $\hat{i} = \frac{\sqrt{2}}{2}\vec{i} + \frac{\sqrt{2}}{2}\vec{j}$ and $\hat{j} = -\frac{\sqrt{2}}{2}\vec{i} + \frac{\sqrt{2}}{2}\vec{j}$ at $t = 0$.
We need to obtain $\ddot{x}$ in order to solve for $\vec{a}_{A/O}$.

**Force Analysis:** At $t = 0$, mass $A$ is not constrained by either wires and the slot is smooth; therefore, the FBD (Figure 11.17(b)) contains only one normal force $N$.

**Governing Equations:** Applying *Newton's Second Law* at $t = 0$, we obtain

$$-\frac{\sqrt{2}}{2} N = m \left( \frac{\sqrt{2}}{2} \ddot{x} \right) \tag{11.13g}$$

$$\frac{\sqrt{2}}{2} N = m \left( \frac{\sqrt{2}}{2} \ddot{x} + \dot{\omega} r \right) \tag{11.13h}$$

**Solving for Unknowns:** We can solve for $N$ and $\ddot{x}$,

$$\ddot{x}(t=0) = -\frac{\dot{\omega} r}{\sqrt{2}} \tag{11.13i}$$

$$N(t=0) = m \frac{\dot{\omega} r}{\sqrt{2}} \tag{11.13j}$$

The negative sign in the above equation indicate that mass $A$ has an acceleration toward point $C$ with respect to the slot. As a result, the wire of $AB$ will be stretched soon after $t = 0$.

Consider the time period $0 < t < t_1$ where $t_1$ is very small so that $\omega \approx 0$. With wire $AB$ restraining mass $A$, we are sure that $\hat{v}_{A/D} = \hat{a}_{A/D} = 0$.

From Equation (11.13d), for $0 < t < t_1$, we have

$$\vec{a}_{A/O} = \dot{\omega} r \vec{j} \tag{11.13k}$$

Based on the FBD of Figure 11.17(c), we have

$$-\frac{\sqrt{2}}{2} N + \frac{\sqrt{2}}{2} T = 0 \tag{11.13l}$$

$$\frac{\sqrt{2}}{2} N + \frac{\sqrt{2}}{2} T = m\dot{\omega} r \tag{11.13m}$$

Solving the above equations, we obtain

$$T = \frac{m\dot{\omega} r}{\sqrt{2}} \tag{11.13n}$$

$$N = \frac{m\dot{\omega} r}{\sqrt{2}} \tag{11.13o}$$

The positive sign of $T$ confirms that wire $AB$ is stretched.

Now, let's consider what would happen after the disk starts to gain an angular velocity some time later. Wire $AB$ is still stretched, therefore, $\ddot{x} = 0$ and $A$ and $D$ are still overlapped. When the disk has rotated to an angular position $\theta$, a polar coordinate $\vec{e}_r - \vec{e}_\theta$ can be defined, as shown in Figure 11.17(d).

The acceleration of $A$ becomes

$$\vec{a}_{A/O} = \vec{a}_{D/O} + \vec{\Omega} \times (\vec{\Omega} \times \hat{r}_{A/D})$$

$$+ \dot{\vec{\Omega}} \times \hat{r}_{A/D} + 2\vec{\Omega} \times \hat{v}_{A/D} + \hat{a}_{A/D}$$

$$= \vec{\omega} \times (\vec{\omega} \times \vec{r}_{D/O}) + \dot{\vec{\omega}} \times \vec{r}_{D/O}$$

$$= -\omega^2 r \vec{e}_r + \dot{\omega} r \vec{e}_\theta \tag{11.13p}$$

In the above equation, we choose to represent $\vec{a}_{A/O}$ in $\vec{e}_r - \vec{e}_\theta$ instead of $\vec{i} - \vec{j}$ for convenience. In addition, we also notice that we should further express $\vec{a}_{A/O}$ based on $\hat{i} - \hat{j}$ so that the governing equations based on *Newton's Second Law* have a simpler form because both forces $T$ and $N$ are aligned with $\hat{i}$ and $\hat{j}$, respectively.

We then have

$$\vec{e}_r = \sin 45° \hat{i} - \cos 45° \hat{j} = \frac{\sqrt{2}}{2} \hat{i} - \frac{\sqrt{2}}{2} \hat{j} \qquad (11.13q)$$

$$\vec{e}_\theta = \cos 45° \hat{i} + \sin 45° \hat{j} = \frac{\sqrt{2}}{2} \hat{i} + \frac{\sqrt{2}}{2} \hat{j} \qquad (11.13r)$$

The governing equations based on *Newton's Second Law* with respect to $\hat{i}$ and $\hat{j}$ now become

$$T = -\frac{\sqrt{2}}{2} \omega^2 r + \frac{\sqrt{2}}{2} \dot{\omega} r \qquad (11.13s)$$

$$N = \frac{\sqrt{2}}{2} \omega^2 r + \frac{\sqrt{2}}{2} \dot{\omega} r \qquad (11.13t)$$

When $\omega = \sqrt{\dot{\omega}}$, Equation (11.13s) indicates that $T = 0$, which is the situation that wire $AB$ will lose its tension, while wire $AC$ will gain tension later to restrain $A$. We can also find out the angular position and time when this would happen.

$$\omega(T = 0) = \sqrt{\dot{\omega}} = \int_0^{t^*} \dot{\omega} \, dt = \dot{\omega} t^*$$

$$\implies t^* = \frac{1}{\sqrt{\dot{\omega}}} \qquad (11.13u)$$

$$\theta(T = 0) = \frac{1}{2} \dot{\omega}(t^*)^2 = \frac{1}{2} = 28.64° \qquad (11.13v)$$

It is interesting to note that the angular position for $AB$ to lose its tension is always the same regardless of the value of $\dot{\omega}$.

**Problem 11.14:** The split ring rotates about the axis $O-O$, which is normal to the plane of the ring, and starts from rest with an initial angular acceleration $\alpha$ and the split gap at the lowest point. Determine the bending moment and shear force at point $O$. The mass per unit length of rod from which the ring is made is $\rho$. Neglect gravitational forces (Figure 11.18).

**Solution:**

**Motion Analysis:** We consider the motion at $t = 0$ when $\omega = 0$. The entire ring is considered as a rigid body. Every element on the

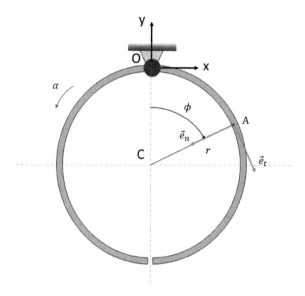

Figure 11.18:   Figures for Problem 14.

ring will have the same angular acceleration $\alpha$. In order to review the bending moment and the forces at point $O$, we will consider the motion of one-half ring, as shown in Figure 11.19(a). The mass center of the half ring is at $G$, which is $\frac{2r}{\pi}$ from $C$.

With $\omega = 0$ at $t = 0$, the acceleration of $G$ is

$$\vec{a}_{G/O} = \vec{\omega} \times (\vec{\omega} \times \vec{r}_{G/O}) + \vec{\alpha} \times \vec{r}_{G/O}$$

$$= \alpha d \cos \beta \vec{i} + \alpha d \sin \beta \vec{j} \qquad (11.14a)$$

where $\tan \beta = \frac{2}{\pi}$ and $d = \sqrt{r^2 + 4r^2/\pi^2}$.

The moment of inertia of the half ring with respect to $G$ is

$$I_G = m_1 r^2 - m_1 \left( \frac{2r}{\pi} \right)^2 = (\rho \pi r) \left( r^2 - \frac{4r^2}{\pi^2} \right)$$

$$= \rho \pi r^3 \left( 1 - \frac{4}{\pi^2} \right) \qquad (11.14b)$$

where $m_1$ is the mass of the half ring.

**Force Analysis:** The FBD for the half ring is shown in Figure 11.19(a). There are two forces and one moment at the cross-section at $O$.

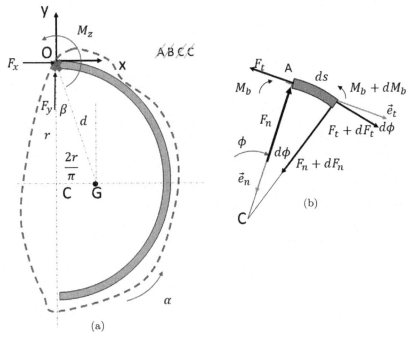

Figure 11.19: (a) FBD for the right-side half ring; and (b) tangential, normal directions, forces, and moments of an infinitesimal element.

**Governing Equations:** Based on *Newton's Second Law*, the governing equations are

$$\left(\sum F\right)_x = F_x = m_1 a_{Gx} = (\rho\pi r)\alpha d\cos\beta$$

$$= \rho\alpha\pi r^2 \tag{11.14c}$$

$$\left(\sum F\right)_y = F_y = m_1 a_{Gy} = (\rho\pi r)\alpha d\sin\beta$$

$$= 2\rho\alpha r^2 \tag{11.14d}$$

$$\left(\sum \tau\right)_G = M_z - F_x r - F_y\left(\frac{2r}{\pi}\right)$$

$$= I_G\alpha = \rho\pi r^3\left(1 - \frac{4}{\pi^2}\right)\alpha \tag{11.14e}$$

**Solving for Unknowns:** Solving the above equations, we obtain

$$F_x = \rho \alpha \pi r^2 \tag{11.14f}$$

$$F_y = 2\rho \alpha r^2 \tag{11.14g}$$

$$M_z = 2\pi \rho \alpha r^3 \tag{11.14h}$$

**Verification:** This problem is not difficult, but let's try to solve for it in a different way. In particular, we would like to invoke the curved beam theory discussed in Chapter 10 to see if we will obtain the same results.

Let's consider cross-section $A$, which is at $\phi$ from $O$. The center of the ring is $C$. At $A$, we define a normal direction $\vec{e}_n$ and a tangential direction $\vec{e}_t$ in Figure 11.19(b), similar to Section 10.6.6.

The acceleration of $A$ can be determined as

$$
\begin{aligned}
\vec{a}_{A/O} &= \vec{a}_{C/O} + \vec{a}_{A/C} \\
&= \vec{\omega} \times (\vec{\omega} \times \vec{r}_{C/O}) + \vec{\alpha} \times \vec{r}_{C/O} \\
&\quad + \vec{\omega} \times (\vec{\omega} \times \vec{r}_{A/C}) + \vec{\alpha} \times \vec{r}_{A/C}
\end{aligned}
\tag{11.14i}
$$

Because at $t = 0$, $\omega = 0$, the above equation is simplified to

$$
\begin{aligned}
\vec{a}_{A/O} &= \vec{\alpha} \times \vec{r}_{C/O} + \vec{\alpha} \times \vec{r}_{A/C} \\
&= \alpha r \vec{i} - \alpha r \vec{e}_t
\end{aligned}
\tag{11.14j}
$$

The relationships of the unit direction vectors are

$$\vec{i} = \cos \phi \vec{e}_t - \sin \phi \vec{e}_n \tag{11.14k}$$

$$\vec{j} = -\sin \phi \vec{e}_t - \cos \phi \vec{e}_t \tag{11.14l}$$

We express $\vec{a}_{A/O}$ in $\vec{e}_n$ and $\vec{e}_t$ as

$$\vec{a}_{A/O} = (\alpha r \cos \phi - \alpha r)\vec{e}_t - \alpha r \sin \phi \vec{e}_n \tag{11.14m}$$

For the little element of Figure 11.19(b), we can also define its moment of inertia with respect to point $C$ as

$$dI_G = \rho(rd\phi)r^2 = \rho r^3 d\phi \tag{11.14n}$$

**Force Analysis:** The FBD of an infinitesimal ring cross-section element is shown in Figure 11.19(b).

**Governing Equations:** Based on *Newton's Second Law*, for the infinitesimal element, the governing equations are

$$\left(\sum F\right)_t = -F_t + (F_t + dF_t)\cos(d\phi) - (F_n + dF_n)\sin(d\phi)$$
$$= dF_t - F_n d\phi = (\rho r d\phi)(a_{A/O})_t$$
$$= (\rho r d\phi)(\alpha r \cos\phi - \alpha r)$$
$$= \rho\alpha r^2(\cos\phi - 1)d\phi \qquad (11.14\text{o})$$

$$\left(\sum F\right)_n = (F_n + dF_n)\cos(d\phi) - F_n + (F_t + dF_t)\sin(d\phi)$$
$$= dF_n + F_t d\phi = (\rho r d\phi)(a_{A/O})_n$$
$$= (\rho r d\phi)(-\alpha r \sin\phi) = -\rho\alpha r^2 \sin\phi d\phi \qquad (11.14\text{p})$$

$$\left(\sum \tau\right)_C = -M_b + (M_b + dM_b) + F_t r - (F_t + dF_t)r$$
$$= dM_b - r\,dF_t = (dI_G)\alpha = \rho\alpha r^3 d\phi \qquad (11.14\text{q})$$

**Solving Equations:** Equations (11.14o) and (11.14p) can be combined into a second-order differential equation of $F_n$ as

$$\frac{d^2 F_n}{d\phi^2} + F_n = \rho\alpha r^2 + -2\rho\alpha r^2 \cos\phi \qquad (11.14\text{r})$$

Solving the above equation and with the introduction of boundary conditions, we have

$$F_n = \rho\alpha r^2(\cos\phi + \pi\sin\phi - \phi\sin\phi + 1) \qquad (11.14\text{s})$$
$$F_t = \rho\alpha r^2(-\pi\cos\phi + \sin\phi + \phi\cos\phi) \qquad (11.14\text{t})$$

At $\phi = 0$, we obtain

$$F_n(\phi = 0) = 2\rho\alpha r^2 \qquad (11.14\text{u})$$
$$F_t(\phi = 0) = -\rho\alpha\pi r^2 \qquad (11.14\text{v})$$

The above results match the results of $F_x$ and $F_y$, but the signs are the opposite due to the definition of $\vec{e}_t$ and $\vec{e}_n$ and the construction of the FBD.

With the functions of $F_t$, we can now solve Equation (11.14q) to obtain $M_b$ as

$$M_b = \rho\alpha r^3(-\pi\cos\phi + \sin\phi + \phi\cos\phi + \phi - \pi) \qquad (11.14\text{w})$$

At $\phi = 0$, we obtain

$$M_b(\phi = 0) = -2\rho\alpha\pi r^3 \qquad (11.14\text{x})$$

Again, the result matches $M_z$ obtained earlier, but the sign is the opposite due to the definition of the unit vectors.

With this solution, we actually extend the curved beam theory to a dynamic situation. Now, we have the shear force, the tangential force, and the bending moment for every cross-section of the split ring.

**Problem 11.15:** The square frame mounted on corner rollers of radius $r$ is moving in a straight line with a velocity $v$ before it contacts the circular path of radius $R$ at $C$. Determine the acceleration of corner $A$ at the instant after the roller at $B$ passes $C$ (Figure 11.20).

**Solution:**

**Motion Analysis:** Basically, we have a rigid body of the square frame which initially moves along a straight line and then when the roller at $B$ passes $C$, it will gain a rotation. The bottom of the

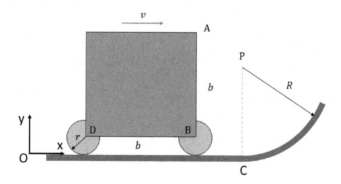

Figure 11.20: Figures for Problem 15.

front rollers, in contact with $C$ will be at zero velocity due to the rolling-without-slipping condition. With the front rollers rolling over the circular path, the motion of $B$ will follow the circular path and become a circular motion. On the other hand, $D$, the center of the rear wheel, will still be on a straight path. With the motion of $B$ and $D$ defined, we can define the rotation of the square frame, which in turn will allow us to determine the motion of every point on the frame, including $A$.

With respect to Figure 11.21(a), the angular velocity and acceleration of the square are defined as $\omega_s$ and $\alpha_s$, respectively. The circular motion of $B$ is defined by its angular velocity and acceleration, $\omega_P$ and $\alpha_P$, respectively. Point $P$ is the center of the circular path of $B$

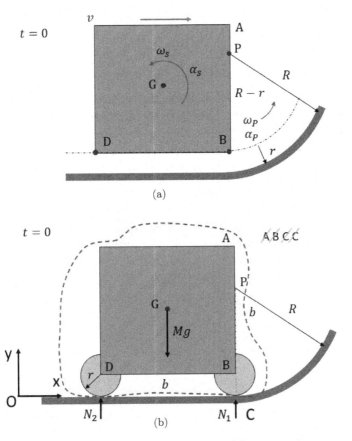

(a)

(b)

Figure 11.21: (a) Motion analysis; and (b) force analysis.

and it is a stationary point in space. The velocity and acceleration of $B$ at $t = 0$ can now be defined as

$$\vec{v}_{B/O} = \vec{\omega}_P \times \vec{r}_{B/P} = \omega_P(R - r)\vec{i} = v\vec{i} \qquad (11.15\text{a})$$

$$\vec{a}_{B/O} = \vec{\omega}_P \times (\vec{\omega}_P \times \vec{r}_{B/P}) + \vec{\alpha}_P \times \vec{r}_{B/P}$$

$$= \omega_P^2(R - r)\vec{j} + \alpha_P(R - r)\vec{i}$$

$$= \frac{v^2}{R - r}\vec{j} + \alpha_P(R - r)\vec{i} \qquad (11.15\text{b})$$

where $\omega_P = v/(R - r)$.

For point $D$, it follows a straight line; therefore, at $t = 0$,

$$\vec{v}_{D/O} = v\vec{i} \qquad (11.15\text{c})$$

$$\vec{a}_{D/O} = \dot{v}\vec{i} \qquad (11.15\text{d})$$

In the above equations, we do not know $\alpha_P$ and $\dot{v}$.

**Force Analysis:** The FBD of the entire system is shown in Figure 11.21(b) at $t = 0$. There are no tangential forces acting on the rollers, but the normal forces $N1$ and $N2$ are no longer the same to result in an angular acceleration, $\alpha_s$, of the square. This assumption is quite reasonable because the rollers have very low rolling friction.

**Governing Equations:** Before establishing governing equations, the unknowns we have include $\alpha_P$, $\dot{v}$, $\omega_s$, and $\alpha_s$. We will at least need four governing equations to determine these unknowns.

First, we will establish the motion relationship between $D$ and $B$.

$$\vec{v}_{D/O} = v\vec{i} = \vec{v}_{B/O} + \vec{v}_{D/B}$$

$$= \vec{v}_{B/O} + \vec{\omega}_s \times \vec{r}_{D/B}$$

$$= v\vec{i} + \omega_s b\vec{j} \qquad (11.15\text{e})$$

$$\vec{a}_{D/O} = \dot{v}\vec{i} = \vec{a}_{B/O} + \vec{a}_{D/B}$$

$$= \vec{a}_{B/O} + \vec{\omega}_s \times (\vec{\omega}_s \times \vec{r}_{D/B}) + \vec{\alpha}_s \times \vec{r}_{D/B}$$

$$= \frac{v^2}{R - r}\vec{j} + \alpha_P(R - r)\vec{i} + \omega_s^2 b\vec{i} - \alpha_s b\vec{j} \qquad (11.15\text{f})$$

**Solving for Unknowns:** Equations (11.15e) and (11.15f) are vector equations, constituting four equations. Solving them, we obtain

$$\omega_s = 0 \tag{11.15g}$$

$$\alpha_s = \frac{v^2}{b(R-r)} \tag{11.15h}$$

$$\dot{v} = \alpha_P(R-r) \tag{11.15i}$$

Equation (11.15i) does not provide us the information of $\dot{v}$ because $\alpha_P$ is unknown, but we know now that the horizontal accelerations of $B$ and $D$ are the same. We need an additional equation to determine $\dot{v}$ or $\alpha_P$.

With $\omega_s$ and $\alpha_s$ known, the acceleration of $G$ can be determined as

$$\vec{a}_{G/O} = \vec{a}_{D/O} + \vec{a}_{G/D}$$

$$= \vec{a}_{D/O} + \vec{\omega}_s \times (\vec{\omega}_s \times \vec{r}_{G/D}) + \vec{\alpha}_s \times \vec{r}_{G/D}$$

$$= \dot{v}\vec{i} + \alpha_s \left(\frac{\sqrt{2}}{2}b\right)\left(-\frac{\sqrt{2}}{2}\vec{i} + \frac{\sqrt{2}}{2}\vec{j}\right)$$

$$= \left(\dot{v} - \frac{1}{2}\frac{v^2}{R-r}\right)\vec{i} + \frac{1}{2}\frac{v^2}{R-r}\vec{j} \tag{11.15j}$$

From the FBD, there are no horizontal forces acting on the system at $t = 0$; therefore, we have

$$\dot{v} = \frac{1}{2}\frac{v^2}{R-r} \tag{11.15k}$$

$$\alpha_P = \frac{1}{2}\frac{v^2}{(R-r)^2} \tag{11.15l}$$

We now have the complete information about the motion of $B$ and we can express the acceleration of $A$ with respect to $B$ as

$$\vec{a}_{A/O} = \vec{a}_{B/O} + \vec{a}_{A/B}$$

$$= \vec{a}_{B/O} + \vec{\omega}_s \times (\vec{\omega}_s \times \vec{r}_{A/B}) + \vec{\alpha}_s \times \vec{r}_{A/B}$$

$$= \frac{v^2}{R-r}\vec{j} + \alpha_P(R-r)\vec{i} - \alpha_s b\vec{i}$$

$$= (\alpha_P(R-r) - \frac{v^2}{R-r})\vec{i} + \frac{v^2}{R-r}\vec{j}$$

$$= (\dot{v} - \frac{v^2}{R-r})\vec{i} + \frac{v^2}{R-r}\vec{j}$$

$$= -\frac{1}{2}\frac{v^2}{R-r}\vec{i} + \frac{v^2}{R-r}\vec{j} \qquad (11.15\text{m})$$

**Verification:** We should know that the results above cannot be extended to later time. A few takeaways for this problem are discussed below. First, we should always recognize that acceleration can change instantly when the force condition changes. However, the velocity does not change instantly unless the acceleration is infinity. Therefore, at $t = 0$, the velocity of the square is still the same and $\omega_s = 0$. It also takes time to gain the angular velocity unless there is a impact moment. Finally, we rely on the assumption of zero horizontal forces at $t = 0$ to construct the last governing equation needed to solve the problem. With the square mounted on rollers, it can be considered as a proper assumption. We should recognize that there will be horizontal forces when $t > 0$ because the normal force on roller $B$ is no longer vertical when it goes on the circular path.

**Problem 11.16:** The uniform $1.2\,\text{m}$ bar, initially at rest on the horizontal surface, has a mass of $15\,\text{kg}$ and is supported at its ends by smaller rollers. Compute the initial angular acceleration $\alpha$ of the bar and the initial reaction under roller $B$ resulting from the application of the $250\,\text{N}$ force (Figure 11.22).

**Solution:**

**Motion Analysis:** We have one rigid body and its angular acceleration is defined as $\alpha$. When the force of $250\,\text{N}$ is applied at $t = 0$, the rigid body bar could develop both linear acceleration and

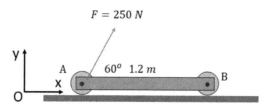

*Figure 11.22: Figures for Problem 16.*

angular acceleration. Because $B$ likely will remain on the surface, its acceleration will be horizontal. We will define it as

$$\vec{a}_{B/O} = a_B \vec{i} \qquad (11.16\text{a})$$

The acceleration of $A$ can then be defined based on $B$ and $\alpha$,

$$\vec{a}_{A/O} = \vec{a}_{B/O} + \vec{a}_{A/B}$$
$$= a_B \vec{i} + \vec{\omega} \times (\vec{\omega} \times \vec{r}_{A/B}) + \vec{\alpha} \times \vec{r}_{A/B}$$
$$= a_B \vec{i} + 1.2\alpha \vec{j} \qquad (11.16\text{b})$$

where $\omega = 0$ at $t = 0$.

Accordingly, the acceleration of the mass center is

$$\vec{a}_{G/O} = \vec{a}_{B/O} + \vec{a}_{G/B}$$
$$= a_B \vec{i} + \vec{\omega} \times (\vec{\omega} \times \vec{r}_{G/B}) + \vec{\alpha} \times \vec{r}_{G/B}$$
$$= a_B \vec{i} + 0.6\alpha \vec{j} \qquad (11.16\text{c})$$

**Force Analysis:** The FBD is shown in Figure 11.23. With rollers, we assume no frictional forces. The normal forces under $A$ and $B$ are defined accordingly.

**Governing Equations:** Based on *Newton's Second Law*, we have

$$\left( \sum F \right)_x = 250 \cos 60° = m a_{Gx} = 15\, a_B \qquad (11.16\text{d})$$

$$\left( \sum F \right)_y = N_A + N_B - mg + 250 \sin 60°$$
$$= m a_{Gy} = 15 \cdot 0.6\alpha = 9\alpha \qquad (11.16\text{e})$$

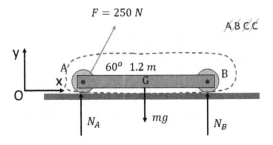

Figure 11.23: The FBD for force analysis

$$\left(\sum \tau\right)_G = -0.6 N_A - 250 \cdot 0.6 \cdot \sin 60° + 0.6 N_B = -I_G \alpha$$

$$= -\frac{1}{12}(15)(1.2)^2 \alpha = -1.8\alpha \qquad (11.16f)$$

**Solving for Unknowns:** We have three equations but four unknowns, $N_A$, $N_B$, $\alpha$, and $a_B$. If we assume that $N_A = 0$, it means that the contact between roller $A$ and the ground is lost when the force is applied.

With this assumption, we proceed to solve for all unknowns.

$$\alpha = 23.8 \text{ rad/s}^2 \qquad (11.16g)$$

$$N_B = 145.0 \text{ N} \qquad (11.16h)$$

$$a_B = 8.33 \text{ m/s}^2 \qquad (11.16i)$$

**Verification:** We need to be cautious not to miss a negative sign for the right side of Equation (11.16f) because we always define counterclockwise rotation as positive. Because we define $\alpha$ clockwise in Figure 11.23, we need to add a negative sign to the right side of Equation (11.16f).

Now, let's consider how realistic is the assumption of $N_A = 0$. If we assume $\alpha = 0$ instead, what will be the value of $N_A$? Resolving Equations (11.16e) and (11.16f), we obtain $N_A = -142.9\,\text{N}$. The negative sign shows that $N_A$ is not compressive but pulling roller $A$ down. This is not possible. Therefore, roller $A$ indeed needs to lose contact.

Let's consider the original solution with $N_B = 145.0\,\text{N}$. This will lead to a weight gain on $B$ of $71.4\,\text{N}$, which is almost equal to $73.6\,\text{N}$, which is the dead weight before $F$ is applied, supporting the assumption again.

In other words, the floor likely will help to support the weight, but it is unlikely to help flipping the rod by pushing it upward.

Finally, if we conduct experiments to verify the above results, we should gradually increase the force to observe when the rod will gain an angular momentum. Without a sufficient force, the rod could remain at equilibrium with weight gradually shifting from $A$ to $B$. It could be true that there will not be an angular acceleration until the contact at $A$ disappears.

**Problem 11.17:** The rotating tube will pump water through the height $h$ and discharge it through its open ends if the rotational speed $\omega$ is sufficiently high. Determine the minimum speed $\omega$ at which pumping will begin, assuming the pump is primed (Figure 11.24).

**Solution:**

**Motion Analysis:** Without the spinning action of the tube, the fluid will not rise up. The spinning action causes the fluid pressure field to change, resulting in the fluid rising up. We are not considering the exit speed of the fluid but only considering the minimum $\omega$ to cause the fluid to rise up by $h$.

**Force Analysis:** We need to model the pressure field distribution and relate it to the spinning action and gravity. The static pressure of the fluid inside the spinning tube is shown in Figure 11.25(a). Between point $E$, the exit, and $A$, the pressure difference is due to gravity because they have the same spinning effect. From $A$ to $B$, the pressure field is due to the spinning action. Finally from $B$ to $O$, it is due to gravity because the spinning is zero at the spinning axis.

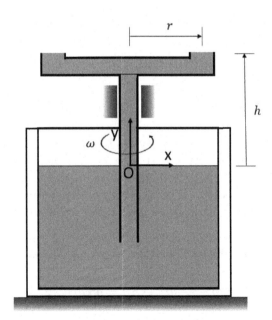

Figure 11.24: Figures for Problem 17.

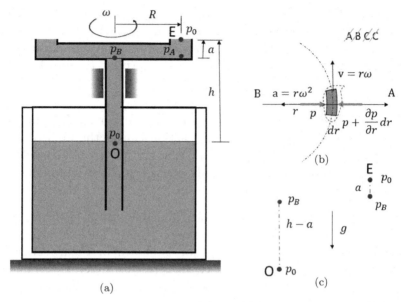

Figure 11.25: (a) Pressure identified at different locations; (b) pressure field due to rotation; and (c) pressure field due to gravity.

We first write down the pressure relationship due to gravity,

$$p_A = p_0 + \rho g a \tag{11.17a}$$

$$P_B = p_0 - \rho g(h - a) \tag{11.17b}$$

where $\rho$ is the density of the fluid. The pressure field due to the spinning action is modeled in Figure 11.25(b). For a small fluid element, the FBD shows that it is subjected to pressure at different radial positions from the center $B$. At $r = R$, the pressure is $P_A$ and at $r = 0$, it is $P_B$. The small fluid element is undergoing a rotation with the tangential speed $r\omega$. The centripetal acceleration is $r\omega^2$. This small element has a little mass $dm = \rho d\vee = \rho S\, dr$, where $S$ is the cross-section area.

**Governing Equations:** Based on *Newton's Second Law*, we have

$$S(p + \frac{\partial p}{\partial r} dr) - Sp = \rho S\, dr(r\omega^2)$$

$$\implies \frac{\partial p}{\partial r} = \rho r\omega^2 \tag{11.17c}$$

**Solving for Unknowns:** We can solve the differential equation of (11.17c) to obtain

$$p(r) = \frac{1}{2}\rho r^2 \omega^2 + c_0 \tag{11.17d}$$

Introducing the boundary condition, when $r = R$, $p = p_A$, we have

$$p(r) = \frac{1}{2}\rho r^2 \omega^2 + p_0 + \rho g a - \frac{1}{2}\rho R^2 \omega^2 \tag{11.17e}$$

The pressure at $B$ with $r = 0$ becomes

$$p_B = p_0 + \rho g a - \frac{1}{2}\rho R^2 \omega^2 \tag{11.17f}$$

The pressure at $O$ can now be expressed alternatively as

$$p_O = p_0 = p_0 + \rho g a - \frac{1}{2}\rho R^2 \omega^2 + \rho g(h - a) \tag{11.17g}$$

From the above equation, we obtain

$$\omega = \frac{\sqrt{2gh}}{R} \tag{11.17h}$$

**Verification:** This problem is similar to the phenomenon in which the fluid free surface inside a cup becomes parabolic when the cup is spinning, as shown in Figure 11.26.

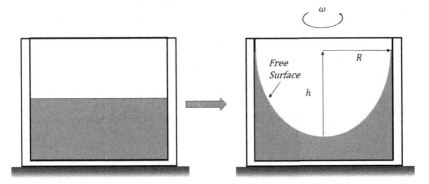

Figure 11.26: Fluid rise in a spinning cup.

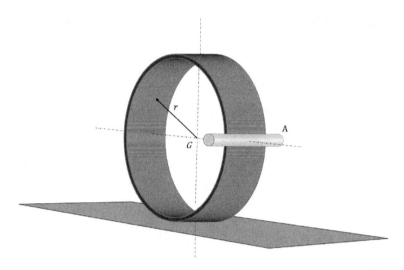

Figure 11.27: Figures for Problem 18.

**Problem 11.18:** The pipe $A$ of mass $m$ is welded to the rim of a circular hoop of radius $r$ and negligible mass which is free to roll on a horizontal plane. If the hoop is released from rest with the pipe initially in a horizontal plane through the center of the hoop, determine the initial angular acceleration $\alpha$ of the hoop if the coefficient of friction between the hoop and the plane is (a) greater than unity; and (b) less than unity. The pipe can be considered as a particle (Figure 11.27).

**Solution:**

**Motion Analysis:** The weld rod is considered a particle with a mass $m$ while the ring is massless. From rest, mass $A$ will be rotating around $G$, while $G$ would move along the horizontal line. At $t = 0$, the acceleration of $A$ can be described as

$$\vec{a}_{A/O} = \vec{a}_{G/O} + \vec{a}_{A/G}$$

$$= a_G \vec{i} + \vec{\omega} \times (\vec{\omega} \times \vec{r}_{A/G}) + \vec{\alpha} \times \vec{r}_{A/G}$$

$$= a_G \vec{i} - r\alpha \vec{j} \tag{11.18a}$$

**Force Analysis:** The FBD at $t = 0$ is shown in Figure 11.28. The normal force $N$ and the frictional force $F$ acting at $C$ are unknowns.

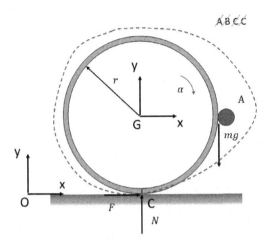

Figure 11.28: The FBD.

The relationship between $F$ and $N$ depends on the condition at the contact point. If the ring can roll without slipping, we know that $F \leq \mu_s N$, where $\mu_s$ is the static friction coefficient. If the ring slips, we have $F = \mu_d N$, where $\mu_d$ is the dynamic friction coefficient. Typically, $\mu_d < \mu_s$.

**Governing Equations:** Based on *Newton's Second Law* for $A$, the governing equation are

$$\left( \sum F \right)_x = F = ma_G \qquad (11.18b)$$

$$\left( \sum F \right)_y = N - mg = -mr\alpha \qquad (11.18c)$$

We can also have a moment equation for the rigid body of the ring and the pipe based on a stationary point within the inertial reference frame. We can choose point $C$ on the ground as this stationary point. The moment equation becomes

$$\left( \sum \vec{\tau} \right)_C = \vec{r}_{A/C} \times (-mg)\vec{j} = -rmg\vec{k}$$

$$= \vec{r}_{A/C} \times m\vec{a}_{A/O} = -m(r^2\alpha + ra_G)\vec{k} \qquad (11.18d)$$

**Solving for Unknowns:** There are four unknowns in Equations (11.18b) and (11.18c). We cannot solve for them. If we assume that the ring is rolling without slipping, we have

$$a_G = r\alpha \tag{11.18e}$$

The angular acceleration and $F$ and $N$ are then solved as

$$\alpha = \frac{g}{2r} \tag{11.18f}$$

$$F = \frac{mg}{2} \tag{11.18g}$$

$$N = \frac{mg}{2} \tag{11.18h}$$

We need to verify if the assumption of rolling without slipping is true. Because the ratio $\frac{F}{N} = 1$, this indicates that the static friction coefficient needs to be larger than one, i.e., $\mu_s > 1$ to render the rolling-without-slipping assumption true.

If the actual static friction coefficient is less than unity, we have a slipping situation. In this case, assuming that we know the dynamic friction coefficient, we have

$$F = \mu_d N \tag{11.18i}$$

Solving Equations (11.18b)–(11.18d) and (11.18i), we obtain

$$\alpha = \frac{g}{r} \tag{11.18j}$$

$$F = 0 \tag{11.18k}$$

$$N = 0 \tag{11.18l}$$

$$a_G = 0 \tag{11.18m}$$

**Verification:** The results above are interesting. If the static friction coefficient is less than unity, the ring will spin without frictional resistance because $N = 0$. Pipe $A$ will have a free fall initially so that $N = 0$.

**Problem 11.19:** The motor $M$ supplies 240 Nm of starting torque to its pinion of 100 mm $(r_p)$ radius in order to elevate the assembly $B$ vertically. The large gear $A$ meshes with the pinion and carries the cable drum of 200 mm radius $(r_d)$. If the gear and the cable drum together have a mass of $m_C = 100$ kg and a combined radius

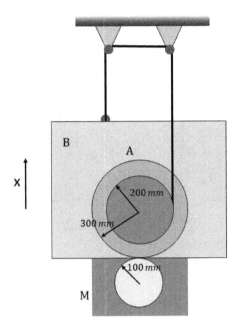

Figure 11.29: Figures for Problem 19.

of gyration of $r_G = 150\,\text{mm}$, and if the mass of the entire assembly is $m = 450\,\text{kg}$, determine the initial upward acceleration of the assembly. Neglect the moment of inertia of the pinion (Figure 11.29).

**Solution:**

**Motion Analysis:** We have an assembly of several rigid bodies. The large gear $A$ and the cable drum are rotating in addition to the upward motion.

**Force Analysis:** The FBD for the entire assembly is shown in Figure 11.30(a).

**Governing Equations:** Based on this FBD and *Newton's Second Law* in the $y$ direction, we obtain

$$\left(\sum F\right)_y = 2T - mg = ma_y \qquad (11.19a)$$

where $T$ is the cable tension, $m$ is the total mass, and $a_y$ is the acceleration of the entire assembly. We cannot solve for $a_y$ because we do not know $T$.

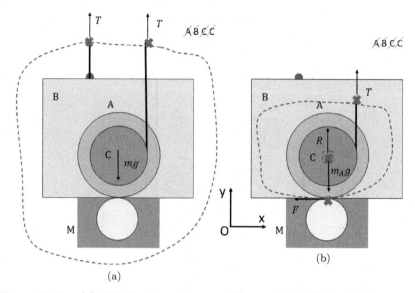

Figure 11.30: (a) FBD for the entire assembly; and (b) FBD for the large gear and cable drum only.

We construct a second FBD as shown in Figure 11.30(b) for the large gear and cable drum only. The large gear receives a driving force $F$ from the pinion. The driving force can be calculated for the motor torque,

$$F = 240 \text{ Nm}/100 \text{ mm} = 2400 \text{ N} \tag{11.19b}$$

Let's construct a moment equation with respect to the center of the large gear $C$,

$$\left(\sum \tau\right)_C = -F \cdot 0.3 + T \cdot 0.2 = -I_C \alpha \tag{11.19c}$$

where $\alpha$ is the angular acceleration of the gear/drum and $I_C$ is its moment of inertia. Because we assign $\alpha$ as clockwise, we need to put a negative sign on the right-hand side of the equation above. The value of $I_C$ is

$$I_C = m_c r_G^2 = 100 \cdot (0.15)^2 = 2.25 \, (\text{kg m}^2) \tag{11.19d}$$

**Solving for Unknowns:** Although we construct an additional equation, we, unfortunately, also introduce an additional unknown $\alpha$.

From the pulley/cable relationship, we know that the tangential speed of the cable at the drum should be twice of the speed of the entire assembly. The acceleration holds the same relationship. We determine that

$$2a_y = r_d\alpha = 0.2\alpha \tag{11.19e}$$

Now, we can solve for all three unknowns,

$$a_y = 4.13 \text{ m/s}^2 \tag{11.19f}$$

$$T = 3136 \text{ N} \tag{11.19g}$$

$$\alpha = 41.3 \text{ rad/s}^2 \tag{11.19h}$$

**Verification:** The moment of inertia of the gear/drum reduces the upward acceleration. If we do not consider this moment of inertia, then the tension is directly proportional to the motor torque. The cable tension would be 3600 N, which results in an upward acceleration at 6.19 m/s$^2$, a 50% overestimate. In a similar situation, the gear trains of an automobile vehicle can reduce the acceleration substantially due to the rotating parts as discussed in Chapter 9. Therefore, it is important to reduce the moment of inertia as long as the strength requirement can be met.

**Problem 11.20:** The large concrete slab of length $l$ is being tilted slowly into position by the winch at $A$. In the position shown the cable at $B$ is horizontal. If the horizontal lower cable breaks at the winch in this position, determine the initial acceleration $a$ of the bottom $C$ of the slab which is free to move on rollers on the horizontal surface (Figure 11.31).

**Solution:**

**Motion Analysis:** We have a rigid body of the slab. Because it was slowly tilted into position before the lower cable at $A$ breaks, we can assume that it is at rest at $t = 0$. $C$ will be in a straight line motion while the entire slab will have an angular acceleration. Note that at

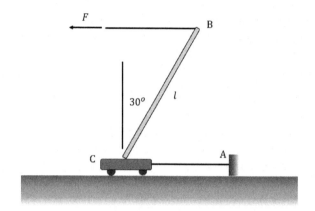

Figure 11.31: Figures for Problem 20.

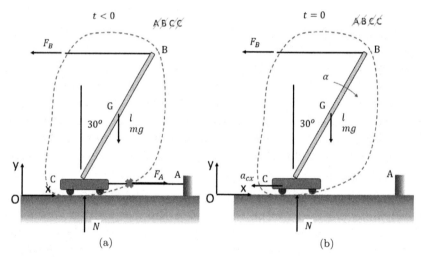

Figure 11.32: (a) the FBD for the slab at equilibrium when $t < 0$; and (b) the FBD for the slab at $t = 0$ when the wire at A breaks.

$t = 0$, there are no velocities of the slab because the forces acting on it are not infinite or the acceleration (angular or linear) is not infinite. At $t = 0$, as shown in Figure 11.32(b), we have

$$\vec{a}_{C/O} = -a_{cx}\vec{i} \tag{11.20a}$$

$$\vec{\alpha} = -\alpha\vec{k} \tag{11.20b}$$

The acceleration of the mass center $G$ becomes

$$\vec{a}_{G/O} = \vec{a}_{C/O} + \vec{a}_{G/C}$$

$$= \vec{a}_{C/O} + \vec{\alpha} \times \vec{r}_{G/C}$$

$$= -a_{cx}\vec{i} + \frac{l}{2}\alpha \sin 60°\vec{i} - \frac{l}{2}\alpha \cos 60°\vec{j}$$

$$= -(a_{cx} - 0.433l\alpha)\vec{i} - 0.25l\alpha\vec{j} \qquad (11.20c)$$

**Force Analysis:** The FBDs of the slab when $t < 0$ and at $t = 0$ are shown in Figure 11.32.

**Governing Equations:** Let's first determine the forces of $F_A$ and $F_B$ when $t < 0$ when the slab is at equilibrium.

$$\left(\sum F\right)_x = F_A - F_B = 0 \qquad (11.20d)$$

$$\left(\sum F\right)_y = N - mg = 0 \qquad (11.20e)$$

$$\left(\sum \tau\right)_{Gz} = \frac{l}{2}F_B \cos 30° + \frac{l}{2}F_A \cos 30°$$

$$- \frac{l}{2}N \sin 30° = 0 \qquad (11.20f)$$

We can solve the above equations to obtain

$$F_{A0} = \frac{\sqrt{3}}{6}mg \approx 0.289\ mg \qquad (11.20g)$$

$$F_{B0} = \frac{\sqrt{3}}{6}mg \approx 0.289\ mg \qquad (11.20h)$$

$$N_0 = mg \qquad (11.20i)$$

At $t = 0$, $F_A = 0$ and the FBD is shown in Figure 11.32(b). We do not know if $F_B$ and $N$ would stay the same.

Based on *Newton's Second Law*, we have

$$\left(\sum F\right)_x = -F_B = ma_{Gx} = -m(a_{cx} - 0.433l\alpha) \quad (11.20j)$$

$$\left(\sum F\right)_y = N - mg = ma_{Gy} = -0.25\,m\,l\alpha \qquad (11.20k)$$

$$\left(\sum \tau\right)_{Gz} = \frac{l}{2}F_B \cos 30° - \frac{l}{2}N \sin 30° = -I_G\alpha \qquad (11.20l)$$

**Solving for Unknowns:** There are four unknowns in the above equations. We will assume that $F_B = F_{B0}$ and solve for the other three unknowns. We obtain

$$a_{cx} = 0.66 \ g \tag{11.20m}$$

$$N = 0.79 \ mg \tag{11.20n}$$

$$\alpha = 0.86\frac{g}{l} \ \text{rad/s}^2 \tag{11.20o}$$

**Verification:** The solutions from Meriam (1975) are

$$a_{cx} = \frac{3\sqrt{3}}{8}g \approx 0.65 \ g \tag{11.20p}$$

$$N = \frac{13}{16} \ mg \approx 0.81 \ mg \tag{11.20q}$$

$$\alpha = \frac{3}{4}\frac{g}{l} \ \text{rad/s}^2 \tag{11.20r}$$

These results are close to the results obtained earlier, but the angular accelerations are quite different. In fact, the Meriam solutions can be obtained by solving the same set of governing equations of (11.20j)–(11.20l) with $F_B = \frac{9}{8}F_{B0}$. This verifies that the governing equations are correct.

With the angular acceleration in the clockwise direction, we can be sure that, at $t = 0$, the acceleration of slab top at $B$ is down and to the right. This indicates that the wire at $B$ will be further stretched at $t = 0$ once the wire at $A$ breaks. The Meriam solution used a higher $F_B$ force which could be more realistic, but we simply do not have a physical law to predict the actual tension of the wire at $B$. An experiment can verify the actual value of the wire tension at $B$ and the wire at $A$ breaks. We can evaluate how the value of $F_B$ affects the results by solving the same set of equations with different $F_B$ values. The results are shown in Figure 11.33.

**Problem 11.21:** One end of a rope of total length $L$ is released from rest with $y = 0$. Determine the velocity $v$ of the falling end of the rope in terms of $y$. Assume that the rope is perfectly flexible in bending but not extensible. Discuss the result when $y \longrightarrow L$ (Figure 11.34).

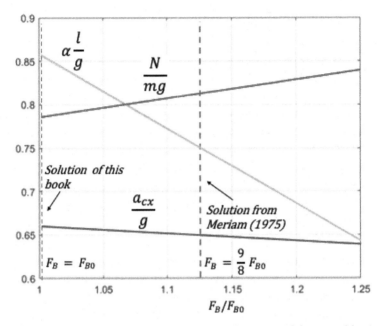

Figure 11.33:  The normalized angular acceleration, normal force, and horizontal acceleration at $C$ with respect to normalized force $F_B/F_{B0}$.

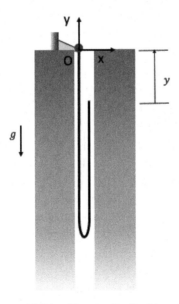

Figure 11.34:  Figures for Problem 21.

## Solution:

**Motion Analysis:** For this problem, due to the rope being totally flexible but not stretchable, we divide the rope into the falling part and the stationary part. The entire falling part has the same velocity $v$.

**Force Analysis:** We construct three FBDs, for the entire rope, the bending portion at the lowest end, and the falling part, as shown in Figure 11.35. For the entire rope (Figure 11.35(a)), the forces involved are the force at $O$ where the rope is tied and gravity. For the lowest bending portion (Figure 11.35(b)), there are gravity and two tensile rope forces, $F_1$ and $F_2$, to the stationary side and the falling side, respectively. For the falling rope (Figure 11.35(c)), it is $F_2$ and gravity.

**Governing Equations:** Because we are interested in finding out the velocity of the falling part, we likely will not use *Newton's Second Law* because it leads to acceleration, not velocity directly. To get velocity, we need to integrate the acceleration over time. Therefore, we will consider either the energy equation or the linear momentum equation. For the linear momentum equation, there is a need for the information related to time, which is not available from the problem

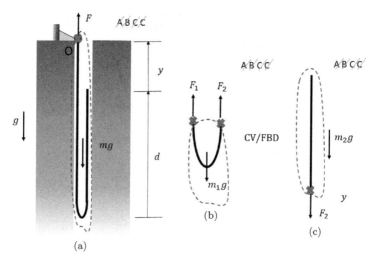

Figure 11.35:   (a) FBD for the entire rope; (b) CV/FBD for the bending portion; and (c) the FBD for the falling rope.

statement. As a result, we examine if we can apply the energy equation first.

Based on Figure 11.35, the only external forces involved are $F$ and $mg$. Because $F$ is applied at the fixed end, it does not do any work. The internal force, i.e., the rope tension, is considered as not doing work because the rope is not stretchable. We conclude that this rope system is conserved in mechanical energy. Based on the datum at $O$, the conservation of mechanical energy for the rope system between $t = 0$ and $t_1$ when the free end is $y$ distance below the datum is

$$T_0 + U_{g0} = 0 - \rho L g \left( \frac{1}{4} L \right) = T_1 + U_{g1}$$

$$= \frac{1}{2} \rho d v_1^2 - \rho d g \left( y + \frac{d}{2} \right)$$

$$- \rho (d + y) g \left( \frac{d + y}{2} \right) \tag{11.21a}$$

where $v_1$ is the falling speed and $d$ is the length of the falling rope. We also have $L = y + 2d$.

**Solving for Unknowns:** There is only one unknown in Equation (11.21a) and it can be readily solved to obtain

$$v_1 = \sqrt{\frac{gy(2L - y)}{L - y}} \tag{11.21b}$$

**Verification:** This problem appears to be simple, but the solution of Equation (11.21b) could be of grave concern when $y \longrightarrow L$, when $L - y$ becomes very small, which results in a very large $v_1$. At $y = L$, $v_1$ is $\infty$. Apparently, this is not possible in reality. We can plot $v_1$ vs. $y$ based on Equation (11.21b) with $L = 1\,\mathrm{m}$ and compare it to the free falling velocity, as shown in Figure 11.36.

In Figure 11.36, we plot the rope speed based on Equation (11.21b) by assuming $L = 1\,\mathrm{m}$. We only plot the velocity up to $y = 0.9\,L$. Initially, the rope is falling similarly to a free-falling object. As $y$ becomes larger, the rope falling speed starts to far outpace the free-falling object. At $y = 0.9\,L$ with $L = 1\,\mathrm{m}$, the rope falling speed is $9.85\,\mathrm{m/s}$ compared with $4.20\,\mathrm{m/s}$ of a free-falling object, a factor of 2.35. A free-falling object will be at a constant acceleration of $g$, while the falling rope, based on Equation (11.21b) will have a higher

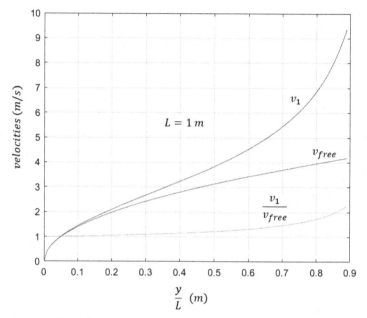

Figure 11.36: The falling velocity of the rope compared with a free falling object.

and higher acceleration. The acceleration will be at $\infty$ as well when $y = L$. This is obviously not possible.

According to Figure 11.35(c), the extra acceleration comes from the rope tension $F_2$ which also becomes larger and larger as $y \longrightarrow L$. In practice, $F_2$ is limited up to the strength of the rope. Therefore, Equation (11.21b) is only true up to a certain point.

We should also examine the assumption we made to invoke the conservation law of mechanical energy. In practice, no rope is perfectly flexible in bending nor is it non-stretchable. The internal force due to the rope tension will consume energy.

Let's examine the control volume for the entire rope (Figure 11.35(a)). Based on this control volume, the linear momentum equation is

$$\left(\sum F\right)_y = F - \rho L g = \frac{d}{dt}(P)_{CV} - \frac{d}{dt}(P)_{in} + \frac{d}{dt}(P)_{out}$$

$$= \frac{d}{dt}(-\rho d v_1) = \frac{d}{dt}\left(-\rho \frac{L-y}{2} v_1\right)$$

$$= -\rho \frac{L-y}{2}\frac{dv_1}{dt} + \frac{1}{2}\rho v_1^2 \qquad (11.21c)$$

From Equation (11.21c), we have

$$\frac{dv_1}{dt} = \frac{1}{L - y}\left(2Lg - 2\frac{F}{\rho} + v_1^2\right) \tag{11.21d}$$

From the above equation, it appears that when $y \longrightarrow L$, $\frac{dv_1}{dt} \longrightarrow \infty$ unless $(2Lg - 2\frac{F}{\rho} + v_1^2) \longrightarrow 0$. After $y = L$, the entire rope is stationary $(v_1 = 0)$ and $F = \rho Lg$, Equation (11.21d) is satisfied. How about before $y = L$? As long as $F = \rho Lg + \rho v_1^2/2$, $\frac{dv_1}{dt}$ may not blow up. In fact, as $v_1$ increases, $F$ will increase as well so that it can bring the falling rope to a stationary state. With the acceleration limited, the final value of $v_1$ should not be $\infty$ as indicated by Equation (11.21b).

Recall Problem 11.4, in which the rope never falls with an acceleration greater than $g$ due to the rope tension. An experiment would verify which equation is more accurate in predicting the fall of the rope for this example. The odds that Equation (11.21d) is more accurate are higher because it does not rely on the assumption of zero energy loss of the rope.

**Problem 11.22:** A rope of length $L$ is whirled around a fixed vertical axis with an angular velocity $\omega$. Determine the shape assumed by the rope if $\omega$ is large enough so that the distance along the rope to any element may be approximated by the horizontal radius to the element (Figure 11.37).

**Solution:**

**Motion Analysis:** The entire rope can be considered as a rigid body once it reaches the steady state, assuming its shape, while rotating.

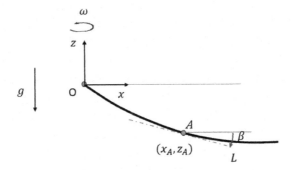

Figure 11.37: Figures for Problem 22.

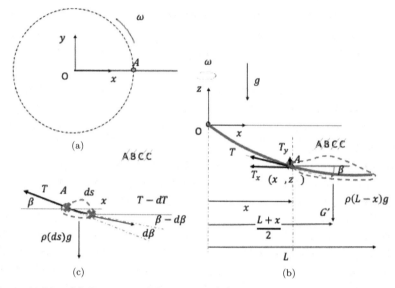

Figure 11.38: (a) Top view of the motion; (b) the FBD for the rope from $x$ to $L$; and (c) the FBD for one infinitesimal rope element.

From the top view, as shown in Figure 11.38(a), each rope element has a circular motion on its own horizontal plane. At an instant of time, the rope is also at the $x–z$ plane. For an infinitesimal element of the rope at $A$, its acceleration is

$$\vec{a}_{A/O} = \vec{\omega} \times (\vec{\omega} \times \vec{r}_{A/O}) = -x\omega^2 \vec{i} \qquad (11.22a)$$

**Force Analysis:** We can treat the entire rope as a rigid body. With the assumption that $\omega$ is large enough so that $x \longrightarrow s$, where $s$ is the actual length of the rope with respect to $O$. For a segment of the rope from $x$ to $x = L$, its mass center is located $(L + x)/2$ horizontally from $O$, as shown in Figure 11.38(b). The FBD of this segment has the rope tension at $A$ and the gravitational force at $G'$. Because the rope tension at $A$ is aligned with the rope tangential direction, we can further define the horizontal and vertical tension components as $T_x$ and $T_z$, respectively.

**Governing Equations:** Applying *Newton's Second Law* for this "rigid" rope segment of Figure 11.38(b), we have

$$-T_x = -\rho(L - x)\left(\frac{L + x}{2}\right)\omega^2 \qquad (11.22b)$$

$$T_y - \rho(L - x)g = 0 \qquad (11.22\text{c})$$

**Solving for unknowns:** Combining the above two equations to define the slope, we have

$$\frac{dz}{dx} = \frac{-T_x}{T_z} = \frac{-2g}{\omega^2(L + x)} \qquad (11.22\text{d})$$

Integrating Equation (11.22d), we obtain

$$z = \frac{2g}{\omega^2} \ln \frac{L}{L + x} \qquad (11.22\text{e})$$

**Verification:** What if $\omega$ is not large enough to render $x \longrightarrow s$?

We can construct an FBD for an infinitesimal rope element, as shown in Figure 11.38(c).

Based on *Newton's Second Law* for the infinitesimal element, we have

$$\left(\sum F\right)_x = -T \cos \beta + (T - dT)\cos(\beta - d\beta)$$
$$= -\rho(ds)x\omega^2 \qquad (11.22\text{f})$$

$$\left(\sum F\right)_z = T \sin \beta - (T - dT)\sin(\beta - d\beta)$$
$$- \rho(ds)g = 0 \qquad (11.22\text{g})$$

where $ds$ is the rope length of the infinitesimal element and $\beta$ is the tangential angle of the rope at $A$.

The above-mentioned two equations can be simplified to

$$- \cos \beta \, dT + T \sin \beta \, d\beta = \rho x \omega^2 \, ds \qquad (11.22\text{h})$$
$$\sin \beta \, dT + T \cos \beta \, d\beta = \rho g \, ds \qquad (11.22\text{i})$$

We can rearrange Equations (11.22h) and (11.22i) as

$$\cos \beta \, dT = dT_x = T \sin \beta \, d\beta - \rho x_A \omega^2 \, ds \qquad (11.22\text{j})$$
$$\sin \beta \, dT = dT_y = -T \cos \beta \, d\beta + \rho g \, ds \qquad (11.22\text{k})$$

Again, we recognize that the rope tension at point $x$ should be also in the tangential direction. Therefore, from Equations (11.22j) and (11.22k), integrating over the segment of Figure 11.38(c), we have

$$\frac{dy}{dx} = \frac{T_y}{T_x} = \frac{\int_{T_{Ay}}^0 dT_y}{\int_{T_{Ax}}^0 dT_x} = \frac{-\int_{\beta_A}^{\beta_L} T\cos\beta\,d\beta + \int_{s_A}^L \rho g\,ds}{\int_{\beta_A}^{\beta_L} T\sin\beta\,d\beta - \int_{s_A}^L \rho x\omega^2\,ds} \quad (11.22l)$$

We also have

$$\tan\beta = \frac{dy}{dx} = \frac{T_y}{T_x} \quad (11.22m)$$

It is possible to pursue a numerical solution to determine the rope profile based on the above two equations. However, let's check if Equation (11.22l) can be simplified and lead to the earlier results if $\omega$ is large enough.

Under this condition, $\beta$ will be at a small angle as well. Let's examine the integration over $\beta$ in Equation (11.22l).

$$\left| \int_{\beta_A}^{\beta_L} T\sin\beta\,d\beta \right| < |-T_{max}(\cos\beta_L - \cos\beta_A)| \quad (11.22n)$$

$$\left| \int_{\beta_A}^{\beta_L} -T\cos\beta\,d\beta \right| < |-T_{max}(\sin\beta_L - \sin\beta_A)| \quad (11.22o)$$

With $\beta$ small when $\omega$ is sufficiently large, the values above two integration will be very small because $\sin\beta_L \approx \sin\beta_A$ and $\cos\beta_L \approx \cos\beta_A$. Equation (11.22l) can now be simplified to

$$\frac{dy}{dx} = \frac{T_y}{T_x} = \frac{\int_{T_y}^0 dT_y}{\int_{T_x}^0 dT_x} \approx \frac{\int_s^L \rho g\,ds}{-\int_s^L \rho x\omega^2\,ds} \quad (11.22p)$$

With $x \longrightarrow s$, Equation (11.22p) can be further simplified and the integration can be carried out as

$$\frac{dy}{dx} = \frac{T_y}{T_x} = \frac{\int_{T_y}^0 dT_y}{\int_{T_x}^0 dT_x} \approx \frac{\int_x^L \rho g\,dx}{-\int_x^L \rho x\omega^2\,dx}$$

$$= \frac{\rho g(L-x)}{-\frac{1}{2}\rho(L^2 - x^2)} = \frac{-2g}{\omega^2(L+x)} \quad (11.22q)$$

We then obtain the same result as earlier. We can plot Equation (11.22q) for a special setting to review the results. Let $L = 2\,\text{m}$

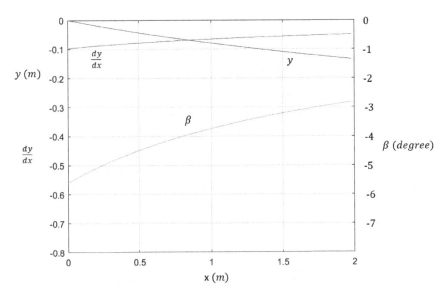

Figure 11.39: The simulated rope shape with $\omega = 10$ rad/s and $L = 2$ m, along with its gradient and the tangential angle.

and $\omega = 10$ rad/s. The rope shape and its tangential are shown in Figure 11.39.

From the tangential angle, we verify that

$$|\cos\beta_0 - \cos\beta_L| = |\cos(-5.60°) - \cos(-2.82°)|$$
$$= 0.004 \qquad (11.22r)$$

$$|\sin\beta_0 - \sin\beta_L| = |\sin(-5.60°) - \sin(-2.82°)|$$
$$= 0.048 \qquad (11.22s)$$

Therefore, neglecting the terms related to $\beta$ to obtain Equation (11.22p) is justified because we will have at most 5% error in the solution.

The rope spinning problem can be further generalized into the spinning rope by cowboys, sometimes noted as the Lasso problem. The readers can check out many videos to see the various rope shapes achieved by spinning.

**Problem 11.23:** The slotted disk rotates in the horizontal plane about its center $O$ at a constant angular speed $\omega$, and the slider of mass $m$ with attached spring slides in the slot with negligible friction.

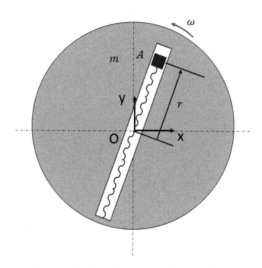

Figure 11.40:   Figures for Problem 23.

The spring has a stiffness $k$ and is un-deformed when the slider is at the center of the disk. During rotation of the disk, the slider is released from rest relative to the disk at the position for which $r = r_0$ where it was initially latched to the disk and the spring is stretched. If $\omega^2 < k/m$, determine (a) the initial value of $\ddot{r}$ just after the slider is released; and (b) the horizontal force $F$ on the slider as it first passes the center of the disk (Figure 11.40).

**Solution:**

**Motion analysis:** First, we establish two reference frames to determine the motion of the slider, identified as A. The inertial reference frame is $O-x-y$, while a rotating reference, $O-\hat{x}-\hat{y}$ is defined with $\hat{y}$ aligned with the slot and rotating with the disk. The frame rotation is $\Omega = \omega$.

At $t = 0$, slider A is at $\hat{y}(t = 0) = r_0\hat{j}$ and $\hat{v}(t = 0) = 0$ when it is just released. The acceleration along the slot direction in $\ddot{\hat{y}}_0$. The acceleration of the slider can be determined as

$$\vec{a}_{A/O}(t = 0) = \vec{\Omega} \times (\vec{\Omega} \times \hat{r}_{A/O}) + \dot{\vec{\Omega}} \times \hat{r}_{A/O} + 2\vec{\Omega} \times \hat{v}_A + \hat{a}_A$$

$$= -\omega^2 r_0\hat{j} + \ddot{\hat{y}}_0\hat{j} \tag{11.23a}$$

After some time, at $t_1$, the slider just passes through the center at $O$. At this time, it has gained a speed $\hat{v}_1$. Note that there is no radial displacement, i.e., $\hat{r}_{A/O}(t_1) = 0$. The acceleration of the slider at $t_1$ becomes

$$\vec{a}_{A/O}(t_1) = \vec{\Omega} \times (\vec{\Omega} \times \hat{r}_{A/O}) + \vec{\dot{\Omega}} \times \hat{r}_{A/O} + 2\vec{\Omega} \times \hat{v}_A + \hat{a}_A$$

$$= 2\omega\hat{v}_1\hat{i} + \ddot{y}_1\hat{j} \qquad (11.23b)$$

where $\ddot{y}_1$ is the acceleration of A at $t_1$ with respect to the slot.

During the time $0 < t < t_1$, the acceleration of $A$ is

$$\vec{a}_{A/O}(t < t_1) = \vec{\Omega} \times (\vec{\Omega} \times \hat{r}_{A/O}) + \vec{\dot{\Omega}} \times \hat{r}_{A/O} + 2\vec{\Omega} \times \hat{v}_A + \hat{a}_A$$

$$= -\omega^2 r\hat{j} + 2\omega\hat{v}\hat{i} + \ddot{y}\hat{j} \qquad (11.23c)$$

The unknowns involved with Equations (11.23a) and (11.23b) are to be calculated later.

**Force analysis:** The corresponding FBD is shown in Figure 11.41. The spring force $F_k = k\hat{y}$.

**Governing equations and solutions:** At $t = 0$, based on *Newton's Second Law*, we have

$$\left(\sum F\right)_{\hat{x}} = -N = 0 \qquad (11.23d)$$

$$\left(\sum F\right)_{\hat{y}} = -F_k = -kr_0 = m(-\omega^2 r_0 + \ddot{y}_0) \qquad (11.23e)$$

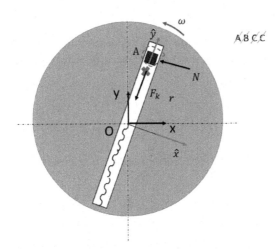

Figure 11.41:   The motion analysis and force analysis for Problem 23.

From Equation (11.23e), we determine, at $t = 0$, the acceleration of the slider with respect to the slot is

$$\ddot{\hat{y}} = \left( \omega^2 - \frac{k}{m} \right) r_0 \tag{11.23f}$$

In the problem statement, the condition was defined as $\omega^2 < \frac{k}{m}$, so that $\ddot{\hat{y}}_0 < 0$, pointing to the center. At $\omega^2 = \frac{k}{m}$, we will have a resonance problem, which is beyond the discussion of this problem. At $t_1$, the governing equations are

$$\left( \sum F \right)_{\hat{x}} = -N = m(2\omega \hat{v}_1) \tag{11.23g}$$

$$\left( \sum F \right)_{\hat{y}} = -F_k = 0 = m(\ddot{\hat{y}}_1) \tag{11.23h}$$

Equation (11.23h) indicates that at $t_1$, $\ddot{\hat{y}}_1 = 0$. However, there are two unknowns in Equation (11.23g). Before, we can solve for $N$, we need to determine the speed of the slider when it passes through the slot.

From Equation (11.23c), we have

$$\ddot{\hat{y}} = \left( \omega^2 - \frac{k}{m} \right) r \tag{11.23i}$$

We can invoke $ads = vdv$ to determine the speed $\hat{v}$ because it is a straight line motion with respect to the rotating reference frame. We obtain

$$\hat{v}_1 = r_0 \sqrt{\omega^2 - \frac{k}{m}} \tag{11.23j}$$

Finally, the normal force of the slot on the slider is

$$N = -2m\omega r_0 \sqrt{\omega^2 - \frac{k}{m}} \tag{11.23k}$$

The negative sign indicates that the normal force is the opposite direction of Figure 11.41.

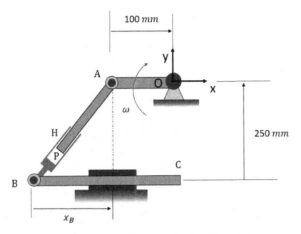

Figure 11.42:   Figures for Problem 24.

**Verification:** This is a fairly straight forward problem as long as we define the reference frames correctly. The inertial reference frame and the rotating reference frame share the same origin.

**Problem 11.24:** The angular velocity of the crank $OA$ is $\omega = 1.5$ rad/s and $\dot{\omega} = 0$ in the horizontal position as shown. Simultaneously, for the control rod $BC$, $x_B = 150$ mm, $\dot{x}_B = -100$ mm/s, and $\ddot{x}_B = 0$. For these conditions, determine the velocity $u$ and $\dot{u}$ of the piston $P$ with respect to the hydraulic cylinder $H$ and calculate the angular acceleration $\alpha$ of the cylinder axis (Figure 11.42).

**Solution:**

**Motion Analysis:** We have a rigid-body assembly. We will define the velocity and acceleration of those points for solutions. For point $A$, we have

$$\vec{v}_{A/O} = \vec{\omega} \times \vec{r}_{A/O} = 0.1\omega\vec{j} = 0.15\vec{j}\,(\text{m/s}) \qquad (11.24a)$$

$$\vec{a}_{A/O} = \vec{\omega} \times (\vec{\omega} \times \vec{r}_{B/P}) + \dot{\vec{\omega}} \times \vec{r}_{A/O}$$

$$= 0.1\omega^2\vec{i} = 0.225\vec{i}\,(\text{m/s}^2) \qquad (11.24b)$$

The motion of $P$ can then be defined as

$$\vec{v}_{P/O} = \vec{v}_{A/O} + \vec{v}_{P/A}$$

$$= 0.15\vec{j} + \vec{\omega}_P \times \vec{r}_{P/A}$$

$$= 0.15\vec{j} + \omega_P \cdot r_{P/A}(0.857\vec{i} - 0.514\vec{j})$$

$$= (0.857\omega_P \cdot r_{P/A})\vec{i}$$

$$+(0.15 - 0.514\omega_P \cdot r_{P/A})\vec{j} \,(\text{m/s}) \tag{11.24c}$$

$$\vec{a}_{P/O} = \vec{a}_{A/O} + \vec{a}_{P/A}$$

$$= 0.225\vec{i} + \vec{\omega}_P \times (\vec{\omega}_P \times \vec{r}_{P/A}) + \dot{\vec{\omega}}_P \times \vec{r}_{P/A}$$

$$= 0.225\vec{i} + \omega_P^2 \cdot r_{P/A}(0.514\vec{i} + 0.857\vec{j})$$

$$+\alpha_P \cdot r_{P/A}(0.857\vec{i} - 0.514\vec{j})$$

$$= (0.225 + 0.514\omega_P^2 \cdot r_{P/A} + 0.857\alpha_P \cdot r_{P/A})\vec{i}$$

$$+(0.857\omega_P^2 \cdot r_{P/A} - 0.514\alpha_P \cdot r_{P/A})\vec{j} \,(\text{m/s}^2) \tag{11.24d}$$

In the mean time, let's consider point $H$, which is at the same position as $P$ but on the cylinder. The velocity and acceleration of $H$, can be defined as

$$\vec{v}_{H/O} = \vec{v}_{B/O} + \vec{v}_{H/B}$$

$$= 0.1\vec{i} + \vec{\omega}_P \times \vec{r}_{H/B}$$

$$= 0.1\vec{i} + \omega_P \cdot r_{H/B}(-0.857\vec{i} + 0.514\vec{j})$$

$$= (0.1 - 0.857\omega_P \cdot r_{H/B})\vec{i}$$

$$+(0.514\omega_P \cdot r_{H/B})\vec{j} \,(\text{m/s}) \tag{11.24e}$$

$$\vec{a}_{H/O} = \vec{a}_{B/O} + \vec{a}_{H/B}$$

$$= 0 + \vec{\omega}_P \times (\vec{\omega}_P \times \vec{r}_{H/B}) + \dot{\vec{\omega}}_P \times \vec{r}_{H/B}$$

$$= \omega_P^2 \cdot r_{H/B}(-0.514\vec{i} - 0.857\vec{j})$$

$$+\alpha_P \cdot r_{H/B}(-0.857\vec{i} + 0.514\vec{j})$$

$$= (-0.514\omega_P^2 \cdot r_{H/B} - 0.857\alpha_P \cdot r_{H/B})\vec{i}$$

$$+(-0.857\omega_P^2 \cdot r_{H/B}$$

$$+ 0.514\alpha_P \cdot r_{H/B})\vec{j} \,(\text{m/s}^2) \tag{11.24f}$$

The velocity and acceleration of $P$ with respect of $H$ can be defined using a rotating reference frame attached to $H$ and the cylinder, as shown in Figure (11.42). This rotating reference frame has the same angular motion as the cylinder. Based on this rotating reference frame, the motion of $P$ will appear as a straight line motion along the axis of the cylinder. We can then write down the following relative motion equations

$$\vec{v}_{P/O} = \vec{v}_{H/O} + \vec{\omega}_P \times \hat{r}_{P/H} + \hat{v}_{P/H}$$
$$= \vec{v}_{H/O} + \hat{v}_{P/H} = \vec{v}_{H/O} + u\hat{i} \qquad (11.24\text{g})$$
$$\vec{a}_{P/O} = \vec{a}_{H/O} + \vec{\omega}_P \times (\vec{\omega}_P \times \hat{r}_{P/H})$$
$$+ \vec{\alpha}_P \times \hat{r}_{P/H} + 2\vec{\omega}_P \times \hat{v}_{P/H} + \hat{a}_{P/H}$$
$$= \vec{a}_{H/O} + 2\vec{\omega}_P \times \hat{v}_{P/H} + \hat{a}_{P/H}$$
$$= \vec{a}_{H/O} + 2\omega_P u\hat{j} + \dot{u}\hat{i} \qquad (11.24\text{h})$$

where $\hat{r}_{P/H} = 0$, $\hat{i} = 0.514\vec{i} + 0.857\vec{j}$, and $\hat{j} = -0.857\vec{i} + 0.514\vec{j}$.
We also have

$$r_{P/A} + r_{H/B} = 0.292\,\text{m} \qquad (11.24\text{i})$$

**Solving for Unknowns:** First, we use Equations (11.24g) to obtain $\omega_P$ and $u$ as

$$\vec{\omega}_P = 0.559\vec{k}\ \text{rad/s} \qquad (11.24\text{j})$$
$$u = 0.0772\ \text{m/s} \qquad (11.24\text{k})$$

Next, we solve Equation (11.24h) to obtain $\alpha_P$ and $\dot{u}$ as

$$\vec{\alpha}_P = -0.958\vec{k}\ \text{rad/s}^2 \qquad (11.24\text{l})$$
$$\dot{u} = 0.207\ \text{m/s}^2 \qquad (11.24\text{m})$$

**Verification:** We can use the values of $\omega_P$ and $\alpha_P$ to calculate the velocity and acceleration of $B$ to verify the above results.

**Problem 11.25:** An aircraft $A$ is flying with a constant speed $v$ in a horizontal circle of radius $r$ at an altitude $h$. Relate the acceleration observed by the spherical coordinate system of the radar tracking antenna located at $O$ to the inertial reference $O-X-Y-Z$. This spherical coordinate is identified as $O-\hat{x}_2-\hat{y}_2-\hat{z}_2$ in Figure (11.43).

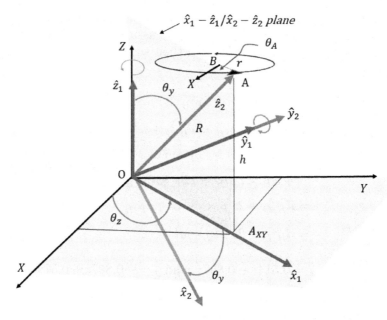

Figure 11.43:   Figures for Problem 25.

**Solution:**

**Motion Analysis:** With respect to the inertial reference frame $O-X-Y-Z$, aircraft $A$ undergoes a perfect circular motion on a horizontal plane. However, for the spherical tracking system with a reference frame $O_2-\hat{x}_2-\hat{y}_2-\hat{z}_2$, it would appear to have only a radial straight line motion along axis $O-\hat{z}_2$.

The radar tracking system has two motors to control the aiming of the antenna. The first motor rotates the platform around axis $Z$. On the platform, two horizontal axes are identified as $x$ and $y$. The reference frame $x-y$ originally aligns with $X-Y$. The second motor is mounded on the platform to rotate the antenna around $y$.

To track aircraft $A$ with a horizontal circular path as shown in Figure 11.43, the radar system first rotate around the $Z$ axis by an angle of $\theta_z$. The resulting little coordinate is $O_1 - \hat{x}_1 - \hat{y}_1 - \hat{z}_1$. This rotation places the $\hat{x}_1$ axis vertically under the aircraft. Therefore, $A$ is now on the $\hat{x}_1 - \hat{z}_1$ plane. The tracking system then active the second motor to rotate around $y(\hat{y}_1)$ so that $\hat{z}_2$ passes through $A$.

Based on Chapters 3 and 4, the relationship between $O-X-Y-Z$ and $O-\hat{x}_2-\hat{y}_2-\hat{z}_2$ can be related as

$$\begin{bmatrix} X \\ Y \\ Z \\ 1 \end{bmatrix} = T_{tz}(\theta_z) \cdot T_{ty}(\theta_y) \cdot \begin{bmatrix} \hat{x}_2 \\ \hat{y}_2 \\ \hat{z}_2 \\ 1 \end{bmatrix} \tag{11.25a}$$

$$\begin{bmatrix} X \\ Y \\ Z \\ 1 \end{bmatrix} = T_{tz}(\theta_z) \cdot \begin{bmatrix} \hat{x}_1 \\ \hat{y}_1 \\ \hat{z}_1 \\ 1 \end{bmatrix} \tag{11.25b}$$

$$\begin{bmatrix} \hat{x}_1 \\ \hat{y}_1 \\ \hat{z}_1 \\ 1 \end{bmatrix} = T_{ty}(\theta_y) \cdot \begin{bmatrix} \hat{x}_2 \\ \hat{y}_2 \\ \hat{z}_2 \\ 1 \end{bmatrix} \tag{11.25c}$$

From the above three equations, we can determine the transformation from $\vec{I} - \vec{J} - \vec{K}$ to $\hat{i}_1-\hat{j}_1-\hat{k}_1$, from $\hat{i}_1-\hat{j}_1-\hat{k}_1$ to $\hat{i}_2-\hat{j}_2-\hat{k}_2$, and from $\vec{I} - \vec{J} - \vec{K}$ to $\hat{i}_2-\hat{j}_2-\hat{k}_2$.

We have

$$\vec{I} = \cos\theta_z \hat{i}_1 - \sin\theta_z \hat{j}_1 \tag{11.25d1}$$

$$\vec{J} = \sin\theta_z \hat{i}_1 + \cos\theta_z \hat{j}_1 \tag{11.25d2}$$

$$\vec{K} = \hat{k}_1 \tag{11.25d3}$$

$$\hat{i}_1 = \cos\theta_y \hat{i}_2 + \sin\theta_y \hat{k}_2 \tag{11.25e1}$$

$$\hat{j}_1 = \hat{j}_2 \tag{11.25e2}$$

$$\hat{k}_1 = -\sin\theta_y \hat{i}_2 + \cos\theta_y \hat{k}_2 \tag{11.25e3}$$

$$\vec{I} = \cos\theta_z \cos\theta_y \hat{i}_2 - \sin\theta_z \hat{j}_2 + \cos\theta_z \sin\theta_y \hat{k}_2 \tag{11.25f1}$$

$$\vec{J} = \sin\theta_z \cos\theta_y \hat{i}_2 + \cos\theta_z \hat{j}_2 + \sin\theta_z \sin\theta_y \hat{k}_2 \tag{11.25f2}$$

$$\vec{K} = -\sin\theta_y \hat{i}_2 + \cos\theta_y \hat{k}_2 \tag{11.25f3}$$

Equations (11.25f1)–(11.25f3) can also be obtained by combining Equations (11.25d1)–(11.25d3) and (11.25e1)–(11.25e3).

With the transformations between unit directional vectors defined, we can derive the acceleration for $A$. With respect to $O-X-Y-Z$, it is

$$\vec{a}_{A/O} = \vec{\omega}_A \times (\vec{\omega}_A \times \vec{r}_{A/B})$$

$$= \left(\frac{v}{r}\vec{K}\right) \times \left(\left(\frac{v}{r}\vec{K}\right) \times \vec{r}_{A/B}\right)$$

$$= \left(\frac{v}{r}\vec{K}\right) \times \left(\left(\frac{v}{r}\vec{K}\right) \times (r\cos\theta_A\vec{I} + r\sin\theta_A\vec{J})\right)$$

$$= -\frac{v^2}{r}\cos\theta_A\vec{I} - \frac{v^2}{r}\sin\theta_A\vec{J} \tag{11.25g}$$

where $\theta_A$ is the angular position of $A$ with respect to the $X$-axis, as shown in Figure 11.43.

Transforming $\vec{I}$ and $\vec{J}$ to $\hat{i}_1$ and $\hat{j}_1$, we have

$$\vec{a}_{A/O} = -\frac{v^2}{r}\cos\theta_A\vec{I} - \frac{v^2}{r}\sin\theta_A\vec{J}$$

$$= \frac{v^2}{r}(-\cos\theta_A\cos\theta_z + \sin\theta_A\sin\theta_z)\hat{i}_1$$

$$+ \frac{v^2}{r}(\cos\theta_A\sin\theta_z - \sin\theta_A\cos\theta_z)\hat{j}_1 \tag{11.25h}$$

We can further transform it to $\hat{i}_2$, $\hat{j}_2$, and $\hat{k}_2$, as

$$\vec{a}_{A/O} = -\frac{v^2}{r}\cos\theta_A\vec{I} - \frac{v^2}{r}\sin\theta_A\vec{J}$$

$$= \frac{v^2}{r}(-\cos\theta_A\cos\theta_z\cos\theta_y - \sin\theta_A\sin\theta_z\cos\theta_y)\hat{i}_2$$

$$+ \frac{v^2}{r}(+\cos\theta_A\sin\theta_z - \sin\theta_A\cos\theta_z)\hat{j}_2$$

$$+ \frac{v^2}{r}(-\cos\theta_A\cos\theta_z\sin\theta_y - \sin\theta_A\sin\theta_z\sin\theta_y)\hat{k}_2$$

$$\tag{11.25i}$$

Sometimes, just like in the $r - \theta$ coordinate in a 2D motion, we want to define acceleration components in $a_r$ and $a_\theta$, we can also define the 3D acceleration components in $a_R$, $a_{\theta_z}$, and $a_{\theta_y}$ even though these components definitions do not have much practical use. Specifically, $a_R$ is the component of $\hat{k}_2$, $a_{\theta_z}$ is the component of $\hat{j}_2$, and $a_{\theta_y}$ is the component of $\hat{i}_2$.

When $\theta_A = 0$, we have

$$a_R = -\frac{v^2}{r} \cos\theta_z \sin\theta_y \tag{11.25j}$$

$$a_{\theta_z} = \frac{v^2}{r} \sin\theta_z \tag{11.25k}$$

$$a_{\theta_y} = -\frac{v^2}{r} \cos\theta_z \cos\theta_y \tag{11.25l}$$

In some conventions, the rotation around $Z$ is defined simply as $\theta$, i.e., $\theta_z = \theta$. The rotation around $\hat{y}_1$ is defined as $\phi$ as the angle from the $\hat{x}_1 - \hat{y}_1$ plane to the $\hat{z}_2$-axis, which leads to $\phi = \pi/2 - \theta_y$ and $a_\phi = -a_{\theta_y}$. Use these conventions, we have

$$a_R = -\frac{v^2}{r} \cos\theta \cos\phi \tag{11.25m}$$

$$a_\theta = \frac{v^2}{r} \sin\theta \tag{11.25n}$$

$$a_\phi = \frac{v^2}{r} \cos\theta \sin\phi \tag{11.25o}$$

**Verification:** Knowing these acceleration components do not help in the control of the radar tracking system. The original problem statement is a good exercise to conduct transformations among different reference frames. We will consider the tracking system for a general 3D airplane motion in the next problem.

**Problem 11.26:** Similar to Problem 11.25, a radar system is used to track the motion of an aircraft. Once locked, the tracking system has the information of $R$, $\dot{R}$, and $\ddot{R}$. In addition, it has $\theta_z$, $\dot{\theta}_z$, and $\ddot{\theta}_z$. Finally, it has $\theta_y$, $\dot{\theta}_y$, and $\ddot{\theta}_y$. We will define $\theta_z = \theta$ and $\theta_y = \phi$ for a simpler expression. Determine the actual velocity and acceleration of the aircraft $A$ with respect to $O-X-Y-Z$ (Figure 11.44).

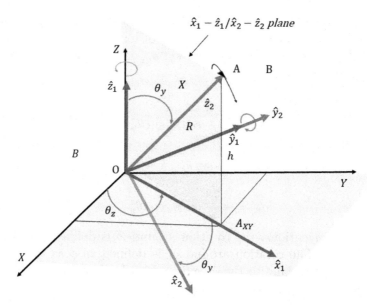

Figure 11.44:   Figures for Problem 26.

## Solution:

**Motion Analysis:** With the tracking system, we are using two reference frames to determine the motion of $A$. The inertial reference is $O-X-Y-Z$, while the rotating reference frame is $O_2-\hat{x}_2-\hat{y}_2-\hat{z}_2$. The frame rotation of $O_2-\hat{x}_2-\hat{y}_2-\hat{z}_2$ can be defined based on the relative motion equations with rotating reference frames, with the help of $O_1-\hat{x}_1-\hat{y}_1-\hat{z}_1$.

The transformation from $\hat{i}_2 - \hat{j}_2 - \hat{k}_2$ to $\vec{I} - \vec{J} - \vec{K}$ are determined by the following equations.

$$
\begin{bmatrix} \hat{x}_2 \\ \hat{y}_2 \\ \hat{z}_2 \\ 1 \end{bmatrix} = T_{ty}(-\phi) \cdot T_{tz}(-\theta) \cdot \begin{bmatrix} X \\ Y \\ Z \\ 1 \end{bmatrix} \tag{11.26a}
$$

$$
\begin{bmatrix} \hat{x}_1 \\ \hat{y}_1 \\ \hat{z}_1 \\ 1 \end{bmatrix} = T_{tz}(-\theta) \cdot \begin{bmatrix} X \\ Y \\ Z \\ 1 \end{bmatrix} \tag{11.26b}
$$

$$\begin{bmatrix} \hat{x}_2 \\ \hat{y}_2 \\ \hat{z}_2 \\ 1 \end{bmatrix} = T_{ty}(-\phi) \cdot \begin{bmatrix} \hat{x}_1 \\ \hat{y}_1 \\ \hat{z}_1 \\ 1 \end{bmatrix} \tag{11.26c}$$

$$\hat{i}_1 = \cos\theta \vec{I} + \sin\theta \vec{J} \tag{11.26d1}$$

$$\hat{j}_1 = -\sin\theta \vec{I} + \cos\theta \vec{J} \tag{11.26d2}$$

$$\hat{k}_1 = \vec{K} \tag{11.26d3}$$

$$\hat{i}_2 = \cos\phi \hat{i}_1 - \sin\phi \hat{k}_1 \tag{11.26e1}$$

$$\hat{j}_2 = \hat{j}_1 \tag{11.26e2}$$

$$\hat{k}_2 = \sin\phi \hat{i}_1 + \cos\phi \hat{k}_1 \tag{11.26e3}$$

$$\hat{i}_2 = \cos\phi\cos\theta \vec{I} + \cos\phi\sin\theta \vec{J} - \sin\phi \vec{K} \tag{11.26f1}$$

$$\hat{j}_2 = -\sin\theta \vec{I} + \cos\theta \vec{J} \tag{11.26f2}$$

$$\hat{k}_2 = \sin\phi\cos\theta \vec{I} + \sin\phi\sin\theta \vec{J} + \cos\phi \vec{K} \tag{11.26f3}$$

The angular velocity of $O_2-\hat{x}_2-\hat{y}_2-\hat{z}_2$ with respect to $O-X-Y-Z$ becomes

$$\vec{\Omega} = \vec{\Omega}_{2/O} = \vec{\Omega}_{1/O} + \vec{\Omega}_{2/1} = \dot{\theta}\vec{K} + \dot{\phi}\hat{j}_1$$

$$= \dot{\theta}(-\sin\phi \hat{i}_2 + \cos\phi \hat{k}_2) + \dot{\phi}\hat{j}_2$$

$$= -\dot{\theta}\sin\phi \hat{i}_2 + \dot{\phi}\hat{j}_2 + \dot{\theta}\cos\phi \hat{k}_2 \tag{11.26g}$$

$$= -\dot{\phi}\sin\theta \vec{I} + \dot{\phi}\cos\theta \vec{J} + \dot{\theta}\vec{K} \tag{11.26h}$$

As discussed in Section 4.2.1, we can represent $\vec{\Omega}$ in $\hat{i}_2-\hat{j}_2-\hat{k}_2$ or in $\vec{I}-\vec{J}-\vec{K}$ for motion analysis.

From Equation (11.26h), we can determine the frame angular acceleration in $\vec{I}-\vec{J}-\vec{K}$ and then transform it to $\hat{i}_2-\hat{j}_2-\hat{k}_2$ if necessary. The frame angular acceleration becomes

$$\dot{\vec{\Omega}} = \dot{\vec{\Omega}}_{2/O} = (-\ddot{\phi}\sin\theta - \dot{\phi}\cos\theta\dot{\theta})\vec{I}$$

$$+ (\ddot{\phi}\cos\theta - \dot{\phi}\sin\theta\dot{\theta})\vec{J}$$

$$+ \ddot{\theta}\vec{K} \tag{11.26i}$$

When the radar tracking locks on the aircraft, $\hat{z}_2$ points at $A$. The radial distance $R$, $\dot{R}$, and $\ddot{R}$ can be determined by the Doppler effect. To $O_2-\hat{x}_2-\hat{y}_2-\hat{z}_2$, the position, velocity, and acceleration vectors are

$$\hat{r}_{A/O_2} = R\hat{k}_2 \tag{11.26j}$$

$$\hat{v}_{A/O_2} = \dot{R}\hat{k}_2 \tag{11.26k}$$

$$\hat{a}_{A/O_2} = \ddot{R}\hat{k}_2 \tag{11.26l}$$

We can now construct the relative motion equation to determine the velocity and acceleration of $A$ with respect to $O-X-Y-Z$. We have

$$\vec{v}_{A/O} = \vec{v}_{O_2/O} + \vec{\Omega} \times \hat{r}_{A/O_2} + \hat{v}_{A/O_2}$$

$$= \vec{\Omega} \times (R\hat{k}_2) + \dot{R}\hat{k}_2$$

$$= R\dot{\phi}\hat{i}_2 + R\dot{\theta}\sin\phi\,\hat{j}_2 + \dot{R}\hat{k}_2 \tag{11.26m}$$

$$= (-R\dot{\theta}\sin\theta\sin\phi + R\dot{\phi}\cos\phi\cos\theta + \dot{R}\sin\phi\cos\theta)\vec{I}$$

$$+ (R\dot{\theta}\sin\phi\cos\theta + R\dot{\phi}\cos\phi\sin\theta + \dot{R}\sin\phi\sin\theta)\vec{J}$$

$$+ (-R\dot{\phi}\sin\phi + \dot{R}\cos\phi)\vec{K} \tag{11.26n}$$

To obtain the acceleration of $A$, we can simply take a derivative of Equation (11.26n) or use the relative acceleration equation as

$$\vec{a}_{A/O} = \vec{a}_{O_2/O} + \vec{\Omega} \times (\vec{\Omega} \times \hat{r}_{A/O_2})$$

$$+ \dot{\vec{\Omega}} \times \hat{r}_{A/O_2} + 2\vec{\Omega} \times \hat{v}_{A/O_2} + \hat{a}_{A/O_2} \tag{11.26o}$$

Readers are welcome to complete the derivation of Equation (11.26o) as an exercise.

**Verification:** Just as a verification, if the radar tracking system is tracking aircraft $A$ of Problem 11.25 but with point $B$ vertically up from point $O$, then from the velocity vector, we have

$$(\vec{v}_{A/O})_Z = 0 = -R\dot{\phi}\sin\phi + \dot{R}\cos\phi$$

$$\implies \dot{\phi} = 0 \tag{11.26p}$$

where $\dot{R} = 0$ because $B$ is vertically up from $O$ and the circular path of $A$ forms a vertical cone with point $O$.

The velocity of $A$ is simplified to

$$\vec{v}_{A/O} = -R\dot{\theta}\sin\theta\sin\phi\vec{I} + R\dot{\theta}\sin\phi\cos\theta\vec{J} \qquad (11.26q)$$

The airplane speed is then determined as

$$v = R\dot{\theta}\sin\phi \qquad (11.26r)$$

Because $v$ is a constant, we can rewrite Equation (11.26q) as

$$\vec{v}_{A/O} = -v\sin\theta\vec{I} + v\cos\theta\vec{J} \qquad (11.26s)$$

Taking a derivative of Equation (11.26s), we obtain the acceleration as

$$\vec{a}_{A/O} = -v\dot{\theta}\cos\theta\vec{I} - v\dot{\theta}\sin\theta\vec{J} \qquad (11.26t)$$

Equation (11.26t) matches Equation (11.25g).

Knowing that $A$ undergoes circular motion at constant speed, we obtain

$$r = \frac{v}{\dot{\theta}} \qquad (11.26u)$$

**Problem 11.27:** The vertical shaft at $O$ rotates about the vertical $Z$-axis with a constant angular velocity $\omega$ and causes the wheel to roll in a horizontal circle of radius $R$. The radius of the wheel is $r$. Determine the acceleration of point $A$ on the rim at the instant represented. The wheel rolls without slipping (Figure 11.45).

**Solution:**

**Motion Analysis:** The shaft and the wheel form one rigid body. It has one angular velocity and one angular acceleration. We define an inertial reference frame $O-X-Y-Z$ and a rotating reference frame $O-\hat{x}-\hat{y}-\hat{z}$ which is attached to the shaft. At this instant of time, $O-X-Y-Z$ and $O-\hat{x}-\hat{y}-\hat{z}$ are aligned. We can use the relative motion analysis to solve this problem but we will present a different way here.

With respect to $O-\hat{x}-\hat{y}-\hat{z}$, the rigid body has a rotation around the negative $\hat{y}$-axis. Due to the no-slip condition, this angular rotation speed is $\hat{\omega} = -R\omega/r\hat{j}$. The frame rotation of $O-\hat{x}-\hat{y}-\hat{z}$ observed by $O-X-Y-Z$ is $\vec{\Omega} = \omega\vec{K} = \Omega\hat{k}$.

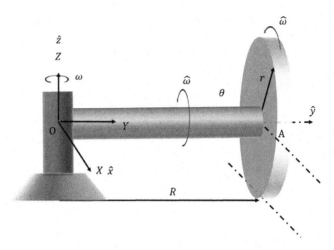

Figure 11.45: Figures for Problem 27.

The total angular velocity of the shaft and the shaft/wheel $\vec{\omega}_{SW}$ is determined as

$$\vec{w}_{SW/O} = -\frac{R}{r}\omega\hat{j} + \omega\vec{K} \tag{11.27a}$$

$$= -\frac{R}{r}\omega\hat{j} + \omega\hat{k} \tag{11.27b}$$

$$= -\frac{R}{r}\omega\vec{J} + \omega\vec{K} \tag{11.27c}$$

where $\hat{i}$, $\hat{j}$, and $\hat{k}$ are unit directional vectors of $O-\hat{x}-\hat{y}-\hat{z}$ and $\vec{I}$, $\vec{J}$, and $\vec{K}$ are unit direction vectors for $O-X-Y-Z$. They are aligned for the instant shown in Figure 11.45.

Based on the expression of Equation (11.27b), the angular acceleration of the shaft/wheel is determined as

$$\dot{\vec{w}}_{SW/O} = \frac{\partial}{\partial t}\left(-\frac{R}{r}\omega\hat{j} + \omega\hat{k}\right) + \vec{\Omega} \times \left(-\frac{R}{r}\omega\hat{j} + \omega\hat{k}\right)$$

$$= \frac{R\omega^2}{r}\hat{i} \tag{11.27d}$$

$$= \frac{R\omega^2}{r}\vec{I} \tag{11.27e}$$

The position vector of $A$ with respect to $O-X-Y-Z$ at this instant is

$$\vec{r}_{A/O} = r\vec{I} + R\vec{J} \tag{11.27f}$$

Because $O$ is on the shaft/wheel rigid body and it has a zero acceleration, we can determine the acceleration of $A$ as if it is undergoing a rotation with respect $O$ at $\vec{\omega}_{SW/O}$ and $\dot{\vec{\omega}}_{SW/O}$,

$$\vec{a}_{A/O} = \vec{\omega}_{SW/O} \times (\vec{\omega}_{SW/O} \times \vec{r}_{A/O}) + \dot{\vec{\omega}}_{SW/O} \times \vec{r}_{A/O}$$

$$= -\frac{\omega^2}{r}(r^2 + R^2)\vec{I} - \omega^2 R\vec{J} \tag{11.27g}$$

**Verification:** This is not a difficult problem, but we must determine $\vec{\omega}_{SW/O}$ and $\dot{\vec{\omega}}_{SW/O}$ correctly using two sets of reference frames defined in the solution.

**Problem 11.28:** A particle of mass $m$ is released from rest at $A$ and slides down the smooth spiral tube of constant helix angle. Determine the force $N$ exerted by the tube on the particle as it completes one full turn at $B$ (Figure 11.46).

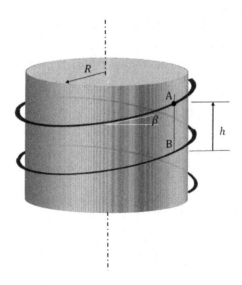

Figure 11.46:   Figures for Problem 28.

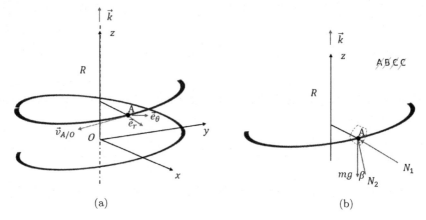

(a)                                   (b)

Figure 11.47:   (a) Motion analysis; and (b) force analysis.

**Solution:**

**Motion Analysis:** We have one particle in the helix motion. The most convenient coordinate to use is the $r-\theta-z$ coordinate, as shown in Figure 11.47(a). The inertial reference frame is $O-x-y-z$ at the same horizontal plane as $B$. The $x$-axis can be aligned to $B$ so that when the particle is released, $\theta = 0$ and it reaches $B$, $\theta = 2\pi$.

At any instant before reaching $B$, particle $A$ has a velocity $\vec{v}_{A/O}$ which is tangential to the tube, as shown in Figure 11.47(a). At this instant, the radial and transverse unit directional vectors, $\vec{e}_r$ and $\vec{e}_\theta$, are defined accordingly. The velocity and acceleration vectors of $A$ can now be defined as

$$\vec{v}_{A/O} = \dot{r}\vec{e}_r - r\dot{\theta}\vec{e}_\theta - \dot{z}\vec{k}$$

$$= -R\dot{\theta}\vec{e}_\theta - \dot{z}\vec{k} \tag{11.28a}$$

$$\vec{a}_{A/O} = (\ddot{r} - r\dot{\theta}^2)\vec{e}_r - (r\ddot{\theta} + 2\dot{r}\dot{\theta})\vec{e}_\theta - (\ddot{z})\vec{k}$$

$$= (-R\dot{\theta}^2)\vec{e}_r - (R\ddot{\theta})\vec{e}_\theta - (\ddot{z})\vec{k} \tag{11.28b}$$

Note that the negative signs for the terms involved with $\vec{\theta}$ and $\vec{k}$ are due to how $O-x-y-z$ is defined based on the right-hand rule and $A$ is sliding downward and in the negative $\theta$ direction.

Because it is a helical motion, the vertical displacement and the angular displacement are related as

$$z = \frac{h}{2\pi}\theta \tag{11.28c}$$

$$\dot{z} = \frac{h}{2\pi}\dot{\theta} \tag{11.28d}$$

$$\ddot{z} = \frac{h}{2\pi}\ddot{\theta} \tag{11.28e}$$

The helix angle is also defined as

$$\tan\beta = \frac{h}{2\pi R} \tag{11.28f}$$

**Force Analysis:** The FBD of $A$ is shown in Figure 11.47(b). Because the tube is smooth, there are no tangential forces. The normal force has two components, one is normal to the tube in the vertical plane ($N_2$) and the other along the $\vec{e}_r$ direction ($N_1$). The total normal force is

$$N = \sqrt{N_1^2 + N_2^2} \tag{11.28g}$$

**Governing Equations:** We need to think about which governing equation(s) to use. Because there is no friction involved and the normal force does not do work, we know that the conservation of mechanical energy holds. We will define point $B$ as the datum for the gravitational potential energy. With respect to $t_0$ when $A$ is released and $t_1$ when it reaches $B$, the energy equation is

$$T_0 + U_{g0} = 0 + mgh = T_1 + U_{g1} = \frac{1}{2}mv_1^2 \tag{11.28h}$$

From Equation (11.28h), we obtain

$$v_1 = \sqrt{2gh} \tag{11.28i}$$

From Equation (11.28a), we also have

$$v_1 = \sqrt{(R\dot{\theta})^2 + (\dot{z})^2} \qquad (11.28\text{j})$$

Combining Equations (11.28h), (11.28j), and (11.28d), we have angular velocity at $t_1$ as

$$\dot{\theta}(t_1) = 2\pi\sqrt{\frac{2gh}{4\pi^2 R^2 + h^2}} \qquad (11.28\text{k})$$

In order to get the information of the normal force, we need to invoke *Newton's Second Law*, which leads to

$$\left(\sum F\right)_r = -N_1 = -mR\dot{\theta}^2 \qquad (11.28\text{l})$$

$$\left(\sum F\right)_\theta = -N_2 \sin\beta = -mR\ddot{\theta} \qquad (11.28\text{m})$$

$$\left(\sum F\right)_z = -mg + N_2 \cos\beta = -m\ddot{z} = -m\frac{h}{2\pi}\ddot{\theta} \qquad (11.28\text{n})$$

**Solving for Unknowns:** At $t_1$, Equations (11.28l)–(11.28n) contain three unknowns, $N_1$, $N_2$, and $\ddot{\theta}$. We can proceed to solve for them to obtain

$$N_1 = 8\pi^2 mR\frac{gh}{4\pi^2 R^2 + h^2} \qquad (11.28\text{o})$$

$$N_2 = \frac{mg\frac{2\pi R}{h}}{\frac{2\pi R}{h}\cos\beta + \sin\beta} \qquad (11.28\text{p})$$

$$\ddot{\theta}_1 = \frac{g\frac{2\pi}{h}\sin\beta}{\frac{2\pi R}{h}\cos\beta + \sin\beta} \qquad (11.28\text{q})$$

Finally, we have

$$N = \sqrt{N_1^2 + N_2^2}$$

$$= mg\frac{4\pi^2 R^2}{4\pi^2 R^2 + h^2}\sqrt{\frac{4\pi^2 R^2 + 16\pi^2 h^2 + h^2}{4\pi^2 R^2}} \qquad (11.28\text{r})$$

**Verification:** Equation (11.28r) only applies for $t_1$. A more general expression of the normal force can be derived when $A$ is at a different vertical position by revising Equation (11.28h) accordingly.

**Problem 11.29:** The bent rod shown rotates about the $z$-axis and starts from rest with an angular acceleration $\alpha = 1,000\,\text{rad/s}^2$ resulting from an applied torque. The material of the rod has a mass of 1.2 kg per meter of the length. Neglect the small effect of gravity and determine the bending moment in the rod at the section near center $O$ (Figure 11.48).

**Solution:**

**Motion Analysis:** The entire rod is rotating around the $z$-axis. We will assume that the bent rod as rigid.

The moment of inertia of the entire bent rod with respect to the $z$-axis can be determined by adding the moment of inertial of each rod section.

$$I_Z = I_{AB} + I_{BC} + I_{AE} + I_{CD} + I_{FE}$$

$$= I_{AB} + 2I_{BC} + 2I_{CD} = 0.0098125\,\text{kg} \cdot \text{m}^2 \quad (11.29\text{a})$$

**Force Analysis:** The FBD for a small section of the rod near $O$ with negligible moment of inertia is shown in Figure 11.49. Because of the negligible moment of inertia of this little portion, we have

$$\tau - 2\tau_1 = I_s\alpha \approx 0 \quad (11.29\text{b})$$

The bending moment at the horizontal rod near $O$ is one half of the applied torque.

$$\tau_1 = \frac{1}{2}\tau \quad (11.29\text{c})$$

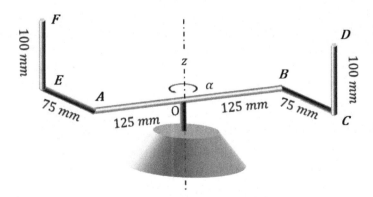

Figure 11.48: Figures for Problem 29.

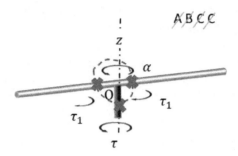

Figure 11.49: The FBD for the center portion of the rod with negligible moment of inertia.

**Governing Equations:** Applying *Newton's Second Law* for the moment around the $z$-axis for the entire rod, we have

$$\left(\sum \tau\right)_z = \tau = I_z \alpha \qquad (11.29\text{d})$$

**Solving for Unknowns:** We obtain

$$\tau = 9.81 \,\text{Nm} \qquad (11.29\text{e})$$

$$\tau_1 = 4.91 \,\text{Nm} \qquad (11.29\text{f})$$

**Verification:** The calculation of the moment of inertia is quite tedious. The moments of inertia for rods $EA$ and $BC$ require integration because the radial distance to $O$ is different along the rod.

**Problem 11.30:** The disk rotates in the fixed $x-y$ plane about the $z$-axis. The link $AB$ is connected to the rim of the disk and to the slider at $A$ by ball joints. Point $A$ moves in the $y-z$ plane and is a distance $r$ from the $z$-axis. Determine the velocity $\dot{z}$ of $A$ in terms of $\theta$ if the disk turns with a constant speed $\omega = \dot{\theta}$. Also, find the components of the angular velocity, $\omega_{AB}$ of $AB$ at the instant when $\theta = \pi/2$.

**Solution:**

**Motion Analysis:** We are dealing with the motion of rigid body $AB$ for this problem. We need to know its angular velocity and angular acceleration in addition to the motion of one point in order to know the motion of every point. The problem only asks for the velocity information.

We know the motion of point $B$ based on the rotation of the disk. We assume that $AB$ has an angular velocity $\vec{\Omega} = \Omega_x \vec{i} + \Omega_y \vec{j} + \Omega_z \vec{k}$. We also know that the velocity of $A$ is along the $z$-axis,

$$\vec{v}_{A/O} = v_A \vec{k} = \dot{z}_A \vec{k} \tag{11.30a}$$

Based on the reference frame $O-x-y-z$ defined in Figure 11.50, we can define the position vectors for $A$ and $B$ as

$$\vec{r}_{A/O} = r\vec{i} + z_A \vec{k} \tag{11.30b}$$

$$\vec{r}_{B/O} = -r\cos\theta\vec{i} + r\sin\theta\vec{j} \tag{11.30c}$$

$$\vec{r}_{A/B} = r(1 + \cos\theta)\vec{i} - r\sin\theta\vec{j} + z_A \vec{k} \tag{11.30d}$$

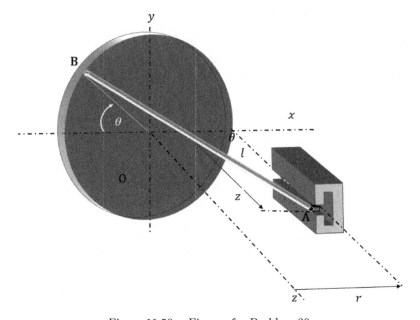

Figure 11.50: Figures for Problem 30.

We can now write down the velocity equation for $A$ with respect to $O$ as

$$\vec{v}_{A/O} = \dot{z}_A \vec{k} = \vec{v}_{B/O} + \vec{v}_{A/B}$$

$$= (-\dot{\theta}\vec{k}) \times \vec{r}_{B/O} + \vec{\Omega} \times \vec{r}_{A/B}$$

$$= (z_A \Omega_y + \Omega_z r \sin\theta + r\dot{\theta}\sin\theta)\vec{i}$$

$$+ (r(1 + \cos\theta)\Omega_z - z_A\Omega_x + r\dot{\theta}\cos\theta)\vec{j}$$

$$+ (-\Omega_x r \sin\theta - r\Omega_y(1 + \cos\theta))\vec{k} \qquad (11.30\text{e})$$

Equation (11.30e) provides three equations, but there are four unknowns, $\dot{z}_A$, $\Omega_x$, $\Omega_y$, and $\Omega_z$.

**Solving for Unknowns:** We need one more equation. From Equation (11.30d), we obtain

$$l = \sqrt{r^2(1 + \cos\theta)^2 + r^2 \sin^2\theta + z_A^2} \qquad (11.30\text{f})$$

Taking a derivative of Equation (11.30f), we have

$$\dot{z}_A = \frac{r^2\dot{\theta}\sin\theta}{\sqrt{l^2 - 2r^2 - 2r^2\cos\theta}} \qquad (11.30\text{g})$$

We can now solve Equation (11.30e) for $\theta = \pi/2$ to obtain

$$\Omega_x = -\frac{r^3\dot{\theta}}{l^2\sqrt{l^2 - 2r^2}} \qquad (11.30\text{h})$$

$$\Omega_y = r\dot{\theta}\frac{r^2 - l^2}{l^2\sqrt{l^2 - 2r^2}} \qquad (11.30\text{i})$$

$$\Omega_z = \frac{-r^2\dot{\theta}}{l^2} \qquad (11.30\text{j})$$

$$z_A = \sqrt{l^2 - 2r^2} \qquad (11.30\text{k})$$

**Verification:** Readers can proceed to conduct acceleration analysis as a practice.

**Problem 11.31:** The uniform circular ring has a mass of 8 kg and a radius of 300 mm. The ring is released in space with a spin velocity

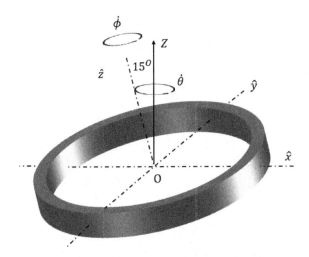

Figure 11.51:   Figures for Problem 31.

$\dot{\phi} = 60\,\text{rpm}$ about its principal $\hat{z}$-axis and with precessional wobble about the $z$-axis corresponding to $\theta = 15°$. Calculate the kinetic energy of rotation of the ring about its center $O$ (Figure 11.51).

**Solution:**

**Motion Analysis:** First recall the torque-free motion discussed in Section 6.3.3.1, where the spinning and the precessing rates are related as

$$\dot{\theta} = \frac{I_{\hat{z}\hat{z}}}{(I_{\hat{y}\hat{y}} - I_{\hat{z}\hat{z}})\cos\theta}\dot{\phi} \qquad (11.31a)$$

We have two sets of reference frames. $O-X-Y-Z$ is the inertial reference frame, while $O-\hat{x}-\hat{y}-\hat{z}$ is attached to the principal axes of the ring and it observes a spinning motion of the ring around $\hat{z}$ as $\dot{\phi}$. With respect to $O-X-Y-Z$, $O-\hat{x}-\hat{y}-\hat{z}$ has a frame rotation as $\vec{\Omega} = \dot{\theta}\vec{K}$.

The angular velocity of the ring can now be defined as

$$\vec{\omega} = \vec{\Omega} + \dot{\phi}\hat{k} = \dot{\phi}\hat{k} + \dot{\theta}\vec{K} \qquad (11.31b)$$

The unit directional vectors between the two observation frames can be defined by

$$\begin{bmatrix} X \\ Y \\ Z \\ 1 \end{bmatrix} = T_{tz}(\theta) \cdot T_{tx}(15°) \cdot \begin{bmatrix} \hat{x} \\ \hat{y} \\ \hat{z} \\ 1 \end{bmatrix} \tag{11.31c}$$

$$\vec{I} = \cos\theta\hat{i} - \cos 15° \sin\theta\hat{j} + \sin 15° \sin\theta\hat{k} \tag{11.31d}$$

$$\vec{J} = \sin\theta\hat{i} + \cos 15° \cos\theta\hat{j} - \sin 15° \cos\theta\hat{k} \tag{11.31e}$$

$$\vec{K} = \sin 15°\hat{j} + \cos 15°\hat{k} \tag{11.31f}$$

We prefer to express $\vec{\omega}$ in terms of $\hat{i}$, $\hat{j}$, and $\hat{k}$ because the moment of inertia matrix expression is simpler.

$$\vec{\omega} = \hat{\omega} = \dot{\phi}\hat{k} + \dot{\theta}(\sin 15°\hat{j} + \cos 15°\hat{k})$$

$$= \dot{\theta}\sin 15°\hat{j} + (\dot{\phi} + \dot{\theta}\cos 15°)\hat{k} \tag{11.31g}$$

Based on $O-\hat{x}-\hat{y}-\hat{z}$, the moment of inertial matrix of the ring is

$$\hat{I}_O = \begin{bmatrix} I_{\hat{x}\hat{x}} & 0 & 0 \\ 0 & I_{\hat{y}\hat{y}} & 0 \\ 0 & 0 & I_{\hat{z}\hat{z}} \end{bmatrix} = \begin{bmatrix} \frac{1}{2}mr^2 & 0 & 0 \\ 0 & \frac{1}{2}mr^2 & 0 \\ 0 & 0 & mr^2 \end{bmatrix} \tag{11.31h}$$

The kinetic energy of the ring becomes

$$T_{\text{ring}} = \frac{1}{2}\hat{\omega}'\hat{I}_O\hat{\omega}$$

$$= \frac{1}{4}mr^2 \sin^2 15°\dot{\theta}^2 + \frac{1}{2}mr^2(\dot{\phi} + \dot{\theta}\cos 15°)^2 \tag{11.31i}$$

From Equation (11.31a), we obtain

$$\dot{\theta} = -\frac{4\pi}{\cos 15°}\dot{\phi} \tag{11.31j}$$

where $\dot{\phi} = 60$ rpm $= 2\pi$ rad/s.

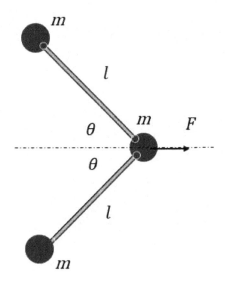

Figure 11.52: Figures for Problem 32.

Introducing all the values into Equation (11.31i), we obtain

$$T_{\text{ring}} = 16.25\,\text{J} \qquad (11.31\text{k})$$

**Verification:** The precession rate was not given in the problem statement and we have to recognize that this is a torque-free motion because the ring is released in space.

**Problem 11.32:** The three identical bodies each of mass $m$ are connected by the two hinged links of negligible mass and equal length $l$. If a force $F$ is applied as shown to the central body of the system at rest in space, determine the initial acceleration of this body (Figure 11.52).

**Solution:**

**Motion Analysis:** We label the three bodies as $A$, $B$, and $C$ as shown in Figure 11.53. With respect to the inertial reference frame $O-x-y$, the acceleration of each body is defined as

$$\vec{a}_{A/O} = \ddot{x}_A \vec{i} \qquad (11.32\text{a})$$
$$\vec{a}_{B/O} = \vec{a}_{A/O} + \vec{a}_{B/A}$$

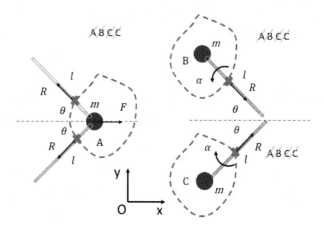

Figure 11.53:   Free-body diagrams for Problem 32.

$$= \ddot{x}_A \vec{i} + \vec{\omega} \times (\vec{\omega} \times \vec{r}_{B/A}) + \vec{\alpha} \times \vec{r}_{B/A}$$

$$= (\ddot{x}_A - l\alpha \sin\theta)\vec{i} - l\alpha \cos\theta\vec{j} \qquad (11.32b)$$

$$\vec{a}_{C/O} = \vec{a}_{A/O} + \vec{a}_{C/A}$$

$$= \ddot{x}_A \vec{i} + \vec{\omega} \times (\vec{\omega} \times \vec{r}_{C/A}) + \vec{\alpha} \times \vec{r}_{C/A}$$

$$= (\ddot{x}_A - l\alpha \sin\theta)\vec{i} + l\alpha \cos\theta\vec{j} \qquad (11.32c)$$

where $\omega = 0$ at $t = 0$ because the system is at rest when $F$ is applied.

**Force Analysis:** The FBDs are shown in Figure 11.53. The force $R$ is along the link axial direction because the link is pinned at both ends.

**Governing Equations:** For body $A$, *Newton's Second Law* leads to

$$\left(\sum F\right)_x = F - 2R\cos\theta = m\ddot{x}_A \qquad (11.32d)$$

$$\left(\sum F\right)_y = R\sin\theta - R\sin\theta = 0 \qquad (11.32e)$$

For body $B$, we have

$$\left(\sum F\right)_x = R\cos\theta = m\ddot{x}_B \qquad (11.32f)$$

$$\left(\sum F\right)_y = -R\sin\theta = m\ddot{y}_B \tag{11.32g}$$

For body $C$, we have

$$\left(\sum F\right)_x = R\cos\theta = m\ddot{x}_C \tag{11.32h}$$

$$\left(\sum F\right)_y = R\sin\theta = m\ddot{y}_C \tag{11.32i}$$

From the above equations, we have

$$\ddot{y}_B = -\frac{R\sin\theta}{m} \tag{11.32j}$$

$$\ddot{y}_C = \frac{R\sin\theta}{m} \tag{11.32k}$$

**Solving for Unknowns:** From the above equations, we obtain

$$\ddot{x}_A = \frac{F}{3m - 2m\sin^2\theta} \tag{11.32l}$$

$$\alpha = \frac{F\sin\theta}{l(3m - 2m\sin^2\theta)} \tag{11.32m}$$

$$R = \frac{F\cos\theta}{3 - 2\sin^2\theta} \tag{11.32n}$$

**Verification:** Let's verify two extreme conditions when $\theta = 0°$ or $\theta = \pi/2$. When $\theta = 0°$, the acceleration, angular acceleration, and force $R$ are found to be

$$\ddot{x}_A = \frac{F}{3m} \tag{11.32o}$$

$$\alpha = 0 \tag{11.32p}$$

$$R = \frac{F}{3} \tag{11.32q}$$

This is correct because all three bodies move together as one and $\alpha = 0$

When $\theta = \pi/2$, the acceleration, angular acceleration, and force $R$ are

$$\ddot{x}_A = \frac{F}{m} \tag{11.32r}$$

$$\alpha = \frac{F}{ml} \qquad (11.32\text{s})$$

$$R = 0 \qquad (11.32\text{t})$$

This is correct because the link force $R$ does not have a component in the $x$ direction and $\alpha = \frac{F}{ml}$, which is the largest. At this instant, both $B$ and $C$ have zero acceleration.

# Index

Printed in the United States
by Baker & Taylor Publisher Services